Oxidative Stress and Antioxidant Defense in Crop Plants, 2nd Edition

Oxidative Stress and Antioxidant Defense in Crop Plants, 2nd Edition

Guest Editors

Grzegorz Chrzanowski
Marta Jańczak-Pieniążek

MDPI

Basel • Beijing • Wuhan • Barcelona • Belgrade • Novi Sad • Cluj • Manchester

Guest Editors
Grzegorz Chrzanowski
Collegium Medicum
University of Rzeszów
Rzeszów
Poland

Marta Jańczak-Pieniążek
Department of Crop
Production
University of Rzeszów
Rzeszów
Poland

Editorial Office
MDPI AG
Grosspeteranlage 5
4052 Basel, Switzerland

This is a reprint of the Special Issue, published open access by the journal *Antioxidants* (ISSN 2076-3921), freely accessible at: https://www.mdpi.com/journal/antioxidants/special_issues/8EYVQQK79A.

For citation purposes, cite each article independently as indicated on the article page online and as indicated below:

Lastname, A.A.; Lastname, B.B. Article Title. *Journal Name* **Year**, *Volume Number*, Page Range.

ISBN 978-3-7258-7819-2 (Hbk)
ISBN 978-3-7258-7820-8 (PDF)
https://doi.org/10.3390/books978-3-7258-7820-8

Contents

About the Editors

Grzegorz Chrzanowski

Grzegorz Chrzanowski is an academic and researcher in the field of natural sciences, specializing in the interdisciplinary area of biology, biotechnology, and agriculture, with a primary focus on the biochemistry of secondary metabolites, their biosynthesis, and role in biological systems. His research is mostly focused on the biosynthesis of secondary metabolites in plants and green microalgal cells under stress conditions, as well as on their insecticidal and antimicrobial activities, particularly against Candida species. On the other hand, he is interested in the generation of oxidative stress in plants and insects and in the development of the antioxidative and detoxifying systems within their tissues, involving both enzymatic and non-enzymatic constituents.

Marta Jańczak-Pieniążek

Marta Jańczak-Pieniążek is an Associate Professor at the Department of Crop Production, Faculty of Technology and Life Sciences, University of Rzeszów, Poland. Her research focuses on crop physiology, antioxidant metabolism in plants, and the biochemical and physiological responses of crops to abiotic stress. She is particularly interested in plant adaptation mechanisms to changing environmental conditions and their role in maintaining crop productivity and quality under stress. She has authored and co-authored numerous peer-reviewed publications in the field of plant stress physiology and crop science. Her scientific work contributes to a better understanding of oxidative stress regulation and antioxidant defense systems in plants, with relevance for sustainable agricultural production and crop improvement strategies.

Preface

This Special Issue is focused on antioxidant processes in cultivated crop plants. The aim is to summarize current research on oxidative stress and antioxidant mechanisms in crop species and their roles in plant development, stress responses, and agricultural productivity. The scope includes physiological, biochemical, and molecular studies addressing plant responses to environmental challenges.

This Reprint is addressed to researchers and professionals in plant science, agronomy, and related fields interested in antioxidant functions in cultivated plants and their relevance for modern agriculture.

Grzegorz Chrzanowski and Marta Jańczak-Pieniążek
Guest Editors

antioxidants

Article

Combined Salt and Heat Stress Aggravates Oxidative Stress and Photosynthetic Damage, Disrupting Carbon and Nitrogen Metabolism and Yield in Rice

Lin Li [1,2,†], Jie Xu [1,2,†], Jinqi Liu [1,2], Wenhao Bi [1,2], Yingjiang Li [1,2], Aibin He [1,3], Xiayu Guo [1,3,*] and Zhiyong Ai [1,3,*]

1 National Center of Technology Innovation for Saline-Alkali Tolerant Rice in Sanya, Sanya 572024, China; lilin18@hainanu.edu.cn (L.L.); xujie11272026@126.com (J.X.); 13755161912@163.com (Y.L.); heaibing@hhrrc.ac.cn (A.H.)

2 College of Breeding and Multiplication (Sanya Institute of Breeding and Multiplication), Hainan University, Sanya 572025, China

3 State Key Laboratory of Hybrid Rice, Hunan Hybrid Rice Research Center, Changsha 410125, China

* Correspondence: wanilybao@163.com (X.G.); aizhiyongw@163.com (Z.A.)

† These authors contributed equally to this work.

Abstract

In the context of global climate change, the co-occurrence of salt and heat stress represents a major constraint to rice production, resulting in greater yield penalties than either stress alone. This study aimed to assess the effects of salt and heat stress on oxidative homeostasis, photosynthetic performance, carbon (C)–nitrogen (N) metabolism, and rice yield. The experiment comprised four treatments, i.e., control (CK), salt (irrigation with 3.9 dS m^{-1} NaCl solution), heat (exposure to 36 °C/30 °C day/night for 5 days at panicle initiation), and combined salt + heat stress. Results showed that combined stress enhanced reactive oxygen species (ROS) accumulation (i.e., H_2O_2 content and O_2^- contents were 1.3 and 1.5 times higher than CK), and the activities of superoxide dismutase (SOD), peroxidase (POD), and catalase (CAT) were increased by 64.6%, 69.5%, and 74.8% higher than CK. At the molecular level, salt + heat stress upregulated antioxidant defense-related genes, i.e., *OsAPX2*, *OsSODCC1*, and *OsAPX1*, while significantly downregulated ion homeostasis-related genes, i.e., *OsSOSs*, *OsHKT1;3*, *OsHKT1;5*, and *OsNHX4*, and photosynthesis-related genes, i.e., *Ospsbo*, *OsRbcS2*, and *OsRbcS3*, compared with CK. Furthermore, salt + heat stress reduced the activities of C-metabolism enzymes (sucrose phosphate synthase, sucrose synthase, and starch synthase) and N-metabolism enzymes (nitrate reductase, glutamine synthetase, and glutamate synthase), leading to 34.3% and 18.6% lower stem-sheath non-structural carbohydrate accumulation in stem sheath and its translocation rate, respectively, while total N accumulation decreased by 42.9%, as compared with CK. Ultimately, these cascading effects inhibited panicle development and reduced yield. The findings provide a theoretical basis for improving rice tolerance to combined abiotic stresses by targeting oxidative stress mitigation, photosynthetic protection, and key stress-responsive gene regulation.

Keywords: antioxidants; stress; photosynthesis; non-structural carbohydrates; panicle development

 https://doi.org/10.3390/antiox15030308

1. Introduction

Rice (*Oryza sativa* L.) is a staple food crop feeding over half of the global population, and its stable productivity is critical for ensuring global food security [1,2]. With the intensification of global climate change, abiotic stresses have emerged as major constraints on rice production, particularly in key rice-growing regions such as the coastal areas of China [3]. Simultaneous occurrence of salt and heat stress during critical reproductive stages of rice, such as panicle initiation, severely disrupts physiological processes and yield formation [4].

Salt stress primarily acts through osmotic stress and ionic toxicity; excessive sodium ion (Na^+) accumulation disrupts ion homeostasis in rice cells, damaging cell membranes and increasing electrolyte leakage [5]. It also induces stomatal closure, reduces chlorophyll content, and inhibits photosynthetic carbon fixation, while triggering overaccumulation of reactive oxygen species (ROS) that further exacerbates cellular damage [6]. For instance, salt stress can reduce the net photosynthetic rate (P_n) in rice and increase hydrogen dioxide (H_2O_2) content compared with normal conditions [7]. Heat stress, on the other hand, mainly targets the photosynthetic system and metabolic enzymes. For instance, high temperatures ($\geq$36 °C) denature key photosynthetic enzymes such as Rubisco, disrupt the stability of photosystem II (PSII), and reduce the maximum photochemical efficiency [8]. Additionally, heat stress inhibits antioxidant enzyme activity, leading to ROS burst and membrane lipid peroxidation [9,10]. Previous studies have shown that the critical windows for heat-induced damage in rice are the heading–flowering and panicle initiation stage [11]. While the mechanisms of high-temperature injury and its impacts on spikelet sterility during heading–flowering have been extensively investigated, the research on the impacts of heat stress during the panicle initiation stage in rice remains limited.

Previous studies have shown that salt and heat stress independently damage plant physiological processes, but their combined effects are more complex and destructive [12,13]. The combined stress exerts synergistic or additive destructive effects and causes more severe impairment to membrane stability, photosynthetic rate, and plant biomass than either single stress [3]. The wheat seedlings under combined heat and salt stress show enhanced Na^+ accumulation, reduced relative water contents, and decreased shoot biomass compared to individual stress [14]. Despite extensive research on the individual effects of salt and heat stress on rice, the mechanisms by which their combined effects on rice physiology remain elusive [15,16]. Notably, existing studies have confirmed that there is significant genetic variability in the response of different rice genotypes to individual and combined salt and heat stress; contrasting tolerant and sensitive genotypes are critical materials for dissecting the metabolic and molecular mechanisms underlying differential stress susceptibility, as well as for mining elite stress-tolerant genes for rice breeding [17,18]. Meanwhile, rice is generally recognized as a moderately salt-tolerant crop, while the breeding of mainstream commercial rice varieties has long been centered on improving yield potential, and the genetic improvement of synergistic tolerance to multiple abiotic stresses (including combined salt and heat stress) has been seriously neglected.

Carbon (C)–nitrogen (N) metabolism serves as the primary energy reserve and carbon source for rice panicle development, and its synthesis and remobilization are strongly linked to photosynthetic efficiency and oxidative balance [17]. ROS overaccumulation inhibits carbon metabolism enzymes (e.g., sucrose synthase, starch hydrolase), while photosynthetic damage reduces nonstructural carbohydrate (NSC) precursor supply [9]. Together, these effects lead to insufficient NSC allocation to panicles, resulting in spikelet abortion and heading delay [18]. However, how combined salt and heat stress disrupts these processes to inhibit panicle development and reduce yield has not been systematically elucidated. Against this background, this study hypothesized that the combined salt and

https://doi.org/10.3390/antiox15030308

heat stress synergistically aggravates oxidative stress, photosynthetic damage, disrupts C-N metabolism and remobilization, interferes with panicle development, and causes yield loss. To verify this hypothesis and address the above research gaps, the core objectives of this study are: (1) to compare the differential effects of single salt, single heat, and combined salt and heat stress on rice oxidative homeostasis, ion balance, and photosynthetic performance; (2) to elucidate the regulatory mechanism of combined salt and heat stress on key enzyme activities of carbon and nitrogen metabolism, as well as the accumulation and remobilization of non-structural carbohydrates and nitrogen in rice; and (3) to clarify the cascade pathway of combined salt and heat stress leading to rice yield loss, and provide a targeted theoretical basis for the genetic improvement of multi-stress tolerance in commercial rice varieties.

2. Materials and Methods

2.1. Plant Materials and Growth Conditions

A pot experiment was conducted at the National Center of Technology Innovation for Saline–Alkali Tolerant Rice (18°35′ N, 109°17′ E). Seeds of a widely cultivated rice cultivar 'Ruanhuayoujinsi' in South China were surface-sterilized with 1% NaClO for 10 min, rinsed with distilled water, soaked in water at 30 °C for 24 h, and allowed to germinate at 32 °C for 48 h. Germinated seeds were sown in 72-hole seedling trays filled with nutrient soil (peat: vermiculite: perlite = 3:1:1). On 14 July 2026, rice seedlings were transplanted at the three-leaf stage (25 d after sowing), with two seedlings per hill and three hills per pot to ensure transplanting survival rate. At seven days after transplanting (seedling recovery period), standard thinning was performed: only one uniformly growing and robust seedling was retained per hill, resulting in a final total of three rice plants per pot for all treatments throughout the whole growth period (Figure S1). The standard normal growth conditions during this period were based on the measured average meteorological data of the experimental site during the rice growth period: natural day/night photoperiod of 13–14 h light per day, average daily maximum temperature of 29.4 °C, average daily minimum temperature of 25.3 °C, with a daily average temperature of 27.3 °C, average daily total solar radiation of 17.1 MJ m^{-2} (corresponding to an average daily photosynthetic photon flux density of ~729.1 μmol m^{-2} s^{-1}, Figure S2), and relative humidity of 60–80%. Based on the confirmed transplanting date of 14 July 2026, key growth stages of rice were clearly defined using days after transplanting (DAT) as the unified time scale throughout the experiment: The full experiment adopted a unified timeline with 14 July 2026 as the transplanting date (0 days after transplanting, DAT). Key growth stages and experimental operations are detailed below: 20 DAT: Salt stress treatment initiated; 49 DAT: 5-day continuous heat stress treatment initiated at the panicle initiation stage; 54 DAT: Heat stress terminated; 73 DAT: Full heading stage; 98 DAT: Physiological maturity stage; plants harvested for yield trait evaluation and final biomass/nutrient accumulation determination.

The pots used were 36 cm (top diameter) × 26 cm (bottom diameter) × 28 cm (height). The soil was air-dried, sieved (2 mm), and filled with 15 kg of mixed soil. The 2.76 g urea per pot (containing 46% N) applied at a ratio of 1:1:1 for basal: tillering: panicle fertilizer; 3.03 g superphosphate (containing 12% P_2O_5) applied once as basal fertilizer; and 1.18 g potassium chloride (containing 60% K_2O) was applied at a 1:1 ratio for basal: panicle fertilizer. This fertilization scheme was based on Li et al. [19] to ensure sufficient nutrient supply and eliminate intraspecific nutrient competition. The normal growth state of rice plants in pots throughout the heading stage is shown in Figure S1. The experimental paddy soil used for potting contained 22.6 g kg^{-1} organic matter, 1.32 g kg^{-1} total N, 45.8 mg kg^{-1}

https://doi.org/10.3390/antiox15030308

available P, 81.5 mg kg^{-1} available K, with an initial pH of 6.8 and electrical conductivity (EC) of 0.12 dS m^{-1}.

2.2. Experimental Design

The experiment was arranged in a completely randomized design (CRD) with four treatments: control (CK), salt stress, heat stress, and combined salt + heat stress. A total of 80 pots were used in the experiment, with 20 pots per treatment. For each treatment, the 20 pots were randomly divided into 4 independent groups, with 5 pots per group. Each group of 5 pots was defined as one independent experimental unit, which also served as one independent biological replicate for subsequent index determination and statistical analysis. All pots were randomly placed in the natural field conditions in a rain-proof screen house. To minimize the effects of environmental gradients (e.g., light, temperature, and air flow), all pots were randomly rotated every three days. CK: No salt or heat stress; irrigated with tap water and grown under normal temperature throughout the growth period. Salt stress: Salt stress treatment was initiated at the tillering stage (20 days after transplanting, DAT), by irrigating with NaCl solution with an electrical conductivity (EC) of 3.9 dS m^{-1}, to maintain a stable target EC of 4.1 $\pm$ 0.2 dS m^{-1} in the root zone (5–10 cm soil depth) throughout the entire growth period. Soil EC was measured in situ directly using a calibrated portable soil conductivity meter (2266FS, Spectrum Technologies, Inc., Aurora, IL, USA). For each pot, three replicate measurement points were randomly selected in the root zone, and the average value was taken as the soil EC of the pot. Monitoring was performed every 3 days; if the measured EC fell below the target range of 4.1 $\pm$ 0.2 dS m^{-1}, supplementary irrigation with the 3.9 dS m^{-1} NaCl solution was performed immediately to restore and maintain the stable target salinity level [19]. Heat stress: Heat stress treatment was conducted at the panicle initiation stage (49 days after transplanting). The plants grown under normal conditions were transferred to a climate chamber at the panicle initiation stage and exposed to 36 °C/30 °C day/night temperature for 5 days [10]. Combined salt + heat stress: plants were exposed to the combination of salt and heat stress treatments. All potted plants from all treatment groups were synchronously transferred to identical programmable artificial climate chambers to eliminate the interference of environmental heterogeneity. To avoid pseudoreplication and eliminate the interference of environmental gradients within the climate chambers, all pots from all treatment groups were completely randomly arranged in the chambers, and their positions were randomly re-rotated every 2 days throughout the entire stress treatment period. To ensure temperature was the only variable in the experiment, all environmental parameters except the day/night temperature regime were set completely consistent across all chambers and treatment groups, including photosynthetic photon flux density of 1200 μmol $m^{-2}s^{-1}$, relative humidity of 75 $\pm$ 5%, and a photoperiod of 12 h/12 h (day/night). The temperature regime was set as follows: the CK and salt were maintained at the field's natural normal temperature regime of 29.35 °C/25.30 °C (day/night). After the heat stress, all potted plants from all treatment groups were moved back to the original natural environment, and grown with conventional management until physiological maturity.

2.3. Measurement Indexes and Methods

2.3.1. Oxidative Stress and Antioxidant Enzyme Activity

At the post-heat stress (exposure to 36 °C/30 °C day/night for 5 days at panicle initiation) and full heading stage (about 25 days after panicle initiation), the top fully expanded leaves were collected and stored at –80 °C to estimate ROS accumulation and antioxidant enzyme activity. The H_2O_2, O_2^-, malondialdehyde (MDA) content, superoxide dismutase (SOD), peroxidase (POD), and catalase (CAT) activities were determined via assay kits provided by Jiangsu Jingmei Biotechnology Co., Ltd. (Yancheng, China). The detailed measurement method was found in Supplementary Material.

2.3.2. Photosynthetic Parameters and Aboveground Biomass

At the post-heat stress and heading stage, photosynthetic parameters of the top fully expanded leaf were measured using a Li-6800XT portable photosynthetic analyzer (LiCor, Lincoln, NE, USA), including P_n, stomatal conductance (g_s), intercellular CO_2 concentration (C_i), and transpiration rate (Tr). At the heading and maturity stages, six representative individual rice plants (hills) were randomly sampled from each treatment. The complete aboveground part of each plant, including the main stem and all effective tillers, was rinsed with distilled water, and then partitioned into stem sheaths, leaves, and panicles.

2.3.3. C and N Metabolism Enzyme Activity

At the post-heat stress and heading stage, the NSC metabolism enzyme activities, such as sucrose phosphate synthase (SPS), sucrose synthase (SS), and starch synthase (SSS), were estimated. Fresh stem-sheath tissue (0.5 g) was determined using the detection kits provided by Jiangsu Jingmei Biotechnology Co., Ltd. The activities of leaf N metabolism enzymes such as nitrate reductase (NR), glutamine synthetase (GS), and glutamate synthase (GOGAT) were all determined using the detection kits provided by Jiangsu Jingmei Biotechnology Co., Ltd. The detailed measurement method was found in Supplementary Material.

2.3.4. NSC Content and N Content

The stem-sheath samples were collected at heading and maturity from Section 2.3.2. Samples were dried at 105 °C for 30 min, then at 65 °C to constant weight, and ground into powder. Soluble sugar and starch were determined following the method of Wei et al. [20]. NSC content = soluble sugar + starch; NSC accumulation in the stem (mg plant^{-1}) = Stem sheath biomass × NSC content; NSC translocation in the stem sheath (mg plant^{-1}) = NSC accumulation at heading − NSC accumulation at maturity; NSC translocation rate (%) = (NSC accumulation at heading − NSC accumulation at maturity)/NSC accumulation at heading × 100%. Additionally, the N concentration in each organ at maturity was quantified by the Kjeldahl procedure according to Wei et al. [20]. Dry samples were digested in concentrated H_2SO_4 to transform total N into ammonium sulfate, which was then distilled and titrated for quantification.

2.3.5. Determination of Sodium (Na^+) and Potassium (K^+) Content

At the post-heat stress and heading stage, leaf and stem-sheath samples were dried and ground into powder. Approximately 0.1 g of dry samples was digested with concentrated HNO_3, and then diluted with deionized water. The K^+ and Na^+ contents were determined via flame photometer (M410, Sherwood Scientific, Cambridge, UK), following the method described by Li et al. [19].

2.3.6. Gene Expression Determination

To quantify transcript-level shifts across stress intensities and durations, representative photosynthetic, ROS-scavenging Na^+ and K^+ transport genes identified in our prior RNA-seq [3] dataset were assayed by qRT-PCR, and the key gene information is shown in Table S1. All primers are listed in Supplementary Table S2. The $2^{-\Delta\Delta CT}$ method, as described by Livak et al. [21], was used to calculate the relative gene expression levels using the mean value of four replicates. The detailed measurement method is found in Supplementary Material.

https://doi.org/10.3390/antiox15030308

2.3.7. Panicle Development and Yield Traits

Fifteen main panicles of uniform size were randomly selected, and the length of the first internode, panicle exsertion, and panicle length were measured. Panicle exsertion (%) = (panicle exsertion length/panicle length) × 100%. Yield and its traits were estimated at maturity. Six plants per treatment were harvested to determine panicle number per plant, grain number per panicle, seed setting rate, 1000-grain weight, and yield per plant (adjusted to 14% moisture content).

2.4. Data Analysis

Data processing was performed with Excel 2021 and SPSS 26.0 (SPSS Inc., Chicago, IL, USA); statistical evaluation included analysis of variance (ANOVA). For multiple pairwise comparisons among treatments, Tukey's Honest Significant Difference (HSD) test was performed at the $p < 0.05$ significance level, to strictly control the Type I error rate in multiple comparisons. To test the main effects of salt, heat stress, and their interaction effect on all measured parameters, a two-way ANOVA with a completely randomized 2 × 2 factorial design was performed. A significant salt × heat interaction effect ($p < 0.05$) was considered the statistical basis for the presence of a synergistic effect between salt and heat stress; for variables with a non-significant interaction ($p > 0.05$), the combined effect was described as an additive or more severe combined effect. Principal component analysis (PCA) was performed using the built-in PCA analysis plug-in of Origin 2023 (OriginLab Corp., Northampton, MA, USA). Given that the variables included in the analysis have different scales, units, and numerical ranges, Z-score standardization (mean centering and variance scaling) was strictly performed on all variables via the plug-in's built-in standard preprocessing function before formal analysis, to eliminate the interference of dimension differences on principal component weighting and analysis results. Graphs were drawn using Origin 2023 (OriginLab Corp., Northampton, MA, USA).

3. Results

3.1. Effects of Salt and Heat Stress on Oxidative Stress

Combined salt + heat stress induced the most severe oxidative stress, with significantly higher ROS accumulation and oxidative damage than single stresses (Figure 1). Post-heat stress, H_2O_2 content, and O_2^- production rate in salt + heat stress were 1.3 and 1.5 times higher than CK, and 14.5% and 22.1% higher than heat stress, and 56.6% and 72.2% higher than salt stress. The MDA content in salt + heat stress was 48.3% higher than CK, indicating extensive membrane lipid peroxidation under combined stress conditions. At heading, the combined heat + salt stress was comparatively lower in individual stress than in the combined. Specifically, H_2O_2 content and O_2^- generation rate in salt + heat stress were 67.9% and 60.8% higher than CK. Individual heat and salt stress also enhanced the H_2O_2 and O_2^- contents more than CK, but their values remained relatively lower than those under the combined salt + heat stress treatment. At the post-heat stress stage, the salt × heat interaction effect was significant for H_2O_2 content and O_2^- ($p < 0.05$). At the heading stage, the salt × heat interaction effect was not significant for all oxidative stress-related parameters ($p > 0.05$).

https://doi.org/10.3390/antiox15030308

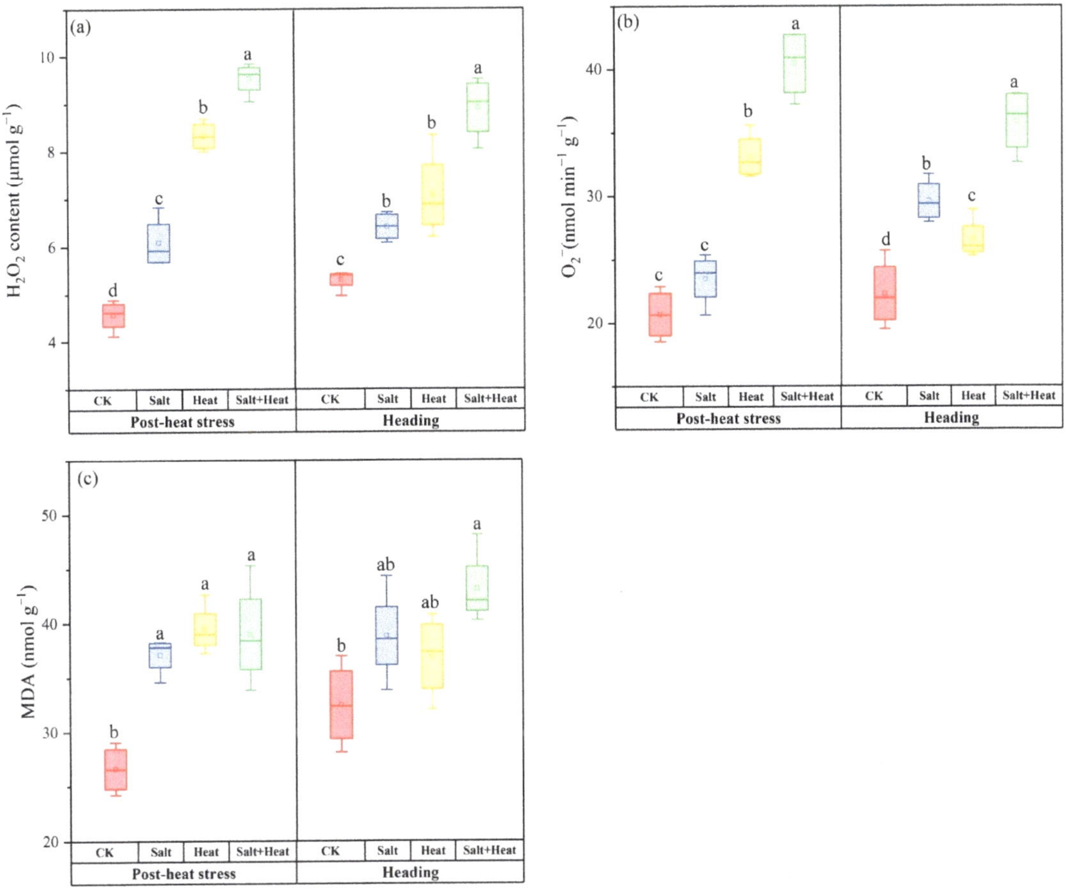

Figure 1. Effect of individual and combined heat and salt stress on (**a**) H_2O_2, (**b**) O_2^-, and (**c**) MDA contents at post-heat stress and heading stage. Data are mean ± SE (n = 4, where n represents the number of independent experimental units (biological replicates) per treatment). Different lowercase letters indicate significant differences among treatments at the $p < 0.05$ level by Tukey's HSD test.

3.2. Effects of Salt and Heat Stress on Antioxidant Enzyme Activities

At post-heat stress, the antioxidant enzyme activities in salt + heat stress were markedly increased. For example, the activities of SOD, POD, and CAT were 64.6%, 69.5%, and 74.8% higher than CK, respectively (Figure 2). At heading, the SOD, POD, and CAT were increased by 58.5%, 52.3%, and 36.2%, respectively. Moreover, heat stress showed lower CAT and POD activities than salt stress, but the difference was not significant. At the post-heat stress stage, the salt × heat interaction effect was significant for SOD and POD activity ($p < 0.05$), showing a synergistic induction effect of combined stress on these two antioxidant enzymes; while the salt × heat interaction effect was not significant for CAT activity ($p > 0.05$), with an additive induction effect under combined stress.

https://doi.org/10.3390/antiox15030308

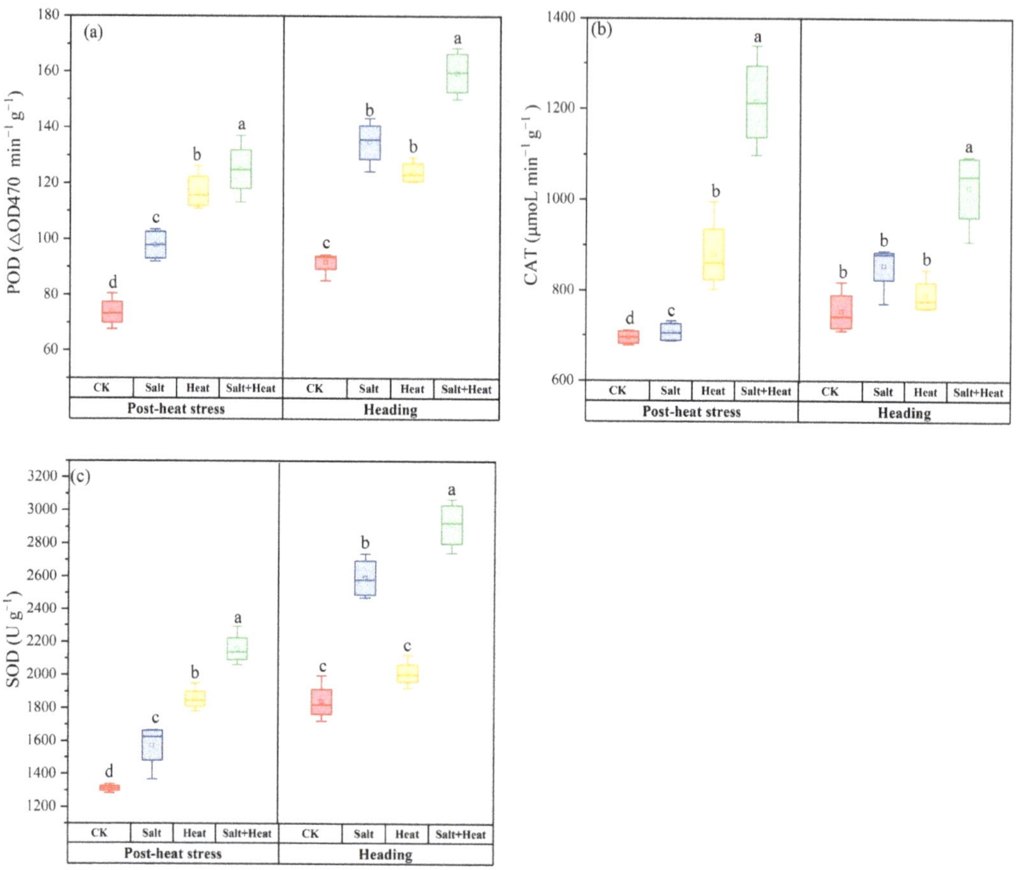

Figure 2. Effect of individual and combined heat and salt stress on (**a**) POD, (**b**) CAT, and (**c**) SOD activities at the post-heat stress and heading stage. Data are mean ± SE (n = 4, where n represents the number of independent experimental units (biological replicates) per treatment). Different lowercase letters indicate significant differences among treatments at the $p < 0.05$ level by Tukey's HSD test.

3.3. Effects of Salt and Heat Stress on Photosynthetic Traits and Aboveground Biomass

The salt + heat stress substantially reduced the net photosynthesis (P_n) and aboveground dry biomass in rice (Figure 3). At post-heat stress, the P_n in salt + heat stress reduced by 45.9% relative to CK. The P_n under heat stress treatment remained statistically similar ($p > 0.05$) to CK but was significantly higher than salt stress. At heading, P_n was 10.0%, 19.7%, and 48.2% lower under heat, salt, and combined salt + heat stress, as compared with CK. Furthermore, the Tr and C_i under heat stress treatment remained statistically similar ($p > 0.05$) to CK but were substantially higher than under salt treatment. For the P_n, the salt × heat interaction effect was extremely significant at both post-heat stress and heading stages ($p < 0.01$).

https://doi.org/10.3390/antiox15030308

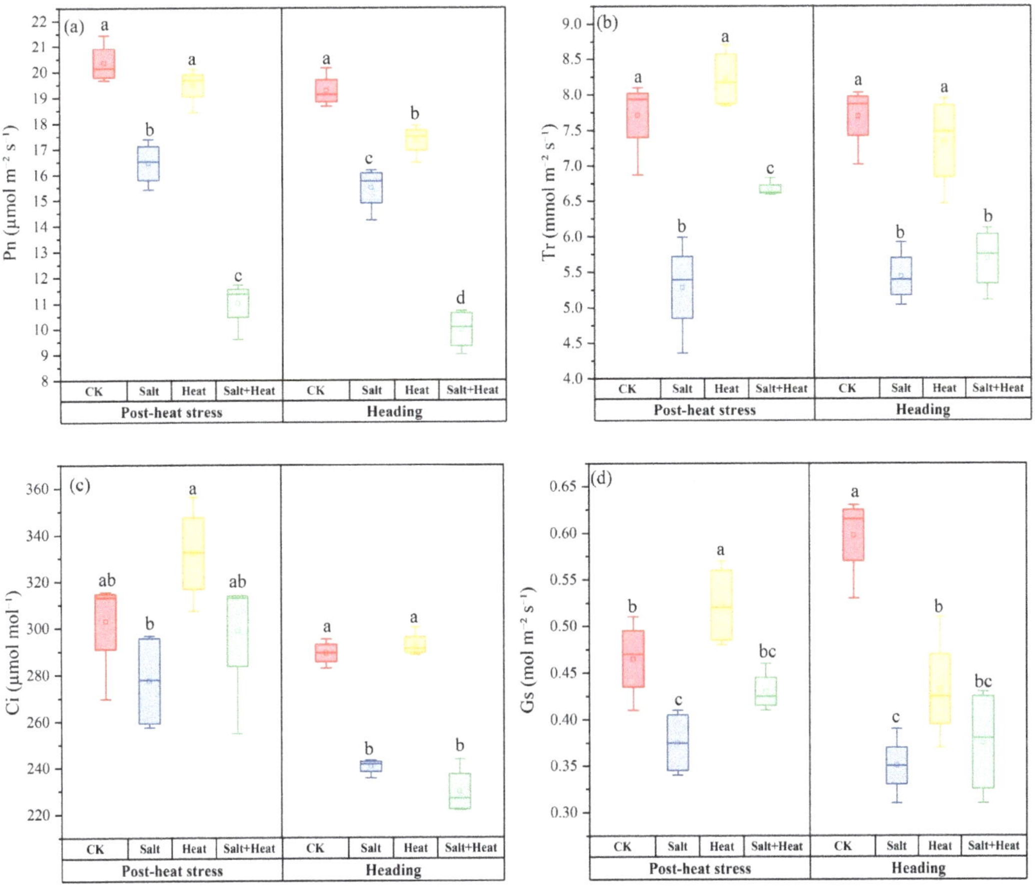

Figure 3. Effect of individual and combined heat and salt stress on (**a**) Pn, (**b**) Tr, (**c**) Ci, and (**d**) Gs at the post-heat stress and heading stage. Data are mean ± SE (n = 4, where n represents the number of independent experimental units (biological replicates) per treatment). Different lowercase letters indicate significant differences among treatments at the $p < 0.05$ level by Tukey's HSD test.

Combined salt and heat stress significantly altered the dry matter accumulation pattern and translocation efficiency in rice (Figure S3). At the heading stage, the leaf biomass of rice under combined salt and heat stress was reduced by 16.3% compared with CK, while the stem sheath and aboveground biomass were decreased by 31.3% and 32.1%, respectively. At maturity, no difference was found in leaf biomass among all treatments, whereas the salt + heat stress resulted in reduced stem sheath and aboveground biomass compared to individual stress. In addition, the leaf dry matter and stem-sheath dry matter translocation of rice under combined salt and heat stress were lower than those under individual stress and the CK. The salt × heat interaction effect was extremely significant for aboveground biomass at maturity, stem-sheath biomass at heading and maturity, and stem-sheath biomass translocation ($p < 0.01$), showing a synergistic inhibition effect.

3.4. Effects of Salt and Heat Stress on N Accumulation and Translocation

The salt + heat stress treatment significantly altered N accumulation and translocation in rice (Figure 4). Total N accumulation in the whole plant at maturity was substantially reduced by salt + heat stress treatment, and the inhibitory effect was more pronounced under combined stress than under individual stress treatment. For organ-specific total N accumulation per plant, leaf N accumulation at the heading stage showed no significant difference among all treatments, whereas leaf N accumulation under salt + heat stress treatment was increased by 12.9% compared with CK at maturity. In contrast, stem-sheath N accumulation at the heading stage was significantly lower under stress conditions, with salt + heat stress treatment showing a 29.6% reduction compared with CK, whereas salt and heat stress treatments decreased by 16.1% and 8.5%, respectively. However, at maturity, no significant difference was found among all treatments in stem-sheath N accumulation. The total N translocation amount from leaf to grain under salt + heat stress treatment was significantly lower than that of CK. Similarly, the stem-sheath total N translocation in salt + heat stress treatment decreased by 54.9% compared with CK, while single salt and heat stress treatment caused reductions in N translocation of up to 31.0% and 18.7%, respectively. The salt × heat interaction effect was significant for total N accumulation at maturity ($p < 0.05$), indicating a synergistic inhibition of total N accumulation under combined stress.

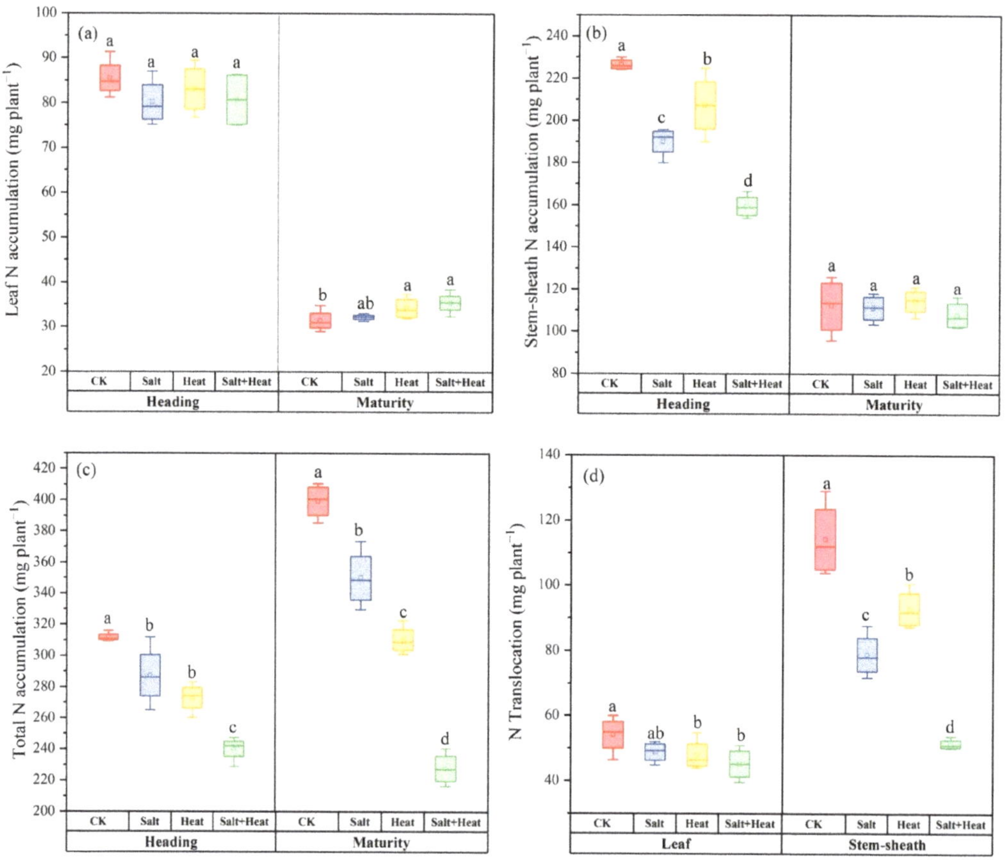

Figure 4. Effect of individual and combined heat and salt stress on (**a**) Leaf N accumulation, (**b**) Stem-sheath accumulation, (**c**) Total N accumulation, and (**d**) N translocation at leaf and stem sheath. Data are mean ± SE ($n = 4$, where n represents the number of independent experimental units (biological replicates) per treatment). Different lowercase letters indicate significant differences among treatments at the $p < 0.05$ level by Tukey's HSD test.

https://doi.org/10.3390/antiox15030308

3.5. Effects of Salt and Heat Stress on Nitrogen-Related Metabolism Enzymes

The salt + heat stress treatment exerted significant and stage-specific effects on the activities of N assimilation enzymes, i.e., NR, GS, and GOGAT (Figure 5). The NR activity in rice leaves under individual heat and salt stress treatment was reduced by 8.1% and 59.3% compared with the CK at post-heat stress. The combined salt + heat stress led to a more severe reduction of 66.8% in NR activity, whereas at the heading stage, the NR activity in all stress treatments was further decreased relative to CK, with the most pronounced inhibition (29% reduction) being observed in the salt + heat stress treatment.

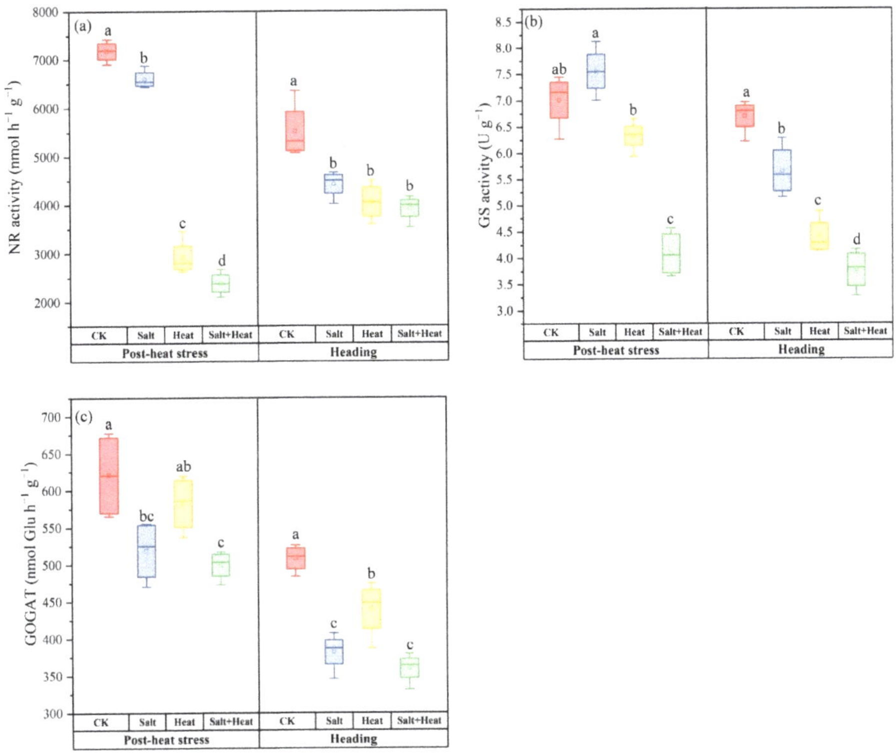

Figure 5. Effect of individual and combined heat and salt stress on nitrogen-related metabolism enzymes, i.e., (**a**) NR, (**b**) GS, and (**c**) GOGAT activity at the post-heat stress and heading stage. Data are mean ± SE (n = 4, where n represents the number of independent experimental units (biological replicates) per treatment). Different lowercase letters indicate significant differences among treatments at the $p < 0.05$ level by Tukey's HSD test.

Furthermore, the GS activity in salt stress treatment was slightly increased by 7.8% relative to CK, while salt + heat stress treatment reduced GS activity by 41.8% at post-heat stress. At the heading stage, the GS activity in CK was the highest, whereas all stress treatments showed a substantial reduction in GS activity. The GOGAT activity in salt + heat stress treatment was reduced by 19.6% compared with CK during post-heat stress. At the heading stage, the GOGAT activity in salt + heat stress treatment was reduced by 29.2%, whereas the reductions in salt and heat stress treatments were 24.9% and 13.5%, as compared to CK. At the post-heat stress stage, the salt × heat interaction effect was extremely significant for GS activity ($p < 0.01$), while it was not significant for NR and GOGAT activities ($p > 0.05$). At the heading stage, the salt × heat interaction effect was significant for NR activity ($p < 0.05$), and was not significant for GS and GOGAT activities ($p > 0.05$)

3.6. Effects of Salt and Heat Stress on K^+ and Na^+ Content, and K^+/Na^+ Ratio in Leaves and Stem Sheaths of Rice

At the post-heat stress, salt + heat stress treatment significantly disrupted ion homeostasis in rice, in terms of K^+, Na^+ content, and K^+/Na^+ ratio between leaves and stem sheaths (Figure 6). For Na^+ accumulation, both single and combined salt + heat stresses induced a significant increase in Na^+ content in leaves and stem sheaths compared with the CK. In contrast to Na^+, K^+ content in leaves and stem sheaths was significantly reduced by salt and heat stress treatment, with the most severe reduction observed under salt + heat stress treatment. The K^+/Na^+ ratio was substantially decreased under stress conditions, with the lowest ratio observed under salt + heat stress treatment. The salt × heat interaction effect was significant for Na^+ content and K^+/Na^+ ratio ($p < 0.05$).

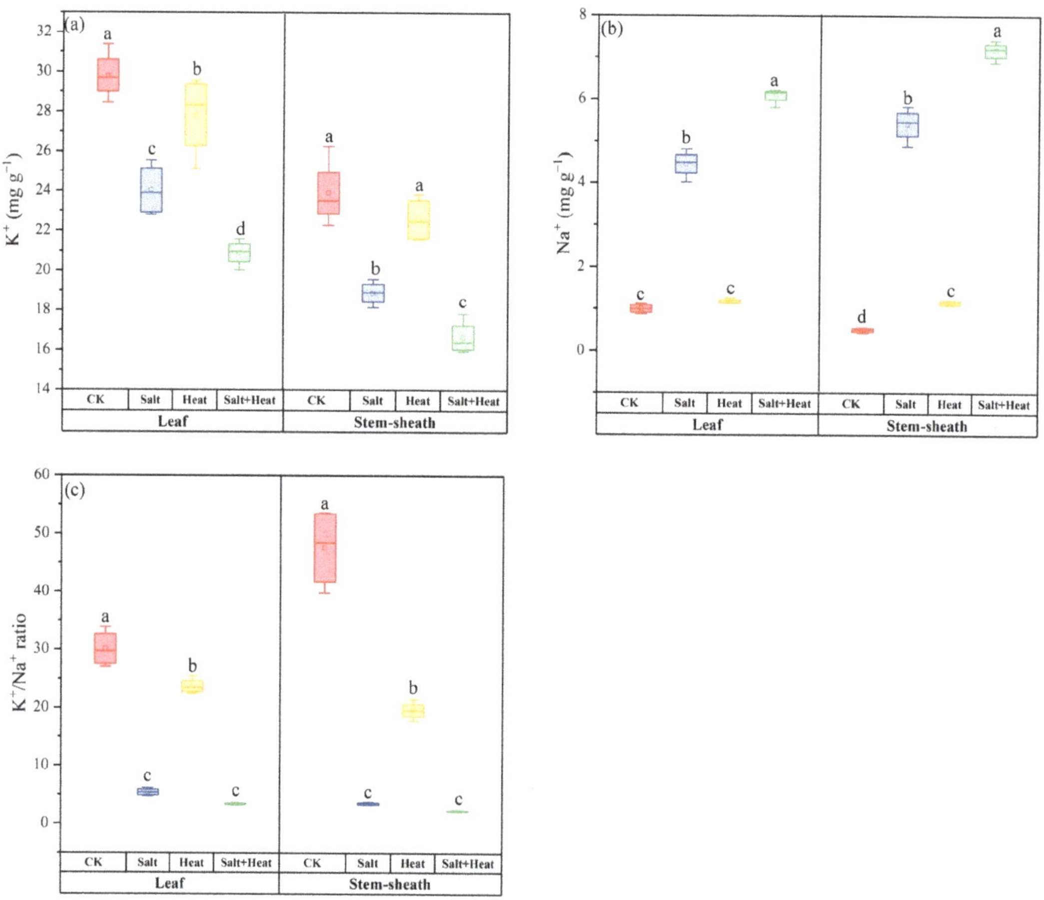

Figure 6. Effect of individual and combined heat and salt stress on (**a**) K^+ content, (**b**) Na^+ content, and (**c**) K^+/Na^+ ratio, and at the post-heat stress. Data are mean ± SE ($n = 4$, where n represents the number of independent experimental units (biological replicates) per treatment). Different lowercase letters indicate significant differences among treatments at the $p < 0.05$ level by Tukey's HSD test.

https://doi.org/10.3390/antiox15030308

3.7. Effects of Salt and Heat Stress on Stem NSC-Related Traits

At the heading stage, both single and combined stresses affected NSC content and accumulation in the stem sheath, but the response patterns differed (Table 1). Relative to CK, single salt stress decreased NSC content by 3.7%, while combined salt + heat stress led to a 4.3% reduction in NSC content. At the heading stage, the NSC accumulation was 9.3%, 10.9%, and 34.3% lower under salt, heat stress, and salt + heat stress treatment than CK. At maturity, stem-sheath NSC content under combined stress treatments was significantly lower than CK, with the most severe reduction observed under salt + heat stress. Furthermore, the NSC translocation rate in CK was 40.24%, which decreased to 33.93%, 35.66%, and 32.75%, respectively, under salt, heat, and salt + heat treatment. The salt × heat interaction effect was extremely significant for NSC accumulation at both heading and maturity stages ($p < 0.01$).

Table 1. Effects of salt and heat stress on stem NSC-related traits.

Treatments	NSC Content at Heading (mg g^{-1})	NSC Content at Maturity (mg g^{-1})	NSC Accumulation at Heading (g Plant^{-1})	NSC Accumulation at Maturity (g Plant^{-1})	NSC Translocation (g Plant^{-1})	NSC Translocation Rate (%)
CK	204.98 ± 3.94 ^{b}	163.91 ± 2.98 ^{a}	6.96 ± 0.36 ^{a}	4.16 ± 0.27 ^{a}	2.80 ± 0.23 ^{a}	40.24 ± 2.51 ^{a}
Salt	197.26 ± 2.49 ^{c}	156.06 ± 3.37 ^{b}	6.31 ± 0.17 ^{b}	4.17 ± 0.19 ^{a}	2.14 ± 0.11 ^{b}	33.93 ± 1.79 ^{b}
Heat	216.37 ± 3.10 ^{a}	163.97 ± 3.73 ^{a}	6.20 ± 0.13 ^{b}	3.99 ± 0.10 ^{a}	2.21 ± 0.13 ^{b}	35.66 ± 1.66 ^{b}
Salt + Heat	196.10 ± 3.39 ^{c}	148.46 ± 6.00 ^{c}	4.57 ± 0.16 ^{c}	3.07 ± 0.03 ^{b}	1.50 ± 0.16 ^{c}	32.75 ± 1.07 ^{c}

Note: Different lowercase letters indicate significant differences among treatments at the $p < 0.05$ level by Tukey's HSD test.

3.8. Effects of Salt and Heat Stress on Carbon-Related Metabolic Enzymes

The salt + heat stress treatment induced stage-specific and enzyme-specific responses SPS, SS, and SSS enzyme activities in the stem sheath (Figure 7). Heat stress decreased SPS activity by 10.3% compared with the CK, while salt stress and salt + heat stress caused 23.9% and 33.3% reduction in SPS activity. At the heading stage, the maximum SPS activity was recorded in CK, whereas all stress treatments showed significant reductions in SPS activity. The SS and SSS activity in salt + heat stress was significantly lower than single salt stress, heat stress, and CK during post-heat stress and heading stage. The salt × heat interaction effect was significant for SS and SPS activities at the heading stage ($p < 0.01$).

3.9. Effects of Salt and Heat Stress on Relative Gene Expression Levels

Antioxidant defense-related genes, i.e., *OsAPX2*, *OsSODCC1*, and *OsAPX1*, were significantly upregulated under both single and combined stresses, whereas the upregulation was more pronounced under salt + heat stress treatment (Figure 8). In contrast, photosynthesis-related genes, i.e., *OsRbcS2* and *OsRbcS3*, were predominantly downregulated under salt stress conditions, and the suppression was more severe under combined stress. The expression of *Ospsbo* and *OsRbcS3* was increased under heat stress compared with CK, while no significant difference was found regarding *OsRbcS2* expression between CK and heat stress.

https://doi.org/10.3390/antiox15030308

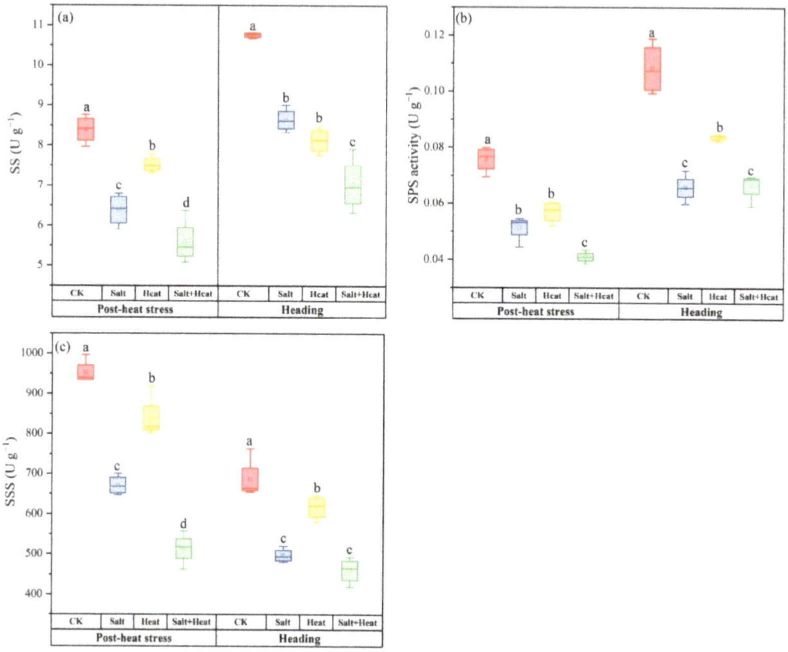

Figure 7. Effect of alt and heat stress on stem carbon-related metabolic enzymes: (**a**) SS, (**b**) SPS, and (**c**) SSS at the post-heat stress and heading stage. Data are mean ± SE ($n = 4$, where n represents the number of independent experimental units (biological replicates) per treatment). Different lowercase letters indicate significant differences among treatments at the $p < 0.05$ level by Tukey's HSD test.

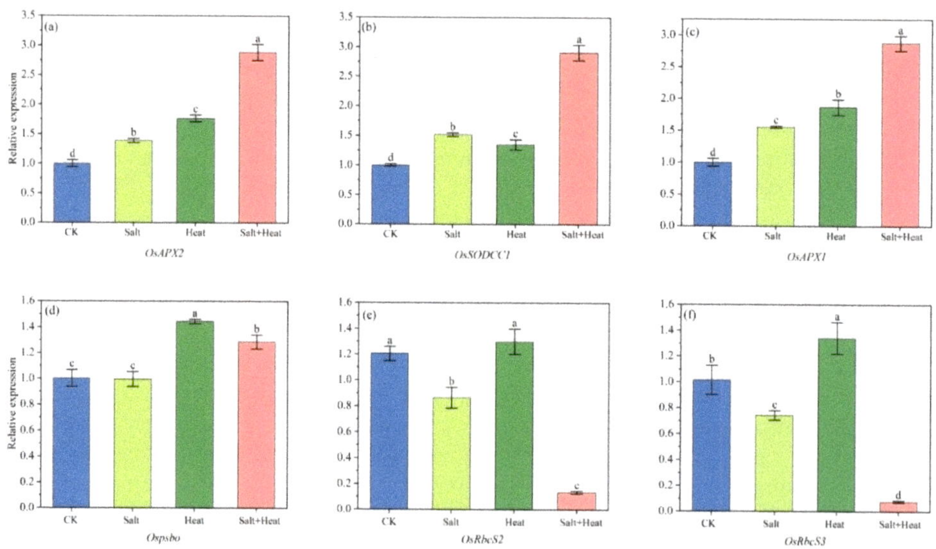

Figure 8. Effect of salt and heat stress on relative expression level of (**a**) *OsAPX2*, (**b**) *OsSODCC1*, (**c**) *OsAPX1*, (**d**) *Ospsbo*, (**e**) *OsRbcS2*, and (**f**) *OsRbcS3* in rice at the post-heat stress using qRT-PCR. Data are mean ± SE ($n = 4$, where n represents the number of independent experimental units (biological replicates) per treatment). Different lowercase letters indicate significant differences among treatments at the $p < 0.05$ level by Tukey's HSD test.

https://doi.org/10.3390/antiox15030308

Notably, all Na^+ and K^+ transport genes, i.e., *OsSOS1, OsSOS2, OsSOS3, OsHKT1;3, OsHKT1;5, OsNHX4,* and *OsNHX5,* were significantly affected under stress treatments (Figure 9). The SOS pathway genes, i.e., *OsSOS1, OsSOS2,* and *OsSOS3,* were significantly downregulated under salt + heat stress, respectively. For the high-affinity K^+ transporter genes, i.e., *OsHKT1;3* and *OsHKT1;5,* and their relative expressions under salt + heat stress were 68.1% and 65.9% lower than CK. Additionally, the vacuolar Na^+/H^+ antiporter genes, i.e., *OsNHX4* and *OsNHX5,* exhibited 59.6% and 78.6% downregulations under salt + heat stress, respectively, which were 2.0- and 2.9-fold higher than those under single salt stress. For antioxidant defense-related genes (*OsAPX2, OsSODCC1,* OsAPX1), photosynthesis-related genes (*OsRbcS2, OsRbcS3*), and ion transport-related genes (*OsSOS1, OsSOS2, OsSOS3, OsHKT1;3, OsHKT1;5, OsNHX4*), the salt × heat interaction effect was extremely significant ($p < 0.01$), showing a synergistic regulatory effect of combined stress on the transcription of these genes.

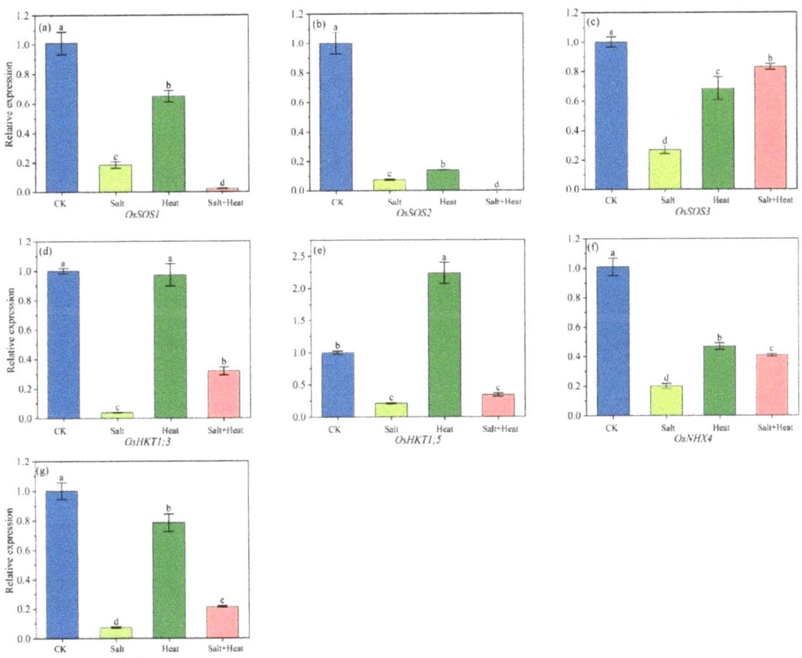

Figure 9. Effect of salt and heat stress on relative expression level of K^+ and Na^+ transport genes: (**a**) *OsSOS1*, (**b**) *OsSOS2*, (**c**) *OsSOS3*, (**d**) *OsHKT1;3*, (**e**) *OsHKT1;5*, (**f**) *OsNHX4*, and (**g**) *OsNHX5* in rice at the post-heat stress using qRT-PCR. Data are mean ± SE ($n = 4$, where n represents the number of independent experimental units (biological replicates) per treatment). Different lowercase letters indicate significant differences among treatments at the $p < 0.05$ level by Tukey's HSD test.

3.10. Effects of Salt and Heat Stress on Grain Yield and Its Components in Rice

Grain yield and its components were substantially reduced under salt + heat stress compared with individual stress (Table 2). Compared with CK, salt treatment reduced effective panicles per plant by 14.2%; however, salt + heat stress led to 28.6% reduction, which was substantially lower than that in single salt and heat stress. Similarly, grains per panicle were decreased by 17.6% (salt stress) and 10.0% (heat stress), while salt + heat stress treatment resulted in a 30.3% reduction compared with CK. Compared with CK, 1000-grain weight and seed-setting rate were significantly decreased by salt and heat stress, while salt + heat stress treatment resulted in a severe reduction. Compared to CK, the grain

https://doi.org/10.3390/antiox15030308

yield was decreased by 34.6% (salt stress) and 20.5% (heat stress), while salt + heat stress led to a 66.4% reduction in grain yield. The salt × heat interaction effect was extremely significant for panicle development traits and final grain yield ($p < 0.01$), significant for 1000-grain weight, seed-setting rate, and productive panicle per plant ($p < 0.05$).

Table 2. Effects of salt and heat stress on rice yield and its components.

Treatments	Panicle Exsertion (%)	Productive Panicle per Plant	Grains per Panicle	Seed-Setting Rate (%)	1000-Grain Weight (g)	Grain Yield (g Plant^{-1})
CK	96.8 ± 1.4 ^{a}	12.3 ± 0.9 ^{a}	128.6 ± 8.2 ^{a}	81.7 ± 4.3 ^{a}	21.1 ± 0.8 ^{a}	27.1 ± 1.6 ^{a}
Salt stress	70.4 ± 4.5 ^{c}	10.5 ± 0.6 ^{b}	106.0 ± 2.6 ^{c}	75.8 ± 1.6 ^{b}	18.2 ± 0.5 ^{b}	17.7 ± 1.0 ^{c}
Heat stress	82.5 ± 5.5 ^{b}	12.0 ± 0.8 ^{a}	115.2 ± 2.6 ^{b}	75.9 ± 4.3 ^{b}	17.6 ± 0.6 ^{b}	21.6 ± 3.0 ^{b}
Salt + heat stress	55.4 ± 10.4 ^{d}	8.8 ± 0.9 ^{c}	89.4 ± 7.1 ^{d}	59.7 ± 3.4 ^{c}	16.0 ± 0.6 ^{c}	9.1 ± 0.8 ^{d}

Note: Data are mean ± SE ($n = 4$, where n represents the number of independent experimental units (biological replicates) per treatment). Different lowercase letters indicate significant differences among treatments at the $p < 0.05$ level by Tukey's HSD test.

3.11. Principal Component Analysis (PCA)

The first two principal components (PC1 and PC2) explained 60.6% and 23.1% of the total variance, respectively, with a cumulative contribution rate of 83.7% (Figure 10a). The four treatment groups showed a clear separation along the PC1 axis, which represented the core dimension of rice stress response: the CK group was distinctly clustered in the positive direction of PC1, the single salt and single heat stress groups were distributed in the intermediate region, and the combined salt + heat stress group was exclusively clustered in the far negative direction of PC1. This distribution pattern directly confirmed that the combined stress caused a more drastic shift in the overall physiological and molecular state of rice than either single stress, showing a significantly more severe combined effect. The loading plot further revealed two core trait clusters with opposite distribution patterns along the PC1 axis, which corresponded to the key physiological and molecular regulatory pathways revealed in this study. Positive PC1 cluster (traits for normal growth and stress resistance): This cluster was dominated by traits related to ion homeostasis (K^+/Na^+ ratio, relative expression of *OsSOS1/2/3*, *OsHKT1;3*, *OsHKT1;5*, *OsNHX4/5*), photosynthetic performance (P_n, G_s, Tr, relative expression of *Ospsbo*, *OsRbcS2*, *OsRbcS3*), C-N metabolic activity (activities of NR, GS, GOGAT, SPS, SS, and SSS), and subsequent NSC and nitrogen accumulation and translocation efficiency. Negative PC1 cluster (stress response traits): This cluster was mainly composed of oxidative stress-related traits, including ROS accumulation (H_2O_2, O_2^- production rate), membrane lipid peroxidation level (MDA content), antioxidant enzyme activities (SOD, POD, CAT), and relative expression of antioxidant defense-related genes (*OsAPX2*, *OsSODCC1*, *OsAPX1*). These traits were highly positively correlated with each other, and showed a significant negative correlation with the positive PC1 cluster traits.

As illustrated in Figure 10b, the first two principal components (PC1 and PC2) explained 68.8% and 13.2% of the total variance, respectively, with a cumulative contribution rate of 82.0%. Similarly, the four treatment groups were clearly separated along the PC1 axis: the CK group was clustered in the positive direction of PC1, characterized by high NSC and total N accumulation, high translocation efficiency from vegetative organs to panicles, and superior yield component traits; while the combined salt + heat stress group was clustered in the negative direction, with significantly inhibited C-N accumulation and translocation, and severe yield loss.

https://doi.org/10.3390/antiox15030308

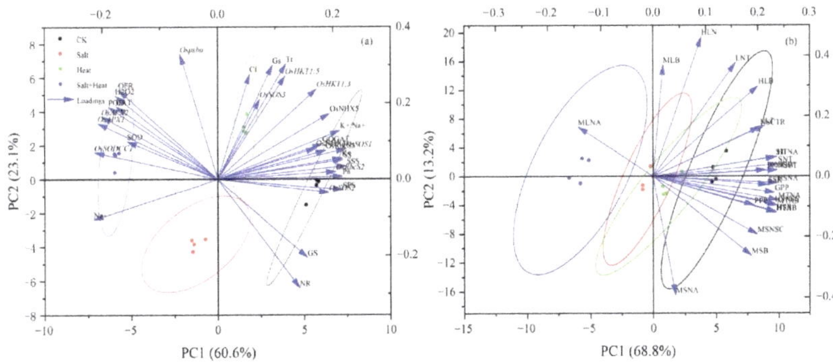

Figure 10. Principal component analysis (PCA) of measured traits: (**a**) PCA of various traits at post-heat stress; (**b**) C and N accumulation and allocation in different plant parts at heading and their translocation to grain.

4. Discussion

4.1. Combined Salt and Heat Stress Aggravates Oxidative Stress and Photosynthetic Damage

Oxidative homeostasis and photosynthesis are important physiological processes in plants prone to abiotic stresses [22–24]. Co-occurrence of two or more stresses may amplify the magnitude of the stress factors, leading to more severe damage in crop plants than a single stress.

In the present study, combined salt + heat stress significantly elevated ROS levels (H_2O_2 and O_2^-) and disrupted antioxidant enzyme balance, which is in line with the findings of Li et al. [3], who found severe damage in tomato plants under salt and heat stress. It was found that salt + heat stress induced extreme ROS accumulation in rice, whereas the H_2O_2 content and O_2^- production rate in salt + heat stress were 1.3 and 1.5 times higher than the CK. This aligns with Rivero et al. [25], who found that combined abiotic stresses often exceed the ROS scavenging capacity of plants, leading to cumulative oxidative damage. Su et al. [26] reported that combined salt + heat stress treatment disrupts the cooperative function of the antioxidant enzyme system, while SOD activity continues to rise, CAT and ascorbate peroxidase (APX) activities were substantially declined, resulting in an imbalance between ROS production and scavenging ability. In the present study, although antioxidant enzymes (SOD, CAT, and POD) were upregulated in salt + heat stress, their activity was insufficient to scavenge excess ROS, resulting in severe membrane damage in terms of lipid peroxidation. Likewise, antioxidant defense-related genes, i.e., *OsAPX2*, *OsSODCC1*, and *OsAPX1*, were upregulated under all stress treatments, but the induction magnitude was most pronounced under salt + heat stress, which indicates that rice plants actively activate the transcriptional expression of antioxidant genes to cope with oxidative stress under combined stress, but the synergistic effect accumulation of ROS under salt + heat stress exceeds the scavenging capacity of the antioxidant system, leading to irreversible cellular damage.

Moreover, ionic imbalance acts as a key mediator through which combined stress amplifies oxidative stress and photosynthetic damage. Na^+ toxicity directly disrupts chloroplast membrane integrity, while K^+ deficiency impairs the structural stability of antioxidant enzymes, leading to insufficient ROS scavenging efficiency [27]. Our results showed that combined salt + heat stress treatment significantly decreased leaf K^+ content, increased Na^+ content, and consequently reduced the K^+/Na^+ ratio. Na^+ and K^+ homeostasis was further confirmed by key ion transport genes involved in Na^+ exclusion, i.e., *OsSOS1*, *OsSOS2*, and *OsSOS3*; Na^+ compartmentalization genes, i.e., *OsNHX4* and *OsNHX5*; and K^+ uptake related genes, i.e., *OsHKT1;3* and *OsHKT1;5*, which were significantly down-

https://doi.org/10.3390/antiox15030308

regulated under salt + heat stress. The downregulation of these genes reduced the ability of rice to maintain ion homeostasis, leading to excessive Na^+ accumulation and K^+ loss. Alzahrani et al. [28] also found that combined stress causes more oxidative damage than either salt or drought stress alone, primarily because salt-induced Na^+ overaccumulation and suppressed K^+ uptake disrupt ion homeostasis, triggering a ROS overproduction and reducing the membrane stability index.

Photosynthetic damage under combined stress is driven by both stomatal and non-stomatal limitations. Salt stress alone induces stomatal closure to reduce ion uptake [5], while heat stress primarily damages the photosynthetic apparatus [16]. High salt (≥100 mM NaCl) combined with heat significantly reduces g_s and P_n, with non-stomatal damage as the dominant factor in tomato [3]. Our results showed similar trends, i.e., salt + heat stress reduced P_n by 45.9% and 48.2% during the post-heat and heading stage, respectively, compared with CK. Studies have shown that high temperature directly disrupts the structural integrity of photosystem II, shortens the crop growth cycle, and ultimately hinders plant growth and development [29,30]. Based on previous research on combined salt and heat stress [3], we hypothesize that the imbalance of the oxidative phosphorylation pathway may be a key mechanism underlying the aggravated photosynthetic damage and oxidative stress observed in this study, which may lead to insufficient ATP synthesis, thereby suppressing both ATP-dependent photosynthetic enzyme activity and the energy supply to the ROS-scavenging system. Combined salt + heat stress inhibited photosynthesis in quinoa more than either single stress, mainly by reducing total chlorophyll content and suppressing stomatal conductance. Heat stress aggravates salt-induced ionic imbalance, jointly triggering oxidative stress that disrupts photosynthetic pigment stability and stomatal regulation [31,32]. We found that leaf Tr and G_s increased under heat stress treatment, whereas salt stress inhibited Tr, G_s, and P_n, indicating fundamental differences in how salt and heat affect photosynthesis. Nevertheless, the photosynthetic inhibition pattern of combined salt + heat stress treatment in alfalfa resembles the effects of salt stress alone [26]. At the molecular level, the photosynthesis-related genes showed divergent responses to different stresses. Specifically, *Ospsbo* and *OsRbcS3* were significantly upregulated under heat stress compared to CK. In contrast, under salt-containing treatments (salt and salt + heat stress), these photosynthetic genes were significantly downregulated, and the suppression was most severe under salt + heat stress. Thus, salt stress may play a dominant role in regulating the photosynthetic physiological processes in rice in response to combined salt + heat stress treatment. It should be noted that chlorophyll fluorescence parameters were not measured in this study, as our work focused on the photosynthetic dark reaction and its downstream carbon–nitrogen metabolic effects. This indicator will be a core focus of our follow-up research to further clarify the photosynthetic damage mechanism of combined salt and heat stress.

4.2. Temporal Dynamics of NSCs: Persistent Pre-Anthesis Synthesis Deficit and Further Failure of Post-Anthesis Remobilization

NSCs stored in rice stem sheath are the core carbon reserve for panicle development and grain filling, whose synthesis, accumulation, and remobilization present strict stage-specific temporal dynamics in the rice growth cycle [9,12,33]. After combined salt and heat stress, the first wave of damage to NSC metabolism occurred, which laid the irreversible foundation for the subsequent continuous NSC deficit (Figure 7). The P_n under salt + heat stress was significantly reduced, which directly and persistently inhibited the de novo synthesis of sucrose and starch, the core components of NSCs. At the molecular level, combined stress significantly downregulated the expression of photosynthesis-related genes, i.e., *OsPsbO*, *OsRbcS2*, and *OsRbcS3* [3]. Rubisco is the rate-limiting enzyme for plant CO_2 fixation, and the reduced expression of its small-subunit genes (*OsRbcS2*, *OsRbcS3*)

https://doi.org/10.3390/antiox15030308

directly inhibits Rubisco activity, leading to insufficient biosynthesis of photosynthates and decreased carbohydrate accumulation in the stem sheath [34,35]. Meanwhile, the synergistic oxidative burst (Figure 1 and Table S3) induced by combined stress further inhibited the activity of key NSC metabolism enzymes: salt stress-induced ROS accumulation has been widely shown to SS and SSS in rice, while heat stress further suppresses these enzymes [36,37]. Our results confirmed that salt + heat stress significantly reduced the activities of SS, SPS, and SSS in the stem sheath, compared with CK after combined salt and heat stress. This persistent inhibition of photosynthetic carbon fixation and key metabolic enzymes directly impaired the basic synthesis capacity of NSCs from the early panicle development stage, which is consistent with previous findings that abiotic stress at panicle initiation reduces pre-anthesis carbohydrate accumulation in rice [9,20].

The present study showed that the combined salt + heat stress had the lowest stem-sheath NSC content and total NSC accumulation at the heading stage, which were significantly lower than CK, respectively, with no transient accumulation observed (Table 1). This significant NSC reserve deficit at the heading stage is the direct phenotypic result of the persistent damage of combined stress to photosynthetic carbon fixation and C metabolism enzymes from panicle initiation to the heading stage. The results were consistent with Wei et al. [20], who found that combined salt and drought stress reduces photosynthetic assimilation and inhibits NSC accumulation in the rice stem sheath. Zhen et al. [9] confirmed that heat-induced photosynthetic damage reduces NSC synthesis in rice.

Our results showed that the NSC translocation amount and translocation rate from the stem sheath to grain under combined stress were significantly lower than CK, respectively, and were also significantly lower than the single salt and heat stress (Table 1). This indicates that the core damage of combined stress to C metabolism is dual, which impairs the NSC "synthesis-remobilization" balance: on the one hand, it causes a significant reduction in pre-anthesis NSC reserve accumulation; on the other hand, it further inhibits the post-anthesis remobilization and translocation of the already limited NSC reserve, which forms a double bottleneck for grain carbon supply.

4.3. Combined Stress Triggers Aberrant Whole-Plant Nitrogen Partitioning Dynamics: From Pre-Anthesis Assimilation and Allocation Imbalance to Post-Anthesis Remobilization Blockage

In this study, the impact of combined salt and heat stress on rice N metabolism led to the systematic reshaping of the whole-plant N partitioning pattern, which was underpinned by reduced total N uptake, and severe blockage of post-anthesis N remobilization from vegetative source organs to reproductive sink organs [26,27]. At the panicle initiation stage, the initial inhibition of N uptake and assimilation established the baseline for subsequent partitioning imbalance. Combined stress suppressed root N acquisition: salt-induced Na^+ overaccumulation competed with inorganic N (NO_3^-, NH_4^+) uptake via root transporters [38], while heat stress reduced root vitality [39], and the concurrent downregulation of ion homeostasis genes (*OsSOS1/2/3, OsHKT1;3, OsHKT1;5, OsNHX4/5*) further weakened mineral nutrient absorption [27,28]. Meanwhile, the activities of key N assimilation enzymes (NR, GS, GOGAT) in leaves were significantly reduced under combined stress, limiting the conversion of inorganic N into metabolically active organic N [40]. These effects ultimately resulted in a 42.9% reduction in total whole-plant N accumulation at maturity, defining the limited size of the total plant N pool.

Notably, N was not uniformly reduced across vegetative organs even at this stage, but showed a clear survival-prioritized redistribution pattern. At the heading stage, stem-sheath total N accumulation was significantly reduced under combined stress compared with CK, while leaf total N accumulation showed no significant difference among treatments. This organ-specific divergence indicated that under the limited total N pool, rice preferentially allocated available N to leaf tissues (core photosynthetic organs) to maintain

https://doi.org/10.3390/antiox15030308

basic carbon assimilation, at the cost of N reserve accumulation in stem sheaths (the primary temporary N storage pool for panicle development) [17,41]. This pre-anthesis allocation imbalance directly reduced the N reserve pool available for post-anthesis remobilization.

The most prominent aberrant N redistribution occurred during the grain filling stage (heading to maturity), characterized by a near-complete blockage of N remobilization from vegetative organs to grains. In rice, the majority of grain nitrogen (N) comes from the remobilization of pre-stored N in vegetative organs after heading [42], but our results showed that combined stress reduced stem-sheath-to-grain N translocation by 54.9% compared with CK, with a similarly significant reduction in leaf-to-grain N translocation. This remobilization blockage drove two key redistribution phenotypes: (1) N that should have been transported to grains was trapped in senescing leaves, resulting in a 12.9% increase in total leaf N accumulation at maturity under combined stress; (2) stem-sheath N accumulation, which was significantly lower at heading, showed no significant difference from CK at maturity, confirming the remobilization failure of pre-stored N.

4.4. C and N Deficiency Inhibits Panicle Development and Reduces Yield

Our results showed that combined salt and heat stress significantly reduced the N remobilization rate in rice, causing severe blockage of N translocation from vegetative organs to grains. This blockage not only led to increased total N retention in mature leaves and stem sheaths but also persistently reduced the availability of metabolically active, remobilizable N in leaves from the heading stage onward (Figures 4 and 5). The reduced leaf N availability further disrupted C-N metabolic balance, inhibited photosynthetic carbon assimilation, and ultimately decreased NSC accumulation and effective substrate supply to developing grains (Table 2 and Figure 4).

These findings are consistent with the established source-sink theory that reproductive sink activity is the primary determinant of rice yield under combined abiotic stress [43], as grain filling depends on both post-anthesis photosynthate allocation and pre-stored C/N remobilization from vegetative organs [9,18]. In this study, the severe limitation of both source-end C/N substrate supply and translocation to the panicle sink under combined salt and heat stress directly explains the inhibited grain filling and yield reduction in rice. The blocked transport of NSCs and N is attributed to the synergistic adverse effects of salt and heat on panicle development. Salt + heat stress reduced grains per panicle and seed-setting rate, leading to reduced sink intensity (Table S3). Under combined stress, the delayed panicle exsertion and sharp reduction in spikelet number transformed vegetative organs into temporary "alternative sinks" that stored excess NSCs and N, further inhibiting the transport of assimilates to reproductive organs [44,45]. Concurrently, the decreased sink capacity conversely suppressed leaf photosynthesis, forming a vicious cycle between source and sink limitation.

Moreover, beyond sink constraints, the activity of C and N metabolism enzymes and the expression of key regulatory genes are involved in mediating NSC and N transport [46,47]. The SS and SSS are key enzymes for converting sucrose to starch in developing panicles [48], and their persistent low activity under salt + heat stress inhibited starch synthesis in spikelets, which was correlated with reduced NSC accumulation in panicles. Similarly, N metabolism enzymes (NR, GS, and GOGAT) are critical for synthesizing amino acids (e.g., glutamine) that regulate panicle filling [41], and their failed recovery after stress reduced the transport efficiency of N and further disrupted the C-N balance. Together, this "substrate-energy double bottleneck" inhibited panicle development at multiple levels, i.e., from spikelet differentiation to grain filling, ultimately leading to a significant yield loss under combined salt and heat stress.

https://doi.org/10.3390/antiox15030308

Notably, this study only used one widely cultivated commercial rice variety as the research material, which has certain limitations. Existing studies have confirmed that rice genotypes with different stress tolerances show significantly different physiological, metabolic, and transcriptional responses to combined stress, and the yield penalty caused by combined stress also varies greatly among different genotypes [18,27,49]. For instance, salt- and heat-tolerant rice genotypes can maintain higher antioxidant enzyme activity, photosynthetic efficiency, and C-N metabolic homeostasis under combined stress, thus showing lower spikelet sterility and yield loss than sensitive genotypes [18,38]. In follow-up studies, we will screen and compare contrasting rice genotypes with significant differences in tolerance to combined salt and heat stress, to further verify the regulatory mechanism revealed in this study, and mine the key stress-responsive genes and core metabolic pathways that determine the differential tolerance of rice to combined salt and heat stress. This will not only deepen the understanding of the genetic basis of rice multi-stress tolerance, but also provide more precise molecular targets and elite germplasm resources for breeding rice varieties with high and stable yield in saline–alkali land under global warming.

5. Conclusions

Combined salt and heat stress exerts a more severe combined effect than individual salt or heat stress alone. Specifically, combined stress intensified oxidative stress, disrupted Na^+/K^+ ion homeostasis, and caused persistent impairment of the photosynthetic apparatus in rice, and these physiological damages are tightly driven by the coordinated misregulation of antioxidant defense, ion transport, and photosynthesis-related key genes identified in this study. These molecular perturbations suppressed carbon and nitrogen metabolism, decreased NSC accumulation in stem-sheath tissues, and resulted in greater yield loss than individual stresses. More precisely, this molecular–physiological cascade induced dual damage to carbon metabolism (persistent pre-anthesis NSC synthesis deficit and post-anthesis remobilization failure) and coupled nitrogen redistribution disorder (inhibited post-anthesis nitrogen translocation to developing grains), which severely disrupted the source-sink coupling of carbon and nitrogen nutrition, and ultimately drove the substantial synergistic yield loss under combined stress. Collectively, this study clarifies the core regulatory role of carbon and nitrogen metabolic dynamics in rice response to combined salt and heat stress and provides a solid theoretical basis for stress-tolerant rice breeding and targeted cultivation management in saline areas.

Supplementary Materials: The following supporting information can be downloaded at: https://www.mdpi.com/article/10.3390/antiox15030308/s1, Table S1. Information on key genes in photosynthesis, ROS, Na^+, and K^+ in response; Table S2. qRT-PCR primer sequence information; Table S3. Significance of rice physiological parameters under salt and heat stress; Figure S1. Photos of rice at the heading stage; Figure S2. The maximum and minimum temperature and average daily solar radiation during the experiment; Figure S3. Aboveground biomass and translocation at the heading and maturity stage. (a) Leaf biomass; (b) Stem sheath biomass; (c) Aboveground biomass; (d) Aboveground biomass translocation.

Author Contributions: L.L.: Investigation, Data curation, Formal analysis, Writing—original draft. J.L.: Investigation, Data curation, Methodology. J.X.: Investigation, Validation. W.B.: Investigation, Software. Y.L.: Investigation, Visualization. A.H.: Resources, Supervision. X.G.: Resources, Project administration. Z.A.: Conceptualization, Funding acquisition, Writing—review & editing. All authors have read and agreed to the published version of the manuscript.

Funding: National Natural Science Foundation of China (32372221); National Innovation Center for Salt-Tolerant and Alkali-Tolerant Rice Project, Innovation and Application of High-Quality Utilization Model of Salt-Tolerant Rice in Coastal Saline Lands (ZDYF2025GCZX001); Project of the 2024 and

2025 Central Guidance Fund for Local Science and Technology Development, Construction of the National Innovation Center for Salt-Tolerant Rice (2024ZYT002, 2025ZYT024).

Institutional Review Board Statement: Not applicable.

Informed Consent Statement: Not applicable.

Data Availability Statement: The original contributions presented in this study are included in the article/Supplementary Material. Further inquiries can be directed to the corresponding authors.

Acknowledgments: We express our sincere gratitude to all individuals who participated in the experiment of this study.

Conflicts of Interest: The authors declare that there are no conflicts of interest.

References

1. Zhao, C.; Liu, B.; Piao, S.; Wang, X.; Lobell, D.B.; Huang, Y.; Huang, M.; Yao, Y.; Bassu, S.; Ciais, P.; et al. Temperature increase reduces global yields of major crops in four independent estimates. *Proc. Natl. Acad. Sci. USA* **2017**, *114*, 9326–9331. [CrossRef] [PubMed]
2. Guo, X.; Huang, B.; Zhang, H.; Cai, C.; Li, G.; Li, H.; Zhang, Y.; Struik, P.C.; Liu, Z.; Dong, M.; et al. T-FACE studies reveal that increased temperature exerts an effect opposite to that of elevated CO_2 on nutrient concentration and bioavailability in rice and wheat grains. *Food Energy Secur.* **2022**, *11*, e336. [CrossRef]
3. Li, Y.; Jiang, F.; He, Z.; Liu, Y.; Chen, Z.; Ottosen, C.; Mittler, R.; Wu, Z.; Zhou, R. Higher Intensity of Salt Stress Accompanied by Heat Inhibits Stomatal Conductance and Induces ROS Accumulation in Tomato Plants. *Antioxidants* **2024**, *13*, 448. [CrossRef]
4. Rodrigues, F.; Sousa, B.; Soares, C.; Moreira, D.; Pereira, C.; Moutinho-Pereira, J.; Cunha, A.; Fidalgo, F. Are tomato plants co-exposed to heat and salinity able to ensure a proper carbon metabolism?—An insight into the photosynthetic hub. *Plant Physiol. Bioch* **2024**, *206*, 108270. [CrossRef]
5. Yang, Y.; Guo, Y. Elucidating the molecular mechanisms mediating plant salt-stress responses. *New Phytol.* **2018**, *217*, 523–539. [CrossRef]
6. Sierla, M.; Waszczak, C.; Vahisalu, T.; Kangasjärvi, J. Reactive Oxygen Species in the Regulation of Stomatal Movements. *Plant Physiol.* **2016**, *171*, 1569–1580. [CrossRef]
7. Liu, C.; Jiang, X.; Yuan, Z. Plant Responses and Adaptations to Salt Stress: A Review. *Horticulturae* **2024**, *10*, 1221. [CrossRef]
8. Huang, J.; Zhao, X.; Bürger, M.; Chory, J.; Wang, X. The role of ethylene in plant temperature stress response. *Trends Plant Sci* **2023**, *28*, 808–824. [CrossRef] [PubMed]
9. Zhen, F.; Zhou, J.; Mahmood, A.; Wang, W.; Chang, X.; Liu, B.; Liu, L.; Cao, W.; Zhu, Y.; Tang, L. Quantifying the effects of short-term heat stress at booting stage on nonstructural carbohydrates remobilization in rice. *Crop J.* **2020**, *8*, 194–212. [CrossRef]
10. Rao, M.J.; Duan, M.; Zhou, C.; Jiao, J.; Cheng, P.; Yang, L.; Wei, W.; Shen, Q.; Ji, P.; Yang, Y.; et al. Antioxidant Defense System in Plants: Reactive Oxygen Species Production, Signaling, and Scavenging During Abiotic Stress-Induced Oxidative Damage *Horticulturae* **2025**, *11*, 477. [CrossRef]
11. Ma, W.; Wang, X.; Gu, C.; Lu, Z.; Ma, R.; Wang, X.; Lu, Y.; Cai, K.; Tang, Z.; Zhou, Z.; et al. Rice heat stress response: Physiological changes and molecular regulatory network research progress. *Plants* **2025**, *14*, 2573. [CrossRef]
12. Lai, H.; Zhang, C.; Ren, H.; Wang, H.; Liang, M.; Zhang, J.; Zhao, B.; Ren, B.; Liu, P. The continuous damage to nitrogen metabolism caused by heat stress limited the young ear growth and yield of maize (*Zea mays* L.). *Eur. J. Agron.* **2025**, *168*, 127627. [CrossRef]
13. Wang, W.; Ji, D.; Yan, X.; Wei, Y.; Han, Y.; Zhang, W.; Liu, L.; Zhang, H.; Wang, Z.; Zhang, Z.; et al. Methyl jasmonate regulates panicle morphogenesis by mediating the negative effects of high temperature stress on carbon and nitrogen allocation and utilization. *Environ. Exp. Bot.* **2025**, *234*, 106150. [CrossRef]
14. Shunkao, S.; Thitisaksakul, M.; Pongdontri, P.; Theerakulpisut, P. Additive effects of combined heat and salt stress is manifested in an enhanced sodium ions accumulation and increased membrane damage in wheat seedlings. *Chil. J. Agric. Res.* **2022**, *82*, 552–563. [CrossRef]
15. Das, K.; Roychoudhury, A. Reactive oxygen species (ROS) and response of antioxidants as ROS-scavengers during environmental stress in plants. *Front. Environ. Sci.* **2014**, *2*, 53. [CrossRef]
16. Balfagón, D.; Zandalinas, S.I.; Mittler, R.; Gómez-Cadenas, A. High temperatures modify plant responses to abiotic stress conditions. *Physiol. Plantarum* **2020**, *170*, 335–344. [CrossRef]
17. Guo, L.; Lu, Y.; Bao, S.; Zhang, Q.; Geng, Y.; Shao, X. Carbon and nitrogen metabolism in rice cultivars affected by salt-alkaline stress. *Crop Pasture Sci.* **2021**, *72*. [CrossRef]

https://doi.org/10.3390/antiox15030308

18. Visakh, R.L.; Anand, S.; Arya, S.N.; Sasmita, B.; Jha, U.C.; Sah, R.P.; Beena, R. Rice Heat Tolerance Breeding: A Comprehensive Review and Forward Gaze. *Rice Sci.* **2024**, *31*, 375–400. [CrossRef]
19. Li, L.; Huang, Z.; Zhou, Z.; Tao, Y.; Zhang, Y.; Mu, Y.; Wu, S.; Nie, L. Foliar application of zinc oxide nanoparticles improved yield and 2-acetyl-1-pyrroline content in fragrant rice under salt stress. *Crop Environ.* **2025**, *4*, 107–117. [CrossRef]
20. Wei, H.H.; Geng, X.Y.; Zhang, X.; Zhu, W.; Zhang, X.B.; Chen, Y.L.; Huo, Z.Y.; Zhou, G.S.; Meng, T.Y.; Dai, Q.H. Grain Yield, Biomass Accumulation, and Leaf Photosynthetic Characteristics of Rice under Combined Salinity-Drought Stress. *Rice Sci.* **2024**, *31*, 118–128. [CrossRef]
21. Livak, K.; Schmittgen, T. Analysis of Relative Gene Expression Data Using Real-Time Quantitative PCR and the $2^{-\Delta\Delta C}{}_{T}$. *Methods* **2001**, *25*, 402–408. [CrossRef]
22. Chauhan, J.; Prathibha, M.D.; Singh, P.; Choyal, P.; Mishra, U.N.; Saha, D.; Kumar, R.; Anuragi, H.; Pandey, S.; Bose, B.; et al. Plant photosynthesis under abiotic stresses: Damages, adaptive, and signaling mechanisms. *Plant Stress.* **2023**, *10*, 100296. [CrossRef]
23. Sood, M. Reactive oxygen species (ROS): Plant perspectives on oxidative signalling and biotic stress response. *Discov. Plants* **2025**, *2*, 187. [CrossRef]
24. Zahra, N.; Hafeez, M.B.; Ghaffar, A.; Kausar, A.; Zeidi, M.A.; Siddique, K.H.M.; Farooq, M. Plant photosynthesis under heat stress: Effects and management. *Environ. Exp. Bot.* **2023**, *206*, 105178. [CrossRef]
25. Rivero, R.M.; Mestre, T.C.; Mittler, R.; Rubio, F.; Garcia-Sanchez, F.; Martinez, V. The combined effect of salinity and heat reveals a specific physiological, biochemical and molecular response in tomato plants. *Plant Cell Environ.* **2014**, *37*, 1059–1073. [CrossRef] [PubMed]
26. Su, L.; Huang, R.; Fu, D.; Chen, Y.; Zhang, X.; Chen, Y.; Liu, C.; Hu, T.; Ma, C. Effects of Combined Salt and Heat Stress on Agronomic Traits, Photosynthetic Parameters, and Physiological Biochemistry in Six Alfalfa (*Medicago sativa* L.) Cultivars. *Plants* **2025**, *14*, 2479. [CrossRef]
27. Nahar, L.; Aycan, M.; Hanamata, S.; Baslam, M.; Mitsui, T. Impact of Single and Combined Salinity and High-Temperature Stresses on Agro-Physiological, Biochemical, and Transcriptional Responses in Rice and Stress-Release. *Plants* **2022**, *11*, 501. [CrossRef] [PubMed]
28. Alzahrani, Y. Evaluation of drought and salinity tolerance potentials of different soybean genotypes based upon physiological, biochemical, and genetic indicators. *Front. Plant Sci.* **2024**, *15*, 1466363. [CrossRef]
29. Mathur, S.; Agrawal, D.; Jajoo, A. Photosynthesis: Response to high temperature stress. *J. Photochem. Photobiol. B Biol.* **2014**, *137*, 116–126. [CrossRef]
30. Kan, Y.; Mu, X.; Gao, J.; Lin, H.; Lin, Y. The molecular basis of heat stress responses in plants. *Mol. Plant* **2023**, *16*, 1612–1634. [CrossRef]
31. Abbas, G.; Areej, F.; Asad, S.A.; Saqib, M.; Anwar-ul-Haq, M.; Afzal, S.; Murtaza, B.; Amjad, M.; Naeem, M.A.; Akram, M.; et al. Differential Effect of Heat Stress on Drought and Salt Tolerance Potential of Quinoa Genotypes: A Physiological and Biochemical Investigation. *Plants* **2023**, *12*, 774. [CrossRef]
32. Jorge, T.F.; Ramalho, J.C.; Alseekh, S.; Pais, I.P.; Leitão, A.E.; Rodrigues, A.P.; Scotti-Campos, P.; Ribeiro-Barros, A.I.; Fernie, A.R.; António, C. Will *Casuarina glauca* Stress Resilience Be Maintained in the Face of Climate Change? *Metabolites* **2021**, *11*, 593. [CrossRef]
33. Liu, H.; Gao, X.; Fan, W.; Fu, X. Optimizing carbon and nitrogen metabolism in plants: From fundamental principles to practical applications. *J. Integr. Plant Biol.* **2025**, *67*, 1447–1466. [CrossRef]
34. Namachivayam, R.; Manickam, G.P.; Eswaran, K.; Loganathan, A.; Kumar, K.K.; Shanmugam, V. Strategies to improve photosynthesis by modifying the RuBisCO system and its limitations. *Mol. Biol. Rep.* **2025**, *52*, 951. [CrossRef] [PubMed]
35. Zhao, L.; Cai, Z.; Li, Y.; Zhang, Y. Engineering Rubisco to enhance CO_2 utilization. *Synth. Syst. Biotechnol.* **2024**, *9*, 55–68. [CrossRef] [PubMed]
36. Ozawa, Y.; Tanaka, A.; Suzuki, T.; Sugiura, D. Sink–source imbalance triggers delayed photosynthetic induction: Transcriptomic and physiological evidence. *Physiol. Plant.* **2023**, *175*, e14000. [CrossRef]
37. Liu, J.; Fu, C.; Li, G.; Khan, M.N.; Wu, H. ROS Homeostasis and Plant Salt Tolerance: Plant Nanobiotechnology Updates. *Sustainability* **2021**, *13*, 3552. [CrossRef]
38. Li, L.; Huang, Z.; Wu, S.; Zhang, Y.C.; Mu, Y.X.; Li, Y.S.; He, A.B.; Ai, Z.Y.; Guo, X.Y.; Nie, L.X. The Mechanisms of Improved Grain Yield and Nitrogen Use Efficiency in Salt-Tolerant Rice Varieties Under Salt Stress. *J. Agron. Crop Sci.* **2025**, *211*, 70064. [CrossRef]
39. González-García, M.P.; Conesa, C.M.; Lozano-Enguita, A.; Baca-González, V.; Simancas, B.; Navarro-Neila, S.; Sánchez-Bermúdez, M.; Salas-González, I.; Caro, E.; Castrillo, G.; et al. Temperature changes in the root ecosystem affect plant functionality. *Plant Commun.* **2023**, *4*, 100514. [CrossRef]
40. Liu, X.; Hu, B.; Chu, C. Nitrogen assimilation in plants: Current status and future prospects. *J. Genet. Genom.* **2022**, *49*, 394–404. [CrossRef]
41. Peng, Y.; Li, C.; Fritschi, F.B. Diurnal dynamics of maize leaf photosynthesis and carbohydrate concentrations in response to differential N availability. *Environ. Exp. Bot.* **2014**, *99*, 18–27. [CrossRef]

42. Marmagne, A.; Masclaux-Daubresse, C.; Chardon, F. Modulation of plant nitrogen remobilization and postflowering nitrogen uptake under environmental stresses. *J. Plant Physiol.* **2022**, *277*, 153781. [CrossRef]
43. Sun, T.; Liu, B.; Hasegawa, T.; Liao, Z.; Tang, L.; Liu, L.; Cao, W.; Zhu, Y. Sink-source unbalance leads to abnormal partitioning of biomass and nitrogen in rice under extreme heat stress: An experimental and modeling study. *Eur. J. Agron.* **2023**, *142*, 126678. [CrossRef]
44. Wang, D.R.; Wolfrum, E.J.; Virk, P.; Ismail, A.; Greenberg, A.J.; McCouch, S.R. Robust phenotyping strategies for evaluation of stem non-structural carbohydrates (NSC) in rice. *J. Exp. Bot.* **2016**, *67*, 6125–6138. [CrossRef]
45. Jin, W.; Li, L.; Fan, W.; Wei, Z. Enhanced non-structural carbohydrate metabolism and transport contribute to yield improvement in salt-tolerant rice under brine irrigation. *Plant Physiol. Bioch* **2025**, *228*, 110285. [CrossRef]
46. Li, G.; Pan, J.; Cui, K.; Yuan, M.; Hu, Q.; Wang, W.; Mohapatra, P.K.; Nie, L.; Huang, J.; Peng, S. Limitation of Unloading in the Developing Grains Is a Possible Cause Responsible for Low Stem Non-structural Carbohydrate Translocation and Poor Grain Yield Formation in Rice through Verification of Recombinant Inbred Lines. *Front. Plant Sci.* **2017**, *8*, 1369. [CrossRef]
47. Vicente, R.; Martínez-Carrasco, R.; Pérez, P.; Morcuende, R. New insights into the impacts of elevated CO_2, nitrogen, and temperature levels on the regulation of C and N metabolism in durum wheat using network analysis. *New Biotechnol.* **2018**, *40*, 192–199. [CrossRef]
48. Li, J.; Hu, Y.; Hu, J.; Xie, Q.; Chen, X.; Qi, X. Sucrose synthase: An enzyme with multiple roles in plant physiology. *J. Plant Physiol.* **2024**, *303*, 154352. [CrossRef] [PubMed]
49. Sakran, R.M.; Ghazy, M.I.; Gaballah, M.M.; Hussein, F.A.; Aamer, S.M.; Ghazy, H.A.; Lamlom, S.F.; Abdelghany, A.M. Genetic variability in physiological and agronomic traits of newly developed rice lines under well-watered and water-deficit conditions. *BMC Plant Biol.* **2025**, *25*, 1291. [CrossRef]

https://doi.org/10.3390/antiox15030308

antioxidants

MDPI

Article

Transcriptome and Hormone Analysis Revealed Jasmonic Acid-Mediated Immune Responses of Potato (*Solanum tuberosum*) to Potato Spindle Tuber Viroid Infection

Eva Marković [1,†], Bernard Jarić [1,†], Jana Oklešťková [2], Jitka Široká [2], Kristina Majsec [1], Jasna Milanović [3], Snježana Kereša [4], Ivanka Habuš Jerčić [4], Ondřej Novák [2] and Snježana Mihaljević [1,*]

1 Division of Molecular Biology, Ruđer Bošković Institute, Bijenička Cesta 54, 10000 Zagreb, Croatia; bernard.jaric@irb.hr (B.J.)

2 Laboratory of Growth Regulators, Faculty of Science, Palacký University & Institute of Experimental Botany, Czech Academy of Sciences, Šlechtitelů 27, 77900 Olomouc, Czech Republic; jana.oklestkova@upol.cz (J.O.); ondrej.novak@upol.cz (O.N.)

3 Centre for Plant Protection, Croatian Agency for Agriculture and Food, Gorice 68b, 10000 Zagreb, Croatia; jasna.milanovic@hapih.hr

4 Division of Plant Science, Faculty of Agriculture, University of Zagreb, Svetošimunska Cesta 25, 10000 Zagreb, Croatia; skeresa@agr.hr (S.K.); ihabus@agr.hr (I.H.J.)

* Correspondence: snjezana.mihaljevic@irb.hr

† These authors contributed equally to this work.

Abstract

Potato is a globally important non-cereal crop in which infection with potato spindle tuber viroid (PSTVd) can cause stunted growth and significantly reduce tuber yield. We previously showed that PSTVd induces accumulation of the plant hormone jasmonic acid (JA) and alters antioxidant responses in potato plants. To clarify the role of JA in response to PSTVd, we analyzed disease development in transgenic JA-deficient *opr3* and JA-insensitive *coi1* lines compared to the wild-type. Transcriptomic analysis using RNA-Seq revealed that most genotype-specific differentially expressed genes (DEGs) in all comparisons were enriched in plant hormone signal transduction, plant-pathogen interaction, and MAPK signaling pathways, although the number of DEGs varied. These differences were confirmed by independent data from RT-qPCR, hormone, and hydrogen peroxide (H_2O_2) analyses. After PSTVd infection, *opr3* plants showed enhanced JA signaling and increased abscisic acid (ABA) and auxin (AUX) content. In contrast, *coi1* plants showed reduced ABA, AUX, and salicylic acid content. Both *opr3* and *coi1* plants showed reduced JA and H_2O_2 content and lower expression of defense-related genes, resulting in milder symptoms but increased viroid accumulation. In addition, treatment with methyl jasmonate alleviated symptoms in infected wild-type plants. Together, these results indicate a modulatory role for JA and JA signaling in basal immune responses and symptom development in the potato-PSTVd interaction.

Keywords: antioxidant responses; basal defense responses; biotic stress; growth–defense trade-off; hormone crosstalk; jasmonic acid; MeJA; potato; PSTVd; RNA-Seq

1. Introduction

Potato (*Solanum tuberosum*) is the fourth most important food crop worldwide, after rice, wheat, and maize [1]. Potato spindle tuber viroid (PSTVd) is a subviral plant pathogen that can cause significant losses in yield and marketable quality of potato tubers. This threat is heightened by the fact that no potato varieties are known to be naturally resistant to this pathogen, and no curative treatment exists for this disease [2].

PSTVd is a small, non-coding, highly structured, circular, single-stranded RNA molecule consisting of approximately 341–364 nucleotides [3]. It is not clear how plants recognize this non-coding biotrophic pathogen, but transcriptomic studies on some viroid-host interactions suggest that viroids can activate pattern-triggered immunity (PTI) and effector-triggered immunity (ETI), as well as general downstream defense responses, including the generation of reactive oxygen species (ROS), mitogen-activated protein kinase (MAPK) cascades, and hormone signaling, which ultimately lead to transcriptional reprogramming of cells [3,4].

The plant hormone jasmonic acid (JA) is generally considered to play a defensive role in plant responses to necrotrophic pathogens, while salicylic acid (SA) is involved in responses to biotrophic pathogens [5,6]. However, increasing evidence supports a positive role for JA and its signaling pathway in defense against certain biotrophic fungal, bacterial, and viral pathogens in various plant species [7–12]. The accumulation of JA during pathogen perception promotes the expression of CORONATINE INSENSITIVE1 (COI1), which binds jasmonate ZIM domain proteins (JAZ) and promotes their degradation. This enables positive-acting transcription factors such as MYC2 to activate the transcription of JA-responsive genes involved in a wide range of defense responses, including maintenance of redox homeostasis, modulation of signaling crosstalk, and stimulation of defense compound production [13–15]. In response to pathogens, JA does not act alone but interacts with other hormones and signaling pathways through the actions of JAZ and MYC2 regulatory proteins [14,15].

Transcriptomic data from plant hosts infected with viroids indicate that the expression of genes involved in JA biosynthesis and signaling pathways is affected, depending on the specific viroid-host interaction [16–23]. However, studies examining whether changes in gene expression coincide with altered accumulation of endogenous JA levels in affected host plants are rare [24,25]. The accumulation of JA in potato leaves during PSTVd infection is also significant because there is little or no accumulation of SA [25], another important defense hormone that, along with JA, determines the outcome of defense responses. To elucidate the molecular mechanisms underlying basal defense responses in potato, this study analyzed the role of JA accumulation in response to PSTVd. We conducted a comparative transcriptomic analysis during PSTVd infection in transgenic lines with downregulated JA biosynthesis or JA signaling and in the corresponding wild-type potato using RNA-Seq. Transcriptome analysis showed that PSTVd infection suppresses the expression of genes related to photosynthesis and primary metabolism while activating genes involved in protein metabolism, signal transduction pathways, and defense responses. In this study, we focused on hormone, MAPK, and ROS signaling pathways. We also analyzed the temporal dynamics of gene expression, hormone levels, and H_2O_2 accumulation to determine how the timing of their expression influences plant susceptibility to PSTVd infection. Furthermore, we tested the effect of exogenous methyl jasmonate (MeJA) on gene expression and the perception of viroid infection. Key genes and pathways involved in basal defense against PSTVd in potato were identified. Results show that JA and its signaling modulate many defense responses to PSTVd in potato, and that their absence increases viroid accumulation but reduces symptom development, revealing new aspects of plant-viroid interactions.

https://doi.org/10.3390/antiox15010086

2. Materials and Methods

2.1. Plant Material, Viroid Inoculation, and Sampling

Potato cv. Désirée wild-type, two transgenic *StOPR3*-RNAi silenced lines (*opr3*A3 and *opr3*Z2) resulting in JA deficiency, and two transgenic *StCOI1*-RNAi silenced lines (*coi1*H1 and *coi1*X5) resulting in JA insensitivity [26], were propagated in stem node tissue culture. The in vitro plantlets were transferred to a soil/perlite mixture (3:1) and grown in a phytochamber at 22 °C, 60% relative humidity, and a 16-h photoperiod with a light intensity of 160 μmol m^{-2} s^{-1}. Healthy potato plants with 3–4 leaves were inoculated with H_2O (mock) or a viroid inoculum (5.7×10^7 viroid copies/μL) prepared from plants infected with PSTVd (GenBank: KF418768) as described [25]. Briefly, the two lower leaves were dusted with Celite 545 (Sigma-Aldrich, St. Louis, MO, USA) and rubbed with 2 μL of inoculum per leaf. After 10 min, the leaves were thoroughly washed with H_2O. After inoculation, plants were grown at 26 °C to enhance viroid replication and symptom development [16]. The first three intact leaves above the inoculated leaves were harvested weekly from 1 to 8 weeks post-inoculation (wpi), immediately frozen in liquid nitrogen, and stored at −80 °C for RNA, hormone, and H_2O_2 quantification. Leaves from at least four mock- and PSTVd-inoculated plants per genotype per time point were harvested. Each biological replicate consisted of leaves from a single plant. The experiment was repeated twice.

For exogenous MeJA treatment, plants were sprayed twice with H_2O or 1 mM MeJA (Sigma-Aldrich, St. Louis, MO, USA): one day before and six days after viroid inoculation (i.e., 24 h before the first sampling). Leaf samples from at least four mock- and PSTVd-inoculated plants per genotype per treatment per time point were harvested weekly from 1 to 6 wpi.

To distinguish effects of viroid inoculation from possible developmentally regulated changes, healthy, mock-inoculated plants served as controls in all experiments. The appearance of symptoms, such as epinasty and pointed upper leaves, was assessed visually, while growth retardation was evaluated by measuring plant height and fresh weight of the apical leaves (Figure S1).

2.2. RNA Extraction and RT-qPCR

Total RNA was extracted using the Direct-zol RNA Mini Prep Kit (Zymo Research, Irvine, CA, USA) according to the manufacturer's instructions. Absolute quantification of PSTVd RNA was performed using the Brilliant III Ultra-Fast qRT-PCR Master Mix (Agilent, Santa Clara, CA, USA) in a duplex reaction with specific primers and probes for PSTVd and *cytochrome oxidase 1* (*COX*) gene as described [27,28].

For plant gene expression analysis, total RNA was reverse transcribed using the SuperScript IV First-Strand Synthesis System (Thermo Fisher Scientific, Waltham, MA, USA). qPCR was performed with the SsoAdvanced Universal SYBR Green Supermix (Bio-Rad, Hercules, CA, USA) in a CFX96 Touch Real-Time PCR Detection System (Bio-Rad, Hercules, CA, USA). The PCR cycling conditions were 95 °C for 30 s, followed by 40 cycles of 10 s at 95 °C and 20 s at 60 °C, then a dissociation curve was generated to verify amplification specificity. Data were normalized using an inter-plate calibrator to correct for run-to-run variability. Relative gene expression was calculated using the $2^{-\Delta\Delta CT}$ method [29], with *elongation factor 1* (*EFα1*) as the reference gene [30]. Primers were designed using Primer3Plus version 3.3.0 or as described in [31–35] (Table S1). Primer efficiency and melt curves were checked and found acceptable for RT-qPCR.

2.3. RNA-Seq Gene Expression Analysis

Total RNA from three independent biological replicates (each an independent plant) per genotype and treatment was used to construct cDNA libraries with the TruSeq Stranded

https://doi.org/10.3390/antiox15010086

mRNA Library Prep Kit (Illumina, San Diego, CA, USA). Sequencing was performed by Macrogen (Seoul, South Korea) using paired-end sequencing on an Illumina NovaSeq 6000 platform. Quality control, adapter trimming, read filtering, mapping, and analysis were performed by Macrogen through their internal pipeline. Reads with mean Phred values below 15 and lengths shorter than 36 bp were removed. Quality statistics are shown in Table S2. Paired-end RNA-Seq reads were mapped to the reference genome SolTub_3.0 [36] with HISAT2 (v2.1.0) using the Bowtie2 (v2.3.4.1) aligner [37]. Mapped reads were assembled with StringTie to generate read counts, which were further analyzed in R with DESeq2 [38]. In DESeq2, size factors were estimated and used to normalize the libraries with the Relative Log Expression method. The negative binomial Wald test was applied to the normalized count data to determine *p*-values for differentially expressed genes (DEGs) between each comparison pair. Macrogen reported expression data with FDR-adjusted *p*-values for all comparisons. The criteria for DEGs were $|\log_2 \text{ fold change}| \geq 2$ and *p*-value < 0.05. Gene ontology (GO) enrichment analysis was performed using g:Profiler (https://biit.cs.ut.ee/gprofiler/gost (accessed on 12 August 2022)), and data were visualized using ChiPlot (https://www.chiplot.online/polarPlot/circle_enrich_plot.html (accessed on 11 January 2025)). MapMan analysis was performed as described [39]. Venn diagrams and Kyoto Encyclopedia of Genes and Genomes (KEGG) analysis were generated using R packages (v4.2.2), including VennDiagram and KEGGREST. Transcription factor (TF) analysis was performed using PlantTFDB v5.0 [40].

2.4. Plant Hormone Analysis

The acidic plant hormones (SA; jasmonoyl-L-isoleucine, JA-Ile; *cis*-(+)-12-oxo-phytodienoic acid, *cis*-OPDA; abscisic acid, ABA; indole-3-acetic acid, IAA) were determined as described [41]. Briefly, samples (10 mg fresh weight) were extracted in cold 10% aqueous methanol with the addition of internal standards (IS) labeled with stable isotopes (20 pmol SA-d_4, Sigma Aldrich, St. Louis, MO, USA; 5 pmol (−)-JA-d_2-Ile, 10 pmol OPDA-d_5 and 10 pmol IAA-$^{13}C_6$ OlChemIm Ltd., Olomouc, Czech Republic; 10 pmol ABA-d_6, National Research Council Canada, Saskatoon, SK, Canada). The extracts were purified on Oasis® HLB solid phase extraction columns (1 cc/30 mg, Waters) as described [42]. Analysis was performed on an Agilent 6490 Triple Quadrupole LC/MS system coupled to a 1290 Infinity LC system (Agilent Technologies, Santa Clara, CA, USA). The amount of SA-d_4 is set higher than that of other IS due to chromatographic issues related to peak shape (peak broadening associated with the history of the analytical column) and detection sensitivity. A level of 20 pmol ensures well-detectable chromatographic peaks of SA-d_4 in most matrices analyzed. Additionally, the higher SA-d_4 peak areas more closely match those of endogenous SA, which is important for accurate quantification. In contrast, the amount of (−)-JA-d_2-Ile is set lower than that of other IS, as 5 pmol is sufficient to obtain well-detectable chromatographic peaks in all tested matrices. Furthermore, for quantification, the lower JA-d_2-Ile peak area is closer to the peak areas of endogenous JA-Ile. The methodology was validated for accuracy and precision as described in [41].

Extraction and quantification of inactive and active endogenous BR were performed as described previously [43,44]. Samples (10 mg fresh weight) were extracted in ice-cold 60% acetonitrile, and 25 pmol deuterium-labeled internal standards of BRs (OlChemIm Ltd., Olomouc, Czech Republic) were added to each sample. After 12 h, samples were centrifuged (36,670 g, 15 min, 4 °C) and supernatants were purified using 50 mg Discovery DPA-6S cartridges (Supelco, Bellefonte, PA, USA). After evaporation to dryness, samples were reconstituted in 40 µL methanol and analyzed by liquid chromatography with tandem mass spectrometry (UHPLC-MS/MS) using an ACQUITY UPLC I-Class System (Waters,

https://doi.org/10.3390/antiox15010086

Milford, MA, USA) with a triple quadrupole mass spectrometer Xevo TQ-S MS (Waters MS Technologies, Manchester, UK).

2.5. Hydrogen Peroxide Analysis

Localization of H_2O_2 in leaf tissue was performed using 3,3-diaminobenzidine (DAB, Sigma-Aldrich, St. Louis, MO, USA) as described [45] and observed under a stereomicroscope (Zeiss SteREO Discovery.V20, Jena, Germany). The H_2O_2 content in leaves was determined spectrophotometrically using the $TiOSO_4$ method [46]. Absorbance was measured at 405 nm.

2.6. Statistical Analysis

The number of biological replicates in each experiment is indicated in the figure captions. When comparing two mean values, a Student's *t*-test was performed after checking the homogeneity of variance between the two samples with the F-test. Differences between the means of more than two groups were statistically assessed using two-way analysis of variance (ANOVA) followed by a post hoc Duncan's multiple range test (DMRT).

3. Results

3.1. Differences in Transcriptomic Responses Between JA-Deficient, JA-Insensitive, and Wild-Type Plants Infected with PSTVd

To identify genes and pathways specifically regulated by JA in response to PSTVd infection, transcriptomic profiles of PSTVd-infected leaves were compared with control leaves in wild-type, JA-deficient *opr3*, and JA-insensitive *coi1* lines. Samples were collected at 5 wpi, when symptoms began to appear (Figures 1A and S1), and viroid load was significantly higher ($p < 0.05$) in *opr3* and *coi1* plants (6.3×10^5 and 5.8×10^5 copies/µg RNA, respectively), compared to wild-type plants (3.6×10^4 copies/µg RNA) compared to wild-type plants (Figure 1B).

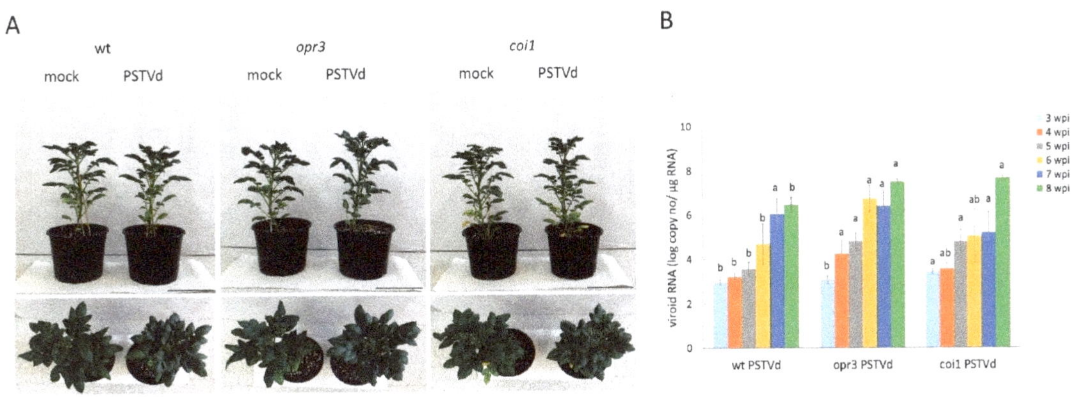

Figure 1. Symptom development and viroid RNA accumulation in *opr3*, *coi1*, and wild-type (wt) plants infected with PSTVd, compared to the corresponding mock-inoculated plants. (**A**) The first mild symptoms, such as stunted stem growth, epinasty, and pointed upper leaves, were observed in all three tested genotypes 5 weeks after inoculation (wpi); bar = 15 cm. (**B**) The dynamics of viroid RNA accumulation were determined by one-step RT-qPCR. At 8 wpi (experimental endpoint), the mean viroid load in *opr3* and *coi1* plants increased by 11-fold (3.15×10^7 copies/µg RNA) and 15-fold

https://doi.org/10.3390/antiox15010086

(4.4×10^7 copies/µg RNA), respectively, compared with wild-type plants (2.8×10^6 copies/µg RNA). Values are means ± SE ($n = 6$) from one representative experiment. Different letters indicate significant differences among the three genotypes, assessed separately at each time point using DMRT ($p < 0.05$).

Overall, twice as many DEGs were detected after PSTVd infection in *coi1* than in *opr3* or wild-type (2509, 906, and 1047 DEGs, respectively) (Figure 2). GO enrichment analysis revealed that the Biological Process (BP) GO terms related to responses to chemical and abiotic stimuli, and photosynthesis were significantly enriched in all examined lines after PSTVd infection (Figure 2, Table S3). However, the BP GO terms response to oxygen-containing compounds and negative regulation of ethylene signaling were enriched only in *coi1*, while response to hormone, phenylpropanoid metabolism, and lipid transport were enriched only in *opr3*. In the Molecular Function (MF) category, GO terms related to catalytic activity, oxidoreductase activity, and ethylene binding were enriched in *coi1* and *opr3*; chlorophyll binding was enriched in *coi1* and wild-type; while unfolded protein binding was enriched in *opr3* and wild-type. Among the Cellular Component (CC) GO terms, those linked to the chloroplast and thylakoid membrane were significantly enriched in all examined lines in response to PSTVd infection.

DEGs from the comparison between PSTVd- and mock-inoculated plants of all tested lines were assigned to MapMan functional classes. In the Biotic Stress overview, the most represented DEGs are related to signaling, proteolysis, and cell wall modification, followed by genes involved in transcriptional regulation, hormone signaling, and redox state (Figure S2, Table S4). Among DEGs encoding transcription factors (TFs), those for ethylene responsive factors (ERFs) and WRKYs were the most abundant in infected plants of all tested lines, as confirmed by PlantTFDB analysis (Figure S3). Additionally, activation of genes from the MYB and bHLH TF families was observed in wild-type and *opr3*, respectively. In *coi1*, a significant number of DEGs for ERF, MYB, and TCP TFs were downregulated compared to the wild-type after PSTVd infection. Besides stress responses and developmental processes, ERFs also participate in ET signaling. Regarding other hormone signaling pathways, more AUX-related genes were upregulated in *opr3* and *coi1* compared to the wild-type after PSTVd infection (Figure S2, Table S4).

3.2. *Comparative Analysis of DEGs*

The overlap between DEGs in PSTVd-infected and control samples of *opr3*, *coi1*, and wild-type plants is shown in Venn diagrams (Figure 3A). Functional KEGG analysis identified only 11 induced genes and 28 repressed genes common to both wild-type and transgenic lines, suggesting regulation by the same or overlapping pathways (Figure 3B, Table S5). Common upregulated genes were enriched in MAPK signaling and plant hormone signaling, including *MAPK7* and ET signaling genes (green ripe-like 1 *GRL1*, ethylene receptors *ETR2*-like and *EBF1*-like, and *ERF1B*-like). Common downregulated genes were enriched in photosynthesis-antenna proteins, glycolysis/gluconeogenesis, and pyruvate metabolism, indicating repression of primary metabolism after PSTVd infection.

https://doi.org/10.3390/antiox15010086

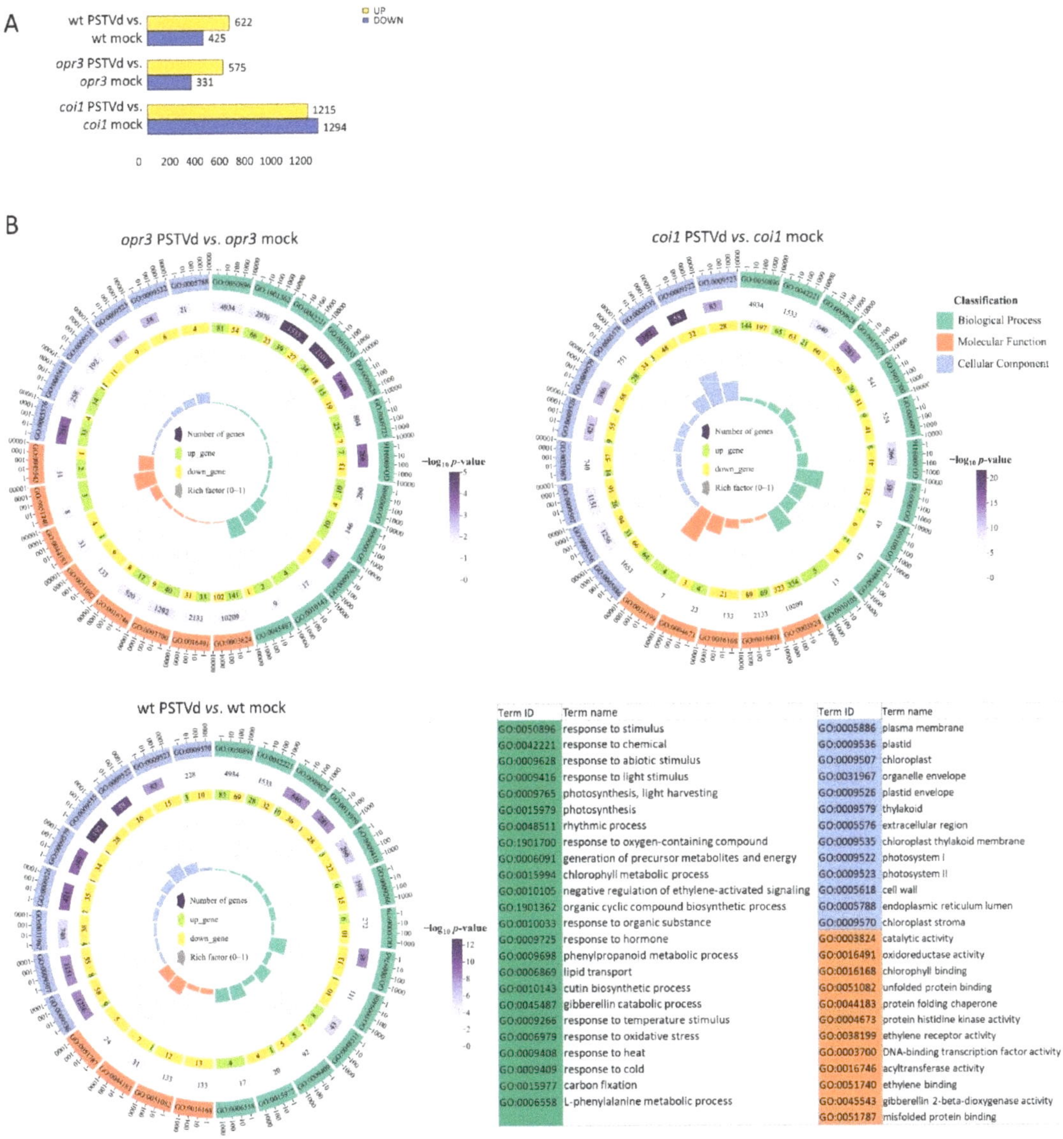

Figure 2. Transcriptional changes associated with PSTVd infection in *opr3*, *coi1*, and wild-type potato plants at 5 wpi. (**A**) The number of upregulated (UP) and downregulated (DOWN) DEGs ($|\log_2 fc| \geq 2$, $p < 0.05$) in three pairwise comparisons between PSTVd-inoculated and corresponding control (mock-inoculated) plants. (**B**) Top 26 enriched gene ontology (GO) terms for DEGs in the three comparisons; all data are in Supplementary Table S3. From the outermost to innermost circles, the first circle represents GO terms, with different colors indicating distinct terms. The second circle shows the total number of genes annotated to each term and the corresponding adjusted p-value, with deeper purple colors indicating smaller p-values. The third circle illustrates the ratio of upregulated to downregulated genes, with green representing the proportion of upregulated genes and yellow representing the proportion of downregulated genes; exact gene counts are labeled below. The fourth circle quantifies the enrichment factor, visualized in increments of 0.1, representing the ratio of DEGs to the total gene set within the given GO term.

https://doi.org/10.3390/antiox15010086

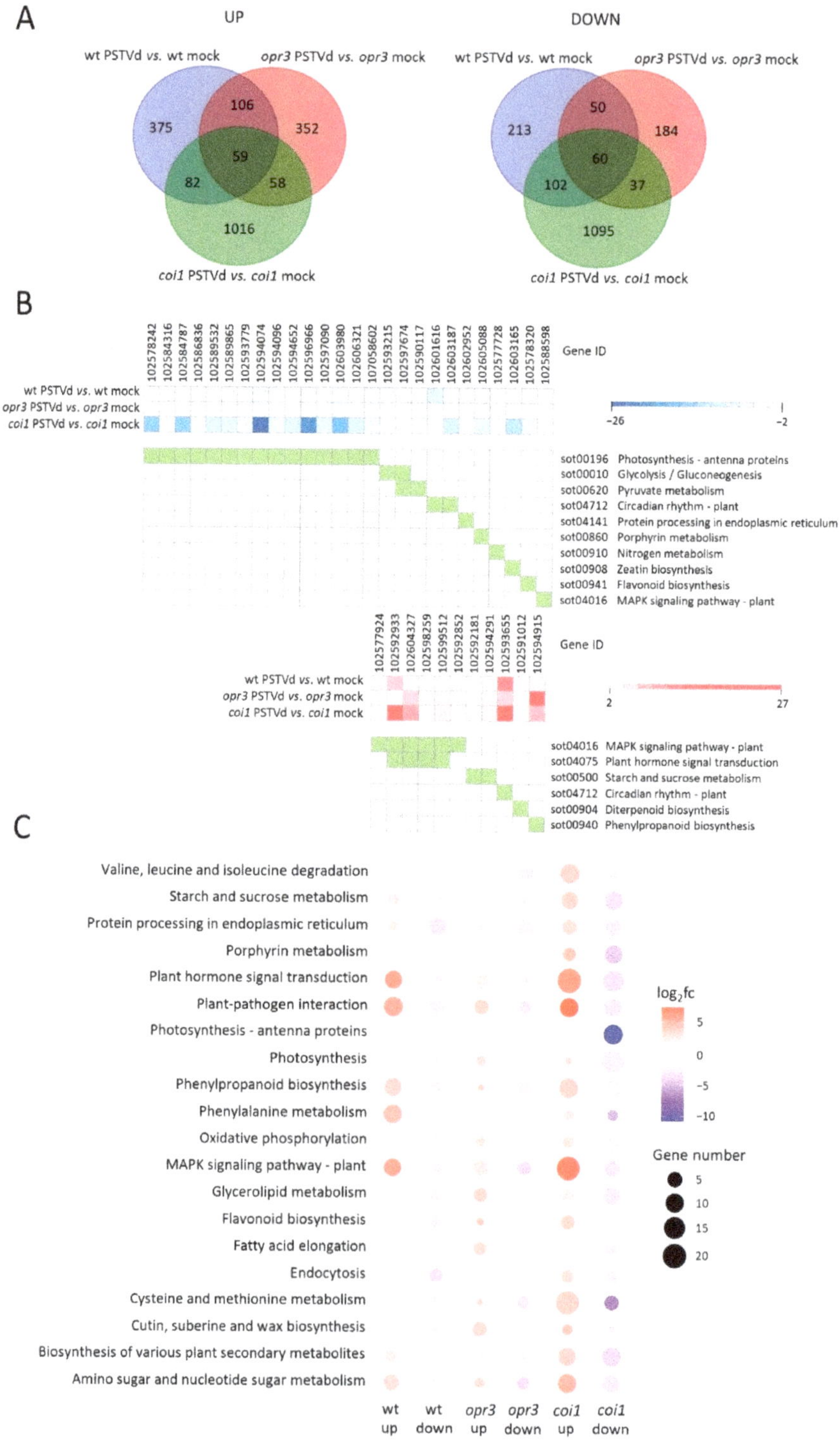

Figure 3. Overview of DEGs in PSTVd-infected *opr3*, *coi1*, and wild-type potato plants. (**A**) Venn diagrams show the overlap of DEGs upregulated or downregulated at 5 wpi in three comparisons

https://doi.org/10.3390/antiox15010086

between infected and control (mock-inoculated) plants. (**B**) Expression heatmaps and KEGG pathway analysis of DEGs shared among the three comparisons. (**C**) Top 20 KEGG pathways enriched with genotype-specific DEGs in the *opr3*, *coi1*, and wild-type comparisons; all data are in Supplementary Table S5.

Most genotype-specific DEGs detected after PSTVd infection in all comparisons were enriched in plant hormone signal transduction, plant-pathogen interaction, and MAPK signaling pathways, although the number of DEGs varied (Figure 3C). The *coi1*-specific DEGs related to hormone signaling were mainly involved in ET signaling (*ERF1B*-like), AUX signaling (Aux/IAAs, SAURs, and AUX-amido synthetase *GH3.6* and *GH3.1* genes), ABA signaling (ABA receptor *PYL4*-like, protein phosphatase 2C (*PP2C*)), GA signaling (*GID1B*-like), and SA signaling (regulatory protein *NPR3*-like and *TGA2*) (Table S5). The *opr3*-specific upregulated DEGs related to hormone signaling were mainly involved in AUX signaling (*GH3.5*, auxin transporter-like protein *LAX3*, and *AUX/IAA14*-like genes), JA signaling (*MYC4*-like), ET signaling (ETHYLENE INSENSITIVE 3-like 3 protein *EIL3* gene), and MAPK signaling (serine/threonine-protein kinase *SAPK3*-like), while genes related to CK biosynthesis and calcium signaling, were downregulated (Table S5). In addition, several more pathways were enriched with *coi1*-specific upregulated DEGs, such as cysteine and methionine metabolism (including ET biosynthesis genes *ACO* and *ACS*) and phenylpropanoid biosynthesis, while photosynthesis and oxidative phosphorylation pathways were enriched with downregulated *coi1*-specific DEGs. In contrast, only infected *opr3* plants showed upregulated DEGs assigned to glycerolipid metabolism, and cutin, suberin, and wax biosynthesis, while *opr3*-specific downregulated DEGs were assigned to protein processing and phenylpropanoid biosynthesis, including class III peroxidases (Figure 3C).

To gain better insight into the opposing regulation of defense responses in PSTVd-infected potato relative to JA-controlled pathways, we analyzed shared genes with contrasting expression between wild-type and transgenic lines (Figure S4A). KEGG analysis showed that DEGs upregulated in *coi1* but downregulated in wild-type were enriched in alanine, aspartate, and glutamate metabolism, as well as phenylpropanoid biosynthesis (Figure S4B, Table S6). In contrast, DEGs downregulated in *coi1* but upregulated in wild-type were enriched in pentose and glucuronate interconversions, amino sugar and nucleotide sugar metabolism, and plant-pathogen interaction pathways, highlighting the role of COI1-dependent JA signaling in the potato-PSTVd interaction. No enriched pathways were detected in the comparison between *opr3* and wild-type.

Identified DEGs were further analyzed using RT-qPCR. Transcript accumulation estimates for 15 selected genes from RNA-Seq closely matched those from RT-qPCR, indicating that the RNA-Seq data are reliable (Figure S5).

3.3. Expression Profiles of Hormone-Related, MAPK, and PR Genes

According to KEGG functional analysis, most genotype-specific DEGs in all comparisons were enriched in plant hormone signal transduction, plant-pathogen interaction, and MAPK signaling pathways. These differences were validated with independent data from RT-qPCR and hormone analyses. The dynamic expression patterns of selected phytohormone-related genes were analyzed at different time points, before and after symptom onset (Figure 4, Table S7). The same tissue samples were used for hormone quantification (Figure 5).

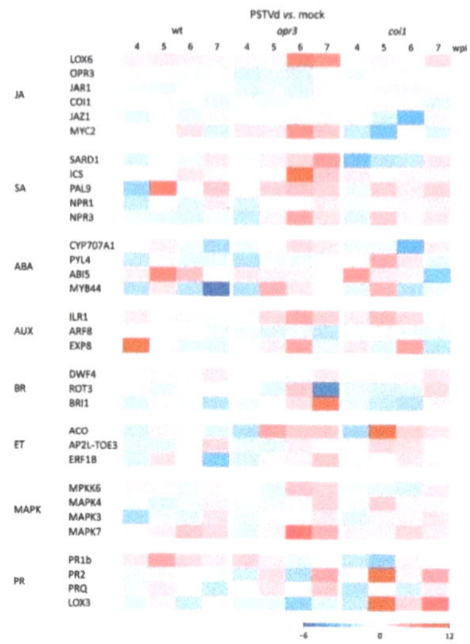

Figure 4. Temporal dynamics of hormone- and defense-related gene expression in leaves of *opr3*, *coi1*, and wild-type (wt) plants in response to PSTVd infection. Heatmaps display changes in relative gene expression at 4, 5, 6, and 7 wpi, as determined by RT-qPCR. The color scale on the heatmap indicates the mean $\log_2$fc from three biological replicates, normalized to the expression of the reference gene *EFα1*, and calculated relative to the corresponding control (mock-inoculated plants). Statistical evaluation of the data is in Table S7. Details on the genes and primers are in Table S1.

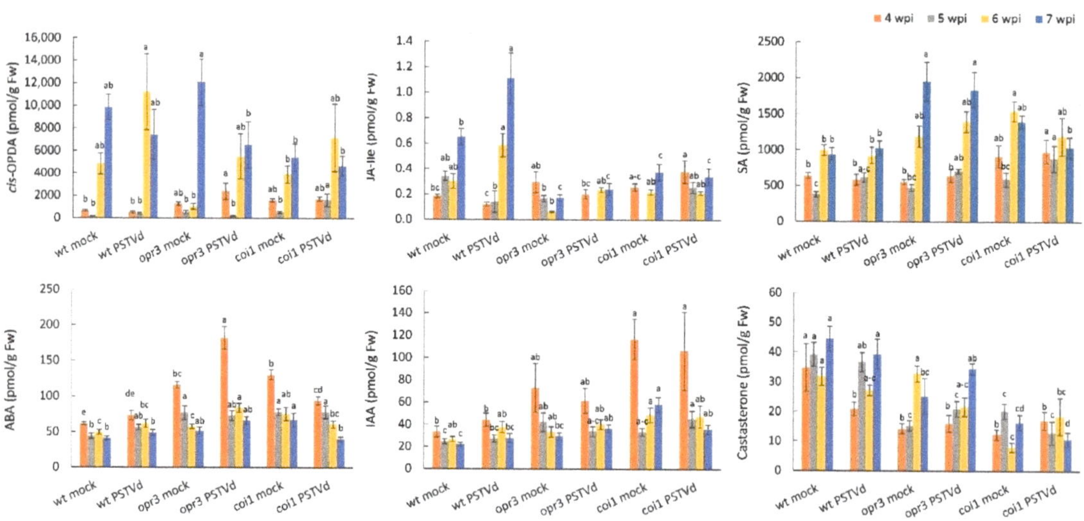

Figure 5. Hormone accumulation in leaves of *opr3*, *coi1*, and wild-type plants during PSTVd infection. Concentrations (pmol/g Fw) of *cis*-(+)-12-oxo-phytodienoic acid (*cis*-OPDA), jasmonyl-isoleucine (JA-Ile), salicylic acid (SA), abscisic acid (ABA), indole-3-acetic acid (IAA), and castasterone (CS) were quantified at different time points. Values are means ± SE (n = 6–8) pooled from two independent experiments. Different letters indicate significant differences among six plant groups (three genotypes; both PSTVd- and mock-inoculated), assessed at each time point separately using DMRT ($p < 0.05$).

After PSTVd infection, earlier activation of JA biosynthesis and perception genes (*LOX6*, *OPR3*, *JAR1*, and *COI1*) was detected in wild-type and *coi1* plants compared to *opr3* plants. Conversely, activation of JA signaling genes (*JAZ1* and *MYC2*) was stronger

https://doi.org/10.3390/antiox15010086

in infected *opr3* plants. Hormone analysis showed that the level of *cis*-OPDA, a precursor of JA-Ile, peaked at 6 wpi and then decreased at 7 wpi in all tested lines, while significant JA-Ile accumulation was detected only in infected wild-type plants. SA biosynthesis (*ICS* and *PAL9*) and SA signaling (*SARD1* and *NPR1*) genes were upregulated in all tested lines (*opr3* > *coi1* > wt); however, SA levels remained unchanged. ABA content and ABA hydroxylase gene expression increased only in *opr3* plants at the early stage of infection, while increased expression of *ABI5* and *MYB44* suggested activation of ABA signaling in all tested lines. The relative expression of auxin-responsive genes (*ILR1*, *ARF8*, and *EXP8*) was stronger in *opr3* and *coi1* compared to wild-type, while IAA content increased slightly in *opr3* and wild-type but decreased in *coi1*, after PSTVd infection. PSTVd-induced activation of the BR receptor gene *BRI1* was higher in *opr3*, coinciding with a slight increase in castasterone (CS) content. The relative expression of ET biosynthesis (*ACO*) and ET signaling (*ERF1B* and *AP2-TOE3*) genes was elevated in all tested lines at 5–7 wpi (*coi1* > *opr3* > wt).

Since cross-talk among MAPK, hormone, and ROS signaling pathways is important for the fine-tuned modulation of plant immunity [47,48], we measured the time course of expression of selected *MAPK* and *PR* protein genes after PSTVd infection (Figure 4, Table S7). Results showed activation of *MPKK6*, *MAPK4*, and *MAPK3* genes in wild-type plants at the early stage, but in *opr3* and *coi1* plants at the late stage of infection. Expression of the gene for MAPK7, an H_2O_2-responsive MAPK [49], was upregulated in all tested lines. Relative *PR1b* gene expression was stronger in wild-type plants, while *PR2* and *PRQ* gene expression was stronger in *opr3* and *coi1* plants, after PSTVd infection. Activation of *LOX3*, a pathogen-responsive lipoxygenase, was strongest in infected *coi1* plants.

3.4. Regulation of Redox Processes in opr3, coi1, and Wild-Type Potato Plants During PSTVd Infection

To investigate the mechanisms underlying altered H_2O_2 accumulation in JA-deficient plants, we analyzed transcript levels of ROS-producing (*RBOH*) and ROS-scavenging genes, including catalase (*CAT2*), ascorbate peroxidase (*APX1*), suberization-associated anionic peroxidase (*POPA*), lignin-forming anionic peroxidase (*LiP*), and peroxidase 12 (*POX12*). After PSTVd infection, wild-type plants accumulated H_2O_2, while *opr3* and *coi1* plants showed reduced H_2O_2 levels at the late stage of infection (Figure 6A,B). RT-qPCR revealed earlier activation of *RBOH*, *CAT2*, and *APX1* genes in *coi1* and *opr3* plants compared to wild-type plants (Figure 6C, Table S7). For Class III peroxidases, a plant-specific family of antioxidant enzymes involved in diverse functions [50], *POPA* expression was upregulated in wild-type plants, *LiP* was upregulated in *coi1* and *opr3* plants, and *POX12* expression was upregulated in *opr3* plants, but downregulated in *coi1* plants.

3.5. Effect of MeJA Treatment on Defense Responses in PSTVd-Infected Potato

Treatment with MeJA alleviated the negative effects of PSTVd infection on apical leaf and stem growth in wild-type plants, but had no significant effect on symptom appearance in *opr3* and *coi1* plants (Figures 7 and S6).

In PSTVd-infected wild-type plants, MeJA caused a transient decrease in the expression of nearly all tested genes involved in JA, SA, MAPK, and ROS signaling (Figure 8, Table S8), along with an increase in H_2O_2 content (Figure S6). In PSTVd-infected *opr3* plants, MeJA initially decreased but then increased the expression of JA-related and SA-related genes, compared to untreated infected *opr3* plants. A decrease in H_2O_2 content was also observed. As expected, MeJA had little effect on the expression of selected genes and H_2O_2 content in JA-insensitive *coi1* plants. Despite its notable impact on gene expression and symptom development, MeJA did not significantly affect the dynamics of PSTVd RNA accumulation in any of the tested lines (Figure S6).

https://doi.org/10.3390/antiox15010086

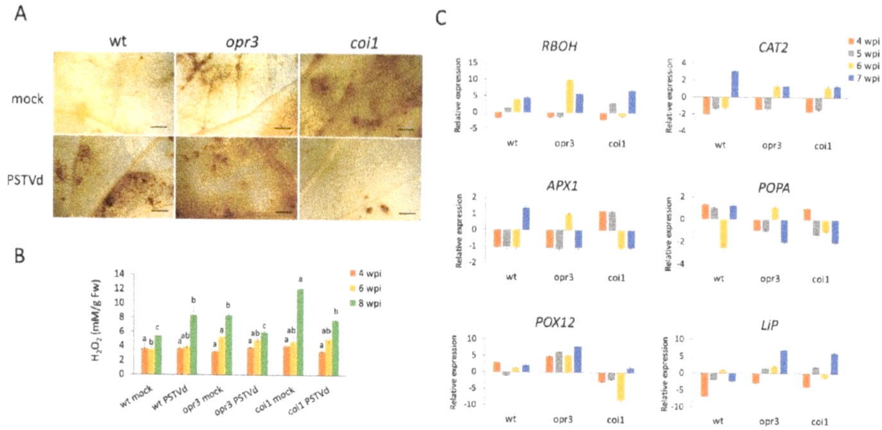

Figure 6. Dynamics of antioxidative responses in *opr3*, *coi1*, and wild-type plants during PSTVd infection. (**A**) H_2O_2 accumulation in leaves of PSTVd- and mock-inoculated plants at 8 wpi was localized by DAB staining, and observed under a stereomicroscope; bar = 2 mm. (**B**) H_2O_2 accumulation (mM/g Fw) at different time points was quantified spectrophotometrically using the $TiOSO_4$ method. Values are means ± SE (n = 4) from one representative experiment. Different letters indicate significant differences among six plant groups (three genotypes; both PSTVd- and mock-inoculated), assessed at each time point separately using DMRT ($p < 0.05$). (**C**) The expression pattern of H_2O_2-producing and scavenging genes was analyzed by RT-qPCR. Values represent the mean $\log_2$fc ± SE (n = 3). Statistical evaluation of the data is in Table S7.

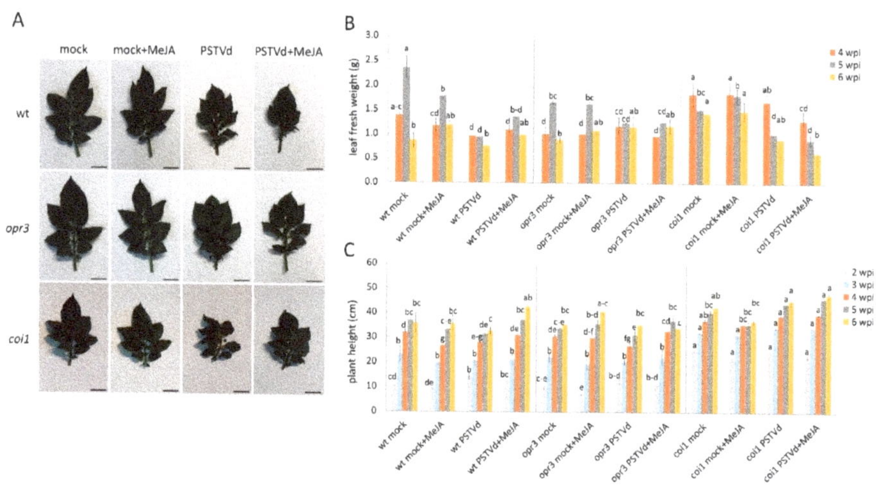

Figure 7. Effects of MeJA pretreatment on symptom appearance in *opr3*, *coi1*, and wild-type plants infected with PSTVd. (**A**) Morphology of upper potato leaves in PSTVd- or mock-inoculated plants in response to MeJA, examined at 6 wpi; bar = 1 cm. (**B**) Growth (g Fw) of upper systemically infected leaves was examined at different time points after MeJA treatment, with or without PSTVd infection. (**C**) Stem elongation was determined at different time points. Values are means ± SE (n = 4–6) from one representative experiment. Different letters indicate significant differences among 12 plant groups (three plant lines; both mock and infected, with or without MeJA), assessed at each time point separately using DMRT ($p < 0.05$). Untreated mock plants (mock); MeJA-treated mock plants (mock + MeJA); PSTVd-infected untreated (PSTVd); MeJA-treated PSTVd-infected (PSTVd + MeJA).

https://doi.org/10.3390/antiox15010086

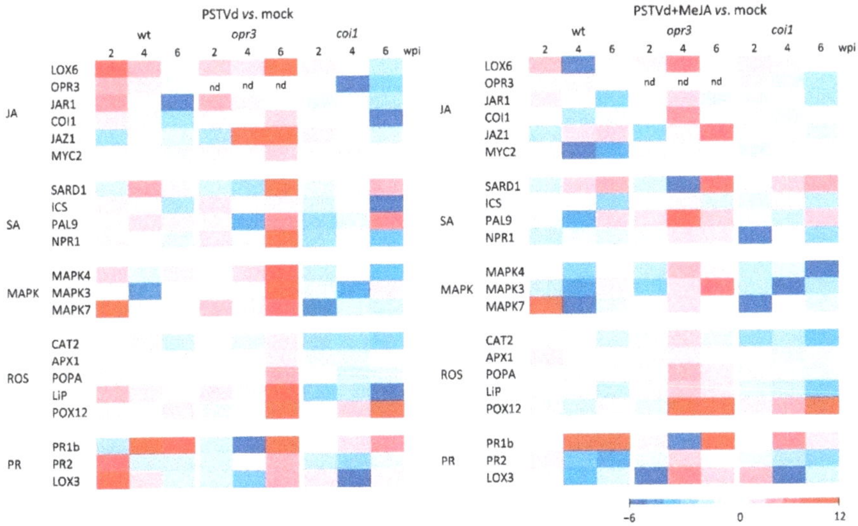

Figure 8. Effects of MeJA pretreatment on hormone- and defense-related gene expression in *opr3*, *coi1*, and wild-type (wt) plants infected with PSTVd. Relative gene expression in leaves was analyzed at different time points by RT-qPCR. The color scale on heatmaps indicates the mean log_2fc from three biological replicates, normalized to the expression of the reference gene *EFα1*, and calculated relative to the corresponding control (mock plants pretreated with H_2O). Statistical evaluation of the data is in Table S8.

4. Discussion

Functional analyses of differentially expressed genes (DEGs) showed that PSTVd infection in all tested potato lines suppressed genes related to photosynthesis and primary metabolism, while activating genes involved in proteolysis, signaling, and defense responses. This indicates a redirection of energy from metabolic processes to defense, a widespread and complex feature of pathogen-infected plants [23,51,52]. The late stage of infection in wild-type plants was characterized by the accumulation of JA-Ile, consistent with our previous results [25]. As expected, the transgenic *opr3* and *coi1* plants produced less JA-Ile than the wild-type, but no significant increase in JA-Ile levels in response to PSTVd was observed. However, the relative expression of the JA signaling gene *MYC2* was higher in *opr3* plants, while the expression of early JA biosynthesis genes such as *LOX* was higher in *coi1* plants. We hypothesize that this may be related to increased production of other jasmonates and oxylipins [10,14]. Both *opr3* and *coi1* transgenic lines had higher viroid loads than the wild-type, suggesting that JA is important for the basal defense response of potato against PSTVd. Transient accumulation of *cis*-OPDA, a precursor of JA-Ile and a JA-independent signaling molecule [53], was observed in infected plants of all tested lines. Therefore, it likely did not contribute to the difference in susceptibility to PSTVd among the tested lines, which requires further investigation.

Consistent with its defensive role, JA accumulation may inhibit plant growth [54]. When infected with PSTVd, JA-deficient *opr3* plants showed milder symptoms in leaf and stem growth, while JA-insensitive *coi1* plants exhibited significantly elongated stems, contrary to the typical PSTVd symptom of stunted stem growth. Detailed analysis of PSTVd-induced transcriptomic, hormone, and physiological responses in transgenic *opr3* and *coi1* lines, compared to the wild-type line, revealed several phenomena possibly associated with altered basal defense responses and symptom development in these lines: (i) activation of other hormone signaling pathways and their interactions with the JA signaling pathway;

https://doi.org/10.3390/antiox15010086

(ii) alteration of MAPK signaling, transcription factors, and *PR* gene expression; and (iii) transcriptional modification of antioxidant responses and metabolic pathways.

4.1. Activation of Other Hormone Signaling Pathways and Their Interactions with the JA Signaling Pathway

An imbalance between JA and AUX pathways can increase plant susceptibility to pathogens because high AUX levels promote cell growth, facilitating pathogen invasion and reducing the plant's ability to activate JA-mediated defense responses [55]. At the transcriptome level, PSTVd-infected *opr3* plants showed increased expression of several AUX metabolism and early AUX signaling genes, including *LAX3*, *SAURs*, and *AUX/IAA* genes [56]. For example, expression of *StGH3.1* (an ortholog of *AtGH3.1* and *OsGH3.8* [57] was upregulated in JA-insensitive and JA-deficient plants compared to wild-type plants, suggesting that JA negatively regulates *GH3.1* expression in response to PSTVd in potato. An interplay between JA and AUX signaling pathways during potato-PSTVd interaction is also suggested by the differential expression of auxin-responsive genes *ARF8* [56] and *EXP8* [58] in *opr3* and *coi1* lines compared to wild-type plants. We hypothesize that impaired JA biosynthesis and signaling affect the regulation of AUX signaling, contributing to symptom alleviation in *opr3* and *coi1* plants to PSTVd infection. However, auxin content remains relatively unchanged after PSTVd infection in all lines tested.

Abscisic acid mediates abiotic stress tolerance and regulates growth and development, but its role in disease resistance is complex and depends on the specific plant-pathogen interaction [5]. Jasmonate deficiency increases stress sensitivity in rice, and ABA biosynthesis or catabolism in response to osmotic stress can be modulated by JA [59]. Similarly, ABA accumulation in the early stage of PSTVd infection and increased expression of genes encoding the transcription factors MYB44 and MYC2 suggest enhanced stress response and stomatal regulation in JA-deficient *opr3* plants. Both MYB44 and MYC2 contribute to stomatal closure under stress conditions; however, MYC2 acts through JA signaling and stimulation of ABA biosynthesis [60], while MYB44 acts through ABA signaling [61]. Multifunctional MYB44 can also act as a negative regulator of ABA signaling in leaf senescence [61]. In contrast, in JA-insensitive *coi1* plants, ABA levels decreased, while the expression of genes for ABA receptors PYR1 and PYL4 [62] and the transcription factor ABI5-like protein 5 increased, suggesting activation of ABA signaling in response to PSTVd infection. ABA receptors and ABI transcription factors can, in turn, influence other signaling pathways, including the JA pathway, to fine-tune downstream stress responses and growth regulation [63].

Ethylene can have both negative and positive effects on plant defense responses and often acts synergistically with JA to activate defense-related genes [6,64]. Coordinated crosstalk between ET and JA converges at the transcriptional activation of *ERF1*. Arabidopsis *ERF1* is an early JA-responsive gene in the ERF branch of the JA defense response pathway, while ethylene promotes a longer-lasting induction of *ERF1* expression [64,65]. The AP2/ERFs, in addition to their role in hormone crosstalk, are also recognized for mediating plant responses to biotic stress, as transcriptional activators or repressors [66,67]. About half of the PSTVd-induced genes shared by wild-type, *opr3*, and *coi1* lines encode ET-responsive proteins (GRL1, ETR2-like, EBF1-like, and ERF1B-like), suggesting that activation of ET signaling is a core response to PSTVd in potato. However, the total number of ET-related DEGs in response to PSTVd was significantly higher in *opr3* and especially in *coi1* plants, which may also suggest an alternative, JA-independent activation of ET signaling and/or that *COI1* is not completely silenced [26]. The involvement of ET signaling in other plant-viroid interactions has also been reported [11,17,68–70], while ET accumulation was mostly associated with symptom development [71]. To clarify the role of ET in the potato response to PSTVd, future studies should analyze ET content.

https://doi.org/10.3390/antiox15010086

Several lines of evidence indicate that suppression of GA biosynthesis genes, which is associated with reduced GA responses, plays a major role in the development of stunting in viroid-infected plants [16,17,24,56,68,72]. Our study showed downregulation of the GA receptor *GID1B* gene, and upregulation of GA-responsive *RSI-1* genes after PSTVd infection in wild-type and *opr3* plants, while in *coi1* plants, the infection was accompanied by opposite responses. *StRSI-1* is an ortholog of *AtGASA5* (Gibberellin-stimulated transcript GAST1 protein homolog 5), a downstream gene of *DELLA*, which acts as a suppressor of gibberellin responses and stem growth in *Arabidopsis thaliana* [73]. In addition, GASA proteins are involved in JA-dependent biotic defense responses [74,75]. Therefore, we hypothesize that the increased expression of *RSI-1* in PSTVd-infected *coi1* plants may be associated with the absence of stem stunting symptoms in *coi1* plants, as part of the crosstalk between growth-promoting GA signaling and defense-promoting JA signaling [54].

Regarding the other hormones analyzed in this study, transcriptomic data showed a limited effect of PSTVd infection on the activation of SA and BR biosynthesis genes. This is consistent with the small changes in endogenous SA and CS content measured in infected compared to uninfected plants across all three lines. However, RT-qPCR analysis detected increased expression of SA biosynthesis (*ICS* and *PAL9*) and SA signaling (*NPR1* and *NPR3*) genes in wild-type plants during the early stage, and in *opr3* and *coi1* plants during the late stage of PSTVd infection. Unlike *Arabidopsis*, in some plant species such as potato and rice, SA levels do not increase significantly after pathogen infection, but the SA signaling pathway remains active and contributes to disease resistance [76,77].

Interestingly, transcriptome analysis also revealed contrasting expression of genes involved in the biosynthesis and metabolism of growth-promoting hormones CK (*IPT*- and *ZOG*-like) and PSK (*phytosulfokines 3-like*) in *opr3* and *coi1* plants infected with PSTVd. This suggests a possible interaction between CK, PSK, and JA signaling [78] during potato-PSTVd interaction, but further investigation is required.

4.2. Alteration of MAPK Signaling, Transcription Factors, and PR Genes in Response to PSTVd

MAPK cascades have been reported to regulate JA biosynthesis and JA-dependent gene expression [79]. JA also regulates both MAPK activity and *MAPK* gene expression, and silencing JA-related genes can affect MAPK activity [48]. After PSTVd infection, late activation of *MAPK3* and *MAPK4* genes was observed in *opr3* and *coi1* plants compared to wild-type plants, while pretreatment with MeJA accelerated gene activation in *opr3* plants. In *Arabidopsis*, *MAPK4* acts as a negative regulator of SAR and SA signaling but as a positive regulator of JA-dependent gene expression [80]. Our transcriptome analysis also revealed *coi1*-specific activation of a *MAP3KA*-like gene [81] and *opr3*-specific activation of *SAPK3*-like and *MAP2K*-like genes [82], suggesting their pathogen-dependent but JA-independent activation in potato. In addition, we detected PSTVd-induced expression of *MAPK7* in all tested potato lines, in all tested lines, which may be related to changes in JA and H_2O_2 status [49] associated with PSTVd infection.

By integrating signals from both hormones and ROS, the MAPK pathways can orchestrate a more precise and efficient plant defense response [47,83]. MapMan analysis showed that most defense-related genes, except miraculin [84], were upregulated in wild-type and opr3 plants compared to coi1 plants after PSTVd infection. Additionally, the expression of *PR1* genes [85] was lower and *PR2* genes [86] higher in *opr3* and *coi1* plants, which may be related to their higher susceptibility to PSTVd compared to wild-type plants, as indicated by viroid RNA accumulation. These results suggest the activation of alternative defense responses in *opr3* and *coi1* plants and highlight the role of JA signaling in regulating *PR* gene expression during PSTVd infection, possibly through interactions with other signaling pathways, including the ET and SA pathways [87].

https://doi.org/10.3390/antiox15010086

Transcription factors (TFs) play a key role in establishing plant defense and symptom development during viroid infections [56,72]. The importance of bHLH TFs, including MYCs that directly interact with JA-responsive genes, in the tomato-PSTVd interaction has also been reported [88]. In our study, the primary TF families showing significant responses to PSTVd in potato were identified, and the potential role of certain TFs in regulating hormone, MAPK, and ROS signaling during the potato-PSTVd interaction is discussed. However, given the numerous TFs and their diverse roles in regulating various defense-related genes, further research on specific TFs in the potato-PSTVd interaction is needed.

4.3. Transcriptional Modification of Antioxidant Responses and Metabolic Pathways in opr3 and coi1 Plants During PSTVd Infection

The physiological effects of JA include activation of ROS signaling pathways and recruitment of the cellular antioxidant defense system to maintain moderate ROS levels [89]. After PSTVd infection, DEGs for enzymes that regulate the cellular redox state, including APX [90], were mainly downregulated in *opr3* and *coi1* compared to wild-type. Since *opr3* and *coi1* plants accumulate less H_2O_2 than wild-type plants after PSTVd infection, their reliance on APX is less pronounced. However, not all antioxidant responses followed this pattern. DEGs for glutathione S-transferases were mainly induced by PSTVd infection in *opr3* and especially in *coi1*, suggesting their involvement in ROS reduction in plants with compromised JA pathways. JA signaling mediated by COI1 is important for controlling ROS accumulation by influencing the activity of MYC proteins, which regulate the expression of genes encoding antioxidant enzymes, including NOX/RBOH and POX12, involved in ROS production during plant defense against pathogens [89,91]. In PSTVd-infected *opr3* plants, one of the most prominent features was increased expression of *MYC2*, *RBOH*, and *POX12* genes, whereas in infected *coi1* plants, both *MYC2* and *POX12* expression were reduced. Treatment with MeJA followed by PSTVd infection induced short-term H_2O_2 production, enhancing *POX12* expression, which in turn reduced H_2O_2 content in *opr3* plants at the late infection stage. High concentrations of MeJA can act pro-oxidatively to stimulate activation of the antioxidant system and some components of basal defense responses [92,93]. Reduced H_2O_2 accumulation may enhance the plant's tolerance to the initial stressor and reduce symptom development, but may also promote pathogen progression [94].

Among other PSTVd-induced *POX* genes, *LiP* and *POPA* encode enzymes that use H_2O_2 as a substrate for lignification and suberinization, respectively. *LiP* and *POPA* showed contrasting expression patterns in wild-type and *opr3* compared with *coi1* plants. Increased expression of *LiP*, along with the lignin biosynthesis genes *4CL* and *CCoAOMT*, suggests increased H_2O_2 consumption during monolignol biosynthesis in infected *coi1* plants. Lignin is a complex polyphenolic polymer that serves as a structural component in plant cell walls, providing rigidity and strength, and acts as a physical and chemical barrier in basal plant immunity [95]. Lignification is negatively regulated by JA [92], which is consistent with the observed increased expression of lignification-related genes in *coi1* plants. In contrast, several *PAL* genes were specifically upregulated in wild-type plants after PSTVd infection, suggesting the production of various other phenolic secondary metabolites [96]. Altered expression of many genes encoding lignin-forming peroxidases and genes from the lignin-specific pathway was also observed in PSTVd-infected tomato [18].

Plants with impaired JA biosynthesis or signaling often activate alternative metabolic pathways, due to a bypass in the JA biosynthesis pathway, allowing accumulation of alternative defense compounds [7]. In PSTVd-infected *opr3* plants, genotype-specific activation of genes related to cutin, suberin, and wax biosynthesis (*CYP94A1* and feruloyl transferase *FHT*), and glycerophospholipid metabolism (glycerol-3-phosphate acyltransferases GPATs),

https://doi.org/10.3390/antiox15010086

was observed. Glycerolipids are important components of plant membranes, which are essential for development, growth, photosynthesis, and stress responses [97]. Cutin and suberin are protective polyester barriers formed by the polymerization of oxygenated fatty acids and glycerol [98]. Because glycerolipids and JA share fatty acids as precursors, inhibition of one pathway enhances regulation of the other, strengthening alternative defense pathways [99,100]. Therefore, we hypothesize that silencing *OPR3* alters glycerolipid and oxylipin metabolism and promotes cutin, wax, and suberin production in potato in response to PSTVd and increased osmotic stress in JA-deficient plants [59].

5. Conclusions

This study showed that silencing JA biosynthesis or JA signaling significantly alters the reprogramming of multiple signaling pathways, including hormone, MAPK, and ROS signaling, which disrupts defense regulation and increases susceptibility to PSTVd in potato. JA deficiency at the late stage of infection reduced symptom development in *opr3* plants, while MeJA treatment at the early stage of infection reduced symptoms in wild-type plants. The weak *coi1* response also supports COI1-dependent MeJA action. However, because only one MeJA dose and timing were used, conclusions about timing and dose–response are limited. Further studies, such as testing different MeJA concentrations, applying stage-specific inhibitory treatments, and conducting more detailed time-course sampling, should be conducted to provide more direct evidence supporting a positive role for JA in the early stage of PSTVd infection. Numerous transcriptional changes caused by PSTVd infection in *coi1* plants indicate the activation of alternative responses that compensate for compromised JA signaling, leading to symptom relief but increased viroid accumulation. This is consistent with the established role of JA as a regulator in the trade-off between growth and defense in plants [101]. Overall, these results enhance our understanding of the role of JA in basal immune responses against PSTVd and provide valuable gene resources for future functional analyses of key genes and regulatory mechanisms underlying immune responses to this pathogen in potato.

Supplementary Materials: The following supporting information can be downloaded at: https://www.mdpi.com/article/10.3390/antiox15010086/s1, Figure S1: Effects of PSTVd infection on potato stem and leaf growth in *opr3* and *coi1* plants compared to wild-type plants; Figure S2: MapMan visualization of DEGs related to biotic stress in *opr3*, *coi1*, and wild-type potato plants during PSTVd infection; Figure S3: Distribution of DEGs in 10 major transcription factor families; Figure S4: Analysis of DEGs with contrasting regulation between opr3, coi1, and wild-type potato plants infected with PSTVd; Figure S5: RT-qPCR validation of RNA-Seq results; Figure S6: Effects of MeJA pretreatment on symptom development, H_2O_2 accumulation, and viroid RNA accumulation in *opr3*, *coi1*, and wild-type plants infected with PSTVd; Table S1: List of primers used in this study; Table S2: Raw and mapped statistics of RNA-Seq data; Table S3: GO enrichment analysis of DEGs; Table S4: MapMan analysis of DEGs associated with biotic stress, in PSTVd-infected versus control samples of wild-type, *opr3*, and *coi1* plants; Table S5: KEGG classification of genotype-specific and common DEGs; Table S6: KEGG classification of contrasting DEGs between different comparisons; Table S7: Expression profiles of genes involved in hormone- and defense-related responses determined by RT-qPCR; Table S8: Expression of hormone-, MAPK-, and ROS-related genes, in response to MeJA treatment and PSTVd infection.

Author Contributions: Conceptualization, S.M.; Formal analysis, I.M., B.J., J.O., J.Š., K.M. and S.M.; Funding acquisition, O.N. and S.M.; Investigation, I.M., B.J., J.O., J.Š., J.M., S.K., I.H.J. and S.M.; Project administration, S.M.; Resources, O.N. and S.M.; Supervision, J.O. and S.M.; Visualization, I.M., B.J. and S.M.; Writing—original draft, S.M.; Writing—review and editing, B.J. and S.M. All authors have read and agreed to the published version of the manuscript.

Funding: This work was supported by the Croatian Science Foundation (HRZZ, IP-2019-04-9915) and the Czech Science Foundation (GACR, 22-17435S).

Institutional Review Board Statement: Not applicable.

Informed Consent Statement: Not applicable.

Data Availability Statement: The RNA-Seq data have been deposited in the NCBI database, with SRA accession code PRJNA1137360. Additional data supporting the conclusions of this article are included in the article and its Supplementary Files. Further inquiries can be directed to the corresponding author.

Acknowledgments: We thank Sabine Roshal (Leibniz Institute of Plant Biochemistry, Halle (Saale), Germany) for providing transgenic Désirée lines *StOPR3*-RNAi and *StCOI1*-RNAi. We also thank Nina Lulić for her help with H_2O_2 analysis.

Conflicts of Interest: The authors declare no conflicts of interest.

References

1. Ferranti, P.; Berry, E.; Anderson, J.R. *Encyclopedia of Food Security and Sustainability*; Elsevier: Amsterdam, The Netherlands, 2018; 1894p, ISBN 9780128126882.
2. Mackie, A.E.; Barbetti, M.J.; Rodoni, B.; McKirdy, S.J.; Jones, R.A.C. Effects of a potato spindle tuber viroid tomato strain on the symptoms, biomass, and yields of classical indicator and currently grown potato and tomato cultivars. *Plant Dis.* **2019**, *103*, 3009–3017. [CrossRef]
3. Navarro, B.; Flores, R.; Di Serio, F. Advances in viroid-host interactions. *Annu. Rev. Virol.* **2021**, *8*, 305–325. [CrossRef] [PubMed]
4. Sano, T. Progress in 50 years of viroid research-molecular structure, pathogenicity, and host adaptation. *Proc. Jpn. Acad. Ser. B Phys. Biol. Sci.* **2021**, *97*, 371–401. [CrossRef] [PubMed]
5. Alazem, M.; Lin, N. Roles of plant hormones in the regulation of host–virus interactions. *Mol. Plant Pathol.* **2014**, *16*, 529–540. [CrossRef]
6. Aerts, N.; Pereira Mendes, M.; Van Wees, S.C.M. Multiple levels of crosstalk in hormone networks regulating plant defense. *Plant J.* **2020**, *105*, 489–504. [CrossRef]
7. Antico, C.J.; Colon, C.; Banks, T.; Ramonell, K.M. Insights into the role of jasmonic acid-mediated defenses against necrotrophic and biotrophic fungal pathogens. *Front. Biol.* **2012**, *7*, 48–56. [CrossRef]
8. Tamaoki, D.; Seo, S.; Yamada, S.; Kano, A.; Miyamoto, A.; Shishido, H.; Miyoshi, S.; Taniguchi, S.; Akimitsu, K.; Gomi, K. Jasmonic acid and salicylic acid activate a common defense system in rice. *Plant Signal Behav.* **2013**, *8*, e24260. [CrossRef]
9. Guerreiro, A.; Figueiredo, J.; Sousa Silva, M.; Figueiredo, A. Linking jasmonic acid to grapevine resistance against the biotrophic oomycete *Plasmopara viticola*. *Front. Plant Sci.* **2016**, *7*, 565. [CrossRef]
10. García-Marcos, A.; Pacheco, R.; Manzano, A.; Aguilar, E.; Tenllado, F. Oxylipin biosynthesis genes positively regulate programmed cell death during compatible infections with the synergistic pair potato virus X-potato virus Y and tomato spotted wilt virus. *J. Virol.* **2013**, *87*, 5769–5783. [CrossRef]
11. Zhang, C.; Ding, Z.; Wu, K.; Yang, L.; Li, Y.; Yang, Z.; Shi, S.; Liu, X.; Zhao, S.; Yang, Z.; et al. Suppression of jasmonic acid-mediated defense by viral-inducible microRNA319 facilitates virus infection in rice. *Mol. Plant* **2016**, *9*, 1302–1314. [CrossRef]
12. Chakraborty, N.; Basak, J. Exogenous application of methyl jasmonate induces defense response and develops tolerance against mungbean yellow mosaic India virus in *Vigna mungo*. *Funct. Plant Biol.* **2018**, *46*, 69–81. [CrossRef]
13. Wasternack, C. Jasmonates: An update on biosynthesis, signal transduction and action in plant stress response, growth and development. *Ann. Bot.* **2007**, *100*, 681–697. [CrossRef] [PubMed]
14. Howe, G.A.; Major, I.T.; Koo, A.J. Modularity in jasmonate signaling for multistress resilience. *Annu. Rev. Plant Biol.* **2018**, *69*, 387–415. [CrossRef] [PubMed]
15. Song, C.; Cao, Y.; Dai, J.; Li, G.; Manzoor, M.A.; Chen, C.; Deng, H. The multifaceted roles of MYC2 in plants: Toward transcriptional reprogramming and stress tolerance by jasmonate signaling. *Front. Plant Sci.* **2022**, *13*, 868874. [CrossRef]
16. Owens, R.A.; Tech, K.B.; Shao, J.Y.; Sano, T.; Baker, C.J. Global analysis of tomato gene expression during Potato spindle tuber viroid infection reveals a complex array of changes affecting hormone signaling. *Mol. Plant Microbe Interact.* **2012**, *25*, 582–598. [CrossRef]
17. Więsyk, A.; Iwanicka-Nowicka, R.; Fogtman, A.; Zagórski-Ostoja, W.; Góra-Sochacka, A. Time-course microarray analysis reveals differences between transcriptional changes in tomato leaves triggered by mild and severe variants of potato spindle tuber viroid. *Viruses* **2018**, *10*, 257. [CrossRef]

18. Góra-Sochacka, A.; Więsyk, A.; Fogtman, A.; Lirski, M.; Zagórski-Ostoja, W. Root transcriptomic analysis reveals global changes induced by systemic infection of *Solanum lycopersicum* with mild and severe variants of potato spindle tuber viroid. *Viruses* **2019**, *11*, 992. [CrossRef]
19. Pokorn, T.; Radišek, S.; Javornik, B.; Štajner, N.; Jakše, J. Development of hop transcriptome to support research into host-viroid interactions. *PLoS ONE* **2017**, *12*, e0184528. [CrossRef]
20. Xia, C.; Li, S.; Hou, W.; Fan, Z.; Xiao, H.; Lu, M.; Sano, T.; Zhang, Z. Global transcriptomic changes induced by infection of cucumber (*Cucumis sativus* L.) with mild and severe variants of hop stunt viroid. *Front. Microbiol.* **2017**, *8*, 2427. [CrossRef]
21. Wang, Y.; Wu, J.; Qiu, Y.; Atta, S.; Zhou, C.; Cao, M. Global transcriptomic analysis reveals insights into the response of 'Etrog' citron (*Citrus medica* L.) to citrus exocortis viroid infection. *Viruses* **2019**, *11*, 453. [CrossRef] [PubMed]
22. Xu, L.; Zong, X.; Wang, J.; Wei, H.; Chen, X.; Liu, Q. Transcriptomic analysis reveals insights into the response to hop stunt viroid (HSVd) in sweet cherry (*Prunus avium* L.) fruits. *PeerJ* **2020**, *8*, e10005. [CrossRef] [PubMed]
23. Joubert, M.; van den Berg, N.; Theron, J.; Swart, V. Transcriptomics advancement in the complex response of plants to viroid infection. *Int. J. Mol. Sci.* **2022**, *23*, 7677. [CrossRef]
24. Takino, H.; Kitajima, S.; Hirano, S.; Oka, M.; Matsuura, T.; Ikeda, Y.; Kojima, M.; Takebayashi, Y.; Sakakibara, H.; Mino, M. Global transcriptome analyses reveal that infection with chrysanthemum stunt viroid (CSVd) affects gene expression profile of chrysanthemum plants, but the genes involved in plant hormone metabolism and signaling may not be silencing target of CSVd-SiRNAs. *Plant Gene* **2019**, *18*, 100181. [CrossRef]
25. Milanović, J.; Oklestkova, J.; Majdandžić, A.; Novák, O.; Mihaljević, S. Organ-specific differences in endogenous phytohormone and antioxidative responses in potato upon PSTVd infection. *J. Plant Physiol.* **2019**, *232*, 107–114. [CrossRef] [PubMed]
26. Halim, V.A.; Altmann, S.; Ellinger, D.; Eschen-Lippold, L.; Miersch, O.; Scheel, D.; Rosahl, S. PAMP-induced defense responses in potato require both salicylic acid and jasmonic jcid. *Plant J.* **2009**, *57*, 230–242. [CrossRef]
27. Boonham, N.; Pérez, L.G.; Mendez, M.S.; Peralta, E.L.; Blockley, A.; Walsh, K.; Barker, I.; Mumford, R.A. Development of a real-time RT-PCR assay for the detection of potato spindle tuber viroid. *J. Virol. Methods* **2004**, *116*, 139–146. [CrossRef]

https://doi.org/10.3390/antiox15010086

28. European and Mediterranean Plant Protection Organization (EPPO). PM 7/138 (1) Pospiviroids (genus *Pospiviroid*). *EPPO Bull* **2021**, *51*, 144–177. [CrossRef]
29. Livak, K.J.; Schmittgen, T.D. Analysis of relative gene expression data using real-time quantitative PCR and the $2^{-\Delta\Delta Ct}$ method *Methods* **2001**, *25*, 402–408. [CrossRef]
30. Expósito-Rodríguez, M.; Borges, A.A.; Borges-Pérez, A.; Pérez, J.A. Selection of internal control genes for quantitative real-time RT-PCR studies during tomato development process. *BMC Plant Biol.* **2008**, *8*, 131. [CrossRef]
31. Alaux, P.L.; Naveau, F.; Declerck, S.; Cranenbrouck, S. Common mycorrhizal network induced JA/ET genes expression in healthy potato plants connected to potato plants infected by *Phytophthora infestans*. *Front. Plant Sci.* **2020**, *11*, 602. [CrossRef]
32. Saubeau, G.; Perrin, F.; Marnet, N.; Andrivon, D.; Val, F. Hormone signalling pathways are differentially involved in quantitative resistance of potato to *Phytophthora infestans*. *Plant Pathol.* **2016**, *65*, 342–352. [CrossRef]
33. Jiang, H.; Wang, Y.; Li, C.; Wang, B.; Ma, L.; Ren, Y.; Bi, Y.; Li, Y.; Xue, H.; Prusky, D. The effect of benzo-(1,2,3)-thiadiazole-7 carbothioic acid S-methyl ester (BTH) treatment on regulation of reactive oxygen species metabolism involved in wound healing of potato tubers during postharvest. *Food Chem.* **2020**, *309*, 125608. [CrossRef]
34. Kuźnicki, D.; Meller, B.; Arasimowicz-Jelonek, M.; Braszewska-Zalewska, A.; Drozda, A.; Floryszak-Wieczorek, J. BABA-induced DNA methylome adjustment to intergenerational defense priming in potato to *Phytophthora infestans*. *Front. Plant Sci.* **2019** *10*, 650. [CrossRef] [PubMed]
35. Weller, S.A.; Elphinstone, J.G.; Smith, N.C.; Boonham, N.; Stead, D.E. Detection of *Ralstonia solanacearum* strains with a quantitative multiplex, real-time, fluorogenic PCR (TaqMan) assay. *Appl. Environ. Microbiol.* **2000**, *66*, 2853–2858. [CrossRef] [PubMed]
36. The Potato Genome Sequencing Consortium. Genome sequence and analysis of the tuber crop potato. *Nature* **2011**, *475*, 189–195 [CrossRef]
37. Pertea, M.; Kim, D.; Pertea, G.M.; Leek, J.T.; Salzberg, S.L. Transcript-level expression analysis of RNA-seq experiments with HISAT, StringTie and Ballgown. *Nat. Protoc.* **2016**, *11*, 1650–1667. [CrossRef] [PubMed]
38. Love, M.I.; Huber, W.; Anders, S. Moderated estimation of fold change and dispersion for RNA-seq data with DESeq2. *Genome Biol.* **2014**, *15*, 550. [CrossRef]
39. Usadel, B.; Poree, F.; Nagel, A.; Lohse, M.; Czedik-Eysenberg, A.; Stitt, M. A guide to using MapMan to visualize and compare omics data in plants: A case study in the crop species, Maize. *Plant Cell Environ.* **2009**, *32*, 1211–1229. [CrossRef]
40. Jin, J.; Tian, F.; Yang, D.-C.; Meng, Y.-Q.; Kong, L.; Luo, J.; Gao, G. PlantTFDB 4.0: Toward a central hub for transcription factors and regulatory interactions in plants. *Nucleic Acids Res.* **2017**, *45*, D1040–D1045. [CrossRef]
41. Široká, J.; Brunoni, F.; Pěnčík, A.; Mik, V.; Žukauskaitė, A.; Strnad, M.; Novák, O.; Floková, K. High-throughput interspecies profiling of acidic plant hormones using miniaturised sample processing. *Plant Methods* **2022**, *18*, 122. [CrossRef]
42. Floková, K.; Tarkowská, D.; Miersch, O.; Strnad, M.; Wasternack, C.; Novák, O. UHPLC–MS/MS based target profiling of stress-induced phytohormones. *Phytochemistry* **2014**, *105*, 147–157. [CrossRef]
43. Tarkowská, D.; Strnad, M. Protocol for extraction and isolation of brassinosteroids from plant tissues. *Methods Mol. Biol.* **2017**, *1564*, 1–7. [CrossRef]
44. Oklestkova, J.; Tarkowská, D.; Eyer, L.; Elbert, T.; Marek, A.; Smržová, Z.; Novák, O.; Fránek, M.; Zhabinskii, V.N.; Strnad, M Immunoaffinity chromatography combined with tandem mass spectrometry: A new tool for the selective capture and analysis of brassinosteroid plant hormones. *Talanta* **2017**, *170*, 432–440. [CrossRef]
45. Daudi, A.; O'brien, J.A. Detection of hydrogen peroxide by dab staining in *Arabidopsis* leaves. *Bio Protoc.* **2012**, *2*, e263. [CrossRef] [PubMed]
46. Mukherjee, S.P.; Choudhuri, M.A. Implications of water stress-induced changes in the levels of endogenous ascorbic acid and hydrogen peroxide in Vigna seedlings. *Physiol. Plant* **1983**, *58*, 166–170. [CrossRef]
47. Xia, X.-J.; Zhou, Y.-H.; Shi, K.; Zhou, J.; Foyer, C.H.; Yu, J.-Q. Interplay between reactive oxygen species and hormones in the control of plant development and stress tolerance. *J. Exp. Bot.* **2015**, *66*, 2839–2856. [CrossRef]
48. Jagodzik, P.; Tajdel-Zielinska, M.; Ciesla, A.; Marczak, M.; Ludwikow, A. Mitogen-activated protein kinase cascades in plant hormone signaling. *Front. Plant Sci.* **2018**, *9*, 1387. [CrossRef]
49. Dóczi, R.; Brader, G.; Pettkó-Szandtner, A.; Rajh, I.; Djamei, A.; Pitzschke, A.; Teige, M.; Hirt, H. The *Arabidopsis* mitogen-activated protein kinase kinase MKK3 is upstream of group C mitogen-activated protein kinases and participates in pathogen signaling. *Plant Cell* **2007**, *19*, 3266–3279. [CrossRef] [PubMed]
50. Almagro, L.; Gómez Ros, L.V.; Belchi-Navarro, S.; Bru, R.; Ros Barceló, A.; Pedreño, M.A. Class III peroxidases in plant defence reactions. *J. Exp. Bot.* **2008**, *60*, 377–390. [CrossRef]
51. Huot, B.; Yao, J.; Montgomery, B.L.; He, S.Y. Growth–defense tradeoffs in plants: A balancing act to optimize fitness. *Mol. Plant* **2014**, *7*, 1267–1287. [CrossRef]
52. Savchenko, T.V.; Rolletschek, H.; Dehesh, K. Jasmonates-mediated rewiring of central metabolism regulates adaptive responses. *Plant Cell Physiol.* **2019**, *60*, 2613–2620. [CrossRef]

https://doi.org/10.3390/antiox15010086

53. Wasternack, C.; Strnad, M. Jasmonate signaling in plant stress responses and development—Active and inactive compounds. *New Biotechnol.* **2016**, *33*, 604–613. [CrossRef]
54. Wasternack, C.; Hause, B. Jasmonates: Biosynthesis, perception, signal transduction and action in plant stress response, growth and development. An update to the 2007 Review in Annals of Botany. *Ann. Bot.* **2013**, *111*, 1021–1058. [CrossRef] [PubMed]
55. Mazzoni-Putman, S.M.; Brumos, J.; Zhao, C.; Alonso, J.M.; Stepanova, A.N. Auxin interactions with other hormones in plant development. *Cold Spring Harb. Perspect. Biol.* **2021**, *13*, a039990. [CrossRef] [PubMed]
56. Bao, D.; Chang, S.; Li, X.; Qi, Y. Advances in the study of auxin early response genes: *Aux/IAA*, *GH3*, and *SAUR*. *Crop J.* **2024**, *12*, 964–978. [CrossRef]
57. Ding, X.; Cao, Y.; Huang, L.; Zhao, J.; Xu, C.; Li, X.; Wang, S. Activation of the indole-3-acetic acid–amido synthetase GH3-8 suppresses expansin expression and promotes salicylate- and jasmonate-independent basal immunity in rice. *Plant Cell* **2008**, *20*, 228–240. [CrossRef] [PubMed]
58. Cosgrove, D.J. Loosening of plant cell walls by expansins. *Nature* **2000**, *407*, 321–326. [CrossRef]
59. Tang, G.; Ma, J.; Hause, B.; Nick, P.; Riemann, M. Jasmonate is required for the response to osmotic stress in rice. *Environ. Exp. Bot.* **2020**, *175*, 104047. [CrossRef]
60. Zhao, W.; Huang, H.; Wang, J.; Wang, X.; Xu, B.; Yao, X.; Sun, L.; Yang, R.; Wang, J.; Sun, A.; et al. Jasmonic acid enhances osmotic stress responses by MYC2-mediated inhibition of protein phosphatase 2C1 and response regulators 26 transcription factor in tomato. *Plant J.* **2022**, *113*, 546–561. [CrossRef]
61. Wang, F.; Yang, F.; Zhu, D.; Saniboere, B.; Zhou, B.; Peng, D. MYB44 plays key roles in regulating plant responses to abiotic and biotic stress, metabolism, and development. *J. Plant Biochem. Biotechnol.* **2023**, *33*, 462–473. [CrossRef]
62. Vieira, J.G.P.; Duarte, G.T.; Barrera-Rojas, C.H.; Matiolli, C.C.; Viana, A.J.C.; Campos, R.d.A.; Canesin, L.E.D.; Vicentini, R.; Nogueira, F.T.S.; Vincentz, M. Regulation of abscisic acid receptor mRNA stability: Involvement of microRNA5628 in *PYL6* transcript decay. *Plant Physiol.* **2024**, *197*, kiae663. [CrossRef]
63. Yu, Q.; Hua, X.; Yao, H.; Zhang, Q.; He, J.; Peng, L.; Li, D.; Yang, Y.; Li, X. Abscisic acid receptors are involves in the jasmonate signaling in *Arabidopsis*. *Plant Signal Behav.* **2021**, *16*, 1948243. [CrossRef]
64. Lorenzo, O.; Piqueras, R.; Sánchez-Serrano, J.J.; Solano, R. ETHYLENE RESPONSE FACTOR1 integrates signals from ethylene and jasmonate pathways in plant defense. *Plant Cell* **2003**, *15*, 165–178. [CrossRef] [PubMed]
65. Wasternack, C.; Song, S. Jasmonates: Biosynthesis, metabolism, and signaling by proteins activating and repressing transciption. *J. Exp. Bot.* **2017**, *68*, 1303–1321. [CrossRef] [PubMed]
66. Coll, A.; Lukan, T.; Stare, K.; Zagorščak, M.; Mahkovec Povalej, T.; Baebler, Š.; Prat, S.; Coll, N.S.; Valls, M.; Petek, M.; et al. The StPti5 ethylene response factor acts as a susceptibility factor by negatively regulating the potato immune response to pathogens. *New Phytol.* **2024**, *244*, 202–218. [CrossRef]
67. Su, Z.-L.; Li, A.-M.; Wang, M.; Qin, C.-X.; Pan, Y.-Q.; Liao, F.; Chen, Z.-L.; Zhang, B.-Q.; Cai, W.-G.; Huang, D.-L. The role of AP2/ERF transcription factors in plant responses to biotic stress. *Int. J. Mol. Sci.* **2025**, *26*, 4921. [CrossRef]
68. Katsarou, K.; Wu, Y.; Zhang, R.; Bonar, N.; Morris, J.; Hedley, P.E.; Bryan, G.J.; Kalantidis, K.; Hornyik, C. Insight on genes affecting tuber development in potato upon potato spindle tuber viroid (PSTVd) infection. *PLoS ONE* **2016**, *11*, e0150711. [CrossRef] [PubMed]
69. Kappagantu, M.; Bullock, J.M.; Nelson, M.E.; Eastwell, K.C. Hop stunt viroid: Effect on host (*Humulus lupulus*) transcriptome and its interactions with hop powdery mildew (*Podospheara macularis*). *Mol. Plant Microbe Interact.* **2017**, *30*, 842–851. [CrossRef]
70. Mishra, A.K.; Kumar, A.; Mishra, D.; Nath, V.S.; Jakše, J.; Kocábek, T.; Killi, U.K.; Morina, F.; Matoušek, J. Genome-wide transcriptomic analysis reveals insights into the response to citrus bark cracking viroid (CBCVd) in hop (*Humulus lupulus* L.). *Viruses* **2018**, *10*, 570. [CrossRef]
71. Prol, F.V.; López-Gresa, M.P.; Rodrigo, I.; Bellés, J.M.; Lisón, P. Ethylene is involved in symptom development and ribosomal stress of tomato plants upon citrus exocortis viroid infection. *Plants* **2020**, *9*, 582. [CrossRef]
72. Aviña-Padilla, K.; Rivera-Bustamante, R.; Kovalskaya, N.Y.; Hammond, R.W. Pospiviroid infection of tomato regulates the expression of genes involved in flower and fruit development. *Viruses* **2018**, *10*, 516. [CrossRef]
73. Zhang, S.; Yang, C.; Peng, J.; Sun, S.; Wang, X. GASA5, a regulator of flowering time and stem growth in *Arabidopsis thaliana*. *Plant Mol. Biol.* **2009**, *69*, 745–759. [CrossRef]
74. An, B.; Wang, Q.; Zhang, X.; Zhang, B.; Luo, H.; He, C. Comprehensive transcriptional and functional analyses of HbGASA genes reveal their roles in fungal pathogen resistance in *Hevea brasiliensis*. *Tree Genet. Genomes* **2018**, *14*, 41. [CrossRef]
75. Herbel, V.; Sieber-Frank, J.; Wink, M. the antimicrobial peptide snakin-2 is upregulated in the defense response of tomatoes (*Solanum lycopersicum*) as part of the jasmonate-dependent signaling pathway. *J. Plant Physiol.* **2017**, *208*, 1–6. [CrossRef]
76. Yu, D.; Liu, Y.; Fan, B.; Klessig, D.F.; Chen, Z. Is the high basal level of salicylic acid important for disease resistance in potato? *Plant Physiol.* **1997**, *115*, 343–349. [CrossRef]
77. Yang, D.-L.; Yang, Y.; He, Z. Roles of plant hormones and their interplay in rice immunity. *Mol. Plant* **2013**, *6*, 675–685. [CrossRef] [PubMed]

https://doi.org/10.3390/antiox15010086

78. Mosher, S.; Kemmerling, B. PSKR1 and PSY1R-mediated regulation of plant defense responses. *Plant Signal Behav.* **2013**, *8*, e2411[9]. [CrossRef] [PubMed]
79. Takahashi, F.; Yoshida, R.; Ichimura, K.; Mizoguchi, T.; Seo, S.; Yonezawa, M.; Maruyama, K.; Yamaguchi-Shinozaki, K.; Shinozaki, K. The mitogen-activated protein kinase cascade MKK3–MPK6 is an important part of the jasmonate signal transduction pathway in *Arabidopsis*. *Plant Cell* **2007**, *19*, 805–818. [CrossRef]
80. Brodersen, P.; Petersen, M.; Bjørn Nielsen, H.; Zhu, S.; Newman, M.; Shokat, K.M.; Rietz, S.; Parker, J.; Mundy, J. Arabidopsis MAP kinase 4 regulates salicylic acid- and jasmonic acid/ethylene-dependent responses via EDS1 and PAD4. *Plant J.* **2006**, *47*, 532–546. [CrossRef]
81. Wu, J.; Wang, J.; Pan, C.; Guan, X.; Wang, Y.; Liu, S.; He, Y.; Chen, J.; Chen, L.; Lu, G. Genome-wide identification of MAPKK and MAPKKK gene families in tomato and transcriptional profiling analysis during development and stress response. *PLoS ONE* **2014**, *9*, e103032. [CrossRef]
82. Ortiz-Masia, D.; Perez-Amador, M.A.; Carbonell, P.; Aniento, F.; Carbonell, J.; Marcote, M.J. Characterization of PsMPK2, the first C1 subgroup MAP kinase from pea (*Pisum sativum* L.). *Planta* **2008**, *227*, 1333–1342. [CrossRef]
83. Zhang, M.; Zhang, S. Mitogen-activated protein kinase cascades in plant signaling. *J. Integr. Plant Biol.* **2022**, *64*, 301–341. [CrossRef]
84. Wang, B.; Wang, Y.; He, W.; Huang, M.; Yu, L.; Cheng, D.; Du, J.; Song, B.; Chen, H. StMLP1, as a kunitz trypsin inhibitor, enhances potato resistance and specifically expresses in vascular bundles during *Ralstonia solanacearum* infection. *Plant J.* **2023**, *116*, 1342–1354. [CrossRef]
85. Gamir, J.; Darwiche, R.; van't Hof, P.; Choudhary, V.; Stumpe, M.; Schneiter, R.; Mauch, F. The sterol-binding activity of PATHOGENESIS-RELATED PROTEIN 1 reveals the mode of action of an antimicrobial protein. *Plant J.* **2017**, *89*, 502–509. [CrossRef]
86. Balasubramanian, V.; Vashisht, D.; Cletus, J.; Sakthivel, N. Plant β-1,3-glucanases: Their biological functions and transgenic expression against phytopathogenic fungi. *Biotechnol. Lett.* **2012**, *34*, 1983–1990. [CrossRef]
87. Leon-Reyes, A.; Du, Y.; Koornneef, A.; Proietti, S.; Körbes, A.P.; Memelink, J.; Pieterse, C.M.J.; Ritsema, T. Ethylene signaling renders the jasmonate response of *Arabidopsis* insensitive to future suppression by salicylic acid. *Mol. Plant Microbe Interact.* **2010**, *23*, 187–197. [CrossRef] [PubMed]
88. Aviña-Padilla, K.; Zambada-Moreno, O.; Jimenez-Limas, M.A.; Hammond, R.W.; Hernández-Rosales, M. Dissecting the role of bHLH transcription factors in the potato spindle tuber viroid (PSTVd)-tomato pathosystem using network approaches. *PLoS ONE* **2025**, *20*, e0318573. [CrossRef] [PubMed]
89. Kadam, S.B.; Barvkar, V.T. COI1 dependent jasmonic acid signalling positively modulates ROS scavenging system in transgenic hairy root culture of tomato. *Plant Physiol. Biochem.* **2024**, *206*, 108229. [CrossRef] [PubMed]
90. Asada, K. Production and scavenging of reactive oxygen species in chloroplasts and their functions. *Plant Physiol.* **2006**, *141*, 391–396. [CrossRef]
91. Zhu, J.; Chen, H.; Liu, L.; Xia, X.; Yan, X.; Mi, X.; Liu, S.; Wei, C. JA-mediated MYC2/LOX/AOS feedback loop regulates osmotic stress response in tea plant. *Hortic. Plant J.* **2024**, *10*, 931–946. [CrossRef]
92. Denness, L.; McKenna, J.F.; Segonzac, C.; Wormit, A.; Madhou, P.; Bennett, M.; Mansfield, J.; Zipfel, C.; Hamann, T. Cell wall damage-induced lignin biosynthesis is regulated by a reactive oxygen species- and jasmonic acid-dependent process in *Arabidopsis*. *Plant Physiol.* **2011**, *156*, 1364–1374, Erratum in *Plant Physiol.* **2015**, *168*, 1181–1182. [CrossRef]
93. Myers, R.J.; Fichman, Y.; Zandalinas, S.I.; Mittler, R. Jasmonic acid and salicylic acid modulate systemic reactive oxygen species signaling during stress responses. *Plant Physiol.* **2022**, *191*, 862–873. [CrossRef]
94. Huang, H.; Ullah, F.; Zhou, D.-X.; Yi, M.; Zhao, Y. Mechanisms of ROS regulation of plant development and stress responses. *Front. Plant Sci.* **2019**, *10*, 800. [CrossRef]
95. Xie, M.; Zhang, J.; Tschaplinski, T.J.; Tuskan, G.A.; Chen, J.-G.; Muchero, W. Regulation of lignin biosynthesis and its role in growth-defense tradeoffs. *Front. Plant Sci.* **2018**, *9*, 1427. [CrossRef]
96. Dong, N.; Lin, H. Contribution of phenylpropanoid metabolism to plant development and plant–environment interactions. *J. Integr. Plant Biol.* **2021**, *63*, 180–209. [CrossRef] [PubMed]
97. Lavell, A.A.; Benning, C. Cellular organization and regulation of plant glycerolipid metabolism. *Plant Cell Physiol.* **2019**, *60*, 1176–1183. [CrossRef]
98. Philippe, G.; Sørensen, I.; Jiao, C.; Sun, X.; Fei, Z.; Domozych, D.S.; Rose, J.K.C. Cutin and suberin: Assembly and origins of specialized lipidic cell wall scaffolds. *Curr. Opin. Plant Biol.* **2020**, *55*, 11–20. [CrossRef]
99. Chanda, B.; Venugopal, S.C.; Kulshrestha, S.; Navarre, D.A.; Downie, B.; Vaillancourt, L.; Kachroo, A.; Kachroo, P. Glycerol-3-phosphate levels are associated with basal resistance to the hemibiotrophic fungus *Colletotrichum higginsianum* in *Arabidopsis*. *Plant Physiol.* **2008**, *147*, 2017–2029. [CrossRef]

https://doi.org/10.3390/antiox15010086

00. Wan, H.; Qiu, H.; Li, Z.; Zhang, X.; Zhang, J.; Jiang, D.; Fernie, A.R.; Lyu, Y.; Cheng, Y.; Wen, W. Transcription factor *CsESE3* positively modulates both jasmonic acid and wax biosynthesis in citrus. *aBIOTECH* **2022**, *3*, 250–266. [CrossRef] [PubMed]
01. Li, C.; Xu, M.; Cai, X.; Han, Z.; Si, J.; Chen, D. Jasmonate signaling pathway modulates plant defense, growth, and their trade-offs. *Int. J. Mol. Sci.* **2022**, *23*, 3945. [CrossRef] [PubMed]

https://doi.org/10.3390/antiox15010086

antioxidants

MDPI

Article

Analyses of Antioxidative Response in Tomato (*Solanum lycopersicum* L.) Grown with Biochar and PGPMs

Silvia Carlo [1], Marta Trazza [1], Luca Pagano [2] and Marta Marmiroli [1,*]

[1] Department of Chemistry, Life Sciences and Environmental Sustainability, University of Parma, 43124 Parma, Italy; silvia.carlo@unipr.it (S.C.); marta.trazza@unipr.it (M.T.)

[2] Consorzio Interuniversitario Nazionale per le Scienze Ambientali (CINSA), University of Parma, 43124 Parma, Italy; luca.pagano@unipr.it

* Correspondence: marta.marmiroli@unipr.it

Abstract

Solanum lycopersicum plants were grown in pots amended with biochar and PGPMs (plant growth-promoting microorganisms: *Pseudomonas fluorescens* and *Azotobacter chroococcum*), applied singularly and in combination, for three months, after which plants and soils were collected, divided into treatment groups based on organs, and analyzed. The following biochemical markers were studied: cellular respiration, shoot fresh and dry weight, root fresh weight, photosynthetic pigments (chlorophyll a, chlorophyll b, and carotenoids), membrane lipid peroxidation, proline content, total antioxidant capacity (DPPH and ABTS assay), hydrogen peroxide, ascorbic acid, total phenolic content, enzymatic activity (SOD, POD, CAT, and APX), total soluble sugar content, and total protein content. Also, soil parameters, such as pH, EC, total enzymatic activity, active carbon, and respiration, were measured. While biochar alone induced root H_2O_2 accumulation, its co-application with PGPMs turned this signal into a systemic trigger for defense, enhancing the antioxidant capacity and the production of proline, phenolics, and ascorbic acid without causing oxidative damage. At the soil level, microorganisms counteracted biochar's inhibitory effects on enzymatic activity and intensified labile carbon use, indicating a more dynamic rhizosphere. Multivariate analysis confirmed that the combined treatment remodulated the plant–soil system, converting a stress factor into a resilience enhancer. This synergy underscores the role of biochar as an effective microbial carrier and PGPM consortia as bioactivators, together providing a powerful tool to prime crops against climate stress while preserving soil health.

Keywords: oxidative stress; antioxidants; plant response; soil health

1. Introduction

Technological advances in agriculture, including the Green Revolution and recent digital innovations, have substantially increased crop yields over the past decades; however, these gains have also come at considerable environmental and social costs, notably through heightened agrochemical use and reduced dietary diversity [1]. Climate variability and projected climate change, including shifts in temperature, precipitation patterns, and the frequency of extreme weather events, are expected to place additional pressure on future crop production and food systems [2]. Furthermore, food systems are facing unprecedented challenges as they attempt simultaneously to accommodate a growing global population and to adapt to the impacts of climate change [1]. In this context, vegetable production is particularly critical, given its importance in balanced diets and its inherently resource-intensive cultivation [3]. However, vegetable crops are highly vulnerable to a wide range of biotic and abiotic stresses, including nutrient deficiencies, which induce genetic, metabolic, physiological, and phenotypic changes in plants, ultimately leading to reduced yield and quality [4]. Moreover, such physiological disruptions have been shown to reduce yield and quality in several vegetable crops, including tomatoes, beans, and leafy greens [5]. Tomato (*Solanum lycopersicum* L.) is a major dietary source of essential macro- and micronutrients. The fruit is especially rich in potassium (K), the most abundant mineral, contributing up to 7.7% of the daily reference intake (DRI) per 100 g serving, depending on the cultivar. Other important minerals include magnesium (Mg), phosphorus (P), calcium (Ca), iron (Fe), and zinc (Zn), although their concentrations can vary according to genotype and cultivation conditions [6,7]. Tomatoes also contain appreciable levels of vitamins (in particular vitamin C), dietary fiber, and bioactive compounds such as carotenoids (e.g., lycopene) and phenolics, which possess recognized antioxidant properties. Nutrient density and composition are influenced by both genetic factors and environmental conditions, with greenhouse-grown tomatoes often exhibiting higher mineral contents than their field-grown counterparts [8,9]. Tomato is, however, a highly water-demanding crop, typically requiring about 500–600 mm of water per growing season, and is sensitive to variability in temperature and rainfall [10]. Its growing season is frequently affected by scarce or excessive precipitation and by extreme temperatures above 40 °C, leading to recurrent water and heat stress during critical phenological stages; such conditions disrupt vegetative and reproductive processes and result in reduced productivity [11]. Between 2023 and 2025, tomato cultivation has shown both global expansion and marked regional volatility. Global production for 2025 is forecast to increase modestly, with an estimated rise of 4.1% over this period, primarily driven by growth in China, India, and Turkey, which are consolidating their dominance in the sector. By contrast, the United States and the European Union are expected to experience a slight decline in total output, largely attributable to water scarcity and heat stress in major producing regions such as California, and to contractions in traditional growing areas such as Spain and Italy, where shortened crop seasons and an increased frequency of extreme weather events are projected to reduce yields and fruit quality [12,13].

Agronomic strategies such as the use of plant growth-promoting microorganisms (PGPM) and novel amendments, including biochar, are increasingly recognized as effective means to mitigate the negative impacts of climate change on crop production [14,15]. Recent research highlights biochar as a promising amendment to enhance soil quality and support sustainable agriculture. Key indicators of good soil quality include favorable texture, nutrient and moisture retention, and the capacity to support robust microbial activity [16]. Evidence suggests that biochar improves soil health and crop yields while preserving overall soil integrity [17]. Achieving higher crop yields without reliance on synthetic fertilizers or additives remains a central challenge in sustainable agriculture [18], particularly as inorganic fertilizer use has previously driven productivity increases at the expense of long-term soil fertility [19]. Biofertilizers, including PGPMs, are eco-friendly alternatives that underpin sustainable agronomic practices. These products contain living microorganisms that enhance soil nutrient status through organic matter decomposition, thereby facilitating mineral nutrient uptake by plants and contributing to improved crop productivity over the longer term [20–22]. Bacterial biofertilizers are pivotal in fixing nitrogen, solubilizing phosphorus, potassium, zinc, and silica, thus aiding the availability of both macro- and micronutrients in the soil [5,23,24]. In addition, biofertilizers promote plant growth by enhancing soil fertility, releasing growth hormones, producing antibiotics, and biodegrading organic matter [5,25]. Scientific literature increasingly indicates that both biochar and plant growth-promoting microbes can enhance soil fertility [26], while current research demonstrates that their combined use significantly modulates soil dynamics and improves productivity [27]. This study aimed to assess the impact of biochar and PGPM application, individually and in combination, on the oxidative status of tomato plants cultivated in soil under controlled greenhouse conditions. Biochar was sourced from plant residues, and a consortium of PGPMs was assembled. The PGPM strains *Pseudomonas fluorescens* DR54 and *Azotobacter chroococcum* LS-132 were selected for their well-characterized nitrogen-fixing and phosphorus-solubilizing capacities, as well as their previously demonstrated rhizosphere-colonization ability. Their combined use has been validated by Tabacchioni et al. [28] as an effective biostimulant consortium capable of enhancing nutrient cycling and modulating plant redox responses. This study provides a novel integrated plant–soil multivariate perspective on how biochar and PGPMs interact to modulate redox signaling. Unlike prior studies, this one focused primarily on growth or nutrient uptake. Our work evaluated the priming-like redox response based on H_2O_2 accumulation, antioxidant capacity, and soil biochemical activity, considering the overall plant-soil system.

2. Materials and Methods

All reagents used in the experiments were purchased from Merck (Darmstadt, Germany) unless specified otherwise. All spectrophotometric analyses were performed with the spectrophotometer Varian Cary 50 (Thermo Fisher, Waltham, MA, USA) and the microplate reader iMark Microplate Absorbance Reader (Bio-Rad Laboratories, Hercules, CA, USA).

2.1. Bacterial Growth and Biochar Functionalization

Two strains of PGPMs (*Pseudomonas fluorescens* DR54 and *Azotobacter chroococcum* LS132) [28] were used in this study, grown in LB medium (tryptone 10 g/L, NaCl 10 g/L, yeast extract 5 g/L) and incubated overnight at 28 °C in a thermostatic orbital shaker (Heidolph unimax 2010, Fisher, Segrate, Italy) set at 200 rpm. After 24 h, cultures were diluted (1:10 ratio), and optical density at 600 nm (OD_{600}) was measured spectrophotometrically to estimate the corresponding colony-forming units per milliliter (CFU/mL). For each strain, an aliquot corresponding to 10^8 CFU/mL was prepared. The two bacterial suspensions were then combined in a 1:1 ratio, resuspended in fresh LB medium, and used for root inoculation and biochar functionalization. Biochar derived from various plant feedstocks, including sunflower residues, was employed as a soil amendment. Detailed biochar characterization is provided in Table S1 and Figure S1. Biochar particles had an average size of 5 mm, obtained by sieving prior to functionalization. Functionalization was carried out under sterile conditions by adsorbing 1 mL of the bacterial suspension (10^8 CFU/mL) per gram of biochar. The biochar was then incubated at room temperature for 24 h to facilitate bacterial adhesion prior to soil application.

2.2. Plant Treatment and Sample Collection

Solanum lycopersicum L. (cv. Ciliegino; BLUMEN Group, Piacenza, Italy) seeds were germinated in standard potting soil (Compo GmbH, Münster, Germany) in seedbeds. Seedlings were maintained in a growth chamber (24 °C, 16 h light/8 h dark photoperiod) until the emergence of the first two fully expanded leaves. After three weeks, the plants were transplanted into pots (25 cm diameter × 25 cm height) containing 3 kg of soil and irrigated with 300 mL of tap water twice a week, with three biological replicates per treatment group. Four experimental conditions were established: (1) untreated control (nt); (2) soil amended with 1% (*w*/*w*) biochar (char); (3) plants inoculated at the root level with a bacterial suspension (10^8 CFU/mL) (PGPM), using the same total bacterial load as that applied during biochar functionalization (30×10^8 CFU per plant); and (4) soil amended with 1% (*w/w*) biochar functionalized with the bacterial suspension (comb). Plants were grown in a greenhouse under controlled conditions (24 ± 2 °C, relative humidity 55–65%, 16 h light/8 h dark photoperiod) for a period of three months before sample collection. No other stress, biotic or abiotic, was applied to plants at the time of the experiment. Soil samples were air-dried for subsequent analysis. Leaves and roots were thoroughly washed, then a part was air dried in an oven at 70 °C for 24 h, and a part was immediately flash-frozen in liquid nitrogen and stored at −80 °C until use for physiological and biochemical assays.

2.3. *Soil Analysis*

Physicochemical analyses of the soil were performed following sieving through a 2 mm mesh.

2.3.1. Determination of Soil pH and Electrical Conductivity (EC)

Soil pH (EPA Method 9045D) and electrical conductivity (EC) (FAO Standard Operating Procedure (SOP) on soil electrical conductivity) were assessed according to the standardized protocols with minor modifications. Specifically, 1 g of soil from each sample was suspended in 20 mL of double-distilled water (ddH_2O) and shaken continuously for two hours. After incubation, pH was measured using a pH meter (Seven Compact Duo, Mettler Toledo, Columbus, OH, USA). For EC determination, the suspension was filtered through Whatman grade 1 filter paper, and EC was measured using a conductivity meter (Seven Compact Duo, Mettler Toledo, Columbus, OH, USA).

2.3.2. Active Carbon (POXC) and Soil Respiration

Active carbon (Permanganate-Oxidizable Carbon, POXC) and soil respiration were measured according to Standard Operating Procedures (SOP) CSH 04 and CSH 06, respectively [29], with minor modifications. For POXC determination, 0.25 g of dried soil was suspended in 20 mL of a 0.02 M potassium permanganate ($KMnO_4$, pH 7.2) solution and shaken for 2 min. The mixture was then allowed to stand for an additional 8 min to enable the oxidation of the active carbon in the soil. The reaction was stopped by diluting the supernatants in ddH_2O (1:100), and the absorbance of the resulting solution was measured at 550 nm to assess the decrease in the characteristic purple color of the permanganate solution. Absorbance values were interpolated against a standard calibration curve prepared from $KMnO_4$ solutions of known concentrations (0.005 M, 0.01 M, and 0.02 M). The concentration of active carbon in the soil was then calculated following the methodology outlined in SOP CSH 04 [29]. Soil respiration rate was assessed by quantifying the carbon dioxide (CO_2) released as an indicator of microbial metabolic activity, using the sealed chamber alkali trap respirometry method: 2 g of air-dried soil was rehydrated and placed in a sealed jar containing a cuvette with 0.9 mL of 0.5 M KOH solution. After a 4-day incubation period, the amount of CO_2 released was estimated by comparing the change in EC of the KOH solution incubated with soil to that of a blank (KOH incubated without soil) as reported in SOP CSH 06 [29].

2.3.3. Total Soil Enzymatic Activity

Microbial enzymatic activity was assessed using the fluorescein diacetate (FDA) hydrolysis assay [30]. Following the preparation step described in ISO 11063:2020 [31] for soil microbial analyses, 2 g of soil was incubated overnight in 20 mL of 60 mM phosphate buffer. Then, 200 µL of FDA solution (1 mg/mL) was added to the suspension and shaken for 1 h at room temperature in the dark. After allowing the mixture to settle, the supernatant was collected and centrifuged to remove soil particles. The absorbance of the clarified solution was then measured at 490 nm.

2.4. Morphological, Physiological, and Biochemical Analysis in Plants

Following plant harvest, the fresh weight of shoots and roots, as well as the dry weight of shoots, was recorded.

2.4.1. Cellular Respiration and Lipid Peroxidation

Cellular respiration in fresh leaves was evaluated using the 2,3,5-triphenyltetrazolium chloride (TTC) assay, while lipid peroxidation in roots and leaves was assessed via the thiobarbituric acid reactive substance (TBAR) assay, according to Pagano et al. [32], with minor modifications. Specifically, a 15 mm diameter leaf disc was excised using a cork borer and incubated in 4 mL of TTC buffer (TTC 0.18 M, 78% $Na_2HPO_4 \cdot H_2O$ 0.05 M, 22% KH_2PO_4 0.05 M) for 15 h at 30 °C, in the dark. After incubation, the discs were rinsed twice with ddH_2O, and the resulting formazan was extracted in 5 mL of 95% ethanol at 80 °C for 10 min. Absorbance of the extract was then measured at 530 nm. Lipid peroxidation of cellular membranes was assessed by quantifying malondialdehyde (MDA) content in flash-frozen leaf and root tissues. Approximately 100 mg of tissue was ground in liquid nitrogen using a mortar and pestle and then homogenized in 0.1% (*w*/*v*) trichloroacetic acid (TCA). The homogenate was centrifuged at 12,000× *g* for 15 min at 4 °C. An aliquot of 500 μL of the resulting supernatant was mixed with 1 mL of reaction solution containing 0.5% TBA dissolved in 20% TCA. The mixture was incubated for 30 min at 95 °C to allow the formation of the TBA–MDA complex. The reaction was stopped by rapid cooling on ice, and absorbance was measured at 532 nm. MDA concentration was determined by interpolation from a standard calibration curve (0–50 μM).

2.4.2. Photosynthetic Activity

Photosynthetic pigments were extracted by incubating 50 mg of powdered leaf tissue with 500 μL of cold 95% (*v*/*v*) acetone on ice for 10 min. The homogenate was then centrifuged to separate plant debris from the pigment-containing acetone extract [33]. Chlorophyll a (Chl a), chlorophyll b (Chl b), and carotenoid contents were quantified spectrophotometrically by measuring absorbance at 662 nm, 647 nm, and 480 nm, respectively, and calculated as reported by Wellburn [34]. Pigment concentrations are expressed as micrograms per gram of fresh leaf weight (μg/g FW).

2.4.3. Proline Content in Leaves

Proline levels were determined in 100 mg of flash-frozen powdered leaf tissue homogenized in 1 mL of 3% (*w*/*v*) sulphosalicylic acid [35]. After centrifugation (10 min, 14,000× *g*, 4 °C), the supernatant was collected and kept on ice. For the colorimetric assay, 50 μL of extract was mixed with 1 mL of reaction solution comprising 3% (*w*/*v*) sulphosalicylic acid, glacial acetic acid, and 2.5% (*w*/*v*) ninhydrin in a 1:1:2 ratio. The mixture was incubated at 100 °C for 15 min, then rapidly cooled on ice. Absorbance was measured at 520 nm. Proline concentrations were calculated via interpolation against a standard curve (0–30 μg/mL) and expressed as μg/g of fresh weight (FW).

2.4.4. Hydrogen Peroxide Content

Hydrogen peroxide (H_2O_2) levels in root and leaf tissues were assessed [36]: a 20 mg aliquot of frozen tissue powder was suspended in 2 mL of reaction mixture containing 0.1% (*w*/*v*) TCA, 10 mM phosphate buffer (PBS), and 1 M potassium iodide (KI) in a 1:1:2 ratio. Samples were sonicated for 10 min at 4 °C (Transsonic T460, Elma Schmidbauer GmbH, Singen, Germany) and then centrifuged (15 min, 12,000× *g*, 4 °C). The supernatant was incubated at room temperature for 20 min, and absorbance was recorded at 390 nm. H_2O_2 concentrations were determined using a standard curve prepared with known H_2O_2 concentrations (0–0.7 mM).

2.4.5. Quantification of Ascorbic Acid

Ascorbic acid content in leaves was determined according to the method of Jagota & Dani [37] adapted for 96-well plate measurements: 50 mg of powdered tissue were extracted in 500 µL of 10% (*w*/*v*) TCA, vortexed, and incubated on ice for 5 min. After centrifugation (5 min, 12,000× *g*, 4 °C), 40 µL of the supernatant was diluted 1:5 in Milli-Q water, 20 µL of 0.2 M Folin–Ciocalteu reagent was added, and the mixture was allowed to react for 10 min. Absorbance was measured at 750 nm. Ascorbic acid concentration was calculated using a standard curve (0–70 µg/mL).

2.4.6. Extraction in Methanol

Antioxidant activity (DPPH, ABTS) and total phenolic content (TPC) were assessed using methanolic plant extracts [33]. Briefly, 100 mg of frozen, powdered tissue was extracted with 1 mL of 75% (*v*/*v*) methanol via 15 min sonication at 35 kHz (Transsonic T460, Elma Schmidbauer GmbH, Germany). After centrifugation (10 min, 4 °C, 10,000× *g*), the supernatant was collected. The pellet was suspended in 1 mL of methanol for another extraction step performed as described above, and both extracts were pooled (total volume: 2 mL) and stored at −80 °C for analysis.

2.4.7. Antioxidant Activity: DPPH Assay

For the DPPH (2,2-diphenyl-1-picrylhydrazyl) radical scavenging activity [33], 50 µL of extract was added to 1.95 mL of 0.06 mM DPPH in methanol and incubated for 30 min at room temperature. The decrease in absorbance at 520 nm indicated antioxidant activity, and the inhibition percentage was calculated by comparing the absorbance of DPPH solution with and without the plant extract, as described by Marmiroli et al. [33].

2.4.8. Antioxidant Activity: ABTS Assay

To have another indication of the antioxidant activity of the plant extract, the ABTS (2,2′-azino-bis(3-ethylbenzotiazoline-6-sulfonic acid)) assay was performed following Mingle & Newsome [38], with minor modifications. The $ABTS^{\bullet+}$ radical was generated by reacting 0.7 mM ABTS with 2.45 mM potassium persulfate in PBS for 30 min in the dark. Plant extract (50 µL) was mixed with 150 µL of ABTS solution (adjusted to an initial absorbance of 0.7 at 734 nm) and incubated for 5-, 15-, and 30-min. Absorbance at 734 nm was measured, and inhibition percentage was calculated as for DPPH.

2.4.9. Total Phenolic Content (TPC)

TPC was determined using the Folin–Ciocalteu method [33], with modifications to allow 96-well plate reading. Briefly, 10 µL of extract was incubated with 75 µL of diluted Folin–Ciocalteu reagent (1:10 in ddH_2O) for 5 min, followed by the addition of 75 µL of 60 g/L sodium carbonate. After 90 min in the dark at room temperature, absorbance was measured at 750 nm. Results are expressed as gallic acid equivalents (GAEs) using a standard curve (0–250 µg/mL).

2.4.10. Protein Extraction and Quantification

For enzyme extraction, frozen leaves (1.0 g) and roots (0.5 g) were ground in liquid nitrogen and homogenized in 5 mL and 2.5 mL of extraction buffer, respectively, containing 50 mM Tris-HCl (pH 7.8), 0.2 mM EDTA, 0.2% (*v*/*v*) Triton X-100, 1 mM PMSF, and 2 mM DTT. Homogenates were centrifuged at 12,000× *g* for 30 min at 4 °C, and the resulting supernatants were collected [36] and stored at −80 °C for total soluble protein quantification and superoxide dismutase (SOD), catalase (CAT), peroxidase (POD), and ascorbate peroxidase (APX) activity.

Total soluble protein content was quantified using the Bradford assay. The Coomassie Brilliant Blue G-250 stock dye (Bio-Rad Laboratories, Hercules, CA, USA) was diluted 1:4 with distilled water, and 40 µL of each extract was added to 2 mL of the diluted dye solution, vortexed for 10 s, and incubated for 5 min at room temperature. Absorbance was measured at 470 and 595 nm to account for spectral contributions from the dye's different ionic forms (red, green, and blue, with absorption maxima at 470, 650, and 590 nm, respectively). Protein concentrations were calculated using a Bovine Serum Albumin (BSA) standard curve (0–1000 µg/mL) and corrected using the equation reported by Ernst & Zor [39].

2.4.11. Determination of SOD Activity

SOD activity was measured by adding 100 µL of extract to 1.4 mL of reaction buffer containing 50 mM of phosphate buffer (pH 7.8), 13 mM of L-methionine, 75 µM of NBT, 0.1 mM of EDTA, and 3.3 µM of riboflavin, added last to initiate the reaction [40,41]. The mixture was exposed to fluorescent white light for 15 min to induce NBT photoreduction. A non-irradiated blank and an irradiated control without extract were included. Absorbance was read at 560 nm, and SOD activity was expressed as the percentage inhibition of NBT reduction.

2.4.12. Determination of POD, CAT, and APX Activities

The activities of POD, CAT, and APX were assessed following the methods described by Shakir et al. [42]. Enzymatic activities were calculated using molar extinction coefficients of 26.6 mM/cm for POD, 39.4 mM/cm for CAT, and 2.8 mM/cm for APX. For each assay, 50 µL of enzyme extract was added to a reaction mixture with a final volume of 1.5 mL. For POD, the reaction mixture consisted of 1.35 mL of 25 mM potassium phosphate buffer (pH 7.0) containing 2 mM EDTA, 50 µL of 300 mM H_2O_2, and 50 µL of 1.5% (*v*/*v*) guaiacol. The increase in absorbance at 470 nm was recorded at 1, 2, and 3 min. For CAT, the mixture included 1.4 mL of 50 mM phosphate buffer (pH 7.0) with 2 mM EDTA and 50 µL of 300 mM H_2O_2. Absorbance at 240 nm was measured over a 30-s interval. For APX, the reaction contained 1.35 mL of 25 mM potassium phosphate buffer (pH 7.0) with 2 mM EDTA, 50 µL of 300 mM H_2O_2, and 50 µL of 7.5 mM ascorbate. The decrease in absorbance at 290 nm was monitored over 1 min.

2.4.13. Determination of Total Soluble Sugars

The total soluble sugar content in root and leaf samples was determined following the procedure described by Shakir et al. [42], with minor modifications. Briefly, 50 mg of frozen tissue powder was homogenized in 1.5 mL of 90% ethanol and incubated for 1 h at 65 °C. The extract was then diluted to a final volume of 12.5 mL with 90% ethanol. A colorimetric reagent was prepared by mixing 5% phenol with concentrated sulfuric acid in a 1:5 ratio. Then, 250 µL of the diluted extract was combined with 1.5 mL of the reagent mixture. Absorbance was measured at 485 nm, and the total soluble sugar concentration was determined by interpolation from a glucose standard curve (20–100 µg/mL) and expressed as µg/mL.

2.5. Statistical Analysis

Normality and homoscedasticity of the data sets were checked with Shapiro–Wilk's and Levene's tests, respectively, utilizing SPSS 27 (https://www.ibm.com/it-it/products/spss-statistics, accessed on 20 November 2025).

Statistical significance among treatments was evaluated using one-way ANOVA followed by Tukey's HSD post hoc test using Past 4.03 (©Copyright Hammer 1999–2020). Multivariate analysis, including principal component analysis (PCA), was performed using R software version 5.1 (R Core Team, 2024). The correlation matrix was created using Pearson's method, and the relative p-values considered significant ($p < 0.05$) were the same as for Tukey's test. The correlation map and the relative chord diagram were both produced using R software.

3. Results and Discussion

3.1. Influence of Biochar–PGPM Combination on Soil Microbial Activity

The determination of soil pH and electrical conductivity showed no significant differences among treatments, indicating that the application of both biochar and PGPMs did not perturb the overall soil physicochemical characteristics (Table 1). In contrast, the analysis of key soil health and microbial activity parameters revealed distinct patterns influenced by the differential application of PGPMs and biochar. The total hydrolytic activity of enzymes such as esterase, lipase, and protease measured by the FDA assay was reduced by about 19% in soil treated with biochar alone compared with the untreated control (Figure 1A), while total respiration (CO_2 release) remained unaffected (Table 1). Conversely, the combination with PGPMs restored enzymatic activity to levels comparable to the control, counteracting the inhibitory effect induced by biochar. This behavior may be linked to its physicochemical properties, including its alkaline pH (9.6) and relatively high copper and nickel contents (1289 mg/kg and 119 mg/kg, respectively), which may suppress microbial activity through pH imbalance [43]. Other hypotheses are that the presence of biochar could also cause the adsorption of bioavailable carbon or could interfere with enzyme activity determination assays [44]. The combination of PGPMs with biochar not only alleviated this inhibition but also increased the consumption of oxidizable carbon (POXC) by about 18% compared to the control (Figure 1B).

Table 1. Mean values (up) and standard deviation (down) obtained for each treatment in the corresponding analyses in soil, along with the p-values from the one-way ANOVA, for cases where the differences were not significant.

Soil Samples	Untreated (nt)	Biochar (char)	PGPM	Biochar + PGPM (comb)	p-Value
Electrical Conductivity (μS/cm^2)	**2817.2** ±1405.5	**2647.2** ±1365.9	**2166.9** ±612.7	**2607.5** ±1125.2	**0.91**
pH	**7.53** ±0.20	**7.35** ±0.74	**7.51** ±0.12	**7.51** ±0.34	**0.95**
Soil respiration (mg CO_2)	**8.91** ±0.47	**7.61** ±1.78	**8.53** ±1.14	**9.30** ±0.13	**0.33**

Bold indicates EM values, normal style indicates SE.

In this context, biochar may act as a microbial carrier, offering porous niches that promote colonization and protection of inoculated strains, which in turn mineralize labile C fractions adsorbed to the matrix. Previous studies support this interpretation, showing that biochar combined with microbial inoculants or organic amendments enhances soil enzymatic activity [45] and promotes the stability and colonization of *Pseudomonas* spp. [46]. Interestingly, soil respiration did not differ significantly among treatments despite clear variations in enzymatic activity and labile C consumption. This apparent decoupling may reflect a microbial shift toward growth and biomass synthesis rather than CO_2 mineralization, or the partial adsorption of CO_2 by biochar [47]. Comparable findings were reported by Qiu et al. [48], who observed reductions in active C pools and enzymatic changes without corresponding increases in CO_2 release. The results indicate that biochar alone reduces microbial metabolic activity, whereas inoculated microorganisms restore enzymatic activity and intensify oxidizable C consumption [49]. In a closed greenhouse pot system, the reduction in POXC does not necessarily signal soil depletion but rather reflects intensive use of available carbon by the microbial community, with potential benefits for nutrient bioavailability. The application of these amendments under field conditions could offer valuable insights into the processes governing POXC reduction, considering the role of carbon balance modulated by external environmental inputs [50].

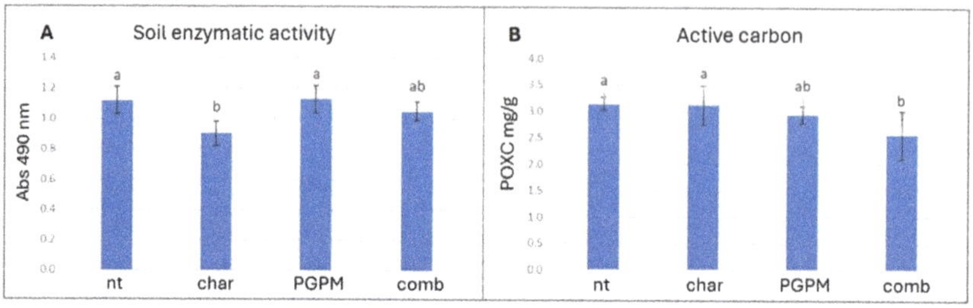

Figure 1. Effects of biochar (char), plant growth-promoting microorganisms (PGPMs), and their combination (comb) on (**A**) soil enzymatic activity (measured by FDA assay) and (**B**) oxidizable carbon (POXC). Different letters indicate significant differences among treatments according to Tukey's HSD post hoc test ($p < 0.05$).

3.2. *Yield of Roots and Above-Ground Parts*

The assessment of plant yield indicated that the applied treatments had no significant effect on plant biomass, either on shoots or roots (Table 2), despite the distinct patterns in soil carbon utilization and enzymatic activity observed under biochar and PGPM treatments. Previous studies have demonstrated that the combined application of biochar and PGPM can enhance nutrient cycling by providing suitable habitats for beneficial microorganisms and improving plant stress resilience, but such effects do not necessarily result in greater biomass accumulation [51,52]. Furthermore, biomass and fresh/dry weight responses may require longer-term observations to become evident, particularly if biochar primarily enhances physiological mechanisms related to stress tolerance or nutrient efficiency during the early stages of growth [53].

Table 2. Mean values (up) and standard deviation (down) obtained for each treatment of roots fresh weight and shoots fresh and dry weight, along with the *p*-values from the one-way ANOVA.

Weight	Untreated (nt)	Biochar (char)	PGPM	Biochar + PGPM (comb)	*p*-Value
Roots fresh weight (g)	**12.5** ±2.78	**8.0** ±0.56	**11.0** ±1.41	**6.0** ±0.74	**0.45**
Shoots fresh weight leaves and stems (g)	**275.0** ±35.36	**212.5** ±95.46	**232.5** ±10.61	**255.0** ±24.04	**0.68**
Shoots dry weight leaves and stems (g)	**26.04** ±8.78	**23.05** ±0.47	**28.50** ±5.71	**21.71** ±1.28	**0.61**

Bold indicates EM values, normal style indicates SE.

3.3. *Roots to Shoots Redox Responses and Oxidative Stress Indicators*

Analyses of root tissues revealed an increase in H_2O_2 content in plants exposed to biochar alone (Figure 2A), with levels rising by 55.4% compared to the control, while H_2O_2 accumulation in leaves remained unchanged (Table 3). Interestingly, this localized rise in root H_2O_2 did not coincide with any significant increase in lipid peroxidation in either roots or leaves, or alteration in leaves' cellular respiration (Table 3). It is well known that moderate accumulation of H_2O_2 appears to function as a key redox signal able to mediate plant stress acclimation: its controlled increase activates ROS-dependent signaling cascades and redox-sensitive proteins, leading to transcriptional and translational adjustments consistent with enhanced tolerance [54]. The lack of oxidative damage in plants treated with biochar suggests that the detected H_2O_2 increase in roots reflects an adaptive rather than damaging response. This mechanism is supported by previous evidence showing that moderate H_2O_2 levels modulate hormonal pathways, particularly abscisic acid signaling, and promote antioxidant reinforcement [55,56]. When biochar was combined with PGPMs, H_2O_2 accumulation in roots was attenuated, suggesting that microbial inoculation contributed to the mitigation of biochar-induced oxidative response. This effect reflects the capacity of several PGPMs, including *Pseudomonas* and *Azotobacter*, to alleviate redox imbalance through the secretion of antioxidant metabolites or by priming host defense pathways [57–59]. The combined biochar–PGPM treatment thus appears to promote a more balanced redox homeostasis in roots, reducing stress intensity while maintaining signaling competence. At the foliar level, antioxidant assays based on radical inhibition revealed distinct patterns: the DPPH assay showed higher inhibition percentages in leaves of plants treated with biochar, either alone or combined with PGPMs (8.9% and 9.4%, respectively), compared to both the PGPM-only and control treatments (4.9% and 4.8%, respectively; Figure 2B). Conversely, the ABTS assay revealed no difference under biochar alone compared to the control, but a significant reduction in radical inhibition under PGPMs alone, dropping from 6.8% in the control to 2.4% in the treated plants (Figure 2C). This divergence between DPPH and ABTS responses may arise from their different chemical sensitivities [60]: DPPH primarily detects lipophilic antioxidants and certain phenolic classes, while ABTS also responds to hydrophilic antioxidants [61,62]. The selective increase in DPPH inhibition suggests that biochar, alone or in combination with PGPMs, stimulated the synthesis and accumulation of lipophilic antioxidant compounds without substantially affecting hydrophilic antioxidant pools. This result may indicate a possible priming mechanism in which biochar-induced H_2O_2 accumulation in roots acts as a mobile redox signal, activating antioxidant metabolism in leaves through hormone-mediated signaling networks involved in abiotic stress tolerance [63,64]. This mechanism would enhance plant resilience against oxidative stress by promoting antioxidant accumulation prior to the induction of oxidative damage [65].

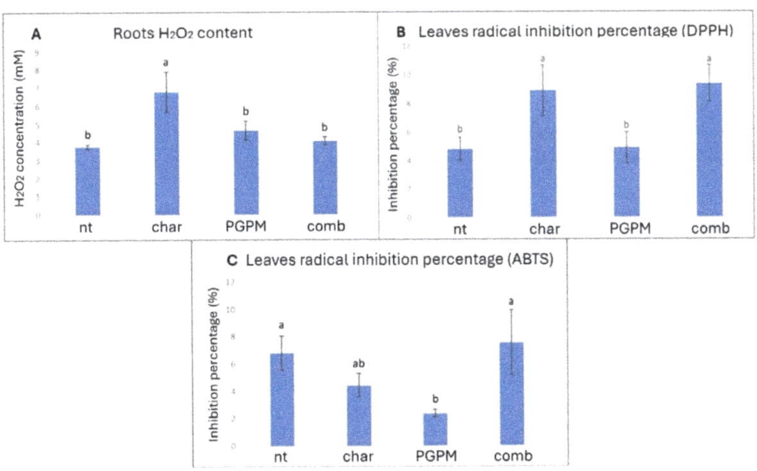

Figure 2. Effects of biochar (char), plant growth-promoting microorganisms (PGPMs), and their combination (comb) on (**A**) roots' H_2O_2 content, (**B**) leaves' radical inhibition percentage measured by DPPH assay, and (**C**) leaves' radical inhibition percentage measured by ABTS assay. Different letters indicate significant differences among treatments according to Tukey's HSD post hoc test ($p < 0.05$).

Table 3. Mean values (up) and standard deviation (down) obtained for each treatment in the corresponding analyses in leaves and roots, along with the *p*-values from the one-way ANOVA, for cases where the differences were not significant.

Oxidative Stress Parameters	Untreated (nt)	Biochar (char)	PGPM	Biochar + PGPM (comb)	*p*-Value
Roots lipidic peroxidation–MDA assay (ng/mL)	**1.15** ±0.13	**1.15** ±0.27	**1.25** ±0.25	**1.17** ±0.15	**0.91**
Leaves lipidic peroxidation–MDA assay (ng/mL)	**10.1** ±2.2	**10.9** ±0.6	**10.2** ±0.9	**12.5** ±1.3	**0.16**
Leaves H_2O_2 content (μM)	**3.95** ±0.33	**4.55** ±0.99	**4.13** ±0.19	**3.45** ±0.62	**0.40**
Leaves cellular respiration–TTC assay (*Abs λ* = 530 nm)	**0.26** ±0.036	**0.23** ±0.028	**0.22** ±0.052	**0.27** ±0.086	**0.65**

Bold indicates EM values, normal style indicates SE.

3.4. Antioxidant Modulation in Roots and Leaves

The assessment of total soluble sugar content in roots and leaves serves as a useful indicator of the plant antioxidant and metabolic response, reflecting their function as osmoprotectants and modulators of ROS dynamics via sugar-derived signaling pathways, particularly in systems affected by PGPM and biochar interactions. At the root level, the inoculation with PGPMs, either alone or in combination with biochar, resulted in a marked reduction (approximately 37%) in soluble sugar concentrations compared with the uninoculated control (Figure 3A). This decrease may reflect enhanced microbial metabolic activity and carbon demand in the rhizosphere, with plant-derived carbohydrates being mobilized to sustain microbial proliferation and root colonization. This interpretation aligns with the observed decline in soil oxidizable carbon (POXC), indicating intensified use of labile carbon fractions by the plant–microbe system. The concurrent depletion of both root sugars and POXC supports the notion that PGPMs promote a dynamic carbon exchange between plant tissues and soil microbial communities, thereby accelerating carbon turnover and improving resource efficiency in the rhizosphere. Comparable responses have been documented in previous studies, where PGPM inoculation stimulated the mobilization of stored carbohydrates in roots, leading to a reduction in sugar reserves that facilitated symbiotic interactions and rhizodeposition processes [58,66]. Beyond this metabolic role, soluble sugars also function as pivotal signaling molecules in root development, energy allocation, and stress-related pathways [67]. Therefore, the observed decline in root sugars may not only represent enhanced microbial consumption but also a PGPM-induced metabolic reprogramming aimed at optimizing carbon fluxes and enhancing the plant's physiological resilience [68]. Indeed, several PGPM strains have been shown to modulate carbon metabolism by adjusting sugar partitioning between shoots and roots, enhancing carbon allocation to metabolically active tissues, and improving carbon use efficiency [69]. The pattern observed in leaves displayed an opposite trend: under biochar alone, soluble sugar concentrations increased significantly by approximately 31.7% compared with the control (Figure 3B), suggesting a possible shift in carbon partitioning between below- and above-ground organs. This behavior may reflect reduced root carbon demand and enhanced allocation of photoassimilates to the shoots, potentially driven by improved soil physical conditions and nutrient availability associated with biochar amendment. Alternatively, the accumulation of soluble sugars in leaves may represent an adaptive strategy to enhance osmotic balance and to provide precursors for secondary metabolite biosynthesis, both of which contribute to stress tolerance [68,69]. Notably, the increase in leaf sugar content coincided with higher root H_2O_2 levels, suggesting a potential signaling link between oxidative stress and carbohydrate metabolism. Exogenous or endogenous H_2O_2 generated in roots is known to act as a systemic signal that triggers sugar accumulation in leaves, thereby priming plants to counteract abiotic stress conditions through the reinforcement of antioxidant and protective metabolic pathways [64]. This increase in root H_2O_2 content observed under biochar alone did not translate into significant changes in the activity of key antioxidant enzymes (SOD, POD, CAT, and APX) in either roots or leaves, nor did it interfere with photosynthetic activity (Table 4). Similarly, total protein content remained statistically unchanged across treatments in both organs (Table 4), indicating that the oxidative signal induced by biochar was not associated with generalized protein degradation or stress-induced metabolic suppression. In contrast, the non-enzymatic antioxidant profile assessed through proline (Figure 3C), ascorbic acid (Figure 3D), and total phenolic content (Figure 3E), revealed a distinct response. Biochar alone had a limited influence on these metabolites, while PGPM inoculation alone resulted in a significant reduction in all three parameters (66.0%, 18.0%, 45.7%, respectively), indicating a downregulation or redistribution of antioxidant and osmolyte metabolism. Interestingly, when PGPMs were combined with biochar, however, metabolite levels were restored to values comparable to, or even exceeding, those of the control, suggesting that the presence of biochar mitigated the reduction induced by PGPMs alone. This pattern supports the hypothesis of a priming mechanism, wherein the moderate H_2O_2 accumulation observed in roots exposed to biochar may function as a systemic signal that activates antioxidant defenses in the leaves through hormonal and redox-mediated signaling pathways associated with abiotic stress tolerance [63,64] prior to the induction of enzymatic defenses or oxidative damage [65]. Furthermore, different species of PGPMs (like *Pseudomonas* spp.) are known to trigger secondary metabolism, stimulating phenolic and flavonoid synthesis through defense pathway activation [57,70]. The enhanced antioxidant response observed under the combined biochar–PGPM treatment, in contrast to the suppression seen with PGPMs alone and the limited effects of biochar alone, suggests a synergistic interaction between the two components: biochar may act as a physical and chemical reservoir for microbial metabolites or signaling molecules, prolonging their bioavailability in the rhizosphere and amplifying their systemic effects on plant metabolism, optimizing redox balance and plant capacity for metabolic adjustment under stress-related conditions.

Table 4. Mean values (up) and standard deviation (down) for each treatment in the corresponding analyses in leaves and roots, along with the *p*-values from the one-way ANOVA, for cases where the differences were not significant.

Antioxidants Parameters	Untreated (nt)	Biochar (char)	PGPM	Biochar + PGPM (comb)	*p*-Value
Chlorophyll A content (mg/g FW)	**0.50** ±0.004	**0.53** ±0.029	**0.55** ±0.030	**0.51** ±0.024	**0.11**
Chlorophyll B content (mg/g FW)	**0.54** ±0.091	**0.69** ±0.068	**0.72** ±0.052	**0.75** ±0.104	**0.06**
Carotenoids content (mg/g FW)	**0.164** ±0.009	**0.174** ±0.004	**0.178** ±0.009	**0.169** ±0.008	**0.21**
Leaves total protein content (BSA µg/mL)	**392.03** ±28.74	**438.13** ±83.61	**387.57** ±38.26	**470.62** ±33.08	**0.22**
Leaves SOD activity (I%)	**54.77** ±13.79	**52.29** ±4.44	**50.56** ±3.88	**55.89** ±5.22	**0.84**
Leaves POD activity (U/mL)	**0.096** ±0.018	**0.127** ±0.033	**0.109** ±0.036	**0.115** ±0.032	**0.68**
Leaves CAT activity (U/mL)	**0.205** ±0.085	**0.086** ±0.047	**0.108** ±0.013	**0.103** ±0.029	**0.36**
Leaves APX activity (U/mL)	**0.272** ±0.071	**0.539** ±0.111	**0.545** ±0.250	**0.401** ±0.128	**0.18**
Roots total protein content (BSA µg/mL)	**223.11** ±50.95	**225.45** ±18.34	**247.83** ±23.98	**251.82** ±9.58	**0.56**
Roots SOD activity (I%)	**45.01** ±3.00	**47.70** ±2.55	**44.31** ±3.63	**48.90** ±6.18	**0.50**
Roots POD activity (U/mL)	**0.634** ±0.10	**0.661** ±0.11	**0.705** ±0.20	**0.739** ±0.08	**0.78**

Table 4. *Cont.*

Antioxidants Parameters	Untreated (nt)	Biochar (char)	PGPM	Biochar + PGPM (comb)	*p*-Value
Roots CAT activity (U/mL)	**0.043** ±0.012	**0.063** ±0.040	**0.048** ±0.031	**0.124** ±0.099	**0.33**
Roots APX activity (U/mL)	**0.925** ±0.080	**0.896** ±0.338	**0.921** ±0.171	**1.354** ±0.291	**0.14**

Bold indicates EM values, normal style indicates SE.

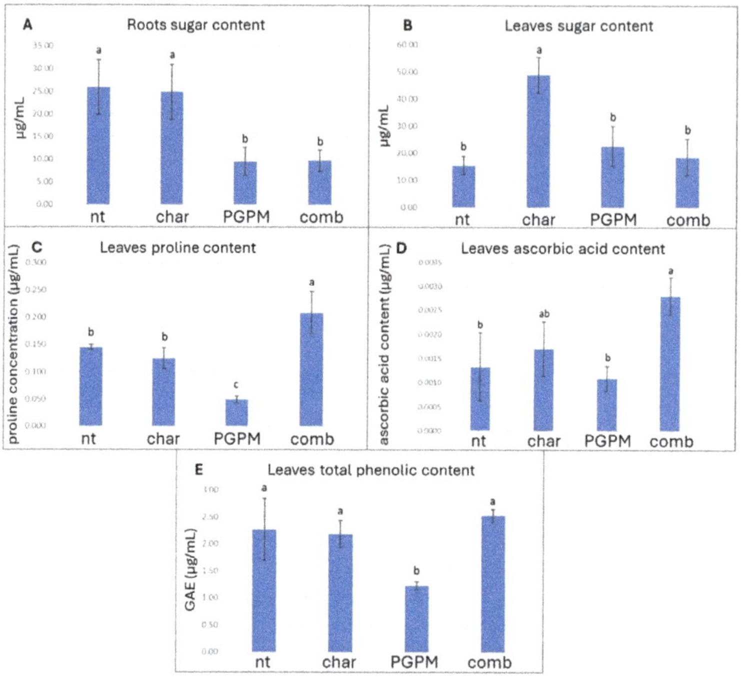

Figure 3. Effects of biochar (char), plant growth-promoting microorganisms (PGPMs), and their combination (comb) on (**A**) roots' total soluble sugar content, (**B**) leaves' total soluble sugar content, (**C**) leaves proline content, (**D**) leaves' ascorbic acid content, and (**E**) leaves' total phenolic content (TPC). Different letters indicate significant differences among treatments according to Tukey's HSD post hoc test ($p < 0.05$).

3.5. Integrated Multivariate Insights into Plant–Soil System: A Decryption Key

Multivariate analysis provided an integrated view of the plant–soil system, revealing consistent patterns across scales. Pearson correlation was applied to assess the correlation among different physiological and biochemical attributes, which were schematized in a correlation matrix and a chord diagram (Figures 4 and S2).

With regard to soil variables, negative correlations were identified between EC, pH, and protein content in roots (r Protein). In addition, pH was positively correlated with soil CO_2 content and negatively correlated with H_2O_2 content in leaf samples (l H_2O_2). Enzymatic activity (s FDA) showed a strong negative correlation with root H_2O_2 (r H_2O_2) and leaf sugar content (l Sugar), with *p* values of 0.009 and 0.001, respectively. A negative correlation was also observed between active C and root catalase (r CAT). Soil respiration (s CO_2) was negatively correlated with l H_2O_2. For plant parameters, a negative correlation was found between lipid peroxidation in roots (r MDA) and certain antioxidant enzymes in leaves (l SOD; l POD). Conversely, leaf lipid peroxidation (l MDA) was positively correlated with total phenolic content (l TPC) and with root protein content (r Protein). Root hydrogen peroxide content exhibited a positive correlation with l POD and was also closely correlated with leaf sugar levels ($p = 0.00003$). Positive correlations among enzymatic parameters were evident, including r SOD with l POD and l vitamin C; r CAT and r APX with cell viability (l TTC); and, in leaves, l APX with one of the photosynthetic pigments (l Car). Regarding pigments, a robust positive correlation ($p = 0.002$) was identified between chlorophyll a (l Chl A) and carotenoid content (l Car), accompanied by a negative correlation with l TPC. Proline levels (l Proline) showed a strong positive correlation with biochemical indicators of antioxidant status, namely l ABTS, l TPC, and l Vit C, with p values of 0.009, 0.003, and 0.008, respectively. The correlation between leaf sugar content and root H_2O_2 suggests that increased hydrogen peroxide in roots is associated with enhanced primary metabolic activity, leading to elevated sugar accumulation in leaves, a metabolic interaction consistent with previous reports [64]. A contrasting relationship between these two parameters (root hydrogen peroxide and leaf sugar content) emerged in the context of soil enzymatic activity, a pattern not yet explored in depth in the literature and potentially offering a starting point for future mechanistic studies. Proline content in leaves was positively correlated with ABTS reduction, total phenolic content, and vitamin C content, indicating a synergistic antioxidant response to oxidative stress. In tomato plants, it is well documented that proline content correlates positively with vitamin C, phenolic content, and antioxidant activity measured by the ABTS assay [71]. Moreover, a positive correlation between Chl A and carotenoids was also observed, highlighting the co-regulation of chlorophyll and carotenoid biosynthesis during plastid development and pigment accumulation, as well as the potential accumulation of photosynthates in the fruit [72].

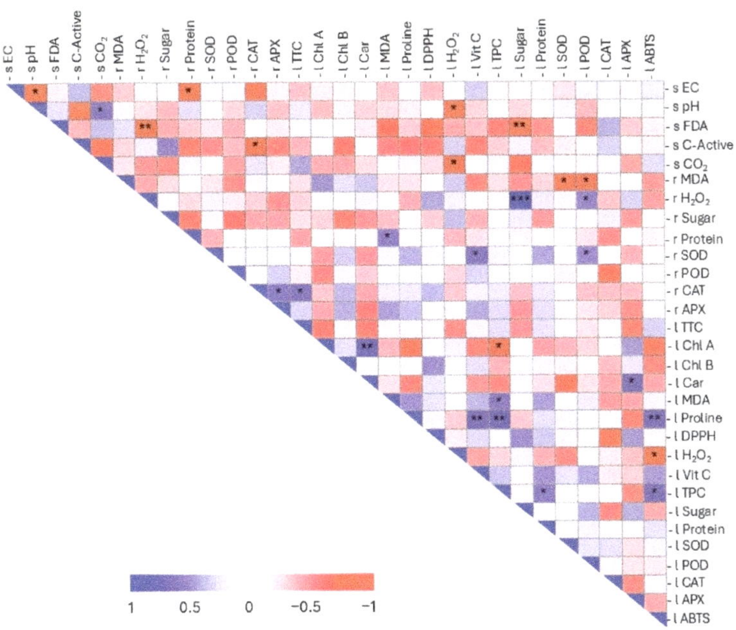

Figure 4. Pearson correlation matrix between the analysed parameters. Positive correlations are shown in blue and negative correlations in red. The colour intensity is proportional to the correlation value, as shown by the scale below (from 1 to −1). Significant correlations are highlighted with asterisks: p-value <0.05 (*); p-value <0.01 (**); and p-value <0.001 (***). The letters s, l, and r were used to represent soil, leaves, and roots, respectively. To distinguish between the various substrates of the matrix under consideration, the parameters were assigned a letter: s for soil, r for roots, and l for leaves.

PCA performed on the different datasets clearly separated the four experimental conditions, identifying the microbial inoculum and the combined biochar strategy as the main drivers of system perturbation. In the soil dataset (Figure 5A), the microbial component, explaining 55.8% of the total variance, was associated with changes in soil-level activity, whereas the second component, related to biochar application and accounting for 19.8% of the total variance, aligned with previous findings attributing such effects to biochar's adsorptive properties or the release of potentially toxic compounds [43,44]. When biochar was combined with PGPMs, microorganisms appeared to exploit biochar surfaces as protective niches, ultimately contributing to the restoration of soil metabolic dynamics and highlighting the effectiveness of this combined strategy [45]. This evidence supports the concept of biochar functioning as an ecological carrier that enhances microbial resilience when co-applied with beneficial microorganisms [73]. Root responses provided further insight (Figure 5B), with the microbial factor (alone or in combination) emerging as a major source of variation (30.2% of the total variance), while biochar alone was projected along axes dominated by H_2O_2, confirming a localized oxidative stress response. In contrast, inoculated treatments, particularly the biochar–PGPM combination, were associated with metabolic and protein-related parameters, indicating a broader metabolic adjustment. This outcome can be explained by the release of phytohormones and metabolites by PGPMs such as *Pseudomonas fluorescens* and *Azotobacter chroococcum*, which contribute to plant carbon demand and redox balance regulation [58,74]. The reduction in root sugars under PGPM treatments, also captured by PCA, may therefore reflect not only microbial consumption but also plant-driven reallocation of carbon towards defense and shoot growth, in line with the signaling role of sugars [68,69]. In leaves (Figure 5C), all treatments were clearly separated, predominantly along gradients defined by antioxidant traits, whereas single-factor treatments were less discriminant, with the first two components together explaining only 51.7% of the total variance. This pattern suggests that the combined biochar–PGPM treatment elicits a systemic antioxidant response, likely aimed at counteracting potential oxidative stress through enhanced ROS scavenging [75].

These findings provide a statistical interpretation of the complex system under investigation and indicate that the biochar–PGPM interaction may trigger systemic priming without measurable oxidative damage, whereby root-derived ROS and hormonal signals induce the accumulation of defensive metabolites in shoots, preparing plants for environmental stress while preserving cellular homeostasis [63,65]. The selective increase in antioxidant activity detected by the DPPH assay, which reflects lipophilic antioxidant pools, supports the activation of secondary metabolic pathways associated with long-term adaptation [61,62]. The absence of these effects under PGPMs alone, together with the limited responses observed under biochar alone, highlights the synergistic nature of the combined treatment in both mitigating root oxidative stress and activating systemic defense responses. Taken together, the three PCA analyses converge on a coherent interpretation: biochar alone introduces potentially detrimental perturbations; PGPMs alone partially improve root status without inducing strong systemic effects; whereas their combination converts biochar-induced stress into a beneficial signal that primes systemic defense. This underscores the importance of evaluating the integrated plant–soil level, while accounting for dynamic environmental interactions [76].

Several biochemical variables showed clear trends; however, a number of parameters did not reach statistical significance ($p > 0.05$), reflecting the inherent variability of pot experiments and the absence of imposed abiotic or biotic stress. In addition, the experiment lasted only three months and did not include measurements of fruit yield or quality, thereby constraining the assessment of physiological and biochemical responses to the vegetative growth stage. These limitations restrict the extrapolation of the findings to broader field conditions and longer cropping cycles. Future work should therefore incorporate defined stress treatments, extended field trials, and metabolomic profiling for higher biochemical resolution, together with yield-level assessments.

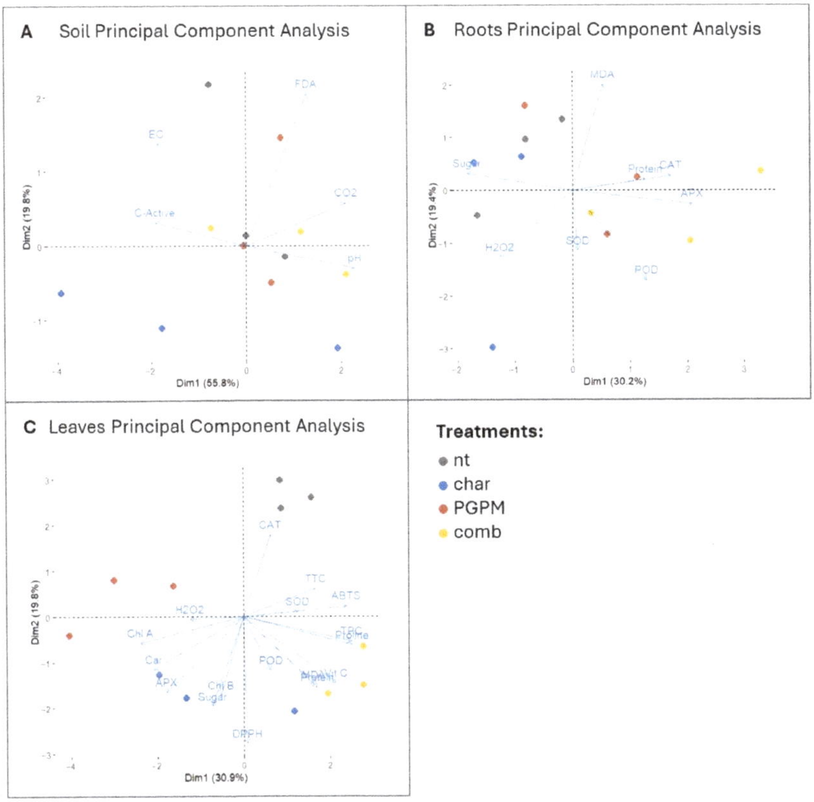

Figure 5. Principal component analysis (PCA) of soil (**A**), root (**B**), and leaf (**C**) parameters under different treatments: untreated control (nt), biochar (char), plant growth-promoting microorganisms (PGPMs), and their combination (comb). Vectors indicate the contribution of individual variables to the variance, while points represent the distribution of treatments across principal components.

4. Conclusions

From an applied perspective, the co-application of biochar and PGPMs emerges as a promising and sustainable strategy that may potentially prime plants toward enhanced defensive capacity, particularly under the increasing frequency of environmental stressors associated with climate change. As global agriculture faces intensified drought, heat, and nutrient imbalances, optimizing beneficial interactions within the plant–soil system will be crucial to improve resource-use efficiency, reduce dependence on chemical inputs, and sustain productivity in degraded or marginal soils. Future field-scale investigations should therefore aim to verify whether these greenhouse results translate into yield benefits and improved stress tolerance under realistic climatic fluctuations and across diverse crop systems, although current evidence strongly supports their integrated use as a tool for future agriculture.

Supplementary Materials: The following supporting information can be downloaded at: https://www.mdpi.com/article/10.3390/antiox14121482/s1, Table S1. Physico-chemical characterization of biochar utilized in the experiments. Details on the methods utilized are reported in Marmiroli et al. (2022) [77]. Figure S1. Structures (A) and EDX spectra of biochar (B). Biochar was assessed using an ESEM FEG2500 FEI Environmental Scanning Electron Microscope (FEI Europe, Eindhoven, Netherlands) with energy-dispersive X-ray spectroscopy (EDX). Methods utilized are described in Marmiroli et al. (2022) [77]. Figure S2. A chord diagram illustrating the significant interactions between the analysed parameters ($p < 0.05$). Positive correlations are indicated by blue chords, while negative correlations are indicated by red chords. Statistically significant correlations with $p < 0.01$ are highlighted with a black border, while those with $p < 0.001$ are highlighted with a thick, dotted black line along the respective chords.

Author Contributions: M.M., S.C., M.T. and L.P.: manuscript draft and revisions, S.C. and M.T.: experimental and statistics, M.M., S.C., M.T. and L.P.: data analyses and revisions, M.M.: first draft revision, M.M., S.C., M.T. and L.P.: second draft revision, final draft approval. All authors have read and agreed to the published version of the manuscript.

Funding: This study was funded under the National Recovery and Resilience Plan (NRRP), Mission 04 Component 2 Investment 1.5—NextGenerationEU, Call for tender n. 3277 dated 30 December 2021 Award Number: 0001052 ECOSISTER.

Institutional Review Board Statement: Not applicable.

Informed Consent Statement: The work has been performed only on plants and microbes; humans or animals were not used during the experiments.

Data Availability Statement: All data are available upon request to the authors.

Conflicts of Interest: The authors declare no conflicts of interest.

References

1. Zhou, R.; Jiang, F.; Liu, Y.; Yu, X.; Song, X.; Wu, Z.; Cammarano, D. Environmental Changes Impact on Vegetables Physiology and Nutrition—Gaps between Vegetable and Cereal Crops. *Sci. Total Environ.* **2024**, *933*, 173180. [CrossRef] [PubMed]
2. Heikonen, S.; Heino, M.; Jalava, M.; Siebert, S.; Viviroli, D.; Kummu, M. Climate Change Threatens Crop Diversity at Low Latitudes. *Nat. Food* **2025**, *6*, 331–342. [CrossRef] [PubMed]
3. Schreinemachers, P.; Simmons, E.B.; Wopereis, M.C.S. Tapping the Economic and Nutritional Power of Vegetables. *Glob. Food Secur.* **2018**, *16*, 36–45. [CrossRef]
4. González Guzmán, M.; Cellini, F.; Fotopoulos, V.; Balestrini, R.; Arbona, V. New Approaches to Improve Crop Tolerance to Biotic and Abiotic Stresses. *Physiol. Plant* **2022**, *174*, e13547. [CrossRef]
5. Chaudhary, S.; Devi, P.; HanumanthaRao, B.; Jha, U.C.; Sharma, K.D.; Prasad, P.V.V.; Kumar, S.; Siddique, K.H.M.; Nayyar, H. Physiological and Molecular Approaches for Developing Thermotolerance in Vegetable Crops: A Growth, Yield and Sustenance Perspective. *Front. Plant Sci.* **2022**, *13*, 878498. [CrossRef]
6. Erika, C.; Griebel, S.; Naumann, M.; Pawelzik, E. Biodiversity in Tomatoes: Is It Reflected in Nutrient Density and Nutritional Yields Under Organic Outdoor Production? *Front. Plant Sci.* **2020**, *11*, 589692. [CrossRef]
7. Ozygit, I.I.; Can, H.; Uyanik, O.L.; Yalcin, I.E.; Demir, G. Fruit Mineral Nutrient Contents of Field and Greenhouse Grown Tomatoes and Comparison with Standard Values. *Not. Bot. Horti Agrobot. Cluj Napoca* **2024**, *52*, 13479. [CrossRef]
8. Ali, M.Y.; Sina, A.A.I.; Khandker, S.S.; Neesa, L.; Tanvir, E.M.; Kabir, A.; Khalil, M.I.; Gan, S.H. Nutritional Composition and Bioactive Compounds in Tomatoes and Their Impact on Human Health and Disease: A Review. *Foods* **2021**, *10*, 45. [CrossRef]
9. Collins, E.J.; Bowyer, C.; Tsouza, A.; Chopra, M. Tomatoes: An Extensive Review of the Associated Health Impacts of Tomatoes and Factors That Can Affect Their Cultivation. *Biology* **2022**, *11*, 239. [CrossRef]
10. Carucci, F.; Gagliardi, A.; Giuliani, M.M.; Gatta, G. Irrigation Scheduling in Processing Tomato to Save Water: A Smart Approach Combining Plant and Soil Monitoring. *Appl. Sci.* **2023**, *13*, 7625. [CrossRef]
11. Khan, Q.; Wang, Y.; Xia, G.; Yang, H.; Luo, Z.; Zhang, Y. Deleterious Effects of Heat Stress on the Tomato, Its Innate Responses, and Potential Preventive Strategies in the Realm of Emerging Technologies. *Metabolites* **2024**, *14*, 283. [CrossRef] [PubMed]
12. Hubab, M.; Lorestani, N.; Al-Awabdeh, R.A.M.; Shabani, F. Climate Change-Driven Shifts in the Global Distribution of Tomato and Potato Crops and Their Associated Bacterial Pathogens. *Front. Microbiol.* **2025**, *16*, 1520104. [CrossRef] [PubMed]
13. Govindasamy, R.; Ceylan, R.F.; Özkan, B. Global Tomato Production: Price Sensitivity and Policy Impact in Mexico, Türkiye, and the United States. *Horticulturae* **2025**, *11*, 84. [CrossRef]

14. Fiodor, A.; Singh, S.; Pranaw, K. The Contrivance of Plant Growth Promoting Microbes to Mitigate Climate Change Impact in Agriculture. *Microorganisms* **2021**, *9*, 1841. [CrossRef]
15. Afshar, M.; Mofatteh, S. Biochar for a Sustainable Future: Environmentally Friendly Production and Diverse Applications. *Results Eng.* **2024**, *23*, 102433. [CrossRef]
16. Allohverdi, T.; Mohanty, A.K.; Roy, P.; Misra, M. A Review on Current Status of Biochar Uses in Agriculture. *Molecules* **2021**, *26*, 5584. [CrossRef]
17. Kabir, E.; Kim, K.H.; Kwon, E.E. Biochar as a Tool for the Improvement of Soil and Environment. *Front. Environ. Sci.* **2023**, *11*, 1324533. [CrossRef]
18. Weber, K.; Quicker, P. Properties of Biochar. *Fuel* **2018**, *217*, 240–261. [CrossRef]
19. Fang, H.; Liu, K.; Li, D.; Peng, X.; Zhang, W.; Zhou, H. Long-Term Effects of Inorganic Fertilizers and Organic Manures on the Structure of a Paddy Soil. *Soil Tillage Res.* **2021**, *213*, 105137. [CrossRef]
20. Aloo, B.N.; Tripathi, V.; Makumba, B.A.; Mbega, E.R. Plant Growth-Promoting Rhizobacterial Biofertilizers for Crop Production: The Past, Present, and Future. *Front. Plant Sci.* **2022**, *13*, 1002448. [CrossRef]
21. Mahmud, A.A.; Upadhyay, S.K.; Srivastava, A.K.; Bhojiya, A.A. Biofertilizers: A Nexus between Soil Fertility and Crop Productivity under Abiotic Stress. *Curr. Res. Environ. Sustain.* **2021**, *3*, 100063. [CrossRef]
22. Khosravi, H.; Khoshru, B.; Nosratabad, A.F.; Mitra, D. Exploring the Landscape of Biofertilizers Containing Plant Growth-Promoting Rhizobacteria in Iran: Progress and Research Prospects. *Curr. Res. Microb. Sci.* **2024**, *7*, 100268. [CrossRef]
23. Nosheen, S.; Ajmal, I.; Song, Y. Microbes as Biofertilizers, a Potential Approach for Sustainable Crop Production. *Sustainability* **2021**, *13*, 1868. [CrossRef]
24. Devi, R.; Kaur, T.; Kour, D.; Yadav, A.; Yadav, A.N.; Suman, A.; Ahluwalia, A.S.; Saxena, A.K. Minerals Solubilizing and Mobilizing Microbiomes: A Sustainable Approach for Managing Minerals' Deficiency in Agricultural Soil. *J. Appl. Microbiol.* **2022**, *133*, 1245–1272. [CrossRef] [PubMed]
25. Wu, X.; Liu, Y.; Jia, B.; Tao, L.; Li, H.; Wang, J.; Yuan, Z.; Sun, X.; Yao, Y. Four Decades of Bacillus Biofertilizers: Advances and Future Prospects in Agriculture. *Microorganisms* **2025**, *13*, 187. [CrossRef] [PubMed]
26. Sarfraz, R.; Hussain, A.; Sabir, A.; Ben Fekih, I.; Ditta, A.; Xing, S. Role of Biochar and Plant Growth Promoting Rhizobacteria to Enhance Soil Carbon Sequestration—A Review. *Environ. Monit. Assess.* **2019**, *191*, 251. [CrossRef]
27. Guan, T.K.; Wang, Q.Y.; Li, J.S.; Yan, H.W.; Chen, Q.J.; Sun, J.; Liu, C.J.; Han, Y.Y.; Zou, Y.J.; Zhang, G.Q. Biochar Immobilized Plant Growth-Promoting Rhizobacteria Enhanced the Physicochemical Properties, Agronomic Characters and Microbial Communities during Lettuce Seedling. *Front. Microbiol.* **2023**, *14*, 1218205. [CrossRef] [PubMed]
28. Tabacchioni, S.; Passato, S.; Ambrosino, P.; Huang, L.; Caldara, M.; Cantale, C.; Hett, J.; Del Fiore, A.; Fiore, A.; Schlüter, A.; et al. Identification of Beneficial Microbial Consortia and Bioactive Compounds with Potential as Plant Biostimulants for a Sustainable Agriculture. *Microorganisms* **2021**, *9*, 426. [CrossRef]
29. Moebius-Clune, B.N.; Moebius-Clune, D.J.; Gugino, B.K.; Idowu, O.J.; Schindelbeck, R.R.; Ristow, A.J.; van Es, H.M.; Thies, J.E.; Shayler, H.A.; McBride, M.B.; et al. *Comprehensive Assessment of Soil Health—The Cornell Framework; Standard Operating Procedures;* Cornell University: Geneva, NY, USA, 2016.
30. Adam, G.; Duncan, H. Development of a Sensitive and Rapid Method for the Measurement of Total Microbial Activity Using Fluorescein Diacetate (FDA) in a Range of Soils. *Soil Biol. Biochem.* **2001**, *33*, 943–951. [CrossRef]
31. *ISO 11063:2020;* Soil Quality—Direct Extraction of Soil DNA. International Organization for Standardization: Geneva, Switzerland, 2020.
32. Pagano, L.; Carlo, S.; Lepore, G.O.; Bonanni, V.; Zizic, M.; Pollastri, S.; Margheri, S.; Orsilli, J.; Puri, A.; Villani, M.; et al. Mechanistic Understanding of Iron Oxide Nanobiotransformation in Zea Mays: A Combined Synchrotron-Based, Physiological and Molecular Approach. *Environ. Sci. Nano* **2025**, *12*, 4107–4121. [CrossRef]
33. Marmiroli, M.; Mussi, F.; Pagano, L.; Imperiale, D.; Lencioni, G.; Villani, M.; Zappettini, A.; White, J.C.; Marmiroli, N. Cadmium Sulfide Quantum Dots Impact Arabidopsis Thaliana Physiology and Morphology. *Chemosphere* **2020**, *240*, 124856. [CrossRef] [PubMed]
34. Wellburn, A.R. The Spectral Determination of Chlorophylls a and b, as Well as Total Carotenoids, Using Various Solvents with Spectrophotometers of Different Resolution. *J. Plant Physiol.* **1994**, *144*, 307–313. [CrossRef]
35. Chen, T.; Zhang, B. Measurements of Proline and Malondialdehyde Content and Antioxidant Enzyme Activities in Leaves of Drought Stressed Cotton. *Bio Protoc.* **2016**, *6*, e1913. [CrossRef]
36. Marmiroli, M.; Mussi, F.; Gallo, V.; Gianoncelli, A.; Hartley, W.; Marmiroli, N. Combination of Biochemical, Molecular, and Synchrotron-Radiation-Based Techniques to Study the Effects of Silicon in Tomato (*Solanum lycopersicum* L.). *Int. J. Mol. Sci.* **2022**, *23*, 15837. [CrossRef]
37. Jagota, S.K.; Dani, H.M. A New Calorimetric Technique for the Estimation of Vitamin C Using Folin Phenol Reagent. *Anal. Biochem.* **1982**, *127*, 178–182. [CrossRef]

38. Mingle, C.E.; Newsome, A.L. An Amended Potassium Persulfate ABTS Antioxidant Assay Used for Medicinal Plant Extracts Revealed Variable Antioxidant Capacity Based upon Plant Extraction Process. *BioRxiv* **2020**.
39. Ernst, O.; Zor, T. Linearization of the Bradford Protein Assay. *J. Vis. Exp.* **2010**, *12*, 1918. [CrossRef]
40. Abedi, T.; Pakniyat, H. Antioxidant Enzyme Changes in Response to Drought Stress in Ten Cultivars of Oilseed Rape (*Brassica napus* L.). *Czech J. Genet. Plant Breed.* **2010**, *46*, 27–34. [CrossRef]
41. Jiang, M.; Zhang, J. Water Stress-Induced Abscisic Acid Accumulation Triggers the Increased Generation of Reactive Oxygen Species and up-Regulates the Activities of Antioxidant Enzymes in Maize Leaves. *J. Exp. Bot.* **2002**, *53*, 2401–2410. [CrossRef]
42. Shakir, S.K.; Irfan, S.; Akhtar, B.; Rehman, S.U.; Daud, M.K.; Taimur, N.; Azizullah, A. Pesticide-Induced Oxidative Stress and Antioxidant Responses in Tomato (*Solanum lycopersicum*) Seedlings. *Ecotoxicology* **2018**, *27*, 919–935. [CrossRef]
43. Dvořáčková, H.; Dvořáček, J.; Záhora, J.; Šimečková, J. Biochar Alone Did Not Increase Microbial Activity in Soils from a Temperate Climate That Had Long-Term Acidity Stress. *Agriculture* **2022**, *12*, 941. [CrossRef]
44. Lehmann, J.; Rillig, M.C.; Thies, J.; Masiello, C.A.; Hockaday, W.C.; Crowley, D. Biochar Effects on Soil Biota—A Review. *Soil Biol Biochem.* **2011**, *43*, 1812–1836. [CrossRef]
45. Bai, M.X.; Situ, G.M.; Li, S.H.; Wu, Q.F.; Liang, C.F.; Qin, H.; Chen, J.H. Effects of Combined Application of Biochar with Organic Amendments on Enzyme Activity and Microbial Metabolic Function of Carbon Sources in Infertile Red Soil. *Ying Yong Sheng Tai Xue Bao* **2022**, *33*, 1283–1290. [CrossRef] [PubMed]
46. Kolton, M.; Graber, E.R.; Tsehansky, L.; Elad, Y.; Cytryn, E. Biochar-Stimulated Plant Performance Is Strongly Linked to Microbial Diversity and Metabolic Potential in the Rhizosphere. *New Phytol.* **2017**, *213*, 1393–1404. [CrossRef]
47. Hagemann, N.; Joseph, S.; Schmidt, H.P.; Kammann, C.I.; Harter, J.; Borch, T.; Young, R.B.; Varga, K.; Taherymoosavi, S.; Elliott K.W.; et al. Organic Coating on Biochar Explains Its Nutrient Retention and Stimulation of Soil Fertility. *Nat. Commun.* **2017** *8*, 1089. [CrossRef]
48. Qiu, H.; Hu, Z.; Liu, J.; Zhang, H.; Shen, W. Effect of Biochar on Labile Organic Carbon Fractions and Soil Carbon Pool Management Index. *Agronomy* **2023**, *13*, 1385. [CrossRef]
49. Wojewódzki, P.; Lemanowicz, J.; Debska, B.; Haddad, S.A. Soil Enzyme Activity Response under the Amendment of Different Types of Biochar. *Agronomy* **2022**, *12*, 569. [CrossRef]
50. Gong, W.; Yan, X.; Wang, J. The Effect of Chemical Fertilizer on Soil Organic Carbon Renewal and CO_2 Emission—A Pot Experiment with Maize. *Plant Soil* **2012**, *353*, 85–94. [CrossRef]
51. Premalatha, R.P.; Poorna Bindu, J.; Nivetha, E.; Malarvizhi, P.; Manorama, K.; Parameswari, E.; Davamani, V. A Review on Biochar's Effect on Soil Properties and Crop Growth. *Front. Energy Res.* **2023**, *11*, 1092637. [CrossRef]
52. Wang, Y.; Li, W.; Du, B.; Li, H. Effect of Biochar Applied with Plant Growth-Promoting Rhizobacteria (PGPR) on Soil Microbial Community Composition and Nitrogen Utilization in Tomato. *Pedosphere* **2021**, *31*, 872–881. [CrossRef]
53. Bolan, S.; Hou, D.; Wang, L.; Hale, L.; Egamberdieva, D.; Tammeorg, P.; Li, R.; Wang, B.; Xu, J.; Wang, T.; et al. The Potential of Biochar as a Microbial Carrier for Agricultural and Environmental Applications. *Sci. Total Environ.* **2023**, *886*, 163968. [CrossRef] [PubMed]
54. Hossain, M.A.; Bhattacharjee, S.; Armin, S.M.; Qian, P.; Xin, W.; Li, H.Y.; Burritt, D.J.; Fujita, M.; Tran, L.S.P. Hydrogen Peroxide Priming Modulates Abiotic Oxidative Stress Tolerance: Insights from ROS Detoxification and Scavenging. *Front. Plant Sci.* **2015** *6*, 141857. [CrossRef] [PubMed]
55. Niu, L.; Liao, W. Hydrogen Peroxide Signaling in Plant Development and Abiotic Responses: Crosstalk with Nitric Oxide and Calcium. *Front. Plant Sci.* **2016**, *7*, 177967. [CrossRef] [PubMed]
56. Anjum, N.A.; Gill, S.S.; Corpas, F.J.; Ortega-Villasante, C.; Hernandez, L.E.; Tuteja, N.; Sofo, A.; Hasanuzzaman, M.; Fujita, M Editorial: Recent Insights Into the Double Role of Hydrogen Peroxide in Plants. *Front. Plant Sci.* **2022**, *13*, 843274. [CrossRef]
57. Ansabayeva, A.; Makhambetov, M.; Rebouh, N.Y.; Abdelkader, M.; Saudy, H.S.; Hassan, K.M.; Nasser, M.A.; Ali, M.A.A.; Ebrahim M. Plant Growth-Promoting Microbes for Resilient Farming Systems: Mitigating Environmental Stressors and Boosting Crops Productivity—A Review. *Horticulturae* **2025**, *11*, 260. [CrossRef]
58. Enebe, M.C.; Babalola, O.O. The Influence of Plant Growth-Promoting Rhizobacteria in Plant Tolerance to Abiotic Stress: A Survival Strategy. *Appl. Microbiol. Biotechnol.* **2018**, *102*, 7821–7835. [CrossRef]
59. Graziano, S.; Caldara, M.; Gullì, M.; Bevivino, A.; Maestri, E.; Marmiroli, N. A Metagenomic and Gene Expression Analysis in Wheat (*T. durum*) and Maize (*Z. mays*) Biofertilized with PGPM and Biochar. *Int. J. Mol. Sci.* **2022**, *23*, 10376. [CrossRef]
60. Lewoyehu, M.; Amare, M. Comparative Evaluation of Analytical Methods for Determining the Antioxidant Activities of Honey: A Review. *Cogent Food Agric.* **2019**, *5*, 1685059. [CrossRef]
61. Floegel, A.; Kim, D.O.; Chung, S.J.; Koo, S.I.; Chun, O.K. Comparison of ABTS/DPPH Assays to Measure Antioxidant Capacity in Popular Antioxidant-Rich US Foods. *J. Food Compos. Anal.* **2011**, *24*, 1043–1048. [CrossRef]
62. Xie, J.; Schaich, K.M. Re-Evaluation of the 2,2-Diphenyl-1-Picrylhydrazyl Free Radical (DPPH) Assay for Antioxidant Activity. *J. Agric. Food Chem.* **2014**, *62*, 4251–4260. [CrossRef]

63. Wang, G.; Hu, C.; Zhou, J.; Liu, Y.; Cai, J.; Pan, C.; Wang, Y.; Wu, X.; Shi, K.; Xia, X.; et al. Systemic Root-Shoot Signaling Drives Jasmonate-Based Root Defense against Nematodes. *Curr. Biol.* **2019**, *29*, 3430–3438.e4. [CrossRef] [PubMed]
64. Ralmi, N.H.A.A.; Khandaker, M.M.; Mohd, K.S.; Majrashi, A.; Fallatah, A.M.; Badaluddin, N.A.; Yusoff, N.; Mahmud, K.; Saifuddin, M.; Osman, N.; et al. Influence of Rhizopheric H_2O_2 on Growth, Mineral Absorption, Root Anatomy and Nematode Infection of Ficus Deltoidea. *Agronomy* **2021**, *11*, 704. [CrossRef]
65. Waszczak, C.; Carmody, M.; Kangasjärvi, J. Reactive Oxygen Species in Plant Signaling. *Annu. Rev. Plant Biol.* **2018**, *69*, 209–236. [CrossRef]
66. Sharma, N.; Mahawar, L.; Mishra, A.; Albrectsen, B.R. Microbial Contributions to Plant Growth and Stress Tolerance: Mechanisms for Sustainable Plant Production. *Plant Stress* **2025**, *17*, 100966. [CrossRef]
67. Supriya, L.; Dake, D.; Woch, N.; Gupta, P.; Gopinath, K.; Padmaja, G.; Muthamilarasan, M. Sugar Sensors in Plants: Orchestrators of Growth, Stress Tolerance, and Hormonal Crosstalk. *J. Plant Physiol.* **2025**, *307*, 154471. [CrossRef] [PubMed]
68. Rolland, F.; Baena-Gonzalez, E.; Sheen, J. Sugar Sensing and Signaling in Plants: Conserved and Novel Mechanisms. *Annu. Rev. Plant Biol.* **2006**, *57*, 675–709. [CrossRef]
69. Saini, S.; Lohani, S.; Khati, P.; Rani, V. PGPR-Mediated Mitigation of Biotic and Abiotic Stress in Plants. In *Advanced Microbial Technology for Sustainable Agriculture and Environment*; Academic Press: Cambridge, MA, USA, 2023; pp. 199–227. [CrossRef]
70. Mehmood, N.; Saeed, M.; Zafarullah, S.; Hyder, S.; Rizvi, Z.F.; Gondal, A.S.; Jamil, N.; Iqbal, R.; Ali, B.; Ercisli, S.; et al. Multifaceted Impacts of Plant-Beneficial *Pseudomonas* Spp. in Managing Various Plant Diseases and Crop Yield Improvement. *ACS Omega* **2023**, *8*, 22296. [CrossRef]
71. Badiaa, O.; Abdelhakim, H.; Yssaad, R.; Topcuoglu, B. Effect of Heavy Metal (Copper and Zinc) on Proline, Polyphenols and Flavonoids Content of Tomato (*Lycopersicon esculentum* Mill.). *Plant Arch* **2020**, *20*, 2125–2137.
72. Shi, Y.; Pang, X.; Liu, W.; Wang, R.; Su, D.; Gao, Y.; Wu, M.; Deng, W.; Liu, Y.; Li, Z. SlZHD17 Is Involved in the Control of Chlorophyll and Carotenoid Metabolism in Tomato Fruit. *Hortic. Res.* **2021**, *8*, 259. [CrossRef]
73. Zou, Q.; Zhao, L.; Guan, L.; Chen, P.; Zhao, J.; Zhao, Y.; Du, Y.; Xie, Y. The Synergistic Interaction Effect between Biochar and Plant Growth-Promoting Rhizobacteria on Beneficial Microbial Communities in Soil. *Front. Plant Sci.* **2024**, *15*, 1501400. [CrossRef]
74. Hayat, R.; Ali, S.; Amara, U.; Khalid, R.; Ahmed, I. Soil Beneficial Bacteria and Their Role in Plant Growth Promotion: A Review. *Ann. Microbiol.* **2010**, *60*, 579–598. [CrossRef]
75. Tattini, M.; Gori, A.; Beatriz, L.; Nascimento, S.; Patanè, C.; Cosentino, S.L.; Romano, D.; Toscano, S. Relative Water Content, Proline, and Antioxidant Enzymes in Leaves of Long Shelf-Life Tomatoes under Drought Stress and Rewatering. *Plants* **2022**, *11*, 3045. [CrossRef]
76. Vandenkoornhuyse, P.; Quaiser, A.; Duhamel, M.; Le Van, A.; Dufresne, A. The Importance of the Microbiome of the Plant Holobiont. *New Phytol.* **2015**, *206*, 1196–1206. [CrossRef]
77. Marmiroli, M.; Caldara, M.; Pantalone, S.; Malcevschi, A.; Maestri, E.; Keller, A.A.; Marmiroli, N. Building a risk matrix for the safety assessment of wood derived biochars. *Sci. Total Environ.* **2022**, *839*, 156265. [CrossRef]

antioxidants

MDPI

Article

Maize Peroxidase ZmPrx25 Modulates Apoplastic ROS Homeostasis and Promotes Seed Germination and Growth Under Osmotic and Drought Stresses

Feixue Zhang [1], Liangjie Niu [1,2], Yingxue Li [1,†], Xiaoli Zhou [1,†], Hui Zhang [1,*], Xiaolin Wu [1], Hui Liu [1] and Wei Wang [1,*]

[1] State Key Laboratory of High-Efficiency Production of Wheat-Maize Double Cropping, College of Life Sciences, Henan Agricultural University, Zhengzhou 450046, China

[2] Dabie Mountain Laboratory, College of Tea and Food Science, Xinyang Normal University, Xinyang 464000, China

[*] Correspondence: zhanghui2607@henau.edu.cn (H.Z.); wangwei@henau.edu.cn (W.W.)

[†] These authors contributed equally to this study.

Abstract

Drought is one of the major abiotic stresses threatening maize production globally. Under drought stress, maize plants produce excessive reactive oxygen species (ROS), leading to oxidative damage. The apoplast, as the site of substance and signal exchange between plant cells and the external environment, is an important location for the production of ROS under drought stress. Elucidating the ROS scavenging mechanisms in the apoplast is crucial for understanding plant stress responses. However, there is still a lack of research on the ROS scavenging enzymes in maize apoplast and their mediated signaling pathways. We verified that maize peroxidase Prx25 (ZmPrx25) is localized in the apoplast, it scan scavenge hydrogen peroxide (H_2O_2), and we systematically investigated the responses of the apoplastic *ZmPrx25*-ROS system to osmotic stress. ROS accumulate in the apoplast of maize mesocotyl in response to osmotic stress and transmit the external osmotic stress signals from the apoplast to the inner cellular compartments. The expression of *ZmPrx25* is highly upregulated in the meristematic regions of maize seedlings under osmotic and oxidative stress. Overexpression of *ZmPrx25* in *Arabidopsis* promoted seed germination and plant growth, significantly enhancing tolerance to osmotic and oxidative stress. This study provides a new perspective on the role of *Prx25* in scavenging ROS under drought stress.

Keywords: reactive oxygen species; osmotic stress tolerance; maize apoplast; peroxidase

1. Introduction

With the intensification of global warming, drought has become one of the major abiotic stresses leading to declining grain production [1]. Maize (*Zea mays* L.), an important food crop and a typical C4 plant, often exhibits phenotypes such as plant wilting and a reduced leaf growth rate under drought stress. Particularly, prolonged drought during the flowering and grain-filling stages can lead to production loss or even complete crop failure in maize [2]. Under drought conditions, maize typically enhances its drought resistance by closing stomata and improving its osmotic adjustment capacity [3]. However, the molecular mechanisms underlying these adaptive strategies remain poorly understood. Therefore, a thorough understanding of the molecular mechanisms underlying maize responses to drought provides a theoretical basis for future strategies aimed at mitigating yield losses.

In various stress environments, plants exhibit a common characteristic during the early response stage—the production of ROS [4]. ROS are important stress signaling molecules in plant cells, and their generation and signal transduction mechanisms have always been a hot and challenging research focus at the forefront of plant stress biology [5]. Plant cells can sense external stress and internal metabolic disturbance signals across various subcellular compartments, and in response, they generate ROS. The main sites of ROS production include chloroplasts [6], mitochondria [7], peroxisomes [8], plasma membranes [9], and the apoplast [10]. Apart from the apoplast—where antioxidant capacity is comparatively low—surplus intracellular ROS are eliminated via antioxidant systems. This process serves to maintain redox homeostasis and ward off oxidative damage [10]. Currently, the processes of ROS production and signaling in intracellular organelles such as chloroplasts [11], mitochondria [12], and peroxisomes [13] are well understood [14]. However, the processes of ROS scavenging and signaling in the apoplast are still poorly understood.

The apoplast is a continuous system that consists of cell walls, intercellular spaces, and dead cell lumens (such as vessels), filled with liquid or air, with cell walls as the main structural component [15]. As a dynamic open region, the apoplast functions as a site where substances and signals are exchanged among plant cells, encompassing many physiological processes, mainly involved in the storage and transport of nutrients and water, cell wall synthesis, and cell growth [10]. As one of the most susceptible areas to environmental stress, the importance of the apoplast in the early perception of environmental stress signals is self-evident [10]. A growing body of evidence demonstrates that apoplastic ROS function is a pivotal early signaling molecule, orchestrating plant growth, development, and adaptive responses to both biotic and abiotic stresses [14,16]. Pathogen infection, drought, salt, low-temperature stress, and soil pollution can all affect physiological processes occurring in the apoplast, leading to increased apoplastic pH and ROS accumulation [16,17]. For example, both drought stress and pathogen attack can trigger the rapid production of ROS in the apoplast, which is essential for stomatal closure [18]. Apoplastic ROS can occur independently of stress-induced abscisic acid (ABA) signaling under water stress, but their levels are also regulated by ABA signaling [19].

Apoplastic peroxidase (Prx) belongs to the plant-specific, Class III peroxidase (EC:1.11.1.7), whcih participates in cell elongation, cell wall component synthesis, camalexin synthesis, auxin metabolism, and plays an indispensable role in the process of clearing ROS [20]. Under biotic or abiotic stress, apoplastic Prx activity generally significantly increases [21,22]. Under biotic stress, gene knockout studies in *Arabidopsis* have demonstrated that the presence of a positive feedback mechanism enhances the defense level of apoplastic ROS signaling [23–25]. In maize, apoplastic *Prx35* is associated with seed responses to insects and mechanical damage, and the resistance mechanisms in maize against corn borers and cutworms that Prx35 is involved in are independent of each other [26]. Under abiotic stress, Arabidopsis plants overexpressing *AtPrx3* exhibit higher salt tolerance, and wheat plants overexpressing *TaPrx109-B1* enhance drought resistance [27,28]. In maize, it has been observed that the apoplast might be the earliest site for the detection of exogenous ABA treatment in leaves [29]. Under water stress, the levels of ROS in the apoplast of the maize root growth zone are spatially regulated, and this response holds a vital position in the adaptation of roots to osmotic stress [30]. However, the key Prx responsible for ROS scavenging in the maize apoplast remains unclear.

This study aims to elucidate the role of the maize apoplastic peroxidase ZmPrx25 in ROS homeostasis and drought tolerance by analyzing its tissue-specific expression, subcellular localization, in vitro enzymatic activity, and functional contribution under osmotic and drought stress.

2. Materials and Methods

2.1. Plant Materials and Growth Conditions

Materials used included hybrid Zhengdan958 and inbred line Chang7-2t. Zhengdan958 is currently widely cultivated in China; Chang7-2t was a drought-tolerant mutant line that was produced by us via ^{60}Co-γ irradiation of Chang7-2 seeds [31]. Maize seeds were surface-disinfected with 1% NaClO for 15 min, rinsed thoroughly, soaked for 12 h, evenly placed in seedling trays, and incubated in darkness at 28 °C in a plant growth chamber. Half of the etiolated seedlings were subjected to 15% PEG-6000 treatment targeting the mesocotyl, and the other half were used as controls. After stress treatment, the mesocotyl was collected for Prx activity, as well as ROS and H_2O_2 levels. Similarly, for gene expression and physiological analysis, etiolated maize seedlings were separated into two groups: one half was exposed to root stress induced by 15% PEG-6000, while the other half served as untreated controls. Tissues were collected for downstream assays. For protein level analysis, seedlings were further divided into three subgroups: (1) one-third received 15% PEG-6000 root stress for 60 min; (2) the remaining two-thirds were pretreated with either 2 mM of NaN_3 or 0.2 mM of DPI on roots for 15 min, followed by 10% PEG-6000 stress on the mesocotyl for 60 min. The stress treatment apparatus is shown in Figure S1.

Arabidopsis lines were Columbia-0 (Col-0) and *ZmPrx25*-overexpressing *Arabidopsis*. Seeds were surface-sterilized first in 70% ethanol for 5 min, then rinsed with sterile water, and finally treated with 1% NaClO for 5 min. The sterilized seeds were sown on MS medium, vernalized in darkness at 4 °C for 48 h, and then transferred to a growth chamber (22 °C, 16 h light/8 h dark). After 7 d, the seedlings were transplanted into soil composed of a 1:1 mixture of vermiculite and nutrient soil. Some plants were normally irrigated for 15 d, followed by 15 d of drought treatment, after which survival rates were recorded. The remaining plants were irrigated for 21 d and then subjected to drought for 10 d to monitor bolting and physiological indices.

2.2. Isolation of Apoplastic Fluid

Apoplast isolation was performed using established methods for maize leaves [32,33]. Mesocotyl segments (about 2.5 cm in length) were washed three times with deionized water, immersed in 10% methanol, and vacuum-infiltrated at −45 kPa for 20 min at room temperature. The segments were bundled together using sealing film. These bundles were placed into 15 mL conical centrifuge tubes equipped with plastic mesh at the bottom and centrifuged at a low speed (1000× *g*, 10 min, 4 °C). The liquid collected at the bottom of the tube was considered apoplastic fluid, which was either used immediately or stored at −80 °C for future use.

2.3. Light Microscopy and ROS In Vivo Visualization

Mesocotyl sections were stained with iodine or safranin-fast green and observed under a light microscope. ROS detection was completed using a H_2DCFDA fluorescent probe (Merck D6883, Rahway, NJ, USA) [34]. Maize seedlings were treated with 50 μM of H_2DCFDA/PBS (pH 7.4) mist for 30 min. After osmotic stress treatment of the mesocotyl with 10% PEG 6000 + 2% agar for varying durations, the mesocotyls were imaged using an IVIS Lumina S5 fluorescence imaging system (excitation 480 nm/emission 520 nm, f/2, exposure time 2 s).

2.4. Bioinformatic Analysis of Prx

Prx family gene sequences were retrieved from the NCBI databases, and 30 maize Prx members were identified that are highly homologous to *ZmPrx25*. Conserved motif analysis was performed using the M; MEME suite. Subcellular localization was predicted based on UniProt annotations combined with Plant-mPLoc analysis. The *Cis*-acting elements in the *ZmPrx25* promoter were recognized through the PlantCARE database.

2.5. Transcriptome Profiling by RNA-Seq

Following the isolation of total RNA from 100 mg of leaf tissue using a Qiagen kit (Qiagen, Hilden, Germany), mRNA was enriched via poly (A) selection and sequenced on a BGIseq500 platform (BGI, Shenzhen, China). Raw reads were quality-controlled by removing adapters and low-quality sequences with SOAPnuke. Alignment to the maize reference genome (GCF_000005005.2_B73_RefGen_v4) was performed using HISAT2 (version 2.0.5) [35]. DESeq2 (version 1.34.0) was applied to identify DEGs based on a threshold of $|\log_2 FC| \geq 1$ and a q-value ≤ 0.05 [36]. The raw data have been deposited in the EBI-ENA (https://www.ebi.ac.uk/ena/browser/home, access data 29 August 2025) under accession E-MTAB-11413.

2.6. Expression Analysis of ZmPrx25 and qRT-PCR Validation

To verify the expression of *ZmPrx25* genes under stress conditions, quantitative reverse transcription PCR (qRT-PCR) was performed, with *ZmUBI* (*GRMZM2G027378*) as the internal standard. The procedures for RNA extraction and qRT-PCR were as described in previous work [37], and the calculation of relative gene expression was performed via the $2^{-\Delta\Delta CT}$ method [38]. The gene-specific primers were listed in Table S1

2.7. Expression and In Vitro Hydrolysis of the Recombinant ZmPrx25 Protein

The coding sequence (CDS) of *ZmPrx25* (GenBank accessions: LOC100192603, Zm00001d028348) was amplified from maize cDNA, with *Bam*HI and *Xho*I restriction enzyme sites introduced during amplification. The recombinant ZmPrx25 protein was expressed in *E. coli* Rosetta (DE3) cells harboring the pET30a (+)−CDS construct. Expression was induced with 0.2 mM or 0.5 mM of isopropyl-β-D-thiogalactopyranoside IPTG and was carried out at 28 °C for 14 h. The cultures were centrifuged (5000× *g*, 5 min, 4 °C) to collect cell pellets. Then, the cultures were washed with PBS, lysed by boiling, and analyzed using SDS-PAGE to verify protein expression. For protein purification, the cell pellets were resuspended in PBS and lysed by sonication. The lysates were centrifuged to collect the supernatants, which were subjected to affinity purification using Ni-NTA agarose. The bound proteins were recovered using an imidazole gradient, and their purities and sizes were assessed by SDS-PAGE. To further verify the recombinant protein, samples were electrophoretically transferred onto PVDF membranes and analyzed by Western blotting with the anti-His antibody. After incubation with a secondary antibody, signals were detected using an appropriate imaging system. Prx activity was determined using the guaiacol assay.

2.8. Transient Expression of ZmPrx25

Transient transformation of tobacco was performed according to the methods of [39]. The CDS of *ZmPrx25* was amplified with *Sal*I/*Bam*HI restriction sites from maize cDNA, cloned, and inserted into the pCAMBIA1302 vector. Tobacco plants were grown for 4–5 weeks under a 16 h light/8 h dark photoperiod at 25 °C. The recombinant *Agrobacterium tumefaciens* carrying the pCAMBIA1302-*ZmPrx25* construct was grown in antibiotic-supplemented LB broth at 28 °C with shaking (200 rpm) for 24 h until turbidity was reached. The bacterial culture (1 mL) was added to 50 mL of antibiotic-containing LB medium and grown to an OD_{600} of 0.5–0.6. Cells were harvested by centrifugation and resuspended in infiltration buffer (10 mM of $MgCl_2$, 10 mM of MES, 200 μM of acetosyringone) to a final OD_{600} of 1.0. Silwet-L77 was then added at a ratio of 1:1000, and the suspension was incubated at room temperature for 2–3 h. Tobacco leaves were infiltrated on the abaxial side with the bacterial suspension using a syringe, and the infiltration sites were marked. After infiltration, the plants were kept in darkness for 24 h and then transferred to 21 °C for an additional 2 d of growth. Fluorescence was observed using confocal laser scanning microscopy.

Transient transformation of onion epidermis was performed according to the method described by [40]. The CDS of *ZmPrx25* was amplified with *Sal*I/*Bam*HI restriction sites from maize cDNA, cloned, and inserted into the pYFPLT-C1 vector. The inner epidermis of onion bulb scales was peeled under sterile conditions and incubated in the dark on MS solid medium for 24 h. The epidermal tissue was immersed in a suspension of recombinant *A. tumefaciens* carrying the pYFPLT-*ZmPrx25* vector (OD_{600} = 1.0) for 30 min, then transferred to MS medium for co-cultivation at 28 °C under a 16 h light/8 h dark photoperiod for 2 d. Plasmolysis was induced by treatment with a 30% sucrose solution for 10–20 min, and YFP fluorescence was detected using confocal laser scanning microscopy.

2.9. Generation of *Arabidopsis* Lines Overexpressing *ZmPrx25*

The pCAMBIA1302-*ZmPrx25* vector was introduced into *A. tumefaciens* strain GV3101 and cultured on LB agar plates containing 50 μg/mL of kanamycin and 25 μg/mL of rifampicin at 28 °C for 48 h. A single positive colony was picked and grown in liquid medium to an OD_{600} of 1.0. Bacterial suspension was collected by centrifugation and resuspended in a solution containing 10% sucrose and 0.01% Silwet-77 (OD_{600} = 1.0), then used to transform *Arabidopsis* at the bolting stage via floral dip. After 24 h of dark incubation, the plants were returned to normal growth conditions, and the infection procedure was repeated 2–3 times at 5–7 d intervals. T0 seeds were harvested, and positive transformants were screened by PCR, followed by the identification of high-expression lines using RT-qPCR. Two lines exhibiting the highest expression of *ZmPrx25* were continuously self-crossed, resulting in the establishment of homozygous transgenic lines by the T3 generation for further analysis.

2.10. Measurement of Plant Phenotype and Physiological-Biochemical Indices

Arabidopsis seeds were surface-sterilized and cultured for 3 d, during which seed germination was monitored. The primary root length of 7 d seedlings was measured. The bolting status of 30 d *Arabidopsis* plants was recorded. Each trait was recorded with at least three biological replicates. The activities of superoxide dismutase (SOD), Prx, catalase (CAT), and the content of malondialdehyde (MDA) in plants were determined using kits from Beijing Solarbio Science & Technology Co., Ltd. (Beijing, China), following the visible spectrophotometry method, with kit numbers BC0170, BC0090, BC0205, and BC0020, respectively. SOD activity was determined based on a xanthine/xanthine oxidase reaction system that generates superoxide anions (O^{2-}), which reduce nitroblue tetrazolium (NBT) to form a blue formazan product with a characteristic absorption at 560 nm. SOD inhibits formazan formation by scavenging O^{2-}, and its activity is negatively correlated with the absorbance at 560 nm. Prx activity was calculated by detecting the change in absorbance at 470 nm resulting from its catalysis of H_2O_2 oxidation of a specific substrate. CAT activity was quantified by monitoring the rate of decrease in the characteristic absorption peak of H_2O_2 at 240 nm due to its catalytic decomposition. MDA content was derived based on its condensation with thiobarbituric acid (TBA) under acidic and high-temperature conditions to produce a brownish-red product with maximum absorption at 532 nm, using the absorbance value for calculation. The specific experimental steps were carried out according to the instruction manual.

3. Results

3.1. Apoplastic ROS Accumulate Under Osmotic Stress

The maize mesocotyl, which connects the coleoptile node and the base of the seed root, plays a critical role in seedling responses to drought, waterlogging, extreme temperatures, and pathogen invasion [41–43], inducing stress-related ROS accumulation. Microscopic observations of both longitudinal and cross sections of the mesocotyl revealed that the outermost layer comprised small epidermal cells with thin walls, enclosing the inner cortex and central vascular tissue. The cortical cells were irregular in shape and exhibited large intercellular spaces (Figure 1A,B). These structural features may facilitate the rapid transmission of environmental stress signals via the apoplastic pathway.

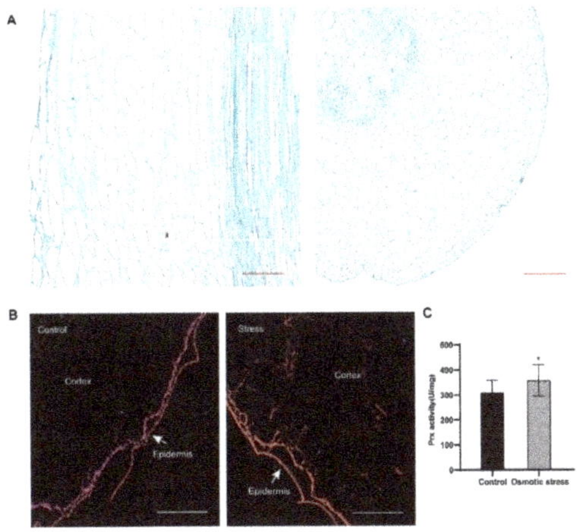

Figure 1. Cellular structure of the maize mesocotyl and ROS fluorescence and detection of apoplastic Prx activity under osmotic stress. (**A**) Cellular structure of the maize mesocotyl. Longitudinal and transverse sections of maize mesocotyl after 132 h of dark treatment, stained with safranin-fast green. Cellulose-rich cell walls appear as green, while lignified or suberized cell walls appear as red. Red lines represent the scale bars, 200 μm. (**B**) Apoplastic ROS fluorescence detection in mesocotyl under 15% PEG 6000 stress. Control, 2% agar; Stress, mesocotyl treated with 2% agar containing 10% PEG 6000 for 30 min. Red fluorescence indicates a significant ROS increase. White lines represent the scale bars, 200 μm. (**C**) Assay of apoplastic Prx activity in maize mesocotyl under osmotic stress. Control, distilled water; Osmotic stress, the mesocotyl was treated with 15% PEG 6000 for 4 h. All experimental data are expressed as mean ± SD. Asterisks indicate statistically significant differences from the control (* $p < 0.05$, analyzed using Student's *t*-test).

To investigate apoplast-specific ROS signaling and transmission under osmotic stress, mesocotyls were treated with 15% PEG 6000. Frozen sections of the mesocotyl stained with dihydroethidium (DHE) and observed under a fluorescence microscope revealed a marked accumulation of ROS in the cell wall and extracellular regions 30 min after osmotic stress treatment, compared to the control, and the ROS signal then progressed inward toward the mesocotyl interior, while intracellular ROS remained at low levels (Figure 1B). This suggests that the observed ROS accumulation is a result of apoplastic sensing of osmotic stress, and that the mesocotyl can sensitively perceive external osmotic stress and transmit the signal inward via apoplastic ROS. Furthermore, Prx activity assays showed a significant increase in apoplastic Prx activity under osmotic stress (Figure 1C), combined with the pattern of ROS accumulation and changes in Prx activity under osmotic stress, suggesting that plants may regulate apoplastic ROS levels by upregulating Prx activity.

3.2. Screening for the Major Prx Responsible for Apoplastic ROS Metabolism

To determine the major Prx responsible for apoplastic ROS metabolism under osmotic stress, we analyzed the changes in the mesocotyl transcriptomes of Chang7-2t plants exposed to osmotic stress. We identified significant changes in the expression of five Prx genes and *Rboh4*. Among these, the expression level of the *ZmPrx25* gene significantly increased, while the expression of *Rboh4* only increased slightly. The expression levels of the other four Prx genes showed varying degrees of increase or decrease, with the corresponding enzymes located in the apoplast or plasma membrane (Figure 2A). Subsequently, qRT-PCR confirmed that the *ZmPrx25* gene was significantly expressed in the roots, mesocotyls, and leaves of maize seedlings, with the highest expression in the mesocotyl (Figure 2B). the expression trend in the mesocotyl was consistent with the RNA-Seq results mentioned above.

A

Gene	Protein	Expression change(Log2 multiple)	Significant differences	Subcellular localization of protein
Zm00001d052653	Rboh4	1.288	*	Plasma membrane
Zm00001d028348	ZmPrx25	3.594	**	Apoplast
Zm00001d040364	ZmPrx18	1.226	*	Apoplast
Zm00001d021965	ZmPrx72	1.030	*	Apoplast
Zm00001d002897	ZmPrx24	−1.781	*	Apoplast
Zm00001d025774	ZmPrx95	−1.710	*	Apoplast

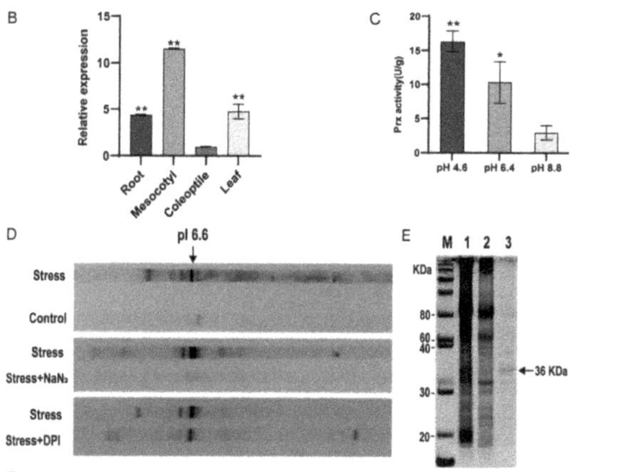

F

Protein	Band identification	Theoretical/Experimental molecular weight	Theoretical/Experimental pI	Sequence coverage	Mascot score	Identified peptides
ZmPrx25	36 kD	36.08/36 kD	7.52/6.6	23.8	364	TCPVVHRVAR; SEKFATPNNNSAR; DQELLSTPGAPTAPIVGR; SDOELLSTPGAPTAPIVGR; DAVKAALEEACPGVVSCADI

Figure 2. Transcriptomic analysis of *ZmPrx* expression and protein-level validation of ZmPrx25. (**A**) RNA-Seq Analysis of Prx and Rboh Gene Expression in Maize mesocotyl under Osmotic Stress. Log2 > 0, upregulated; Log2 < 0, downregulated. Protein localization based on UniProt annotation. (**B**) The expression of *ZmPrx25* was assessed using qRT-PCR. Chang 7-2t roots grown in darkness for 108 h were treated with 10% PEG 6000 for 90 min, after which *ZmPrx25* expression levels were examined in different organs. All experimental data are expressed as mean ± SD. Asterisks indicate statistically significant differences from the control (* $p < 0.05$, ** $p < 0.01$, analyzed using Student's t-test). (**C**) *Prx* activity in the mesocotyl apoplast at different pH levels. (**D**) Detection of mesocotyl apoplast Prx activity in the presence of specific inhibitors. Prx activity is sensitive to NaN_3 but not to DPI. Stress, 60 min of Mannitol stress on the mesocotyl; Control, distilled water; Stress + NaN_3, roots were pretreated with 2 mM of NaN_3 for 15 min, followed by 10% PEG 6000 stress applied to the mesocotyl for 60 min. Stress + DPI, roots were pretreated with 0.2 mM of DPI for 15 min, followed by 10% PEG 6000 stress applied to the mesocotyl for 60 min, apoplastic proteins from the mesocotyl were separated by IEF. Treatment with stress + DPI indicated that Prx activity was predominantly located at pI 6.6. (**E**) SDS-PAGE analysis of the protein. Lane M, protein marker; Lane 1, proteins extracted from the upper 1 cm segment of the mesocotyl; Lane 2, proteins extracted from the lower 1 cm segment of the mesocotyl; Lane 3, proteins extracted around the pI 6.6 of the gel strips. (**F**) The results of mass spectrometry identification for the 36 kD band.

Upon isolating the extracellular fluid from the mesocotyl, Prx activity was measured at different pH values, and it was found that Prx activity was higher under acidic conditions (Figure 2C). Based on the sensitivity of Prx to NaN_3 and its insensitivity to DPI, roots were pretreated with 2 mM of NaN_3 and 0.2 mM of DPI, followed by osmotic stress using 10% PEG 6000. Subsequently, apoplastic proteins were separated by isoelectric focusing (IEF), successfully identifying the position of the major Prx induced by osmotic stress on the IEF strip (arrow; Figure 2D). The protein band at pI 6.6 was isolated, and proteins were extracted and analyzed via SDS-PAGE followed by Coomassie blue staining. A band at 36 kDa showed a molecular weight similar to ZmPrx25. The 36 kDa band (Figure 2E,F) was excised for mass spectrometry analysis, which identified ZmPrx25 as the major protein. Analysis of both transcriptomic and protein level demonstrated that ZmPrx25 may be the key enzyme in the apoplast of maize mesocotyl in response to osmotic stress.

3.3. Transient Expression and Promoter Analysis of ZmPrx25

To determine the transient expression of *ZmPrx25*, we constructed pCAMBIA1302-*ZmPrx25* and pYFPLT-*ZmPrx25* vectors and transiently expressed them in *Nicotiana benthamiana* and onion epidermal cells, respectively. Fluorescence observation in *N. benthamiana* indicated that *ZmPrx25* was localized to the plasma membrane or cell wall (Figure 3A). Further plasmolysis experiments in onion epidermal cells confirmed that *ZmPrx25* specifically localized to the cell wall (Figure 3B), which was consistent with the prediction in the UniProt database. Additionally, bioinformatic analysis revealed that the promoter region of *ZmPrx25* contains multiple *cis*-elements that are responsive to drought, low temperatures, and hormones, including ABA, gibberellins (GA), and IAA (Figure 3C), suggesting that this gene may be involved in drought stress response and developmental regulation in maize.

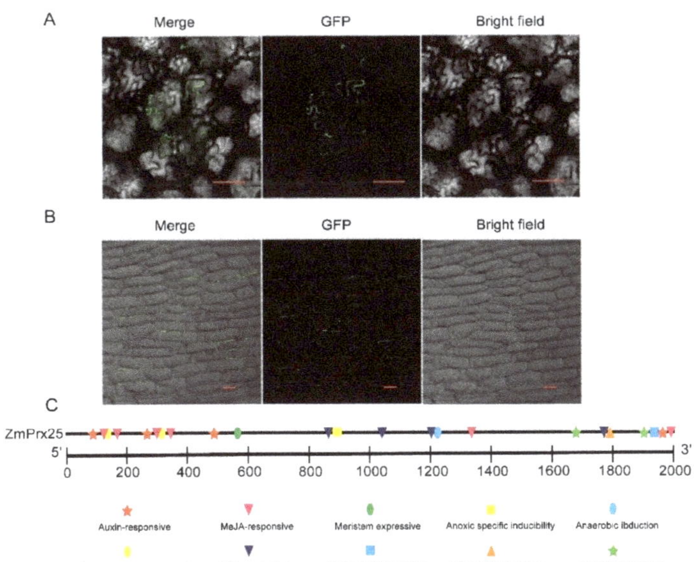

Figure 3. Transient expression and promoter *cis*-element analysis of *ZmPrx25*. (**A**) Transient expression of ZmPrx25 in *N. benthamiana* epidermal cells. (**B**) Transient expression of ZmPrx25 in onion epidermal cells after plasmolysis. Merge, overlay of two kinds of signals; GFP, green fluorescence signal; Bright field, image under light field. Red lines represent the scale bars, 100 µm. (**C**) *Cis*-acting elements in the *ZmPrx25* promoter.

3.4. Enzyme Activity of the Recombinant ZmPrx25 Protein

To determine whether ZmPrx25 can scavenge ROS, the prokaryotic expression vector pET30a (+)−*ZmPrx25* was transformed into *E. coli* competent cells (Figure 4A), which expressed recombinant ZmPrx25. Upon induction with 0.2 mM of IPTG for 12 h, 14 h, and 16 h, a distinct protein band of approximately 41 kDa was observed (Figure 4B), consistent with the predicted molecular weight of the recombinant protein, indicating the successful expression of ZmPrx25. Subsequent purification using imidazole concentrations revealed a single protein band under 100 mM of imidazole elution (Figure 4C). Western blot analysis confirmed that the purified protein was the recombinant ZmPrx25 (Figure 4D). Enzyme activity assays revealed that ZmPrx25 possessed Prx activity, and was capable of efficiently decomposing the substrate H_2O_2 (Figure 4E), indicating its enzymatic function in scavenging ROS.

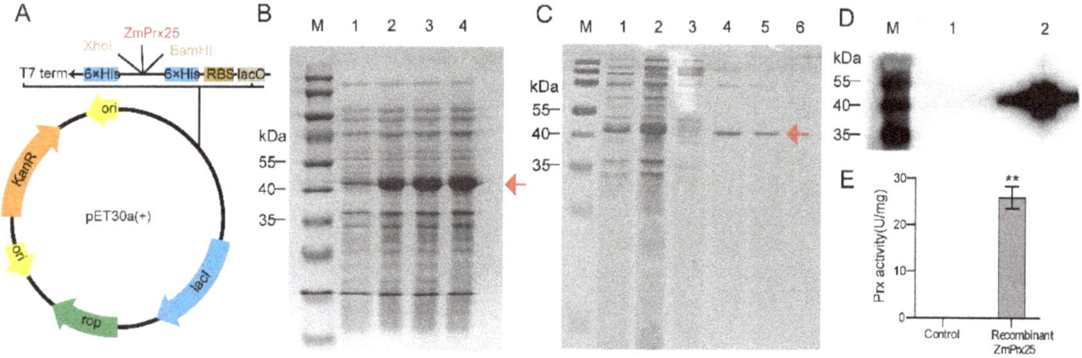

Figure 4. Prokaryotic expression, purification, and enzymatic activity of the recombinant ZmPrx25. (**A**) Establishment of the prokaryotic expression vector pET30a (+)−*ZmPrx25*. (**B**) Expression of the recombinant ZmPrx25 protein. Lane M, protein marker; Lane 1, recombinant ZmPrx25 was not induced; Lane 2, the recombinant ZmPrx25 was induced by 0.2 mM of IPTG for 12 h; Lane 3, induction for 14 h; Lane 4, induction for 16 h. The red arrow indicates the ZmPrx25 protein. (**C**) SDS-PAGE analysis of the recombinant ZmPrx25 purified by Ni-NTA affinity chromatography. After induction with 0.2 mM of IPTG for 12 h, the recombinant ZmPrx25 was purified using different imidazole concentration gradients. The red arrow indicates the ZmPrx25 protein. Lane M, protein marker; Lane 1, PBS buffer; Lane 2, 0 mM of imidazole; Lane 3, 20 mM of imidazole; Lane 4, 50 mM of imidazole; Lane 5, 100 mM of imidazole; Lane 6, 200 mM of imidazole. (**D**) Western blot analysis of the recombinant ZmPrx25. Lane M, protein marker; Lane 1, pET30a (control); Lane 2, the recombinant ZmPrx25 (purified target protein). (**E**) Prx activity assay. All experimental data are expressed as mean ± SD. Asterisks indicate statistically significant differences from the control (** $p < 0.01$, analyzed using Student's *t*-test).

3.5. Tissue-Specific Expression Patterns of ZmPrx25 in Maize

To investigate the function of *ZmPrx25* in maize growth and development, we performed a spatial expression analysis. The results indicated that *ZmPrx25* was significantly expressed in the root meristematic zone (Figure 5A) and at the one cm ends of the mesocotyl (Figure 5B), regions characterized by active cell division. These findings suggest that *ZmPrx25* may play a role in regulating cell proliferation during early developmental stages.

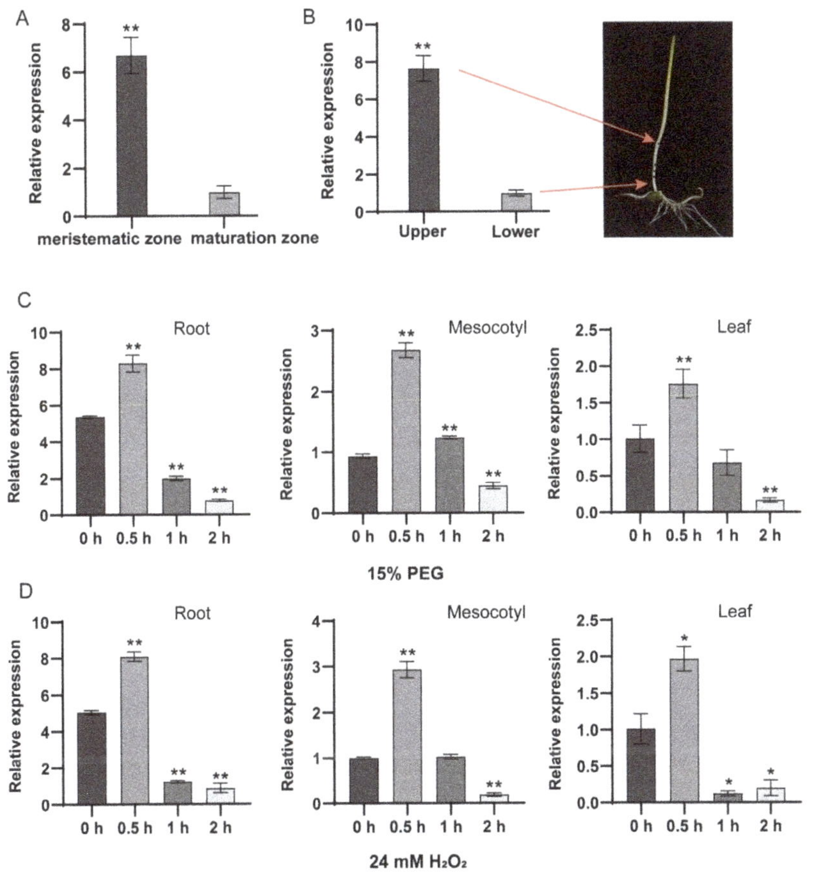

Figure 5. Expression analysis of *ZmPrx25* in different tissues of etiolated maize seedlings and under osmotic and oxidative stress conditions. Different tissues were collected from 108 h etiolated maize seedlings grown in darkness. (**A**) Gene expression of *ZmPrx25* in the meristematic and maturation zones of maize roots. (**B**) Gene expression of *ZmPrx25* in the one cm ends of the mesocotyl. (**C**) Gene expression of *ZmPrx25* in maize roots, mesocotyl, and leaves under 15% PEG 6000 stress. (**D**) Gene expression of *ZmPrx25* in maize roots, mesocotyl, and leaves under 24 mM of H_2O_2 stress. All experimental data are expressed as mean ± SD. Asterisks indicate statistically significant differences from the control (* $p < 0.05$, ** $p < 0.01$, analyzed using Student's *t*-test).

To investigate the role of *ZmPrx25* in stress responses, we treated plants with 15% PEG 6000 and 24 mM of H_2O_2. Temporal expression analysis revealed that *ZmPrx25* was rapidly induced by both stress conditions, with transcript levels in roots, mesocotyls, and leaves peaking at 0.5 h. As the duration of stress extended (1–2 h), its expression levels gradually declined (Figure 5C,D). This dynamic expression profile implies a potential functional significance of ZmPrx25 during the initial phase of stress response in maize.

3.6. Phenotypic and Drought Tolerance Analysis of ZmPrx25 Overexpression Lines

The construct pCAMBIA1302-*ZmPrx25* vector was successfully introduced into *Arabidopsis* (Figure 6A). Transgenic lines were screened using RT-qPCR, and two lines with high *ZmPrx25* expression, OE-5 and OE-7, were selected for propagation. T3 generation plants were obtained and used for subsequent analyses (Figure 6B).

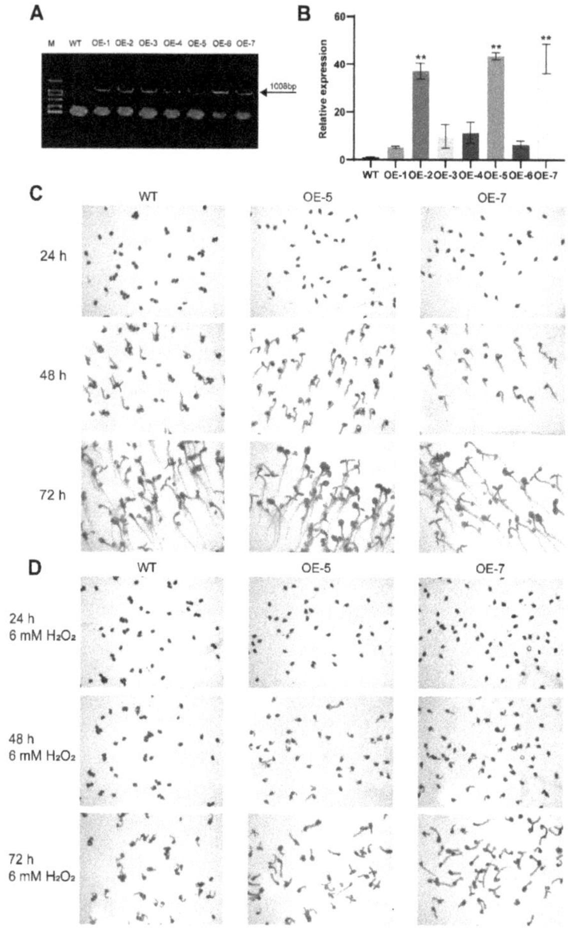

Figure 6. Germination phenotype of *ZmPrx25* overexpressing *Arabidopsis*. (**A**) PCR amplification results of leaf DNA. The arrow points to the location of the target amplicon. Lanes M, WT, and OE-1 to OE-7 correspond to the DNA marker, wild-type, and ZmPrx25-overexpressing lines, respectively. (**B**) RT-qPCR amplification analysis of leaf RNA. Data were obtained from three biological replicates, and significant differences were assessed via Tukey's test (** $p < 0.01$). (**C**) Seed germination of *Arabidopsis* (1–3 d). (**D**) Seed germination of *Arabidopsis* under oxidative stress conditions (1–3 d).

In the seed germination assay, the germination rates of wild-type (WT) and *ZmPrx25* overexpression (OE-5, OE-7) *Arabidopsis* seeds were assessed after 24 h of cultivation. The results showed that the *ZmPrx25*-OE lines exhibited a germination rate of 98%, whereas the WT seeds displayed only 38% germination at the same time point. Complete germination of WT seeds was observed at 72 h (Figure 6C). To further evaluate the role of *ZmPrx25* in the oxidative stress response, seed germination was examined on MS medium supplemented with various concentrations of H_2O_2. The results indicate that *Arabidopsis* overexpressing *ZmPrx25* maintains a high germination rate under oxidative stress conditions (Figure 6D). Notably, under the 6 mM H_2O_2 treatment, the germination rate of transgenic lines exceeded 73% within 24 h, while the WT seeds exhibited only 4% germination (Figure S2). These findings indicate that the overexpression of *ZmPrx25* effectively enhances oxidative stress tolerance and significantly promotes seed germination in *Arabidopsis*.

Under both osmotic stress (100 and 200 mM of mannitol) and oxidative stress (0.6 and 1.2 mM o H_2O_2), root growth in the *ZmPrx25*-OE Arabidopsis lines was significantly less inhibited compared to the WT plants (Figure 7A–D). Notably, under the 200 mM mannitol treatment, root length in the OE lines was approximately 36–48% greater than that of WT (Figure 7A). These results suggest that the overexpression of *ZmPrx25* enhances tolerance to both osmotic and oxidative stress, thereby promoting root growth under environmental stress conditions.

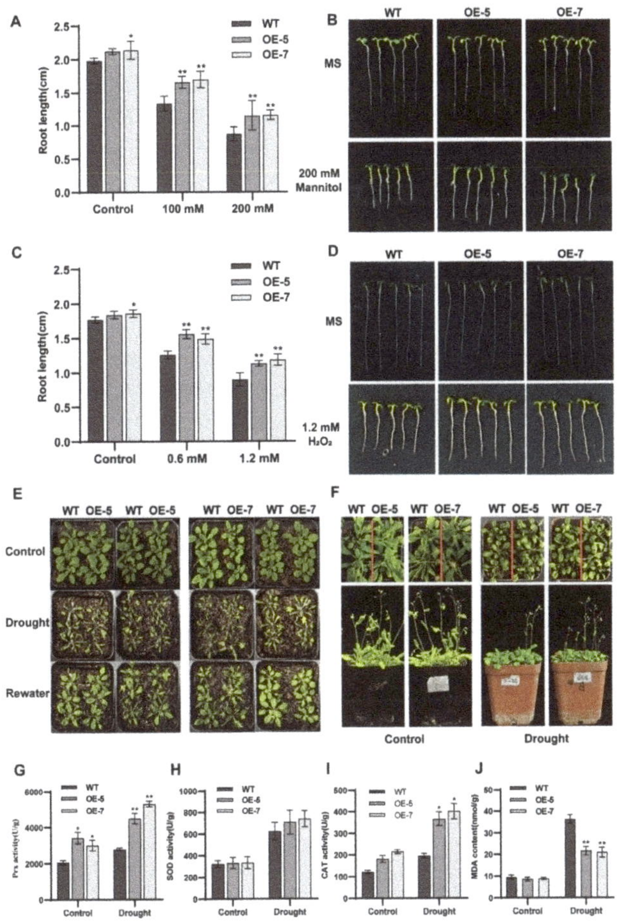

Figure 7. Phenotypic and drought tolerance analysis of transgenic *Arabidopsis* roots under stress. (**A**) Statistical analysis of root length under 100 mM and 200 mM of mannitol stress for 7 d. (**B**) Root phenotype of seedlings grown on 200 mM of mannitol for 7 d. (**C**) Statistical analysis of root length under 0.6 mM and 1.2 mM of H_2O_2 stress for 7 d. (**D**) Root phenotype of seedlings grown on 1.2 mM of H_2O_2 for 7 d. (**E**) Plant survival of 15 d seedlings under drought stress. Control, normal growth; Drought, water withholding for 15 d; Rewater, water withholding for 15 d and 2 d of re-watering. (**F**) Bolting performance of plants under drought stress. Control, normal growth of 21 d seedlings; Drought, 21 d seedlings after 10 d of water withholding. The red line is used to distinguish WT seedlings from OE seedlings. (**G–I**) Activities of antioxidant enzymes (SOD, CAT), and Prx in plants under drought stress. (**J**) Content of MDA in plants under drought stress. All conditions were identical to those of F. All experimental data are expressed as mean ± SD. Asterisks indicate statistically significant differences from the control (* $p < 0.05$, ** $p < 0.01$, analyzed using Student's *t*-test).

Under drought treatment, there was no significant difference in survival rates between *ZmPrx25*-OE lines and WT plants after withholding water for 15 d in seedlings (Figures 7E and S2B). However, in 21 d plants subjected to a 10 d drought period, *ZmPrx25*-OE lines exhibited earlier bolting compared to WT (Figure 7F), indicating that *ZmPrx25* may facilitate drought adaptation by accelerating developmental transitions without compromising rosette leaf development. Further physiological analyses revealed that, under drought stress, *ZmPrx25*-OE lines displayed significantly higher Prx and CAT activities, along with lower MDA accumulation, compared to WT plants (Figure 7G–J). These results suggest that *ZmPrx25* enhances drought tolerance by mitigating oxidative damage through increased antioxidant enzyme activity.

Overexpression of *ZmPrx25* enhances stress tolerance in *Arabidopsis* through a dual mechanism: by directly promoting root growth under short-term osmotic and oxidative stress, and by improving plant survival under prolonged drought stress via the activation of enzymatic antioxidant defenses and the regulation of developmental timing.

4. Discussion

Plants rapidly generate ROS in multiple cellular compartments as an integral early response to stress [10]. The apoplast, a vital conduit for intercellular signaling, is one of the primary sites where plants perceive and respond to environmental stress [15]. Studies have shown that under stress conditions (e.g., pathogen interaction), the apoplast generated ROS and participated in the early signal transduction process [10]. This study found that the mesocotyl could sensitively detect changes in external osmotic stress and effectively transmit stress signals to internal tissues through apoplastic ROS. Under osmotic stress, the activity of apoplastic Prx was significantly upregulated, combined with the specific accumulation and signaling of ROS in the apoplast, suggesting that plants may regulate the activity of peroxidases like Prx to maintain intracellular ROS homeostasis and prevent oxidative damage [44]. Although the apoplast isolation method used for detecting apoplastic Prx activity in this study meets the needs of basic research, it is difficult to completely exclude cytoplasmic contamination during the isolation process, which may affect the rigor of the experimental results. The existing studies support the role of Prx in apoplastic ROS regulation [44]. Further investigations requiring accurate dissection of apoplastic Prx functions would benefit from improved isolation methodologies to minimize the influence of cytoplasmic contaminants on enzymatic activity measurements.

The production of apoplastic ROS depends on the activation of Prx and Rboh activities [24,25]. Among them, Prx, which is unique to plants, plays crucial roles in various physiological and developmental processes [45], including the regulation of ROS levels, cell wall production and maintenance, lignification, fruit growth, seed germination, hormone metabolism, and protection against various stresses (such as drought and high temperatures). Studies have found that the overexpression of the *GsPrx9* gene enhances plant salt tolerance, promotes antioxidant responses, and reduces ROS accumulation [46]. Furthermore, heterologous expression of *CaPrx2* in *Arabidopsis* resulted in a significant increase in H_2O_2 in transgenic plants [47]. Mechanistic validation revealed that sodium azide (NaN_3) (a Prx inhibitor) effectively suppressed H_2O_2 accumulation, whereas diphenyleneiodonium chloride (DPI) (an NADPH oxidase inhibitor) showed no such inhibitory effect, indicating that *CaPrx2* may be involved in the H_2O_2 production process [48]. This study confirmed that Prx played a dual function: they mitigated oxidative damage by scavenging excess ROS and participated in ROS-mediated signal transduction processes [49]. This dual function makes Prx a key regulator in maintaining cellular redox balance [50]. Numerous Prx family members have been identified in *Arabidopsis*, rice, and maize, but functional studies are mainly focused on dicots like *Arabidopsis* [51,52], with less research on the function and mechanisms of Prx in important cereal crops like maize.

Prx family proteins are characterized by highly conserved structural domains with a molecular weight of about 28 to 60 kDa [53]. The ZmPrx25 protein identified in this study matched the characteristics of the aforementioned members and showed significant upregulation under osmotic and oxidative stress conditions, similar to the expression patterns of known stress-responsive homologous proteins such as *AtPrx3* [27] and *CrPrx1* [54], suggesting that *ZmPrx25* may play an important role in plant water stress responses. Prx family proteins are typically localized to the cell wall, where they regulate cell wall dynamics by catalyzing the production, scavenging, and oxidation of ROS, proteins, and phenolic compounds [20]. This study confirmed that *ZmPrx25* was localized to the cell wall and is highly expressed in young, dividing tissues. Previous studies have reported that some peroxidases (Prxs) can promote cell elongation by generating hydroxyl radicals (•OH) that mediate the oxidative cleavage of cell wall polysaccharides [55]. However, although ZmPrx25 shows high expression in rapidly dividing young tissues, there is currently no experimental evidence indicating whether it regulates cell elongation through a similar "•OH-polysaccharide oxidation" mechanism. The direct causal relationship between its high expression and cell elongation phenotypes remains to be further verified.

Functional studies indicate that *ZmPrx25* regulates seed germination through a dual mechanism: on the one hand, this is done by inhibiting the synthesis of ABA and promoting the accumulation of GA to balance hormone metabolism [56]; on the other hand, under oxidative stress induced by H_2O_2, overexpressed lines can clear excessive intracellular H_2O_2 to maintain ROS homeostasis, thereby accelerating germination and increasing germination rates, consistent with the function of cell wall-localized *AtPrx36* [55]. In root development, the overexpression of *ZmPrx25* catalyzed hydroxyl radicals (•OH) to promote cell wall loosening, significantly elongating the primary root of Arabidopsis under normal conditions and maintaining a growth advantage under osmotic/oxidative stress [57]. The overexpression of *ZmPrx25* may catalyze the production of hydroxyl radicals (•OH), promoting the relaxation of the cell wall and significantly elongating the primary roots of *Arabidopsis* under normal conditions, while maintaining a growth advantage under osmotic/oxidative stress [57]. This is consistent with findings on *Arabidopsis AtPrx33/34*, where the overexpression of *AtPrx34* resulted in root elongation, whereas *AtPrx33* mutants exhibited root shortening. The analysis of gene expression level of *ZmPrx25* and the protein level of ZmPrx25 in maize were performed under PEG-6000 stress. Considering that PEG does not reproduce the dynamics of soil drought and may yield phenotypes that differ from soil conditions, we employed a soil-drying assay to verify the drought-resistance function of ZmPrx25. Twenty-one-day-old Arabidopsis transgenic plants were subjected to water withholding for 15 days to evaluate drought-stress phenotypes. By integrating both approaches—PEG for rapid gene-expression screening and soil drought for functional confirmation—we provide a more comprehensive assessment of ZmPrx25's role in drought tolerance. During drought stress, overexpressing lines utilized a "developmental escape" strategy by bolting early to avoid prolonged stress, a process accompanied by increased Prx activity and the synergistic activation of other antioxidant enzymes like CAT, forming a multi-level enzyme-driven antioxidant defense system [58]. However, this study focused solely on *ZmPrx25* and cannot fully rule out potential functional redundancy among other members of the Prx family. Subsequent studies need to conduct more systematic functional analyses of the maize Prx family members.

Under PEG-6000 osmotic stress, transcripts of *ZmPrx25*, *ZmPrx18*, and *ZmPrx72* were significantly upregulated (log_2FC = 3.594, 1.226 and 1.030, respectively), whereas *ZmPrx24* and *ZmPrx95* were markedly downregulated (Figure 2A). Among the three induced genes,

ZmPrx25 exhibited the strongest induction, providing the rationale for selecting it as the primary candidate for drought/osmotic-stress tolerance in this study. ZmPrx25, ZmPrx18, and ZmPrx72 may exhibit functional redundancy. Future work will employ CRISPR/Cas9-mediated multiplex editing to generate single, double, and triple knock-out lines to dissect their individual and overlapping contributions.

5. Conclusions

This study systematically elucidates the central role of the maize peroxidase ZmPrx25 in apoplastic ROS signaling and plant drought/oxidative stress tolerance. In response to osmotic stress, the mesocotyl generates ROS and transmits external osmotic stress signals internally. This process is mediated by the ZmPrx25. This protein maintains ROS homeostasis by scavenging extracellular H_2O_2, thereby mitigating oxidative stress and enhancing drought resistance. Overexpression of *ZmPrx25* promotes seed germination and root growth, strengthens the antioxidant system, and induces a drought escape strategy via early bolting. This research provides a theoretical foundation and a valuable gene resource for drought-resistance breeding in crops.

Supplementary Materials: The following supporting information can be downloaded at https://www.mdpi.com/article/10.3390/antiox14091067/s1: Figure S1: Mesocotyl osmotic stress apparatus. Figure S2: Seed germination rates of Arabidopsis under oxidative stress and survival rates under drought stress. Table S1: Primers used in this study.

Author Contributions: W.W. and H.Z. conceived of and designed the study. F.Z. prepared the plant material for the omics work and conducted the experiments. L.N., Y.L., and X.Z. assisted with the experiments and revised the manuscript. X.W. and H.L. conducted data analysis and collected the references. All authors have read and agreed to the published version of the manuscript.

Funding: This research was supported by the National Natural Science Foundation of China (No. 32272026 to WW) and the Key Scientific Research Project of Colleges and Universities in Henan Province (No. 24A180011 to HL).

Data Availability Statement: The original contributions presented in this study are included in the article/Supplementary Material. Further inquiries can be directed to the corresponding author(s).

Conflicts of Interest: The authors declare no conflicts of interest.

References

1. Tang, H.; Zhang, L.; Xie, X.; Wang, Y.; Wang, T.; Liu, C. Resilience of maize to environmental stress: Insights into drought and heat tolerance. *Int. J. Mol. Sci.* **2025**, *26*, 5274. [CrossRef]
2. Liu, S.; Qin, F. Genetic dissection of maize drought tolerance for trait improvement. *Mol. Breed.* **2021**, *41*, 8. [CrossRef]
3. Dong, A.; Wang, N.; Zenda, T.; Zhai, X.; Zhong, Y.; Yang, Q.; Xing, Y.; Duan, H.; Yan, X. ZmDnaJ-ZmNCED6 module positively regulates drought tolerance via modulating stomatal closure in maize. *Plant Physiol. Biochem.* **2025**, *218*, 109286. [CrossRef]
4. Ali, S.; Tyagi, A.; Bae, H. ROS interplay between plant growth and stress biology: Challenges and future perspectives. *Plant Physiol. Biochem.* **2023**, *203*, 108032. [CrossRef]
5. Fedoreyeva, L.I. ROS as signaling molecules to initiate the process of plant acclimatization to abiotic stress. *Int. J. Mol. Sci.* **2024** *25*, 11820. [CrossRef] [PubMed]
6. Li, M.; Kim, C. Chloroplast ROS and stress signaling. *Plant Commun.* **2022**, *3*, 100264. [CrossRef] [PubMed]
7. Gandin, A.; Dizengremel, P.; Jolivet, Y. Integrative role of plant mitochondria facing oxidative stress: The case of ozone. *Plant Physiol. Biochem.* **2021**, *159*, 202–210. [CrossRef] [PubMed]
8. Huang, L.; Liu, Y.; Wang, X.; Jiang, C.; Zhao, Y.; Lu, M.; Zhang, J. Peroxisome-mediated reactive oxygen species signals modulate programmed cell death in plants. *Int. J. Mol. Sci.* **2022**, *23*, 10087. [CrossRef]
9. Qi, J.; Song, C.P.; Wang, B.; Zhou, J.; Kangasjärvi, J.; Zhu, J.K.; Gong, Z. Reactive oxygen species signaling and stomatal movement in plant responses to drought stress and pathogen attack. *J. Integr. Plant Biol.* **2018**, *60*, 805–826. [CrossRef]
10. Waszczak, C.; Carmody, M.; Kangasjärvi, J. Reactive oxygen species in plant signaling. *Annu. Rev. Plant Biol.* **2018**, *69*, 209–236 [CrossRef]

1. Leister, D. Piecing the puzzle together: The central role of reactive oxygen species and redox hubs in chloroplast retrograde signaling. *Antioxid. Redox Signal* **2019**, *30*, 1206–1219. [CrossRef]
2. Huang, S.; Van Aken, O.; Schwarzländer, M.; Belt, K.; Millar, A.H. The roles of mitochondrial reactive oxygen species in cellular signaling and stress response in plants. *Plant Physiol.* **2016**, *171*, 1551–1559. [CrossRef] [PubMed]
3. Del Río, L.A.; López-Huertas, E. ROS generation in peroxisomes and its role in cell signaling. *Plant Cell Physiol.* **2016**, *57*, 1364–1376. [CrossRef]
4. Zhang, H.; Zhu, J.; Gong, Z.; Zhu, J.K. Abiotic stress responses in plants. *Nat. Rev. Genet.* **2022**, *23*, 104–119. [CrossRef] [PubMed]
5. Miller, G.; Shulaev, V.; Mittler, R. Reactive oxygen signaling and abiotic stress. *Physiol. Plant* **2008**, *133*, 481–489. [CrossRef]
6. Sierla, M.; Waszczak, C.; Vahisalu, T.; Kangasjärvi, J. Reactive oxygen species in the regulation of stomatal movements. *Plant Physiol.* **2016**, *171*, 1569–1580. [CrossRef] [PubMed]
7. Noctor, G.; Mhamdi, A.; Foyer, C.H. The roles of reactive oxygen metabolism in drought: Not so cut and dried. *Plant Physiol.* **2014**, *164*, 1636–1648. [CrossRef]
8. Qi, J.; Wang, J.; Gong, Z.; Zhou, J.M. Apoplastic ROS signaling in plant immunity. *Curr. Opin. Plant Biol.* **2017**, *38*, 92–100. [CrossRef]
9. Hua, D.; Wang, C.; He, J.; Liao, H.; Duan, Y.; Zhu, Z.; Guo, Y.; Chen, Z.; Gong, Z. A plasma membrane receptor kinase, GHR1, mediates abscisic acid- and hydrogen peroxide-regulated stomatal movement in Arabidopsis. *Plant Cell* **2012**, *24*, 2546–2561. [CrossRef]
10. Francoz, E.; Ranocha, P.; Nguyen-Kim, H.; Jamet, E.; Burlat, V.; Dunand, C. Roles of cell wall peroxidases in plant development. *Phytochemistry* **2015**, *112*, 15–21. [CrossRef]
11. Shigeto, J.; Tsutsumi, Y. Diverse functions and reactions of class III peroxidases. *New Phytol.* **2016**, *209*, 1395–1402. [CrossRef]
12. Survila, M.; Davidsson, P.R.; Pennanen, V.; Kariola, T.; Broberg, M.; Sipari, N.; Heino, P.; Palva, E.T. Peroxidase-generated apoplastic ROS impair cuticle integrity and contribute to damp-elicited defenses. *Front. Plant Sci.* **2016**, *7*, 1945. [CrossRef]
13. Bindschedler, L.V.; Dewdney, J.; Blee, K.A.; Stone, J.M.; Asai, T.; Plotnikov, J.; Denoux, C.; Hayes, T.; Gerrish, C.; Davies, D.R.; et al. Peroxidase-dependent apoplastic oxidative burst in Arabidopsis required for pathogen resistance. *Plant J.* **2006**, *47*, 851–863. [CrossRef]
14. Daudi, A.; Cheng, Z.; O'Brien, J.A.; Mammarella, N.; Khan, S.; Ausubel, F.M.; Bolwell, G.P. The apoplastic oxidative burst peroxidase in Arabidopsis is a major component of pattern-triggered immunity. *Plant Cell* **2012**, *24*, 275–287. [CrossRef]
15. O'Brien, J.A.; Daudi, A.; Finch, P.; Butt, V.S.; Whitelegge, J.P.; Souda, P.; Ausubel, F.M.; Bolwell, G.P. A peroxidase-dependent apoplastic oxidative burst in cultured Arabidopsis cells functions in MAMP-elicited defense. *Plant Physiol.* **2012**, *158*, 2013–2027. [CrossRef]
16. López-Castillo, L.M.; González-Leyzaola, A.; Diaz-Flores-Rivera, M.F.; Winkler, R.; Wielsch, N.; García-Lara, S. Modulation of aleurone peroxidases in kernels of insect-resistant maize (*Zea mays* L.; Pob84-C3R) after mechanical and insect damage. *Front. Plant Sci.* **2020**, *11*, 781. [CrossRef] [PubMed]
17. Llorente, F.; López-Cobollo, R.M.; Catalá, R.; Martínez-Zapater, J.M.; Salinas, J. A novel cold-inducible gene from Arabidopsis, RCI3, encodes a peroxidase that constitutes a component for stress tolerance. *Plant J.* **2002**, *32*, 13–24. [CrossRef]
18. Jiao, Y.; Lv, W.; Teng, W.; Li, L.; Lan, H.; Bai, L.; Li, Z.; Lian, Y.; Wang, Z.; Xin, Z.; et al. Peroxidase gene TaPrx109-B1 enhances wheat tolerance to water deficit via modulating stomatal density. *Plant Cell Environ.* **2024**, *47*, 2954–2970. [CrossRef] [PubMed]
19. Hu, X.; Jiang, M.; Zhang, A.; Lu, J. Abscisic acid-induced apoplastic H_2O_2 accumulation up-regulates the activities of chloroplastic and cytosolic antioxidant enzymes in maize leaves. *Planta* **2005**, *223*, 57–68. [CrossRef] [PubMed]
20. Voothuluru, P.; Sharp, R.E. Apoplastic hydrogen peroxide in the growth zone of the maize primary root under water stress. I. Increased levels are specific to the apical region of growth maintenance. *J. Exp. Bot.* **2013**, *64*, 1223–1233. [CrossRef]
21. Zhang, Q.; Liu, H.; Wu, X.; Wang, W. Identification of drought tolerant mechanisms in a drought-tolerant maize mutant based on physiological, biochemical and transcriptomic analyses. *BMC Plant Biol.* **2020**, *20*, 315. [CrossRef]
22. Gentzel, I.; Giese, L.; Zhao, W.; Alonso, A.P.; Mackey, D. A simple method for measuring apoplast hydration and collecting apoplast contents. *Plant Physiol.* **2019**, *179*, 1265–1272. [CrossRef]
23. Witzel, K.; Shahzad, M.; Matros, A.; Mock, H.P.; Mühling, K.H. Comparative evaluation of extraction methods for apoplastic proteins from maize leaves. *Plant Methods* **2011**, *7*, 48. [CrossRef]
24. Fichman, Y.; Miller, G.; Mittler, R. Whole-plant live imaging of reactive oxygen species. *Mol. Plant* **2019**, *12*, 1203–1210. [CrossRef] [PubMed]
25. Kim, D.; Langmead, B.; Salzberg, S.L. HISAT: A fast spliced aligner with low memory requirements. *Nat. Methods* **2015**, *12*, 357–360. [CrossRef] [PubMed]
26. Love, M.I.; Huber, W.; Anders, S. Moderated estimation of fold change and dispersion for RNA-seq data with DESeq2. *Genome Biol.* **2014**, *15*, 550. [CrossRef]
27. Li, Y.; Niu, L.; Zhou, X.; Liu, H.; Tai, F.; Wang, W. Modifying the expression of cysteine protease gene PCP affects pollen development, germination and plant drought tolerance in maize. *Int. J. Mol. Sci.* **2023**, *24*, 7406. [CrossRef]

38. Livak, K.J.; Schmittgen, T.D. Analysis of relative gene expression data using real-time quantitative PCR and the 2(-Delta Delta C(T)) Method. *Methods* **2001**, *25*, 402–408. [CrossRef] [PubMed]
39. Liu, H.; Song, S.; Liu, M.; Mu, Y.; Li, Y.; Xuan, Y.; Niu, L.; Zhang, H.; Wang, W. Transcription factor ZmNAC20 improves drought resistance by promoting stomatal closure and activating expression of stress-responsive genes in maize. *Int. J. Mol. Sci.* **2023**, *24*, 4712. [CrossRef]
40. Xu, K.; Huang, X.; Wu, M.; Wang, Y.; Chang, Y.; Liu, K.; Zhang, J.; Zhang, Y.; Zhang, F.; Yi, L.; et al. A rapid, highly efficient and economical method of agrobacterium-mediated in planta transient transformation in living onion epidermis. *PLoS ONE* **2014**, *9*, e83556. [CrossRef]
41. Christensen, S.A.; Nemchenko, A.; Park, Y.S.; Borrego, E.; Huang, P.C.; Schmelz, E.A.; Kunze, S.; Feussner, I.; Yalpani, N.; Meeley, R.; et al. The novel monocot-specific 9-lipoxygenase ZmLOX12 is required to mount an effective jasmonate-mediated defense against Fusarium verticillioides in maize. *Mol. Plant Microbe Interact.* **2014**, *27*, 1263–1276. [CrossRef] [PubMed]
42. Zuo, W.; Chao, Q.; Zhang, N.; Ye, J.; Tan, G.; Li, B.; Xing, Y.; Zhang, B.; Liu, H.; Fengler, K.A.; et al. A maize wall-associated kinase confers quantitative resistance to head smut. *Nat. Genet.* **2015**, *47*, 151–157. [CrossRef]
43. Niu, L.; Liu, L.; Wang, W. Digging for stress-responsive cell wall proteins for developing stress-resistant maize. *Front. Plant Sci.* **2020**, *11*, 576385. [CrossRef]
44. Li, S.; Zheng, H.; Sui, N.; Zhang, F. Class III peroxidase: An essential enzyme for enhancing plant physiological and developmental process by maintaining the ROS level: A review. *Int. J. Biol. Macromol.* **2024**, *283*, 137331. [CrossRef]
45. Almagro, L.; Gómez Ros, L.V.; Belchi-Navarro, S.; Bru, R.; Ros Barceló, A.; Pedreño, M.A. Class III peroxidases in plant defence reactions. *J. Exp. Bot.* **2009**, *60*, 377–390. [CrossRef]
46. Jin, T.; Sun, Y.; Zhao, R.; Shan, Z.; Gai, J.; Li, Y. Overexpression of peroxidase gene GsPRX9 confers aalt tolerance in soybean. *Int. J. Mol. Sci.* **2019**, *20*, 3745. [CrossRef]
47. Choi, H.W.; Kim, Y.J.; Lee, S.C.; Hong, J.K.; Hwang, B.K. Hydrogen peroxide generation by the pepper extracellular peroxidase $CaPO_2$ activates local and systemic cell death and defense response to bacterial pathogens. *Plant Physiol.* **2007**, *145*, 890–904. [CrossRef]
48. Wally, O.; Punja, Z.K. Enhanced disease resistance in transgenic carrot (*Daucus carota* L.) plants over-expressing a rice cationic peroxidase. *Planta* **2010**, *232*, 1229–1239. [CrossRef] [PubMed]
49. Gokce, A.; Sekmen, A.H. Exploring the regulatory roles of AtGLR3.4 receptors in mitochondrial stress and ROS management in Arabidopsis. *Plant Cell Rep.* **2025**, *44*, 164. [CrossRef] [PubMed]
50. Kang, S.W.; Rhee, S.G.; Chang, T.S.; Jeong, W.; Choi, M.H. 2-Cys peroxiredoxin function in intracellular signal transduction: Therapeutic implications. *Trends Mol. Med.* **2005**, *11*, 571–578. [CrossRef]
51. Shigeto, J.; Itoh, Y.; Hirao, S.; Ohira, K.; Fujita, K.; Tsutsumi, Y. Simultaneously disrupting AtPrx2, AtPrx25 and AtPrx71 alters lignin content and structure in Arabidopsis stem. *J. Integr. Plant Biol.* **2015**, *57*, 349–356. [CrossRef]
52. Jin, J.; Hewezi, T.; Baum, T.J. Arabidopsis peroxidase AtPRX53 influences cell elongation and susceptibility to Heterodera schachtii. *Plant Signal Behav.* **2011**, *6*, 1778–1786. [CrossRef]
53. Freitas, C.D.T.; Costa, J.H.; Germano, T.A.; de O. Rocha, R.; Ramos, M.V.; Bezerra, L.P. Class III plant peroxidases: From classification to physiological functions. *Int. J. Biol. Macromol.* **2024**, *263*, 130306. [CrossRef]
54. Kumar, S.; Jaggi, M.; Sinha, A.K. Ectopic overexpression of vacuolar and apoplastic Catharanthus roseus peroxidases confers differential tolerance to salt and dehydration stress in transgenic tobacco. *Protoplasma* **2012**, *249*, 423–432. [CrossRef]
55. Kunieda, T.; Shimada, T.; Kondo, M.; Nishimura, M.; Nishitani, K.; Hara-Nishimura, I. Spatiotemporal secretion of PEROXIDASE36 is required for seed coat mucilage extrusion in Arabidopsis. *Plant Cell.* **2013**, *25*, 1355–1367. [CrossRef] [PubMed]
56. Schopfer, P.; Plachy, C.; Frahry, G. Release of reactive oxygen intermediates (superoxide radicals, hydrogen peroxide, and hydroxyl radicals) and peroxidase in germinating radish seeds controlled by light, gibberellin, and abscisic acid. *Plant Physiol.* **2001**, *125*, 1591–1602. [CrossRef]
57. Passardi, F.; Tognolli, M.; De Meyer, M.; Penel, C.; Dunand, C. Two cell wall associated peroxidases from Arabidopsis influence root elongation. *Planta* **2006**, *223*, 965–974. [CrossRef] [PubMed]
58. Duan, J.; Zhang, M.; Zhang, H.; Xiong, H.; Liu, P.; Ali, J.; Li, J.; Li, Z. OsMIOX, a myo-inositol oxygenase gene, improves drought tolerance through scavenging of reactive oxygen species in rice (*Oryza sativa* L.). *Plant Sci.* **2012**, *196*, 143–151. [CrossRef] [PubMed]

antioxidants

MDPI

Article

Different Light Wavelengths Differentially Influence the Progression of the Hypersensitive Response Induced by Pathogen Infection in Tobacco

Bao Quoc Tran [1,†], Anh Trung Nguyen [1,†] and Sunyo Jung [1,2,*]

1 BK21 FOUR KNU Creative BioResearch Group, School of Life Science and Biotechnology, Kyungpook National University, Daegu 41566, Republic of Korea; baotran509@gmail.com (B.Q.T.); nguyenanh18190@knu.ac.kr (A.T.N.)

2 Institute of Life Science and Biotechnology, College of Natural Sciences, Kyungpook National University, Daegu 41566, Republic of Korea

* Correspondence: sjung@knu.ac.kr

† These authors contributed equally to this work.

Abstract

Using light-emitting diodes (LEDs), we examined how different light wavelengths influence the hypersensitive response (HR) in tobacco plants infected with *Pseudomonas syringae* pv. *tomato* (Pst). Pst-infiltrated plants exhibited greater resistance to Pst infection under green and blue light compared to white and red light, as indicated by reduced HR-associated programmed cell death, lower H_2O_2 production, and up to 64% reduction in membrane damage. During the late stage of HR, catalase and ascorbate peroxidase activities peaked under green and blue LEDs, with 5- and 10-fold increases, respectively, while superoxide dismutase activity was higher under white and red LEDs. Defense-related genes *CHS1*, *PALa*, *PR1*, and *PR2* were more strongly induced by white and red light. The plants treated with green or blue LEDs during Pst infection prompted faster degradation of phototoxic Mg-porphyrins and exhibited smaller declines in F_v/F_m, electron transport rate, chlorophyll content, and *LHCB* expression compared to those treated with white or red LEDs. By contrast, the induction of the chlorophyll catabolic gene *SGR* was 54% and 77% lower in green and blue LEDs, respectively, compared to white LEDs. This study demonstrates that light quality differentially affects Pst-mediated HR, with green and blue light more effectively suppressing HR progression, mainly by reducing oxidative stress through enhanced antioxidative capacity and mitigation of photosynthetic impairments.

Keywords: chlorophyll metabolism; hypersensitive response; oxidative stress; *Pseudomonas syringae*; ROS detoxification; tobacco

1. Introduction

Biotic stress results in the overaccumulation of reactive oxygen species (ROS) and triggers defense responses, including the expression of phenyl ammonia lyase (PAL), chalcone synthase (CHS), and pathogenesis-related (PR) proteins [1–3]. Infection of plants by *Pseudomonas syringae* can rapidly activate programmed cell death (PCD) in association with disease resistance [4]. Chloroplasts play a crucial role as sources of signals that trigger plant PCD [5–7], and damage to the chloroplast membrane can lead to ROS accumulation [5,6]. In the presence of light, chloroplasts are the primary source of ROS, whereas mitochondria are considered the main source in non-photosynthetic tissues [7].

Chloroplast-derived molecules involved in the initiation of PCD, chloroplast-to-nucleus retrograde signaling, or stress responses include hydrogen peroxide (H_2O_2), singlet oxygen ($^1O_2$), and Mg-protoporphyrin IX (Mg-Proto IX) [8,9].

Both chlorophyll and light-harvesting chlorophyll-binding proteins (LHCs) are essential structural components in the photosynthetic apparatus [10,11]. Chlorophyll biosynthesis shares early biosynthetic steps with other porphyrins such as heme and siroheme in plants [12,13]. After the conversion of 5-aminolevulinic acid into Proto IX, an Mg^{2+} ion is inserted into Proto IX to produce Mg-Proto IX, the first intermediate in the chlorophyll branch of the porphyrin biosynthetic pathway, namely the Mg branch. The biosynthesis and degradation of porphyrins are highly regulated to prevent the accumulation of photoreactive porphyrin intermediates that can undergo photooxidation, resulting in the production of ROS and subsequent promotion of cell death [5,14]. The accumulation of protochlorophyllide (Pchlide) in *Arabidopsis flu* mutants [15] results in cell death phenotypes.

The absorption of excess light energy leads to the generation of ROS within photosynthetic machinery [16]. The conversion of light energy into chemical energy during photosynthesis results in a reduction in chlorophyll *a* fluorescence, a process known as photochemical quenching (q_P) [17]. Excess ROS are scavenged by ROS-detoxifying enzymes [18,19]. In plants, the primary enzyme responsible for ROS detoxification is superoxide dismutase (SOD), which transforms superoxide (O_2^-) into H_2O_2, thus protecting cells from oxidative stress [20,21]. Catalase (CAT) and ascorbate peroxidase (APX) catalyze the elimination of the toxic product of SOD detoxification [16,22]. High ROS concentrations can alter cells' antioxidant status, leading to cell death [7].

Light wavelengths are a key environmental factor regulating plant growth and morphogenesis, as demonstrated by numerous studies [23–25]. However, the impact of different light spectra on the progression of pathogen-mediated hypersensitive response (HR) remains unexplored. We examined HR, oxidative stress, and photosynthetic function in tobacco plants infected with *P. syringae* pv. *tomato* (Pst) under various light wavelengths. We analyzed changes in ROS detoxification activity and gene expression patterns associated with defense mechanisms during HR progression. Additionally, we assessed how different light wavelengths influence the levels of photosensitizing porphyrin intermediates upon Pst infection.

2. Materials and Methods

2.1. Plant Growth, Light Treatment, and Pathogen Inoculation

Tobacco (*Nicotiana tabacum* var. Samsun NN) plants were grown in a growth chamber set to 24 °C with a 16 h light and 8 h dark cycle and a photosynthetic photon flux density (PPFD) of 300 μmol m^{-2} s^{-1}. To produce the bacterial inoculant, *Pseudomonas syringae* pathovar *tomato* strain DC3000 (Pst DC3000) was grown in King's B medium at 28 °C with continuous agitation. The bacteria culture was centrifuged at 3500× *g*, and the pellet was resuspended in 10 mM $MgSO_4$ and adjusted to an OD_{600} of 0.008. Six-week-old tobacco plants were acclimated for two days under one of four different types of light-emitting diodes (LEDs) (Union LED Electronic Co., Siheung-si, Republic of Korea): broad-spectrum white LED (420–680 nm), green LED (520–550 nm), blue LED (460–490 nm), or red LED (620–650 nm). Then, the undersides of fully expanded leaves were inoculated with the suspension via the infiltration method using a needless 1 mL syringe. Each LED treatment was carried out in individually controlled chambers to prevent spectral interference between treatments. The LED array chambers were set to provide a 16 h light and 8 h dark cycle, with a PPFD of 150 μmol m^{-2} s^{-1}. Leaf disks from infiltrated areas or whole leaves were collected at specific time points up to 54 h post-infiltration (hpi) for analysis. Specifically, leaf disks collected at 54 hpi indicate the late stage of PCD.

2.2. Conductivity Measurement

Leaf disks were excised from the Pst-infiltrated zones at specified time points and placed in 5 mL miliQ water for 10 min at 20 °C under gyratory agitation (50 rpm). Cellular leakage into the bathing solution was subsequently measured by detecting electrolyte leakage with a conductivity meter (Cole-Parmer Instruments, Vernon Hills, IL, USA).

2.3. Lipid Peroxidation

To estimate lipid peroxidation, the production of malondialdehyde (MDA) was measured using a modified thiobarbituric acid method [26]. Leaf tissues were ground in a solution containing 0.5% thiobarbituric acid and 20% trichloroacetic acid. After centrifugation, supernatants were heated in a boiling water bath for 25 min, recentrifuged, and utilized for spectrophotometric analysis at 532 nm.

2.4. In Vivo Detection of H_2O_2

For the visual detection of H_2O_2 in leaves, 3,3-diaminobenzidine (DAB) staining was used [27]. The leaves were excised and incubated in a 1 mg mL^{-1} DAB solution (pH 3.8) for 4 h under light at 25 °C. Subsequently, the reaction was stopped by boiling the leaves in ethanol for 10 min. This treatment decolorizes leaves, leaving only the deep brown polymerization product of the reaction between DAB and H_2O_2.

2.5. Measurement of Antioxidant Enzyme Activities

Total soluble proteins were extracted by homogenizing leaf tissues in 0.1 M potassium phosphate buffer (pH 7.5) containing 1% PVP-40, 1 mM phenylmethylsulfonyl fluoride, and 2 mM EDTA, followed by centrifugation at 15,000× *g* for 20 min at 4 °C. SOD activity was spectrophotometrically measured by the xanthine oxidase-cytochrome *c* method [28]. The assay was performed at 25 °C in a 3-mL volume containing 50 mM sodium carbonate buffer (pH 10.0), 0.05 mM xanthine, 0.1 mM EDTA, and 15 μM ferricytochrome *c* by adding xanthine oxidase. Enzyme activity (units/mg protein) is proportional to (V/v − 1)/mg protein, where V and v represent the change in absorbance (550 nm) per min in the absence and presence of SOD, respectively. An assay for APX activity was performed spectrophotometrically [29]. The 3 mL reaction volume contained 0.1 M potassium phosphate buffer (pH 7.5), 0.2 mM H_2O_2, and 0.5 mM ascorbate at 25 °C. The reaction was initiated with the addition of H_2O_2. APX activity was performed by measuring the decrease in absorbance of ascorbate at 290 nm at 25 °C as ascorbate was oxidized. CAT activity was measured spectrophotometrically in a 3 mL volume containing 50 mM potassium phosphate buffer (pH 7.0) and 20 mM H_2O_2 by monitoring the destruction of H_2O_2 at 240 nm [30].

2.6. RNA Extraction and RT-qPCR Analysis

Total RNA was extracted from leaf tissues using TRIzol Reagent (Invitrogen, Carlsbad, CA, USA) in accordance with the manufacturer's instructions. Five μg of RNA from each sample was utilized to synthesize cDNA via the reverse transcription reaction (*ImProm-II*™ Reverse Transcription System, Promega, Madison, WI, USA). Subsequently, qPCR analyses were performed with a StepOnePlus™ Real-Time PCR system (Applied Biosystems, Waltham, MA, USA) using cDNA from each sample, Power SYBR™ Green PCR Master Mix (Applied Biosystems), and gene-specific primers (Supplementary Table S1). The RT-qPCR program consisted of 2 min at 50 °C and 10 min at 95 °C, and then 40 cycles of 15 s at 95 °C and 1 min at 60 °C. The β-*Tubulin* gene was used as an internal control. For analysis, the white LEDs–0 hpi sample served as a calibrator, with its expression level normalized to 1.

2.7. Chlorophyll *a* Fluorescence Measurement

Chlorophyll *a* fluorescence parameters were assessed with a pulse amplitude modulation fluorometer (JUNIOR-PAM, Walz, Effeltrich, Germany). Plants from each treatment were kept in darkness for 30 min to adapt. The minimum fluorescence (F_o) was recorded using a weak measuring beam emitted by a pulse light-emitting diode, while the maximum fluorescence (F_m) was determined after exposing the leaf surface to a saturating pulse of white light. The F_v/F_m value, indicating photosystem II (PSII) activity, was utilized to evaluate functional damage in the plants. The electron transport rate (ETR) is calculated based on the photochemical quantum yield of PSII. The q_P value, used to calculate photochemical quenching, was characterized as $q_P = (F_m' - F')/(F_m' - F_o')$ [31].

2.8. Determination of Porphyrin Contents

Chlorophyll content was measured using a spectrophotometer following the procedure by Lichtenthaler [32]. For measurement of porphyrin contents, leaf tissue was homogenized in a mixture of methanol/acetone/0.1 N NaOH (9/10/1) and then centrifuged at 10,000× *g* and 4 °C [33]. Porphyrin intermediates were analyzed using high-performance liquid chromatography (HPLC) with a Novapak C_{18} column (4 μm particle size, 4.6 × 250 mm, Waters, Milford, MA, USA) at a flow rate of 1 mL min^{-1}. A gradient solvent system was initiated with 0.1 M ammonium phosphate (pH 5.8) and methanol (20:80, *v*/*v*) for 10 min, followed by a linear gradient to 100% methanol over 8 min, which was then maintained for 17 min. The eluate was monitored using a fluorescence detector (2474, Waters) at excitation and emission wavelengths of 440 and 630 nm for protochlorophyllide (Pchlide), and 415 and 595 nm for MgProto IX and MgProto IX methyl ester (MgProto ME), respectively [34].

2.9. Statistical Analysis

All data are shown as means ± standard error (SE). Statistically significant differences were determined by Duncan's multiple range test (at $p < 0.05$) using SPSS 27 software (SPSS Inc., Chicago, IL, USA).

3. Results

3.1. The Impact of Different Light Wavelengths on HR Induced by *P. syringae* Infection

The effects of different light qualities on HR triggered by *P. syringae* infection were examined using LEDs of different wavelengths. Six-week-old tobacco plants were inoculated with Pst to induce HR and exposed to white, green, blue, or red LED light. At 30 hpi, leaf areas infected by Pst started exhibiting mild necrosis, i.e., HR, under the white and red LED treatments, which progressed to severe necrosis by 54 hpi (Figure 1). Plants under green and blue LEDs did not exhibit leaf necrosis at 30 hpi but developed noticeable necrosis by 54 hpi. During Pst infection, plants exposed to white LEDs exhibited the most severe necrosis, followed by those under red, green, and blue LED treatments, in that order.

To determine whether different light wavelengths influence ROS levels during HR, untreated control and Pst-infected leaves were incubated with DAB to detect H_2O_2 generation. Pst-infected leaves began to exhibit faint brown spots as early as 6 hpi under the white and red LEDs, and the spots became stronger by 30 and 54 hpi (Figure 2A). However, under green and blue LEDs, brown spots were not noticeable in Pst-infected leaf areas at 6 hpi but became visible at 30 and 54 hpi. The more pronounced spots observed under white and red LEDs indicated that Pst infection led to significantly higher H_2O_2 production under these conditions.

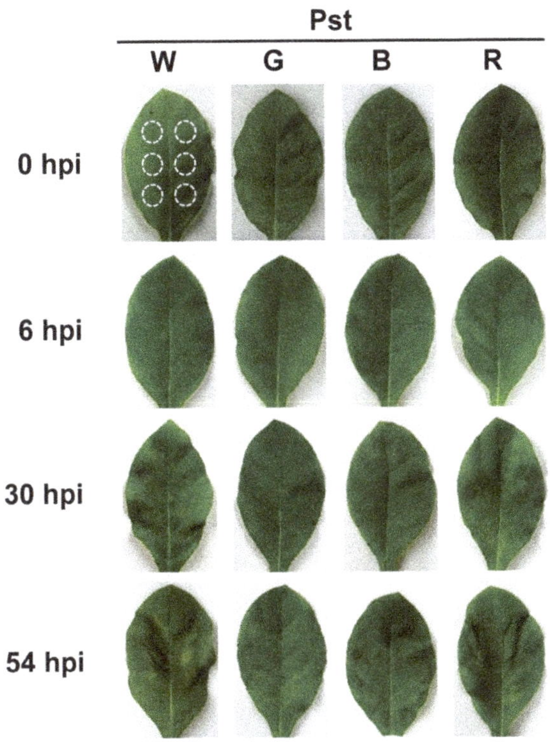

Figure 1. Effects of different light wavelengths on HR in tobacco leaves infected with *P. syringae* pv. *tomato* (Pst). Leaves of 6-week-old tobacco plants were inoculated with Pst and exposed to different light-emitting diodes (LEDs). The treatments included control plants (uninoculated at 0 hpi) and Pst-inoculated plants under broad spectrum white (W; 420–680 nm), green (G; 520–550 nm), blue (B; 460–490 nm), and red (R; 620–650 nm) LEDs. Images of leaves were taken at different time points following Pst inoculation (hpi, hours post-infiltration). The dotted circles indicate the sites of Pst infiltration.

Conductivity, a marker for cellular leakage, was assessed during HR development under the different LED treatments. It continuously increased in Pst-infected leaves from 6 to 54 hpi under all LED treatments (Figure 2B), indicating impaired membrane integrity. Plants exposed to white or red LEDs exhibited a greater increase in conductivity to Pst infection compared to those under blue or green LEDs, with the highest increase in white LED treatment, reaching a 14-fold increase. To further verify membrane integrity, we evaluated MDA content, an indicator of lipid peroxidation. The MDA levels began to increase in response to Pst infiltration under all LED treatments at 30 hpi, with greater increases under white and red LEDs than green and blue LEDs (Figure 2C). At 54 hpi, white and red LED treatments resulted in MDA levels that were up to twice those observed with green and blue LED treatments, confirming greater membrane damage under white and red LEDs.

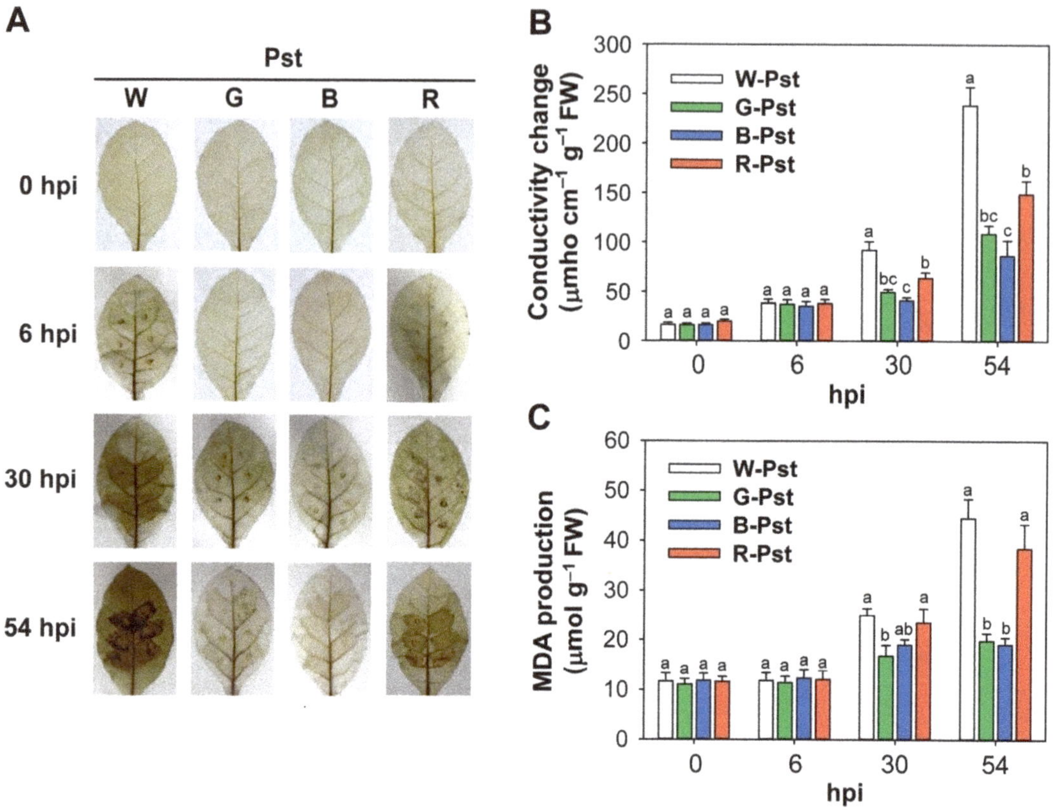

Figure 2. Effects of different light wavelengths on oxidative stress and membrane integrity in leaves infected with Pst. (**A**) DAB staining of H_2O_2 in leaves. The brown spots indicate H_2O_2 production. (**B**) Conductivity change. (**C**) MDA accumulation. Plants were exposed to the treatments described in the caption of Figure 1. W-Pst, G-Pst, B-Pst, and R-Pst represent plants inoculated with Pst under white, green, blue, and red LEDs, respectively. Data are expressed as means ± SE of nine replicates from three independent experiments. Means with the same letter are not significantly different at $p < 0.05$ according to Duncan's Multiple Range Test.

3.2. The Impact of Light Quality on the Regulation of ROS Detoxification and Defense Responses During Pst-Induced HR

Following Pst infection, we monitored how LED lights of different wavelengths influenced the regulation of major ROS-detoxifying enzymes and defense-related genes. In response to Pst infection, the activities of SOD increased under all LED treatments at 6 hpi compared to the untreated control (Figure 3A). At 30 and 54 hpi, SOD activity continued to increase, but the increase was greater under white and red LEDs than under green and blue LEDs. APX activity markedly increased at 6 hpi under white and red LED treatments but decreased by 54 hpi, although it remained higher than that of the control (Figure 3B). Under green and blue LEDs, the activities of APX increased gradually over 54 h. Plants exposed to blue LEDs showed the highest APX activity among all LED treatments at 30 and 54 hpi. During Pst infection, the activities of CAT steadily increased until 54 hpi in all LED treatments (Figure 3C). At 30 and 54 hpi, CAT activity was higher under green LEDs compared to other LEDs, with the highest activity observed at 54 hpi.

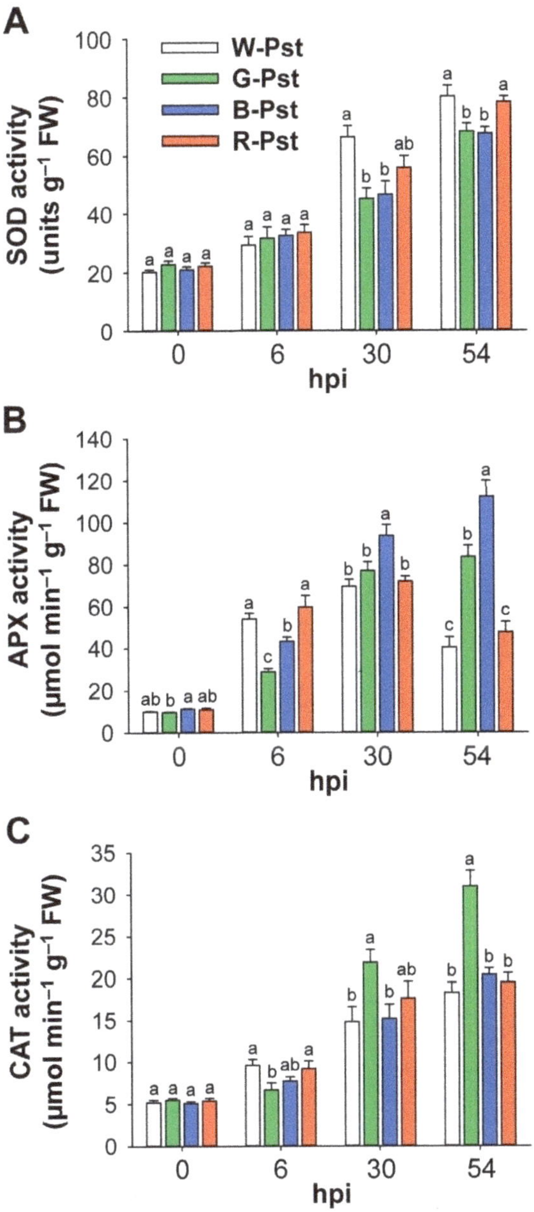

Figure 3. Effects of different light wavelengths on the activity of ROS-detoxifying enzymes in leaves infected with Pst. (**A**) SOD. (**B**) APX. (**C**) CAT. Data are expressed as means ± SE of nine replicates from three independent experiments. Plants were exposed to the treatments described in the caption of Figure 1. The treatment notations are the same as those in Figure 2. Means with the same letter are not significantly different at $p < 0.05$ according to Duncan's Multiple Range Test.

The expression of *CHS1*, a key component in plant defense, was higher under white and blue LEDs at 6 hpi compared to the control and was noticeably upregulated under white and red LEDs at 30 hpi (Figure 4A). By 54 hpi, the expression continued to increase, with greater upregulation under white and red LEDs than under green and blue LEDs.

Within 6 h of Pst infection, at which time no necrosis was visible, *PALa* was dramatically upregulated under all LED treatments, with the greatest increase under red LEDs at 54 hpi (Figure 4B). The *PR1* and *PR2* genes were noticeably upregulated in response to Pst infection at 30 hpi, with their expression further increasing by 54 hpi under all LED treatments (Figure 4C,D). However, the increase was more pronounced under white and red LEDs compared to green and blue LEDs (Figure 4). These results show that Pst-induced HR leads to light quality-dependent activation of ROS detoxification and defense responses under different light wavelengths.

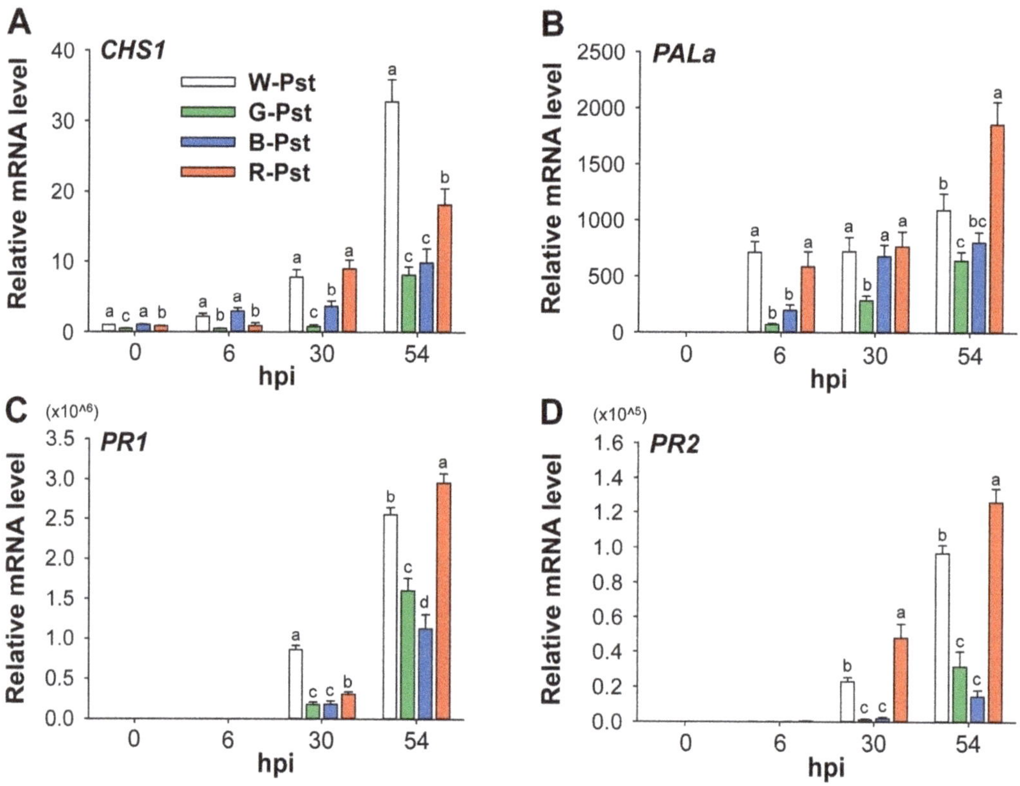

Figure 4. Effects of different light wavelengths on the expression profile of defense-involved genes in leaves infected with Pst. (**A**) *CHS1*. (**B**) *PALa*. (**C**) *PR1*. (**D**) *PR2*. β-*Tubulin* was used as an internal control, and the white LEDs–0 hpi sample was used for normalization, with its expression level set to 1. Plants were exposed to the treatments described in the caption of Figure 1. The treatment notations are the same as those in Figure 2. Data are expressed as means ± SE of nine replicates from three independent experiments. Means with the same letter are not significantly different at $p < 0.05$ according to Duncan's Multiple Range Test.

3.3. The Impact of Light Quality on Regulating Porphyrin Levels During Pst-Induced HR

To assess the effect of light quality on intermediates in the chlorophyll branch of the porphyrin pathway during Pst infection, we examined the porphyrin status in Pst-infected plants subjected to various LED light wavelengths. Mg-Proto IX levels greatly decreased at 6 hpi under white, blue, and red LEDs; Mg-Proto IX was not detected under green LEDs (Figure 5A). Mg-Proto IX methyl ester (ME) levels were also markedly decreased under all LED treatments at 6 hpi, with negligible amounts detected under green LEDs (Figure 5B). By 30 hpi, both Mg-Proto IX and Mg-Proto IX ME further decreased in white and red LED treatments and were undetectable in green and blue LED treatments. Pchlide levels exhibited a decreasing trend similar to that of Mg-Proto IX ME (Figure 5C). By 54 hpi, Mg-Proto IX, Mg-Proto IX ME, and Pchlide were no longer observed under any of the LED treatments.

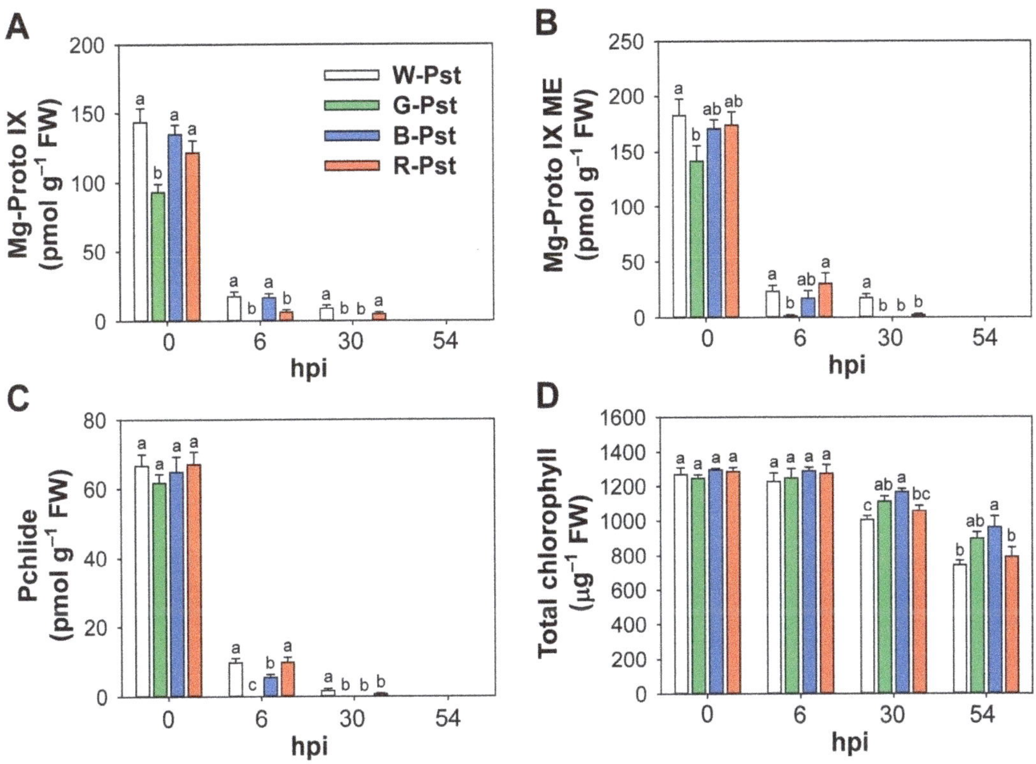

Figure 5. Effects of different light wavelengths on metabolites in the chlorophyll branch of the porphyrin biosynthetic pathway in leaves infected with Pst. (**A–C**) Mg-porphyrin intermediates. (**D**) Total chlorophyll. Plants were exposed to the treatments described in the caption of Figure 1. The treatment notations are the same as those in Figure 2. Data are expressed as means ± SE of nine replicates from three independent experiments. Means with the same letter are not significantly different at $p < 0.05$ according to Duncan's Multiple Range Test.

The levels of chlorophyll, the final product of the chlorophyll branch, were not significantly different among various LEDs at 6 hpi (Figure 5D). However, by 30 hpi, chlorophyll levels decreased by 21%, 18%, 11%, and 10% under white, red, green, and blue LEDs, respectively, compared to the control. By 54 hpi, these decreases reached 42%, 38%, 28%, and 26%, showing greater reductions under white and red LEDs. During Pst-mediated HR, blue and green LEDs induced faster declines in Mg-porphyrin intermediate levels but sustained chlorophyll levels more effectively than white and red LEDs.

3.4. The Impact of Light Quality on Photosynthetic Systems During Pst-Induced HR

We assessed the organization of the photosynthetic machinery in Pst-infected leaves through the quantitative analysis of *LHCB*. Expression levels of *LHCB*, a marker of photosynthetic gene expression patterns, began decreasing at 6 hpi in all LED treatments and continued to decrease at 30 and 54 hpi, with greater reductions under white and red LEDs (Figure 6A). Conversely, the expression of the stay-green gene (*SGR*), an indicator of chlorophyll degradation, increased continuously from 6 to 54 hpi in all LED treatments, with greater increases under white and red LEDs at 30 and 54 hpi (Figure 6B).

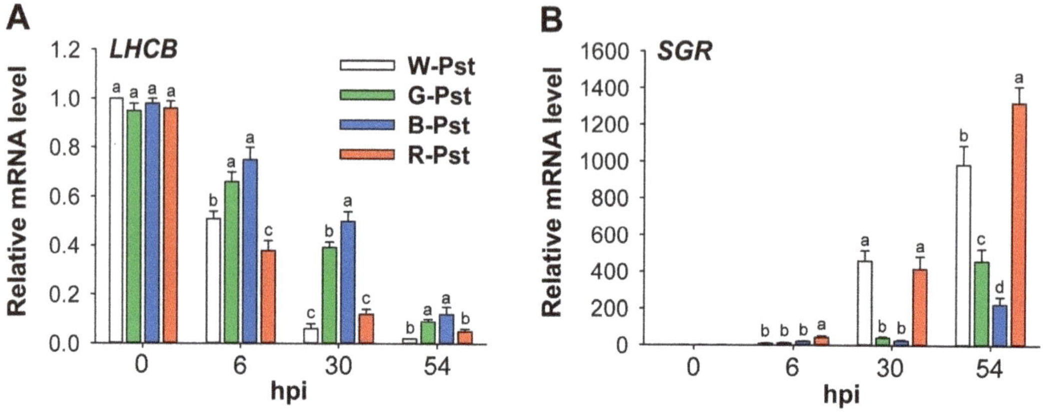

Figure 6. Effects of different light wavelengths on the integrity of the photosystem in leaves infected with Pst. The expression of *LHCB* (**A**) and *SGR* (**B**) genes. β-*Tubulin* was used as an internal control, and the white LEDs–0 hpi sample was used for normalization, with its expression level set to 1. Plants were exposed to the treatments described in the caption of Figure 1. The treatment notations are the same as those in Figure 2. Data are means ± SE of nine replicates from three independent experiments. Means with the same letter are not significantly different at $p < 0.05$ by Duncan's Multiple Range Test.

The effects of different light wavelengths on photosynthetic performance were also examined using the F_v/F_m value, a measure of the photochemical quantum efficiency of PSII, during Pst-mediated HR. After Pst infection, F_v/F_m in plants treated with white or red LEDs continuously decreased from 6 to 54 hpi, with declines of 52% and 34% induced by white and red LEDs, respectively, at 54 hpi (Table 1). Under green LEDs, F_v/F_m remained constant until 30 hpi and then decreased by 10% at 54 hpi compared to the controls, while under blue LEDs, it showed an 8% decline at 54 hpi. The ETR began to decrease in all LED treatments at 6 hpi, continuing to decrease at 30 and 54 hpi, with the greatest reduction observed under white LEDs, followed by red, green, and blue LEDs in decreasing order (Table 1). The q_P values were used to estimate the initiation of photoinhibition in PSII. The q_P started to decline at 6 dpi and continued to decrease at 30 and 54 hpi, with the magnitude of declines in the order of white, red, green, and blue LEDs (Table 1). These results indicate that different light wavelengths caused differential impairments in chloroplast function in response to Pst infection.

Table 1. Effects of different light wavelengths on photosynthetic function in leaves infected with Pst. Plants were exposed to the treatments described in the caption of Figure 1. The treatment notations are the same as those in Figure 2. Data are expressed as means ± SE of nine replicates from three independent experiments. Means with the same letter within a column are not significantly different at $p < 0.05$ based on Duncan's Multiple Range Test.

LED Treatment		*Fv/Fm*	ETR	q_P
0 hpi	White	0.813 ± 0.007 [a]	89.3 ± 2.1 [a]	0.667 ± 0.021 [a]
	Green	0.809 ± 0.004 [a]	85.5 ± 3.7 [a]	0.651 ± 0.013 [a]
	Blue	0.810 ± 0.004 [a]	89.2 ± 2.0 [a]	0.659 ± 0.012 [a]
	Red	0.811 ± 0.003 [a]	86.8 ± 2.0 [a]	0.644 ± 0.014 [a]

Table 1. *Cont.*

LED Treatment		Fv/Fm	ETR	qP
6 hpi	White	0.763 ± 0.017 [b]	70.9 ± 1.7 [b]	0.623 ± 0.011 [a]
	Green	0.801 ± 0.004 [a]	78.4 ± 1.8 [a]	0.638 ± 0.016 [a]
	Blue	0.807 ± 0.003 [a]	79.4 ± 1.8 [a]	0.637 ± 0.020 [a]
	Red	0.779 ± 0.007 [ab]	70.1 ± 2.7 [b]	0.619 ± 0.013 [a]
30 hpi	White	0.678 ± 0.026 [c]	53.5 ± 3.0 [c]	0.466 ± 0.021 [c]
	Green	0.791 ± 0.009 [a]	66.4 ± 2.3 [ab]	0.532 ± 0.011 [ab]
	Blue	0.755 ± 0.011 [ab]	75.0 ± 3.6 [a]	0.584 ± 0.024 [a]
	Red	0.735 ± 0.018 [b]	59.5 ± 5.4 [bc]	0.499 ± 0.016 [bc]
54 hpi	White	0.389 ± 0.039 [c]	18.5 ± 5.5 [c]	0.229 ± 0.027 [d]
	Green	0.730 ± 0.017 [a]	42.7 ± 3.5 [ab]	0.399 ± 0.013 [b]
	Blue	0.749 ± 0.016 [a]	55.3 ± 3.6 [a]	0.503 ± 0.030 [a]
	Red	0.535 ± 0.035 [b]	31.7 ± 4.3 [bc]	0.326 ± 0.011 [c]

4. Discussion

The HR of plants to pathogen infection appears to involve the induction of PCD and other defense responses [35], but little is known about the effects of light wavelength on the mechanisms underlying pathogen-induced HR. Our study demonstrates that different light wavelengths significantly impact HR-associated PCD, primarily by differentially regulating chloroplast-related events during Pst infection, emphasizing the critical role of the chloroplasts in plant HR. Importantly, in tobacco plants infected with Pst, white and red LED treatments triggered more marked necrosis at 30 and 54 hpi compared to blue and green LED treatments (Figure 1). Supporting this observation, irradiation with blue [36], green [37], or red light [36,38], compared to white light, has been shown to suppress disease symptoms in plants infected with pathogens. The suppression of the HR under blue and green LEDs points out the impact of these wavelengths in delaying the progression of PCD in plants.

Some early events associated with HR relate to the rapid generation of ROS in the form of H_2O_2 [39,40]. In Pst-infected leaves, treatments with white or red LEDs led to early H_2O_2 detection at 6 hpi, followed by a marked increase by 30 hpi, with the highest H_2O_2 levels under white LEDs, whereas lower H_2O_2 was formed under green and blue LEDs (Figure 2). The greater production of H_2O_2 under white and red LEDs corresponded to marked necrosis occurring at 54 hpi (Figures 1 and 2). White and red LED treatments also significantly increased conductivity compared to blue and green LEDs, with the highest increase under white LEDs (Figure 2). Lipid peroxidation is known to be linked to HR cell death [41]. At 54 hpi, white and red LEDs produced MDA levels that were twice as high as those seen under green and blue LEDs. These results indicate that green and blue light mitigate lipid peroxidation and membrane disruption in the late stage of the HR. In a previous study, levels of MDA and conductivity decreased under red light but increased under blue light compared to white light in *Camptotheca acuminata* [25]; however, this study did not involve pathogen treatment. During Pst infection, lower H_2O_2 levels and reduced membrane damage under green and blue LEDs indicate a wavelength-dependent impact on oxidative stress and membrane integrity, potentially contributing to the suppression of HR.

In plant cells, ROS detoxification under oxidative stress relies on the efficiency of enzymatic antioxidant components [18,19]. Although numerous studies have explored antioxidant activities under different light wavelengths [24,25,42,43], the impact of light wavelength on antioxidant responses during pathogen infections has been rarely examined. In the late stage of HR, white and red LEDs enhanced SOD activity more than green and blue LEDs (Figure 3). Previous studies without pathogen treatment showed that SOD activity increased under blue and red light compared to white light [42,44]. At 54 hpi, blue LEDs triggered the highest activity of APX, which primarily scavenges harmful H_2O_2 in the chloroplast [21], while CAT activity was highest under green LEDs among all treatments (Figure 3). Green and blue light may help delay HR by lowering H_2O_2 levels through the efficient activation of H_2O_2-detoxifying enzymes during the late stage of the HR. This wavelength-specific impact is consistent with a previous study in which blue light produced higher *APX1* expression than red and white light in broccoli florets [43]. Similarly, blue LED radiation has been shown to enhance APX activity compared to white light in plants infected with *Botrytis cinerea* [45], and plants overexpressing *APX* exhibited reduced visible HR symptoms [35].

Redox signals, such as H_2O_2 accumulation, play a key role in activating plant defenses during the initial phases of HR [46]. CHS and PAL are known to be activated in response to various stress conditions, including fungal or bacterial infection and UV light [2,3]. Under Pst infection, expression of *CHS1* and *PALa* continuously increased under all LED treatments, with notably higher increases under white and red LEDs at 54 hpi (Figure 4). Pathogen-responsive genes *PR-1* and *PR-2* [1,47] were noticeably upregulated at 30 hpi, with even higher levels under white and red LEDs by 54 hpi. Notably, red LEDs induced higher *PALa*, *PR1*, and *PR2* expression than white LEDs at 54 hpi. This result is in accordance with previous observations in *Pseudomonas cichorii*-infected tomato, where *PAL* and *PR-1a* were upregulated more under red light than under white light [48]. Overall, green and blue light led to lower induction of defense-related genes, accompanied by reduced H_2O_2 production (Figures 2 and 4), likely due to the alleviation of HR compared to white and red light.

Porphyrin intermediates may act as an endogenous cell death trigger through the production of $^1O_2$ [9,35]. The chlorophyll branch of the porphyrin pathway begins with the formation of Mg-Proto IX by Mg-chelatase [10,12]. Levels of potent photosensitizers, including Mg-Proto IX, Mg-Proto IX ME, and Pchlide, began to decrease rapidly in response to Pst infection, disappearing by 54 hpi under all LED conditions (Figure 5). The fast disappearance of porphyrin intermediates following Pst infection was suggested to result from photodynamic degradation and reduced porphyrin biosynthesis [49]. Similar declines in porphyrin intermediates have also been previously noted under salt stress, drought, and iron deficiency [34,50,51]. In the late stage of HR, faster scavenging of Mg-porphyrins under green and blue LEDs appears to reduce phototoxicity under Pst infection, thereby helping to delay HR. These results demonstrate that the spectral quality of light regulates porphyrin levels during light quality-dependent HR.

Despite significant declines in the Mg-porphyrin levels, the chlorophyll levels exhibited small reductions under all LED treatments during prolonged Pst infection (Figure 5), indicating limited degradation of chlorophyll. Particularly, green and blue LEDs caused smaller declines in chlorophyll levels compared to white and red LEDs, corresponding to less downregulation of *LHCB* under green and blue LEDs at 30 and 54 hpi (Figures 5 and 6). Previous studies without pathogen infection have also shown that chlorophyll content is higher under blue light compared to white light [24,52], and under green light instead of partial red light compared to red-blue light [53]. Disruption of LHCs can lead to the removal of the porphyrin moiety from chlorophyll [11,40]. Our results indicate that green and blue light induced less photodamage to the photosynthetic apparatus, possibly due to a lower photodegradation rate. This is supported by a lower upregulation of *SGR* under green and blue LEDs (Figure 6B), indicating reduced chlorophyll degradation. The SGR protein extracts Mg^{2+} from chlorophyll *a* to synthesize pheophytin *a* [54] and plays a key role in initiating chlorophyll degradation [39].

Chloroplasts generate contrasting signals that can either inhibit or facilitate cell death [55,56]. In Pst-infected plants, the photosynthetic performance of chloroplasts, as indicated by the F_v/F_m and ETR values, declined more under white or red LEDs compared to green or blue LEDs, with the greatest decline observed under white LEDs (Table 1). This indicates a greater impairment of PSII activity under white and red LEDs, possibly as a result of excessive reduction of the electron transport system. Similar negative impacts of red light on photosynthesis were observed in other plants that were not experiencing pathogen stress [57,58], whereas green light facilitated more efficient photosynthesis than red light [59]. A decline in q_P values indicates increased excitation pressure on PSII [60]. At 30 and 54 hpi, the larger declines in both q_P and chlorophyll levels under white and red LEDs, compared to green and blue LEDs (Figure 5 and Table 1), suggest that PSII is more extensively damaged by white and red light. In contrast, the relatively small decline in q_P observed under blue LEDs indicates a reduced excitation pressure on PSII, allowing for more efficient maintenance of excitation energy for photochemistry. These results show how specific LED wavelengths differentially modulate the integrity of photosystem during Pst infection, highlighting the protective effects of green and blue light, in contrast to the exacerbating effects of white and red light. The observed benefits of green and blue light suggest potential applications of specific light spectra in controlled agricultural environments to reduce pathogen-induced damage by mitigating oxidative stress and preserving photosynthetic efficiency.

5. Conclusions

Our findings demonstrate that green and blue light prominently delay Pst-induced HR, as evidenced by diminished necrosis and lower levels of membrane damage and oxidative stress, compared to white and red light. Green and blue LEDs resulted in the highest activities of CAT and APX, respectively, likely contributing to reduced HR by effectively detoxifying H_2O_2 during the late stage of Pst infection. As HR progressed, green and blue LEDs caused faster declines in Mg-porphyrin levels than white and red LEDs. This may play a protective role in mitigating photodamage to PSII by reducing photosensitivity, which is partly related to a lower susceptibility to photosynthetic impairment under green and blue LEDs. The suppression of HR under green and blue light may be attributed to the tight regulation of chloroplast-related events and enhanced antioxidant properties, aiding plants in coping with oxidative stress during Pst infection. This study provides new insights into how different light wavelengths influence the progression of HR, making a significant advancement in understanding their impact on plant processes related to PCD. Further studies on other economically important crop species and pathogen systems will clarify the broader applicability of light quality-based HR modulation in crop protection.

Supplementary Materials: The following supporting information can be downloaded at https://www.mdpi.com/article/10.3390/antiox14080954/s1, Table S1: Primer used for RT-qPCR assays.

Author Contributions: Conceptualization, B.Q.T. and S.J.; methodology, B.Q.T. and A.T.N.; validation, A.T.N.; formal analysis, B.Q.T. and A.T.N.; investigation, B.Q.T. and A.T.N.; data curation, B.Q.T., A.T.N. and S.J.; writing—original draft preparation, S.J.; writing—review and editing, B.Q.T., A.T.N. and S.J.; supervision, S.J.; project administration, S.J.; funding acquisition, S.J. All authors have read and agreed to the published version of the manuscript.

Funding: This research was supported by the Basic Science Research Program through the National Research Foundation of Korea (NRF), funded by the Ministry of Education (NRF-2020R1I1A3073932), and by the G-LAMP Program of the NRF grant funded by the Ministry of Education (No. RS-2023-00301914).

Data Availability Statement: Data will be made available on request.

Conflicts of Interest: The authors declare no conflicts of interest.

Abbreviations

The following abbreviations are used in this manuscript:

APX	Ascorbate peroxidase
CAT	Catalase
hpi	hours post-infiltration
HR	Hypersensitive response
LEDs	Light-emitting diodes
LHC	Light-harvesting chlorophyll-binding proteins
MDA	Malondialdehyde
PCD	Programmed cell death
Proto IX	Protoporphyrin IX
Proto IX ME	Protoporphyrin IX methyl ester
Pchlide	Protochlorophyllide
Pst	*Pseudomonas syringae* pathovar *tomato*
SOD	Superoxide dismutase

References

1. van Loon, L.C.; Rep, M.; Pieterse, C.M.J. Significance of inducible defense-related proteins in infected plants. *Annu. Rev. Phytopathol.* **2006**, *44*, 135–162. [CrossRef]
2. Dao, T.T.H.; Linthorst, H.J.M.; Verpoorte, R. Chalcone synthase and its functions in plant resistance. *Phytochem. Rev.* **2011**, *10*, 397–412. [CrossRef] [PubMed]
3. Kim, D.S.; Hwang, B.K. An important role of the pepper phenylalanine ammonia-lyase gene (PAL_1) in salicylic acid-dependent signalling of the defence response to microbial pathogens. *J. Exp. Bot.* **2014**, *65*, 2295–2306. [CrossRef] [PubMed]
4. Krzymowska, M.; Konopka-Postupolska, D.; Sobczak, M.; Macioszek, V.; Ellis, B.E.; Hennig, J. Infection of tobacco with different *Pseudomonas syringae* pathovars leads to distinct morphotypes of programmed cell death. *Plant J.* **2007**, *50*, 253–264. [CrossRef] [PubMed]
5. Ishikawa, A.; Okamoto, H.; Iwasaki, Y.; Asahi, T. A deficiency of coproporphyrinogen III oxidase causes lesion formation in Arabidopsis. *Plant J.* **2001**, *27*, 89–99. [CrossRef] [PubMed]
6. Landoni, M.; De Francesco, A.; Bellatti, S.; Delledonne, M.; Ferrarini, A.; Venturini, L.; Pilu, R.; Bononi, M.; Tonelli, C. A mutation in the FZL gene of Arabidopsis causing alteration in chloroplast morphology results in a lesion mimic phenotype. *J. Exp. Bot.* **2013**, *64*, 4313–4328. [CrossRef]
7. Ye, C.; Zheng, S.; Jiang, D.; Lu, J.; Huang, Z.; Liu, Z.; Zhou, H.; Zhuang, C.; Li, J. Initiation and execution of programmed cell death and regulation of reactive oxygen species in plants. *Int. J. Mol. Sci.* **2021**, *22*, 12942. [CrossRef]
8. Koussevitzky, S.; Nott, A.; Mockler, T.C.; Hong, F.; Sachetto-Martins, G.; Surpin, M.; Lim, J.; Mittler, R.; Chory, J. Signals from chloroplasts converge to regulate nuclear gene expression. *Science* **2007**, *316*, 715–719. [CrossRef]
9. Gadjev, I.; Stone, J.M.; Gechev, T.S. Programmed cell death in plants: New insights into redox regulation and the role of hydrogen peroxide. *Int. Rev. Cell Mol. Biol.* **2008**, *270*, 87–144. [CrossRef]
10. Tanaka, R.; Tanaka, A. Tetrapyrrole biosynthesis in higher plants. *Annu. Rev. Plant Biol.* **2007**, *58*, 321–346. [CrossRef]
11. Terry, M.J.; Smith, A.G. A model for tetrapyrrole synthesis as the primary mechanism for plastid-to-nucleus signaling during chloroplast biogenesis. *Front. Plant Sci.* **2013**, *4*, 14. [CrossRef]
12. Beale, S.I.; Weinstein, J.D. Tetrapyrrole metabolism in photosynthetic organisms. In *Biosynthesis of Heme and Chlorophyll*; Daily, H.A., Ed.; McGraw-Hill: New York, NY, USA, 1990; pp. 287–391.
13. Kobayashi, K.; Masuda, T. Transcriptional regulation of tetrapyrrole biosynthesis in *Arabidopsis thaliana*. *Front. Plant Sci.* **2016**, *7*, 1811. [CrossRef] [PubMed]
14. Bruggeman, Q.; Raynaud, C.; Benhamed, M.; Delarue, M. To die or not to die? Lessons from lesion mimic mutants. *Front. Plant Sci.* **2015**, *6*, 24. [CrossRef]
15. Meskauskiene, R.; Nater, M.; Goslings, D.; Kessler, F.; Camp, R.O.D.; Apel, K. FLU: A negative regulator of chlorophyll biosynthesis in *Arabidopsis thaliana*. *Proc. Natl. Acad. Sci. USA* **2001**, *98*, 12826–12831. [CrossRef] [PubMed]
16. Demmig-Adams, B.; Cohu, C.M.; Adams III, W.W. Dealing with the hazards of harnessing sunlight. *Nat. Educ. Knowl.* **2012**, *4*, 18.
17. Murchie, E.H.; Lawson, T. Chlorophyll fluorescence analysis: A guide to good practice and understanding some new applications. *J. Exp. Bot.* **2013**, *64*, 3983–3998. [CrossRef] [PubMed]

18. Das, K.; Roychoudhury, A. Reactive oxygen species (ROS) and response of antioxidants as ROS-scavengers during environmental stress in plants. *Front. Environ. Sci.* **2014**, *2*, 53. [CrossRef]
19. Akram, N.A.; Shafiq, F.; Ashraf, M. Ascorbic acid—A potential oxidant scavenger and its role in plant development and abiotic stress tolerance. *Front. Plant Sci.* **2017**, *8*, 613. [CrossRef]
20. Bose, J.; Rodrigo-Moreno, A.; Shabala, S. ROS homeostasis in halophytes in the context of salinity stress tolerance. *J. Exp. Bot.* **2014**, *65*, 1241–1257. [CrossRef]
21. Kesawat, M.S.; Satheesh, N.; Kherawat, B.S.; Kumar, A.; Kim, H.U.; Chung, S.M.; Kumar, M. Regulation of reactive oxygen species during salt stress in plants and their crosstalk with other signaling molecules—Current perspectives and future directions. *Plants* **2023**, *12*, 864. [CrossRef]
22. Foyer, C.H.; Noctor, G. Redox signaling in plants. *Antioxid. Redox Signal.* **2013**, *18*, 2087–2090. [CrossRef]
23. Li, Q.; Kubota, C. Effects of supplemental light quality on growth and phytochemicals of baby leaf lettuce. *Environ. Exp. Bot.* **2009**, *67*, 59–64. [CrossRef]
24. Ye, S.; Shao, Q.; Xu, M.; Li, S.; Wu, M.; Tan, X.; Su, L. Effects of light quality on morphology, enzyme activities, and bioactive compound contents in *Anoectochilus roxburghii*. *Front. Plant Sci.* **2017**, *8*, 857. [CrossRef] [PubMed]
25. Yu, W.; Liu, Y.; Song, L.; Jacobs, D.F.; Du, X.; Ying, Y.; Shao, Q.; Wu, J. Effect of differential light quality on morphology, photosynthesis, and antioxidant enzyme activity in *Camptotheca acuminata* seedlings. *J. Plant Growth Regul.* **2017**, *36*, 148–160. [CrossRef]
26. Buege, J.A.; Aust, S.D. Microsomal lipid peroxidation. *Methods Enzymol.* **1978**, *52*, 302–310. [CrossRef] [PubMed]
27. Thordal-Christensen, H.; Zhang, Z.; Wei, Y.; Collinge, D.B. Subcellular localization of H_2O_2 in plants. H_2O_2 accumulation in papillae and hypersensitive response during the barley–powdery mildew interaction. *Plant J.* **1997**, *11*, 1187–1194. [CrossRef]
28. Baum, J.A.; Scandalios, J.G. Isolation and characterization of the cytosolic and mitochondrial superoxide dismutases of maize. *Arch. Biochem. Biophys.* **1981**, *206*, 249–264. [CrossRef]
29. Chen, G.-X.; Asada, K. Ascorbate peroxidase in tea leaves: Occurrence of two isozymes and the differences in their enzymatic and molecular properties. *Plant Cell Physiol.* **1989**, *30*, 987–998. [CrossRef]
30. Beers, R.F.; Sizer, I.W. A spectrophotometric method for measuring the breakdown of hydrogen peroxide by catalase. *J. Biol. Chem.* **1952**, *195*, 133–140. [CrossRef]
31. Schreiber, U.; Schliwa, U.; Bilger, W. Continuous recording of photochemical and non-photochemical chlorophyll fluorescence quenching with a new type of modulation fluorometer. *Photosynth. Res.* **1986**, *10*, 51–62. [CrossRef]
32. Lichtenthaler, H.K. Chlorophylls and carotenoids: Pigments of photosynthetic biomembranes. *Methods Enzymol.* **1987**, *148*, 350–382. [CrossRef]
33. Lermontova, I.; Grimm, B. Reduced activity of plastid protoporphyrinogen oxidase causes attenuated photodynamic damage during high-light compared to low-light exposure. *Plant J.* **2006**, *48*, 499–510. [CrossRef]
34. Phung, T.-H.; Jung, H.-I.; Park, J.-H.; Kim, J.-G.; Back, K.; Jung, S. Porphyrin biosynthesis control under water stress: Sustained porphyrin status correlates with drought tolerance in transgenic rice. *Plant Physiol.* **2011**, *157*, 1746–1764. [CrossRef] [PubMed]
35. Yao, N.; Greenberg, J.T. *Arabidopsis* Accelerated Cell Death2 modulates programmed cell death. *Plant Cell* **2006**, *18*, 397–411. [CrossRef] [PubMed]
36. Ahn, S.-Y.; Kim, S.A.; Baek, K.-H.; Yun, H.K. Inhibiting wildfire and inducing defense-related gene expression by LED treatment on *Nicotiana benthamiana*. *J. Plant Pathol.* **2013**, *95*, 477–483. [CrossRef]
37. Kudo, R.; Ishida, Y.; Yamamoto, K. Effects of green light irradiation on induction of disease resistance in plants. *Acta Hortic.* **2011**, *907*, 251–254. [CrossRef]
38. Moyano, L.; Lopéz-Fernández, M.P.; Carrau, A.; Nannini, J.M.; Petrocelli, S.; Orellano, E.G.; Maldonado, S. Red light delays programmed cell death in non-host interaction between *Pseudomonas syringae* pv tomato DC3000 and tobacco plants. *Plant Sci.* **2020**, *291*, 110361. [CrossRef]
39. Mur, L.A.; Aubry, S.; Mondhe, M.; Kingston-Smith, A.; Gallagher, J.; Timms-Taravella, E.; James, C.; Papp, I.; Hörtensteiner, S.; Thomas, H.; et al. Accumulation of chlorophyll catabolites photosensitizes the hypersensitive response elicited by *Pseudomonas syringae* in Arabidopsis. *New Phytol.* **2010**, *188*, 161–174. [CrossRef]
40. Ambastha, V.; Tripathy, B.C.; Tiwari, B.S. Programmed cell death in plants: A chloroplastic connection. *Plant Signal. Behav.* **2015**, *10*, e989752. [CrossRef]
41. Shen, Q.; Naqvi, N.I. The Ferroptosis landscape of biotic interactions in plants. *Curr. Opin. Plant Biol.* **2024**, *77*, 102499. [CrossRef]
42. Manivannan, A.; Soundararajan, P.; Halimah, N.; Ko, C.H.; Jeong, B.R. Blue LED light enhances growth, phytochemical contents, and antioxidant enzyme activities of *Rehmannia glutinosa* cultured in vitro. *Hortic. Environ. Biotechnol.* **2015**, *56*, 105–113. [CrossRef]
43. Setiawan, C.K.; Supriyadi, S.; Santoso, U.; Ma, G.; Kato, M. Effect of light-emitting diode (Led) light on the gene expression related with ascorbate biosynthesis and metabolism in broccoli florets. *KnE Life Sci.* **2017**, *2*, 529. [CrossRef]

44. Kook, H.-S.; Park, S.-H.; Jang, Y.-J.; Lee, G.; Kim, J.S.; Kim, H.; Oh, B.; Chae, J.; Lee, K.-J. Blue LED (light-emitting diodes)-mediated growth promotion and control of *Botrytis* disease in lettuce. *Acta Agric. Scand. Sect. B Soil Plant Sci.* **2013**, *63*, 271–277. [CrossRef]
45. Kim, K.; Kook, H.; Jang, J.; Lee, W.; Kamala-Kannan, S.; Chae, J.-C.; Lee, K.-J. The effect of blue-light-emitting diodes on antioxidant properties and resistance to *Botrytis cinerea* in tomato. *J. Plant Pathol. Microb.* **2013**, *4*, 203. [CrossRef]
46. Mou, Z.; Fan, W.; Dong, X. Inducers of plant systemic acquired resistance regulate NPR1 function through redox changes. *Cell* **2003**, *113*, 935–944. [CrossRef] [PubMed]
47. dos Santos, C.; Franco, O.L. Pathogenesis-related proteins (PRs) with enzyme activity activating plant defense responses. *Plants* **2023**, *12*, 2226. [CrossRef]
48. Nagendran, R.; Lee, Y.H. Green and red light reduces the disease severity by *Pseudomonas cichorii* JBC1 in tomato plants via upregulation of defense-related gene expression. *Phytopathology* **2015**, *105*, 412–418. [CrossRef]
49. Tran, B.Q.; Jung, S. Modulation of chloroplast components and defense responses during programmed cell death in tobacco infected with *Pseudomonas syringae*. *Biochem. Biophys. Res. Commun.* **2020**, *528*, 753–759. [CrossRef]
50. Nguyen, A.T.; Tran, L.H.; Jung, S. Salt stress-induced modulation of porphyrin biosynthesis, photoprotection, and antioxidant properties in rice plants (*Oryza sativa*). *Antioxidants* **2023**, *12*, 1618. [CrossRef]
51. Tran, L.H.; Kim, J.-G.; Jung, S. Expression of the Arabidopsis Mg-chelatase H subunit alleviates iron deficiency-induced stress in transgenic rice. *Front. Plant Sci.* **2023**, *14*, 1098808. [CrossRef]
52. Xiaoying, L.; Shirong, G.; Taotao, C.; Zhigang, X.; Tezuka, T. Regulation of the growth and photosynthesis of cherry tomato seedlings by different light irradiations of light emitting diodes (LED). *Afr. J. Biotechnol.* **2012**, *11*, 6169–6177. [CrossRef]
53. Chen, Y.; Liu, W. Substituting green light for partial red light promoted the growth and quality, and regulated the nitrogen metabolism of *Medicago sativa* grown under red-blue LEDs. *Environ. Exp. Bot.* **2024**, *220*, 105623. [CrossRef]
54. Thomas, H.; Ougham, H. The stay-green trait. *J. Exp. Bot.* **2014**, *65*, 3889–3900. [CrossRef]
55. Galvez-Valdivieso, G.; Mullineaux, P.M. The role of reactive oxygen species in signalling from chloroplasts to the nucleus. *Physiol Plant.* **2010**, *138*, 430–439. [CrossRef] [PubMed]
56. Karpiński, S.; Szechyńska-Hebda, M.; Wituszyńska, W.; Burdiak, P. Light acclimation, retrograde signalling, cell death and immune defences in plants. *Plant Cell Environ.* **2013**, *36*, 736–744. [CrossRef] [PubMed]
57. Miao, Y.-X.; Wang, X.-Z.; Gao, L.-H.; Chen, Q.-Y.; Qu, M. Blue light is more essential than red light for maintaining the activities of photosystem II and I and photosynthetic electron transport capacity in cucumber leaves. *J. Integr. Agric.* **2016**, *15*, 87–100. [CrossRef]
58. Zheng, L.; Steppe, K.; Van Labeke, M.-C. Spectral quality of monochromatic LED affects photosynthetic acclimation to high-intensity sunlight of Chrysanthemum and Spathiphyllum. *Physiol. Plant.* **2020**, *169*, 10–26. [CrossRef]
59. Terashima, I.; Fujita, T.; Inoue, T.; Chow, W.S.; Oguchi, R. Green light drives leaf photosynthesis more efficiently than red light in strong white light: Revisiting the enigmatic question of why leaves are green. *Plant Cell Physiol.* **2009**, *50*, 684–697. [CrossRef]
60. Acosta-Motos, J.R.; Diaz-Vivancos, P.; Álvarez, S.; Fernández-García, N.; Sánchez-Blanco, M.J.; Hernández, J.A. NaCl-induced physiological and biochemical adaptative mechanisms in the ornamental *Myrtus communis* L. plants. *J. Plant Physiol.* **2015**, *183*, 41–51. [CrossRef]

antioxidants

MDPI

Article

Phosphorus Supplementation Enhances Growth and Antioxidant Defense Against Cadmium Stress in Cotton

Asif Iqbal [1,2,3,†], Huiping Gui [1,†], Cangsong Zheng [1], Xiangru Wang [1], Xiling Zhang [1], Meizhen Song [1,2,*] and Xiaoyan Ma [1,*]

1 Institute of Cotton Research of Chinese Academy of Agricultural Sciences/Zhengzhou Research Base, State Key Laboratory of Cotton Bio-Breeding and Integrated Utilization, School of Agricultural Sciences, Zhengzhou University, Anyang 455000, China; asif173aup@gmail.com (A.I.); huiping.828@163.com (H.G.); zhengcangsong@163.com (C.Z.); wxr_z4317@163.com (X.W.); hainan1571@163.com (X.Z.)

2 Xinjiang Key Laboratory of Crop Gene Editing and Germplasm Innovation, Institute of Western Agricultural of CAAS, Changji 831100, China

3 Department of Agriculture, Hazara University, Khyber Pakhtunkhwa, Mansehra 21120, Pakistan

* Correspondence: songmeizhencaas@163.com (M.S.); maxy_caas@126.com (X.M.); Tel.: +86-15837223836 (X.M.)

† These authors contributed equally to this work.

Abstract: Cadmium (Cd) contamination in agricultural soils is increasing due to anthropogenic activities, posing a significant threat to plant growth and productivity. Phosphorus (P) has been suggested as a potential mitigator of Cd toxicity, yet the role of cotton genotypes with contrasting low-P tolerance in contaminated soils remains largely unexplored. A hydroponic experiment was conducted to assess the effects of Cd stress (5 μM) on Jimian169 (strong-low-P tolerant) and DES926 (weak-low-P tolerant) cotton genotypes under low-P (0.01 mM KH_2PO_4) and normal P (1 mM KH_2PO_4) conditions. The results revealed that Cd stress, especially under low-P, significantly reduced plant growth, dry matter, photosynthetic rate, and P use efficiency (PUE), while increasing oxidative damage through increased malonaldehyde levels and reactive oxygen species accumulation. These adverse impacts were very much evident in DES926 compared to Jimian169. In contrast, Jimian169 demonstrated greater resilience to Cd stress by mitigating oxidative damage through enhanced antioxidant enzyme activity, improved photosynthetic performance, and increased accumulation of osmoprotectants. These findings indicate that Jimian169 can better withstand Cd toxicity by enhancing photosynthesis, antioxidant defense mechanisms, and osmotic adjustment. This makes them a promising candidate for cultivation in Cd-contaminated, P-deficient soils.

Keywords: cotton; phosphorus; cadmium stress; reactive oxygen species; antioxidant system; osmotic adjustment

1. Introduction

Cadmium (Cd) is recognized as a highly toxic heavy metal with significant environmental and agricultural implications due to its high solubility in water and rapid uptake by plants [1]. It is one of the most readily translocated heavy metals within the plant tissues, leading to severe physiological disruptions [2]. Cd accumulation inhibits cell division in meristematic tissues [3], disrupts photosynthesis, and interferes with key metabolic processes, leading to excessive production of reactive oxygen species (ROS) and resulting in oxidative stress [4]. These detrimental effects necessitate a robust defense system to mitigate oxidative damage and maintain cellular homeostasis.

In order to protect their photosynthetic structures from the negative impact of ROS and photoinhibition, plants have developed a variety of approaches. One of these approaches is the adjustment of the light-harvesting antenna size, which diminishes light absorption when CO_2 assimilation is insufficient [5]. Moreover, both the photochemical and the non-photochemical quenching mechanisms efficiently dissipate excess excitation energy from the photosynthetic process, thereby preventing the accumulation of ROS [6]. The detoxification of ROS and maintenance of cellular redox balance are facilitated by enzymatic antioxidants, such as superoxide dismutase (SOD), peroxidase (POD), catalase (CAT), as well as non-enzymatic antioxidant systems including glutathione, ascorbate, and carotenoids [7]. SOD catalyzes the dismutation of superoxide radicals (O_2^-), while POD and CAT scavenge hydrogen peroxide (H_2O_2) in mitochondria, collectively mitigating oxidative stress [8].

Cd, a redox-inactive heavy metal induces the generation of ROS, including O_2^-, H_2O_2 and hydroxyl radical [9]. In chloroplast, O_2^- is primarily produced due to electron leakage at photosystem II (PSII), either at the acceptor side or through incomplete water oxidation on the donor side [10]. Cd disrupts PSII functions by replacing calcium in the oxygen-evolving complex and impairing electron transport between the primary quinone acceptor and secondary quinone acceptor [11]. Although O_2^- itself causes minimal damage, it acts as a precursor for more reactive species, such as singlet oxygen ($1O_2$) and OH, which can lead to lipid peroxidation and protein disfunction. To counteract this, SOD plays a critical role in ROS scavenging by converting O_2^- into less harmful molecules like H_2O_2, making it a key player in ROS scavenging [12].

Phosphorus (P) is a key major nutrient that regulates several metabolic processes, including leaf photosynthesis, carbohydrate metabolism, and energy transfer [13]. Besides its agronomic significance, P application in soil is gaining attention for its role in altering heavy metals bioavailability [14]. P enhances soil pH, leading to the immobilization of heavy metals such as Cd, thus mitigating their toxic effects on plants [15]. Additionally, P plays a vital role in reinforcing antioxidant defenses during abiotic stress conditions [16]. Previously, it was reported that P supplementation alleviates the toxicity of Cd in wheat [17], copper toxicity in soybean [18], and aluminum in eucalyptus [19] and Camellia oleifera [20]. Under Cd stress, P enhances chlorophyll contents and activates antioxidant enzymes, thereby improving tolerance in wheat [17]. Likewise, P mitigates toxicity of manganese (Mn) in soybean [21], potato [22], and pasture species such as perennial ryegrass and white clover [23]. High P doses also enhance photosynthetic efficiency and reduce ROS production under Mn toxicity [24], suggesting its broader role in mitigating oxidative stress.

Despite these benefits, P supplementation can enhance heavy metals accumulation like Cd in plants [25]. Studies indicate that P enhances Cd retention in cell walls by increasing cell wall pectin content in mangrove roots under Cd exposure [26]. P also facilitates Cd sequestration by forming Cd-phosphate complexes, that are subsequently deposited in vacuoles and root surfaces [27]. Additionally, P plays a critical role in antioxidant defenses by stimulating the synthesis of radical scavengers such as proline, and various antioxidants both enzymatic and non-enzymatic [25]. P also contributes to carbohydrate synthesis, which supports antioxidative processes [28]. Research on *Myracrodruom urundeuva* reveals that P boosts glutathione biosynthesis, thereby enhancing enzymatic antioxidant activity [29]. In poplar, P deficiency inhibits the ascorbate-glutathione cycle, leading to elevated oxidative stress [25]. Moreover, P reduces malonaldehyde levels and alleviates oxidative damage in Cd-exposed mangrove seedlings by promoting antioxidants and phytochelatin production [30].

Based on the above discussion, this study was designed to investigate whether exogenous P supplementation enhances Cd tolerance in cotton, a globally important cash crop. Specifically, the study investigates how P influences growth, photosynthesis, phosphorus use efficiency, and antioxidant enzyme activities under Cd stress. To explore this, a hydroponic experiment was conducted using two Cd concentrations (0, and 5 μM) with and without 1 mM of P supplementation. We hypothesized that P application, along with low-P tolerant cotton genotypes, can enhance Cd stress tolerance by improving plant morphology and physiology such as photosynthesis, osmotic regulation, and antioxidant system. Consequently, the current study was designed to assess the responses of cotton genotypes and their capacity to mitigate Cd stress with P supplementation.

2. Materials and Methods

2.1. Plant Materials and Growth Conditions

A hydroponic experiment was carried out to assess the impact of Cd stress on Jimian169 (strong-low-P tolerant) and DES926 (weak-low-P tolerant) under low and normal P [31]. The seeds for these genotypes were obtained from the Cotton Research Institute of Chinese Academy of Agricultural Sciences, with official authorization. To ensure uniform germination, healthy seeds were initially sown in sterilized sand and kept under controlled conditions for one week. Once germinated, seedlings with uniform growth were transplanted into 7 L plastic containers filled with a nutrient solution. The plants were maintained under a regulated environment with a 16 h light and 8 h dark cycle, a temperature of 28 °C, and a relative humidity of 60%. Initially, a half-strength solution for the first week was applied, which was then replaced with a full-strength Hoagland solution for the remainder of the experiment [32,33]. Upon reaching the three-leaf stage, the seedlings were categorized into four treatment groups:

(i) LP + Cd (0.01 mM KH_2PO_4 + 5 μM $CdCl_2$)
(ii) LP + CK (0.01 mM KH_2PO_4 + 0 μM $CdCl_2$)
(iii) NP + Cd (1 mM KH_2PO_4 + 5 μM $CdCl_2$)
(iv) NP + CK (1 mM KH_2PO_4 + 0 μM $CdCl_2$)

To ensure proper aeration, an electric pump was used, and nutrient solutions were refreshed weekly. Additionally, to eliminate position bias, the placement of the containers was rotated during solution renewal. After two weeks of treatment, noticeable morphological differences emerged between the genotypes which were then harvested to study various morphological and physiological traits.

2.2. Growth Measurements and Root Analysis

Randomly, six plants were selected from each treatment to assess shoot length using the scale [34]. At the end, plants were harvested, and the roots and shoots were separately dried at 105 and 80 °C for 1 and 48 h, respectively. The root, shoot, and total biomass were then measured using an electric balance. To evaluate root morphology, root samples were scanned through EPSON and the analysis was performed in WinRhizo [32].

2.3. Leaf Physiological Assessments

Leaf gaseous exchange parameters that included net photosynthetic rate, transpiration rate, stomatal conductance, and intercellular CO_2 concentration were recorded from the third fully expanded leaf by using Li-Cor 6800 from morning 9–11 AM [35]. Similarly, leaf chlorophyll and carotenoid contents were measured using the protocol mentioned in our previous study [36].

2.4. Phosphorus Concentration and Use Efficiency

The root and shoot P concentration was quantified by the Kjeldahl method [37]. For this analysis, amounts of 0.2 g of ground samples were digested in a mixture of sulphuric acid and hydrogen peroxide and the resulting P concentrations were determined using Auto Analyzer III. The PUE traits were then calculated using the formulas highlighted in the previous study [38].

2.5. Malonaldehyde and Reactive Oxygen Species Determination

Root and shoot malondialdehyde concentrations were evaluated by thiobarbituric acid (TBA) reactivity [39]. A 0.2 g sample was extracted in 2 mL (0.25% TBA) dissolved in 10% trichloroacetic acid (TCA) and incubated at 95 °C for 30 min. The homogenate was centrifuged at 10,000× *g* for 10 min, and the data were calculated at 532 nm [40].

Hydrogen peroxide (H_2O_2) concentration was determined using the standard protocol [41]. Around 0.1 g of tissue samples were each mixed in 5 mL of 0.1% TCA and centrifuged for 15 min at 12,000× *g*. The amount of H_2O_2 was then measured spectrophotometrically at 390 nm following the standard protocol [31].

Superoxide (O_2^-) concentration was measured using a protocol adapted from the previous study [41]. Briefly, a 0.2 g sample was mixed in 1 mL of 65 mM phosphate buffer (pH 7.8) and then centrifuged at 10,000× *g* for 10 min. The absorbance of the final solution was calculated at 450 nm using spectrophotometer.

2.6. Antioxidant Enzyme Activity Assays

To measure enzymatic activity, 0.5 g of root and shoot tissues were ground using liquid nitrogen and mixed in 10 mL of 50 mM sodium phosphate buffer containing 1% polyvinylpyrrodine, 0.2 mm per L EDTA, and 19 mmol per L magnesium chloride. The mixture was centrifuged and kept at 4 °C for subsequent enzyme analysis.

Peroxidase activity was assessed using a method adopted from our previous study [42]. Superoxide dismutase (SOD) activity was measured using a photochemical nitroblue tetrazolium (NBT) assay and was monitored at 560 nm. The CAT activity was determined by observing the reduction at 240 nm due to H_2O_2 degradation as mentioned in our earlier study [43].

2.7. Osmoprotectants Quantification

Free amino acids were quantified using the protocol of previously published work with some modifications [44]. The extraction buffer was composed of acetic acid/sodium acetate (pH 5.4), and the final data were collected at 580 nm via a spectrophotometer [42].

Total soluble protein content was measured following the method of Bradford [45] with bovine serum albumin as the standard and the absorbance was recorded at 595 nm [46].

Total soluble sugars were analyzed based on a modified protocol adopted from a previous study [47]. A 0.5 g sample was mixed in 3 mL of ethanol (90%), incubated at 70 °C and adjusted to a final volume of 25 mL with 90% ethanol. From this, 1 mL of the supernatant was mixed with 5 mL each of anthrone solution and sulfuric acid. Absorbance was measured at 485 nm, with glucose used as the standard [48].

Proline content was measured by mixing 0.1 g of the sample in 1 mL of sulfosalicylic acid solution. The reaction mixture included sample, glacial acetic acid, and ninhydrin each one with 0.25 mL, and final data were determined at 520 nm [49].

2.8. Statistical and Multivariate Analysis

A two-way ANOVA with a split-plot design was applied, where P and Cd were the main plot factors, and cotton genotypes were the subplot factors using Statistix 8.1 (Analytical Software, Tallahassee, FL, USA) software. Mean comparisons were performed using the least significant difference test at a 5% probability level. Graphs were generated with Graphpad Prism 8 (San Diego, CA, USA) and expressed as mean ± standard error based on three biological and technical replications. Principal component and correlation analysis were calculated in OriginPro (2015) (b9.2.214, OriginLab Corporation, Northampton, MA, USA).

3. Results

3.1. Plant Growth

Various growth traits were evaluated to assess the variation in morphological traits of cotton genotypes. Compared to the control, Cd stress significantly reduced shoot length, root dry weight, shoot dry weight, and total dry weight under low P condition. However, P supplementation alleviated the negative effects of Cd by improving growth parameters (Figure 1). Regardless of the treatment, Jimian169 consistently demonstrated greater shoot length, root dry weight, shoot dry weight, and total dry weight compared to DES926. Additionally, Cd stress negatively impacted root morphological traits, including root length, surface area, volume and diameter under both P conditions (Figure 2). The reductions were more pronounced in DES926, whereas Jimian169 maintained higher values for root morphological traits across both P conditions (Figure 2).

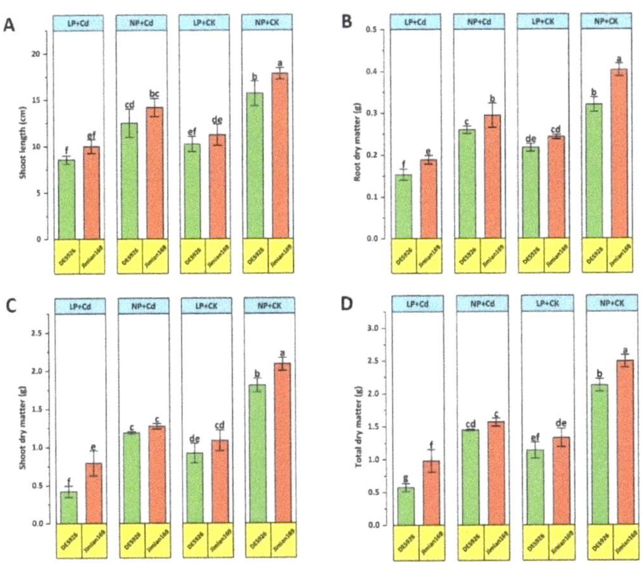

Figure 1. Effect of P supplementation on shoot length (**A**), root dry matter (**B**), shoot dry matter (**C**), and total dry matter (**D**) of Jimian169 and DES926 under Cd stress and control conditions. The values are presented as mean ± standard error. Error bars with different letters show significant differences ($p < 0.05$).

3.2. Leaf Chlorophyll and Gaseous Exchange

Cd stress significantly reduced leaf photosynthesis, transpiration, and stomatal conductance along with chlorophyll and carotenoid contents of contrasting low-P tolerant cotton genotypes under both P conditions, with the exception of intercellular CO_2 concentration

(Figures 3 and 4). Regardless of treatments, Jimian169 exhibited superior photosynthetic performance compared to DES926 (Figures 3 and 4). Under Cd stress, DES926 showed higher intercellular CO_2 concentration, indicating reduced carboxylation efficiency despite the availability of CO_2.

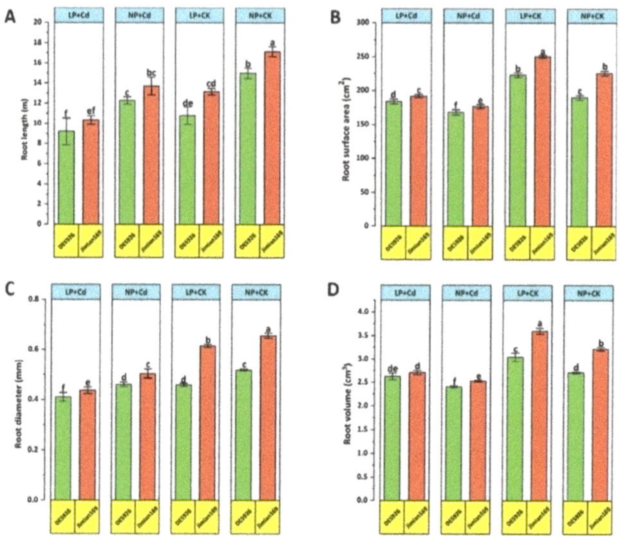

Figure 2. Effect of P supplementation on root length (**A**), root surface area (**B**), root diameter (**C**), and root volume (**D**) of Jimian169 and DES926 under Cd stress and control conditions. The values are presented as mean ± standard error. Error bars with different letters show significant differences ($p < 0.05$).

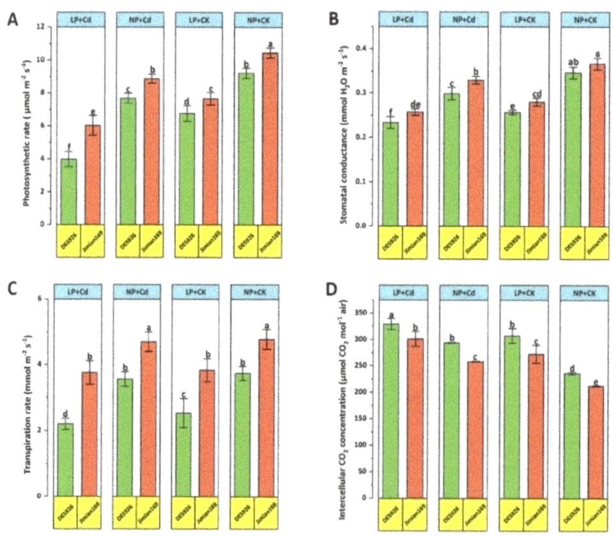

Figure 3. Effect of P supplementation on photosynthetic rate (**A**), stomatal conductance (**B**), transpiration rate (**C**), and intercellular CO_2 concentration (**D**) of Jimian169 and DES926 under Cd stress and control conditions. The values are presented as mean ± standard error. Error bars with different letters show significant differences ($p < 0.05$).

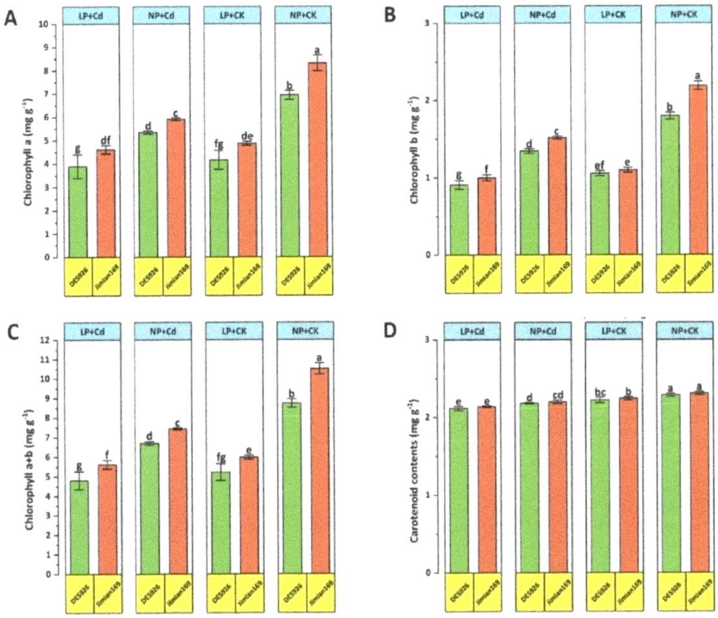

Figure 4. Effect of P supplementation on chlorophyll a (**A**), chlorophyll b (**B**), total chlorophyll (**C**), and carotenoid content (**D**) of Jimian169 and DES926 under Cd stress and control conditions. The values are presented as mean ± standard error. Error bars with different letters show significant differences ($p < 0.05$).

3.3. Phosphorus Concentration and Use Efficiency

Cd stress led to a significant reduction in root and shoot P concentration, particularly under low P conditions. However, Jimian169 maintained higher root and shoot P concentrations compared to DES926 (Figure 5A,B). A notable reduction in P accumulation of root and shoot was recorded in response to low and normal P relative to control. Jimian169 demonstrated greater root and shoot P accumulation than DES926 (Figure 5C,D). Furthermore, Cd stress reduced PUpE and PUtE, especially under low P conditions. Despite this, Jimian169 displayed significantly higher PUpE and PUtE than DES926 across both P conditions (Figure 5E,F).

3.4. Malondialdehyde Content and Reactive Oxygen Species

Cd stress markedly increased root and shoot malondialdehyde (MDA) content of cotton genotypes under low P. However, P supplementation significantly reduced MDA content, particularly in Jimian169 (Figure 6A,B). Regardless of the treatment, DES926 exhibited higher MDA content in roots and shoots compared to Jimian169 (Figure 6A,B). Similarly, Cd stress evaluated reactive oxygen species (ROS) levels, especially under low P conditions, whereas P supplementation effectively mitigated ROS accumulation, with Jimian169 showing the pronounced reduction. Compared to the control, Cd stress increased H_2O_2 concentration levels in roots and shoots under both low and normal P conditions with DES926 displaying higher H_2O_2 accumulation (Figure 6C,D). Similarly, Cd stress raised O_2^- concentration levels in roots and shoots under both P conditions, with DES926 consistently exhibiting higher O_2^- concentration levels than Jimain169 (Figure 6E,F).

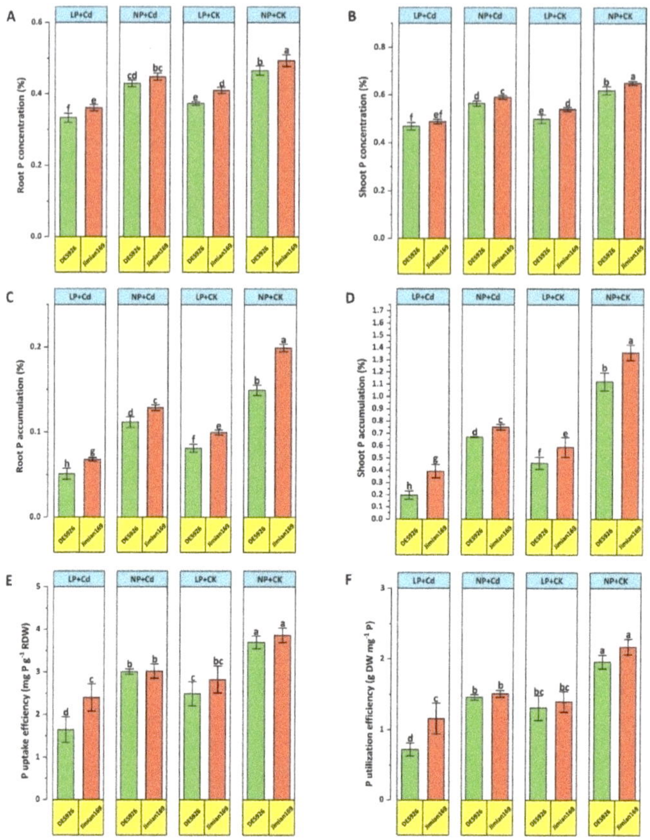

Figure 5. Effect of P supplementation on root P concentration (**A**), shoot P concentration (**B**), Root P accumulation (**C**), shoot P accumulation (**D**), P uptake efficiency (**E**), and P utilization efficiency (**F**) of Jimian169 and DES926 under Cd stress and control conditions. The values are presented as mean ± standard error. Error bars with different letters show significant differences ($p < 0.05$).

3.5. Antioxidant Enzymatic Activities

Antioxidant enzymatic activities were measured to assess the tolerance of cotton genotypes against Cd stress across P concentration levels (Figure 7). Cd stress significantly suppressed the activities of SOD, POD, and CAT in both roots and shoots, particularly under low P conditions compared to control (Figure 7). However, P supplementation enhanced the activities of these enzymes, especially in Jimian169 (Figure 7). Comparative analysis revealed that Jimian169 exhibited higher SOD, POD, and CAT activities in roots and shoots than DES926 (Figure 7).

3.6. Osmo-Protectants

The levels of osmo-protectants were also evaluated. Cd stress significantly reduced free amino acids in roots and shoots under low P conditions. However, Jimian169 maintained significantly higher free amino acid concentration levels compared to DES926 (Figure 8A,B). Total soluble protein and total soluble sugar contents of roots and shoots were greatly improved under Cd stress across both P conditions. Regardless of treatment, Jimian169 showed higher concentration levels of total soluble protein and more total soluble sugars than DES926 (Figure 8C–F).

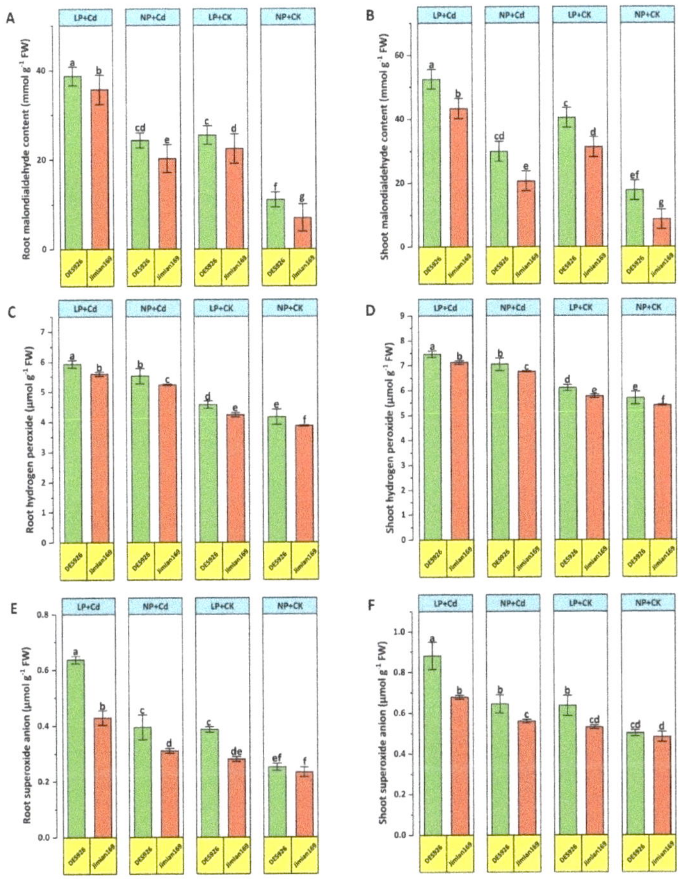

Figure 6. Effect of P supplementation on root MDA (**A**), shoot MDA (**B**), root H_2O_2 (**C**), shoot H_2O_2 (**D**), root O_2^- (**E**), and shoot O_2^- (**F**) of Jimian169 and DES926 under Cd stress and control conditions. The values are presented as mean ± standard error. Error bars with different letters show significant differences ($p < 0.05$).

3.7. Multivariate Analysis

The principal component analysis (PCA) effectively discriminated among samples based on the two principal components, PC1 and PC2, highlighting the influence of Cd stress, P conditions, and cotton genotypes on plant responses. PC1 accounted for a significant portion of the total variation (80.63%) and primarily separated the samples based on external stress factors, particularly Cd exposure and P availability. Traits with high loading on PC1 included key growth parameters, leaf gaseous exchange, PUE, and antioxidant enzyme activities, indicating that these variables are major determinants of the plants' response to Cd and P conditions. PC2, explaining 13.81% of the variation, provided additional discrimination among samples based on genotypic differences. Traits contributing strongly to PC2 included osmo-protectants and some antioxidant enzymes, suggesting that these characteristics are more genotype-dependent and play a secondary but notable role in stress adaptation. Overall, the PCA clearly separated treatments and genotypes, with PC1 representing environmental stress and nutrient factors, while PC2 reflected the inherent genetic variability response mechanism. The factor loading and scores together revealed that growth-related traits and physiological traits underlie the primary axis of variation, while biochemical stress markers define the secondary axis, offering insight into the differential tolerance mechanism among cotton genotypes (Figure 9 and Table S1).

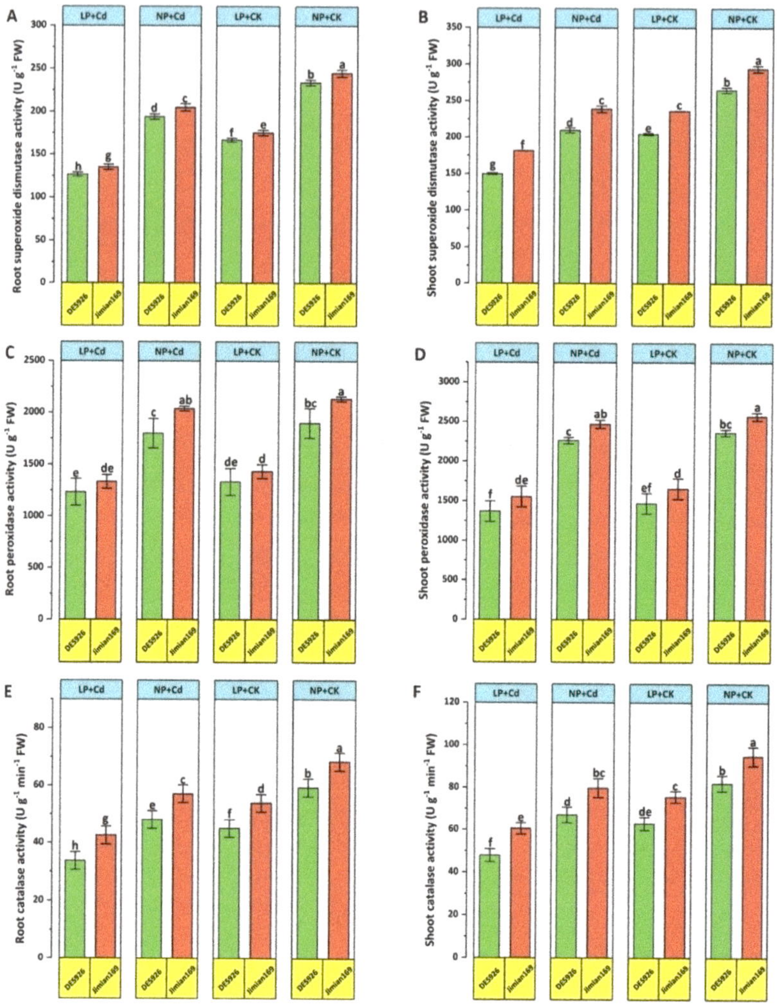

Figure 7. Effect of P supplementation on root SOD (**A**), shoot SOD (**B**), root POD (**C**), shoot POD (**D**), root CAT (**E**), and shoot CAT (**F**) of Jimian169 and DES926 under Cd stress and control conditions. The values are presented as mean ± standard error. Error bars with different letters show significant differences ($p < 0.05$).

The correlation between plant growth, leaf gaseous exchange (excluding intercellular CO_2 concentration), PUE traits, and antioxidant enzyme activities was strongly positive. Additionally, root and shoot growth, along with dry matter, showed strong positive correlations with most traits except for MDA, osmo-protectants, and ROS accumulation (Figure 10). Among root morphological traits, root length exhibited a strong positive correlation with PUE traits and antioxidant enzymes, highlighting the role of P in fostering an improved root system to enhance P uptake and tolerance to Cd stress (Figure 10).

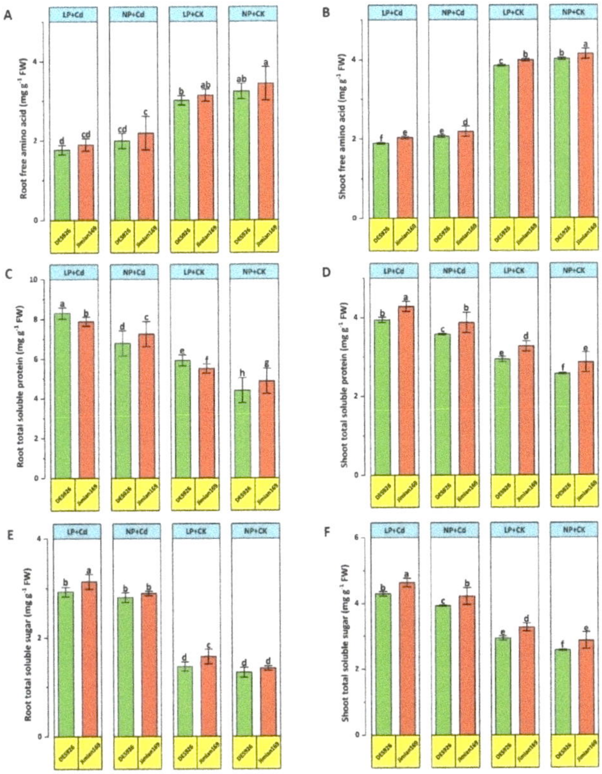

Figure 8. Effect of P supplementation on root-free amino acids (**A**), shoot-free amino acids (**B**), root total soluble protein (**C**), shoot total soluble protein (**D**), root total soluble sugar (**E**), and shoot total soluble sugars (**F**) of Jimain169 and DES926 under Cd stress and control conditions. The values are presented as mean ± standard error. Error bars with different letters show significant differences ($p < 0.05$).

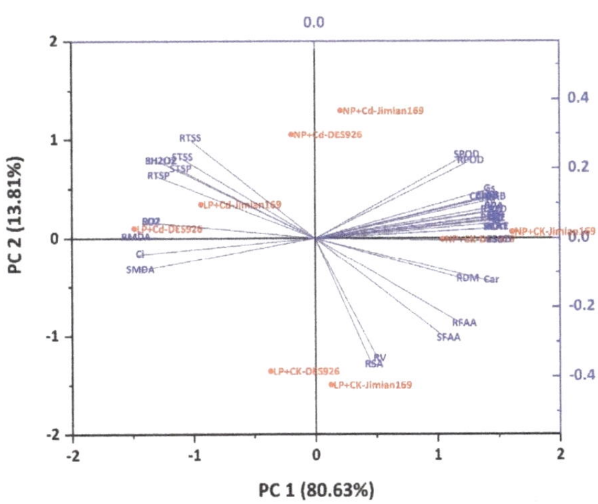

Figure 9. Principal component analysis (PCA) biplot of various studied traits of Jimian169 and DES926 under Cd stress and control supplemented with low and normal P conditions. SL; shoot

length, RDM; root dry matter, SDM; shoot dry matter, TDM; total dry matter, RL; root length, RSA; root surface area, RD; root diameter, RV; root volume, Pn; photosynthetic rate, gs; stomatal conductance, E; transpiration rate, Ci; intercellular CO_2 concentration, Chl a; chlorophyll a content, Chl b; chlorophyll b content, chlorophyll a + b, CAR; carotenoid contents, RP; root phosphorus concentrations, SP; shoot phosphorus concentration, RPA; root phosphorus accumulation, SPA; shoot phosphorus accumulation, PUpE; phosphorus uptake efficiency, PUtE; phosphorus utilization efficiency, RMDA; root malondialdehyde content, SMDA; shoot malondialdehyde content, RO_2; root superoxide anion, SO_2; shoot superoxide anion, RH_2O_2; root hydrogen peroxide, SH_2O_2; shoot hydrogen peroxide, RSOD; root superoxide dismutase, SSOD, shoot superoxide dismutase, RPOD; root peroxidase, SPOD; shoot peroxidase, RCAT; root catalase, SCAT; shoot catalase, RFAA; root free amino acids, SFAA; shoot free amino acids, RTSP; root total soluble proteins, STSP; shoot total soluble proteins, RTSS; root total soluble sugars, and STSS; shoot total soluble sugars.

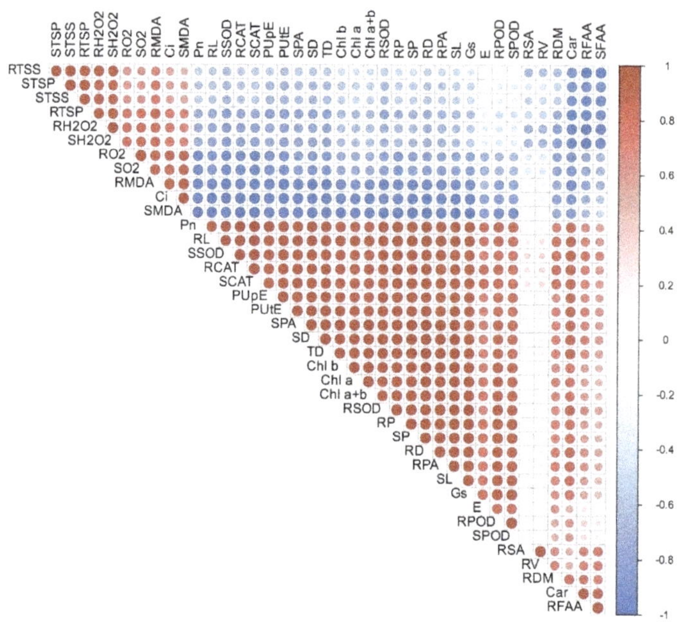

Figure 10. Relationships of various studied traits of Jimian169 and DES926 under Cd stress and control supplemented with low and normal P conditions. Where red color shows a positive correlation and blue color shows a negative correlation, while size of the circle shows the strength of correlation. SL; shoot length, RDM; root dry matter, SDM; shoot dry matter, TDM; total dry matter, RL; root length, RSA; root surface area, RD; root diameter, RV; root volume, Pn; photosynthetic rate, gs; stomatal conductance, E; transpiration rate, Ci; intercellular CO_2 concentration, Chl a; chlorophyll a content, Chl b; chlorophyll b content, chlorophyll a + b, CAR; carotenoid contents, RP; root phosphorus concentrations, SP; shoot phosphorus concentration, RPA; root phosphorus accumulation, SPA; shoot phosphorus accumulation, PUpE; phosphorus uptake efficiency, PUtE; phosphorus utilization efficiency, RMDA; root malondialdehyde content, SMDA; shoot malondialdehyde content, RO_2; root superoxide anion, SO_2; shoot superoxide anion, RH_2O_2; root hydrogen peroxide, SH_2O_2; shoot hydrogen peroxide, RSOD; root superoxide dismutase, SSOD, shoot superoxide dismutase, RPOD; root peroxidase, SPOD; shoot peroxidase, RCAT; root catalase, SCAT; shoot catalase, RFAA; root free amino acids, SFAA; shoot free amino acids, RTSP; root total soluble proteins, STSP; shoot total soluble proteins, RTSS; root total soluble sugars, and STSS; shoot total soluble sugars.

4. Discussion

4.1. Phosphorus Enhances Morphophysiological Tolerance of Cotton Genotypes Under Cadmium Stress

Exposure to heavy or trace elements can lead to toxic effects on plant growth parameters, although the severity of these effects depends on plant species, genotype and tissue type. These growth attributes serve a sensitive indicators for assessing plant responses to metal toxicity [50]. Among heavy metals, cadmium (Cd) is a major pollutant in agricultural soils worldwide, significantly reducing plant growth and productivity [51]. The extent of Cd toxicity in plants varies based on growth conditions, experimental setup, Cd availability, exposure duration, and plant age [52].

In this study, Cd stress (5 μM) markedly reduced root and shoot length, plant height, and dry matter accumulation in two cotton genotypes. The results indicated that Jimian169 exhibited greater growth and dry matter accumulation than DES926 (Figure 1), suggesting a higher tolerance to Cd stress. In contrast, DES926 had low values for such parameters, indicating greater sensitivity, particularly under low P conditions. The observed growth reduction under Cd stress may be attributed to decreased chlorophyll contents, impaired photosynthesis, and related metabolic disruptions, ultimately leading to reduced carbohydrate production and biomass accumulation. These findings align with previous studies that reported significant declines in root and shoot length, as well as dry matter accumulation, under Cd stress [53,54]. Similar reductions were found in *Phyllanthus amarus* grown under high Cd stress [55].

A key early indicator of Cd toxicity in plants is visual symptomology, particularly leaf chlorosis, which results from structural damage to chloroplast [56]. It is generally accepted that Cd concentrations exceeding 5–10 mg/kg in leaf tissue are toxic to plants [57]. In this study, chlorosis was more evident than necrosis in both genotypes, with symptoms intensifying in a dose-dependent manner. DES926 exhibited more severe chlorosis than Jimian169, further confirming its higher sensitivity to Cd stress. However, some degree of natural chlorosis and necrosis were also observed in control plants. Based on visual observations, Jimian169 appeared more tolerant to Cd than DES926. Previous studies suggest that Cd-induced chlorosis may result from disruptions in essential nutrient accumulation in aerial plant parts [58].

The genotypic differences observed in growth responses and visual symptoms under Cd stress highlight distinct tolerance mechanisms in cotton genotypes. These findings are consistent with reports of Cd-induced chlorosis in various plant species, including chickpea [59], *Sedum alfredii* [60], rice [61], and mungbean [62]. The differential responses observed in this study provide valuable insights into morphological and physiological adaptations of cotton genotypes under Cd toxicity.

4.2. Phosphorus Enhances Antioxidant Defense System of Cotton Genotypes Under Cadmium Stress

Enzymatic activities showed different responses against Cd stress. The initial increase in antioxidant enzyme activity under Cd stress may be attributed to the activation of the plant defense mechanism in response to metal stress [63]. These antioxidant enzymes play a crucial role in mitigating oxidative damage by reducing MDA contents under metal stress [64]. However, at higher stress, the reduction in the antioxidant system suggests a disruption in ROS homeostasis, leading to elevated concentration levels of H_2O_2, MDA, and electrolyte leakage. Similar trends in antioxidant enzyme activity and ROS production have been observed in various plants under Cd stress [65,66].

Phosphorus (P) supplementation further enhanced antioxidant enzyme activities, consequently reducing oxidative stress as evidenced by lower MDA and H_2O_2 concentration levels in both roots and shoots under Cd stress (Figure 6). This P-induced enhancement of the antioxidant defense system has been reported under various abiotic stress conditions [65]. The observed inhibition of MDA and H_2O_2 accumulation with P application highlights its role in mitigating Cd-induced stress in cotton plants. Furthermore, the enhanced antioxidant enzyme activity in P-treated plants may be associated with reduced Cd in plants. These findings suggest that P-mediated stimulation of the antioxidant defense system and suppression of ROS production may constitute an essential tolerance mechanism against Cd-induced oxidative stress in cotton.

Under stress conditions, plants accumulate MDA, a marker of lipid peroxidation, which significantly damages cell membranes [67]. Thus, MDA serves as a reliable indicator of oxidative stress. In the current study, Cd stress led to elevated MDA concentration levels and antioxidant enzyme activities (Figures 6 and 7). The increase in MDA content in Cd-stressed plants was likely due to excessive ROS accumulation. Previous studies have reported that abiotic stress disrupts ROS equilibrium, leading to enhanced photooxidative damage to the photosynthetic apparatus and membrane lipid peroxidation [68]. To maintain normal photosynthesis, plants must regulate ROS concentration levels [69] and maintain redox equilibrium in the cells [70]. One of the most critical mechanisms to achieve this balance is the antioxidant defense system [71].

In this study, cotton genotype Jimian169 exhibited a strong capacity to counter oxidative stress damage (Figure 6). Elevated antioxidant defense system and reduced ROS and MDA accumulation under normal P indicate that Jimian169 possesses a robust redox defense system against stress. Similarly, nitrogen application has been shown to enhance maize's soluble protein and antioxidant enzyme activities under stress [72]. Thus, while Cd stress regulates antioxidant enzyme activity, its effectiveness largely depends on the presence of P.

Excessive ROS production and accumulation can lead to severe cellular damage resulting in cell death [7]. In this study, Cd-stressed plants exhibited significantly higher ROS concentration levels compared to control plants. Furthermore, combined low P and Cd stress exacerbated ROS and MDA accumulation, indicating that P application plays a crucial role in suppressing ROS production under Cd stress. The observed increase in ROS levels under Cd stress may be associated with reduced photosynthetic activity (Figure 3), consistent with previous findings [73]. To counteract oxidative stress, plants activate antioxidant defense mechanisms that help alleviate ROS-induced damage. Notably, Jimian169 exhibited higher antioxidant enzyme activities under Cd stress, suggesting a more effective antioxidant defense system compared to DES926 genotype. These findings indicate that DES926 has a weaker antioxidant response to Cd stress, whereas Jimian169 demonstrates a stronger capability to mitigate oxidative damage.

In response to Cd stress, plants accumulate various osmolytes to enhance osmotic potential, thereby maintaining cellular turgidity and physiological processes [74]. Cd stress also affects osmo-protectants, including free amino acids, total soluble proteins, and soluble sugars, which are critical for osmotic adjustment under stress conditions. Cotton seedlings modulate these osmo-protectants to maintain cellular osmoregulation, an essential trait for stress tolerance [49]. Changes in free amino acid levels in response to stress show its production, acting as a stress indicator [75]. Among these osmo-protectants, amino acids play a crucial role in sustaining cell turgidity and preventing dehydration [76]. The higher concentration levels of osmo-protectants in Jimian169 tissues contributed to osmotic adjustment, reducing osmotic potential and maintaining cell turgidity. These findings suggest that Jimian169 has a superior capacity to maintain free amino acid levels, thereby enhancing Cd stress tolerance.

Soluble proteins and sugars play a crucial role as cellular components that enhance plant resilience to stress [77]. Research has consistently shown that various plant species accumulate osmo-protectants when exposed to stressful conditions [36,78,79]. Consistent with these findings, our study revealed that Jimian169 seedlings subjected to Cd stress exhibited a significant increase in osmo-protectant production and accumulation. This response contributed to improved osmotic potential, sustained cell turgidity, and reduced tissue dehydration. As a result, Jimian169 demonstrates a more efficient osmo-protective mechanism, enhancing its tolerance to Cd stress.

5. Conclusions

In conclusion, Cd-induced stress had a profound impact on the growth and metabolic processes of contrasting low-P tolerant genotypes, as evidenced by reduced dry matter accumulation, photosynthetic activity, chlorophyll content, PUE, and antioxidant enzyme activity, along with an increase in MDA concentration levels. However, Jimian169 exhibited greater tolerance to Cd stress across both P conditions compared to DES926. Its enhanced growth, well-developed root system, higher photosynthetic efficiency, improved PUE, stronger antioxidant defense, and greater osmo-protectant accumulation contributed to its superior stress resilience. These findings highlight the critical role of P in alleviating Cd-induced toxicity and suggest that Jimian169 is a more suitable genotype for cultivation in Cd-contaminated soils with normal P.

Supplementary Materials: The following supporting information can be downloaded at https://www.mdpi.com/article/10.3390/antiox14060686/s1, Table S1. Principal component analysis (PCA) biplot of various studied traits of Jimian169 and DES926 under LP + Cd (0.01 mM KH_2PO_4 + 0.5 mM $CdCl_2$), LP + CK (0.01 mM KH_2PO_4 + 0 mM $CdCl_2$), NP + Cd (1 mM KH_2PO_4 + 0.5 mM $CdCl_2$), NP + CK (1 mM KH_2PO_4 + 0 mM $CdCl_2$). SL; shoot length, RDM; root dry matter, SDM; shoot dry matter, TDM; total dry matter, RL; root length, RSA; root surface area, RD; root diameter, RV; root volume, Pn; photosynthetic rate, gs; stomatal conductance, E; transpiration rate, Ci; intercellular CO_2 concentration, Chl a; chlorophyll a content, Chl b; chlorophyll b content, chlorophyll a + b, CAR; carotenoid contents, RP; root phosphorus concentrations, SP; shoot phosphorus concentration, RPA; root phosphorus accumulation, SPA; shoot phosphorus accumulation, PUpE; phosphorus uptake efficiency, PUtE; phosphorus utilization efficiency, RMDA; root malondialdehyde content, SMDA; shoot malondialdehyde content, RO_2; root superoxide anion, SO_2; shoot superoxide anion, RH_2O_2; root hydrogen peroxide, SH_2O_2; shoot hydrogen peroxide, RSOD; root superoxide dismutase, SSOD, shoot superoxide dismutase, RPOD; root peroxidase, SPOD; shoot peroxidase, RCAT; root catalase, SCAT; shoot catalase, RFAA; root free amino acids, SFAA; shoot free amino acids, RTSP; root total soluble proteins, STSP; shoot total soluble proteins, RTSS; root total soluble sugars, and STSS; shoot total soluble sugars.

Author Contributions: Conceptualization, X.M. and A.I.; methodology, H.G. and C.Z.; software, H.G.; validation, A.I. and H.G.; formal analysis, X.W. and C.Z.; investigation, X.Z.; resources, H.G.; data curation, A.I.; Writing—original draft, A.I.; review and editing, X.W.; visualization, H.G. and C.Z.; supervision, X.Z. and X.M.; project administration, X.M. and M.S.; funding acquisition, X.M. All authors have read and agreed to the published version of the manuscript.

Funding: We greatly appreciate the financial support from the Xinjiang Uygur Autonomous Region Tianchi Talent Introduction Program; Xinjiang Key Laboratory of Crop Gene Editing and Germplasm Innovation, Institute of Western Agricultural of CAAS, Changji, 831100, Xinjiang, China; and Xinjiang Science and Technology Major Project of China (2024A02002).

Institutional Review Board Statement: Not applicable.

Informed Consent Statement: Not applicable.

Data Availability Statement: All data generated or analyzed during this study are included in this published article.

Acknowledgments: We are thankful to the Cotton Research Institute of Chinese Academy of Agricultural Sciences for providing the seeds for this experiment.

Conflicts of Interest: The authors declare no conflicts of interest.

References

1. El Rasafi, T.; Oukarroum, A.; Haddioui, A.; Song, H.; Kwon, E.E.; Bolan, N.; Tack, F.M.G.; Sebastian, A.; Prasad, M.N.V.; Rinklebe, J. Cadmium stress in plants: A critical review of the effects, mechanisms, and tolerance strategies. *Crit. Rev. Environ. Sci. Technol.* **2022**, *52*, 675–726. [CrossRef]
2. Thakur, S.; Singh, L.; Wahid, Z.A.; Siddiqui, M.F.; Atnaw, S.M.; Din, M.F.M. Plant-driven removal of heavy metals from soil: Uptake, translocation, tolerance mechanism, challenges, and future perspectives. *Environ. Monit. Assess.* **2016**, *188*, 206. [CrossRef] [PubMed]
3. Bertels, J.; Huybrechts, M.; Hendrix, S.; Bervoets, L.; Cuypers, A.; Beemster, G.T.S. Cadmium inhibits cell cycle progression and specifically accumulates in the maize leaf meristem. *J. Exp. Bot.* **2020**, *71*, 6418–6428. [CrossRef] [PubMed]
4. Adil, M.F.; Sehar, S.; Han, Z.; Lwalaba, J.L.W.; Jilani, G.; Zeng, F.; Chen, Z.-H.; Shamsi, I.H. Zinc alleviates cadmium toxicity by modulating photosynthesis, ROS homeostasis, and cation flux kinetics in rice. *Environ. Pollut.* **2020**, *265*, 114979. [CrossRef]
5. Niyogi, K.K. Photoprotection revisited: Genetic and molecular approaches. *Annu. Rev. Plant Biol.* **1999**, *50*, 333–359. [CrossRef]
6. Takahashi, S.; Badger, M.R. Photoprotection in plants: A new light on photosystem II damage. *Trends Plant Sci.* **2011**, *16*, 53–60. [CrossRef]
7. Apel, K.; Hirt, H. Reactive oxygen species: Metabolism, oxidative stress, and signal transduction. *Annu. Rev. Plant Biol.* **2004**, *55*, 373–399. [CrossRef]
8. Shigeoka, S.; Ishikawa, T.; Tamoi, M.; Miyagawa, Y.; Takeda, T.; Yabuta, Y.; Yoshimura, K. Regulation and function of ascorbate peroxidase isoenzymes. *J. Exp. Bot.* **2002**, *53*, 1305–1319. [CrossRef]
9. Asad, S.A.; Farooq, M.; Afzal, A.; West, H. Integrated phytobial heavy metal remediation strategies for a sustainable clean environment-a review. *Chemosphere* **2019**, *217*, 925–941. [CrossRef]
10. Zhang, H.; Li, X.; Zhang, S.; Yin, Z.; Zhu, W.; Li, J.; Meng, L.; Zhong, H.; Xu, N.; Wu, Y. Rootstock alleviates salt stress in grafted mulberry seedlings: Physiological and PSII function responses. *Front. Plant Sci.* **2018**, *9*, 1806. [CrossRef]
11. Sharma, A.; Kumar, V.; Shahzad, B.; Ramakrishnan, M.; Singh Sidhu, G.P.; Bali, A.S.; Handa, N.; Kapoor, D.; Yadav, P.; Khanna, K. Photosynthetic response of plants under different abiotic stresses: A review. *J. Plant Growth Regul.* **2020**, *39*, 509–531. [CrossRef]
12. Gallie, D.R.; Chen, Z. Chloroplast-localized iron superoxide dismutases FSD2 and FSD3 are functionally distinct in Arabidopsis. *PLoS ONE* **2019**, *14*, e0220078. [CrossRef] [PubMed]
13. Kayoumu, M.; Li, X.; Iqbal, A.; Wang, X.; Gui, H.; Qi, Q.; Ruan, S.; Guo, R.; Dong, Q.; Zhang, X. Genetic variation in morphological traits in cotton and their roles in increasing phosphorus-use-efficiency in response to low phosphorus availability. *Front. Plant Sci.* **2022**, *13*, 1051080. [CrossRef] [PubMed]
14. Gupta, D.K.; Chatterjee, S.; Datta, S.; Veer, V.; Walther, C. Role of phosphate fertilizers in heavy metal uptake and detoxification of toxic metals. *Chemosphere* **2014**, *108*, 134–144. [CrossRef]
15. Seshadri, B.; Bolan, N.S.; Choppala, G.; Kunhikrishnan, A.; Sanderson, P.; Wang, H.; Currie, L.D.; Tsang, D.C.W.; Ok, Y.S.; Kim, G. Potential value of phosphate compounds in enhancing immobilization and reducing bioavailability of mixed heavy metal contaminants in shooting range soil. *Chemosphere* **2017**, *184*, 197–206. [CrossRef]
16. Saleem, M.H.; Ali, S.; Rehman, M.; Rana, M.S.; Rizwan, M.; Kamran, M.; Imran, M.; Riaz, M.; Soliman, M.H.; Elkelish, A. Influence of phosphorus on copper phytoextraction via modulating cellular organelles in two jute (*Corchorus capsularis* L.) varieties grown in a copper mining soil of Hubei Province, China. *Chemosphere* **2020**, *248*, 126032. [CrossRef]
17. Arshad, M.; Ali, S.; Noman, A.; Ali, Q.; Rizwan, M.; Farid, M.; Irshad, M.K. Phosphorus amendment decreased cadmium (Cd) uptake and ameliorates chlorophyll contents, gas exchange attributes, antioxidants, and mineral nutrients in wheat (*Triticum aestivum* L.) under Cd stress. *Arch. Agron. Soil Sci.* **2016**, *62*, 533–546. [CrossRef]
18. Zhang, W.; Zou, C.; Chen, X.; Liu, Y.; Liu, D.; Yang, H.; Deng, Y.; Chen, X. Phosphorus application decreased copper concentration but not iron in maize grain. *Agronomy* **2020**, *10*, 1716. [CrossRef]
19. Teng, W.; Kang, Y.; Hou, W.; Hu, H.; Luo, W.; Wei, J.; Wang, L.; Zhang, B. Phosphorus application reduces aluminum toxicity in two Eucalyptus clones by increasing its accumulation in roots and decreasing its content in leaves. *PLoS ONE* **2018**, *13*, e0190900. [CrossRef]
20. Zhou, J.-Q.; Ai, Z.-Z.; Wang, H.; Niu, G.-H.; Yuan, J. Phosphorus alleviates aluminum toxicity in *Camellia oleifera* seedlings. *Int. J. Agric. Biol.* **2019**, *21*, 237–243.

21. Nogueira, M.A.; Magalhães, G.C.; Cardoso, E.J.B.N. Manganese toxicity in mycorrhizal and phosphorus-fertilized soybean plants. *J. Plant Nutr.* **2004**, *27*, 141–156. [CrossRef]
22. Sarkar, D.; Pandey, S.K.; Sud, K.C.; Chanemougasoundharam, A. In vitro characterization of manganese toxicity in relation to phosphorus nutrition in potato (*Solanum tuberosum* L.). *Plant Sci.* **2004**, *167*, 977–986. [CrossRef]
23. Rosas, A.; Rengel, Z.; Ribera, A.; de La Luz Mora, M. Phosphorus nutrition alleviates manganese toxicity in *Lolium perenne* and Trifolium repens. *J. Plant Nutr. Soil Sci.* **2011**, *174*, 210–219. [CrossRef]
24. Berríos, G.A.; Escobar, A.L.; Alberdi, M.R.; Nunes-Nesi, A.; Reyes-Díaz, M.M. Manganese toxicity amelioration by phosphorus supply in contrasting Mn resistant genotypes of ryegrass. *Plant Physiol. Biochem.* **2019**, *144*, 144–156. [CrossRef] [PubMed]
25. Yang, C.; Qiu, W.; Chen, Z.; Chen, W.; Li, Y.; Zhu, J.; Rahman, S.U.; Han, Z.; Jiang, Y.; Yang, G. Phosphorus influence Cd phytoextraction in Populus stems via modulating xylem development, cell wall Cd storage and antioxidant defense. *Chemosphere* **2020**, *242*, 125154. [CrossRef]
26. Dai, M.; Liu, W.; Hong, H.; Lu, H.; Liu, J.; Jia, H.; Yan, C. Exogenous phosphorus enhances cadmium tolerance by affecting cell wall polysaccharides in two mangrove seedlings *Avicennia marina* (Forsk.) Vierh and *Kandelia obovata* (S., L.) Yong differing in cadmium accumulation. *Mar. Pollut. Bull.* **2018**, *126*, 86–92. [CrossRef]
27. Jiang, H.M.; Yang, J.C.; Zhang, J.F. Effects of external phosphorus on the cell ultrastructure and the chlorophyll content of maize under cadmium and zinc stress. *Environ. Pollut.* **2007**, *147*, 750–756. [CrossRef]
28. Gan, H.; Jiao, Y.; Jia, J.; Wang, X.; Li, H.; Shi, W.; Peng, C.; Polle, A.; Luo, Z.-B. Phosphorus and nitrogen physiology of two contrasting poplar genotypes when exposed to phosphorus and/or nitrogen starvation. *Tree Physiol.* **2016**, *36*, 22–38. [CrossRef]
29. Gomes, M.P.; Soares, A.M.; Garcia, Q.S. Phosphorous and sulfur nutrition modulate antioxidant defenses in *Myracrodruom urundeuva* plants exposed to arsenic. *J. Hazard. Mater.* **2014**, *276*, 97–104. [CrossRef]
30. Dai, M.; Lu, H.; Liu, W.; Jia, H.; Hong, H.; Liu, J.; Yan, C. Phosphorus mediation of cadmium stress in two mangrove seedlings *Avicennia marina* and *Kandelia* obovata differing in cadmium accumulation. *Ecotoxicol. Environ. Saf.* **2017**, *139*, 272–279. [CrossRef]
31. Iqbal, A.; Dong, Q.; Wang, X.; Gui, H.; Zhang, H.; Zhang, X.; Song, M. Low phosphorus tolerance in cotton genotypes is regulated by root morphology and physiology. *J. Plant Growth Regul.* **2023**, *42*, 3677–3695. [CrossRef]
32. Iqbal, A.; Gui, H.; Zhang, H.; Wang, X.; Pang, N.; Dong, Q.; Song, M. Genotypic variation in cotton genotypes for phosphorus-use efficiency. *Agronomy* **2019**, *9*, 689. [CrossRef]
33. Kayoumu, M.; Iqbal, A.; Muhammad, N.; Li, X.; Li, L.; Wang, X.; Gui, H.; Qi, Q.; Ruan, S.; Guo, R. Phosphorus availability affects the photosynthesis and antioxidant system of contrasting low-P-tolerant cotton genotypes. *Antioxidants* **2023**, *12*, 466. [CrossRef] [PubMed]
34. Iqbal, A.; Dong, Q.; Wang, Z.; Wang, X.; Gui, H.; Zhang, H.; Pang, N.; Zhang, X.; Song, M. Growth and nitrogen metabolism are associated with nitrogen-use efficiency in cotton genotypes. *Plant Physiol. Biochem.* **2020**, *149*, 61–74. [CrossRef]
35. Iqbal, A.; Dong, Q.; Wang, X.; Gui, H.P.; Zhang, H.; Pang, N.; Zhang, X.; Song, M. Nitrogen preference and genetic variation of cotton genotypes for nitrogen use efficiency. *J. Sci. Food Agric.* **2020**, *100*, 2761–2773. [CrossRef]
36. Iqbal, A.; Dong, Q.; Wang, X.; Gui, H.; Zhang, H.; Zhang, X.; Song, M. High nitrogen enhance drought tolerance in cotton through antioxidant enzymatic activities, nitrogen metabolism and osmotic adjustment. *Plants* **2020**, *9*, 178. [CrossRef]
37. Villar-Salvador, P.; Peñuelas, J.L.; Jacobs, D.F. Nitrogen nutrition and drought hardening exert opposite effects on the stress tolerance of *Pinus pinea* L. seedlings. *Tree Physiol.* **2013**, *33*, 221–232. [CrossRef]
38. Iqbal, A.; Dong, Q.; Wang, Z.; Gui, H.; Zhang, H.; Zhang, X.; Song, M. Genotypic variation in cotton genotypes for low phosphorus tolerance and efficiency under different growth conditions. *Gesunde Pflanz.* **2023**, *75*, 1975–1993. [CrossRef]
39. Heath, R.L.; Packer, L. Photoperoxidation in isolated chloroplasts: I. Kinetics and stoichiometry of fatty acid peroxidation. *Arch. Biochem. Biophys.* **1968**, *125*, 189–198. [CrossRef]
40. Iqbal, A.; Dong, Q.; Wang, X.; Gui, H.; Zhang, H.; Zhang, X.; Song, M. Phosphorus and carbohydrate metabolism contributes to low phosphorus tolerance in cotton. *BMC Plant Biol.* **2023**, *23*, 97. [CrossRef]
41. Zhang, Y.-J.; Zhang, X.; Chen, C.-J.; Zhou, M.-G.; Wang, H.-C. Effects of fungicides JS399-19, azoxystrobin, tebuconazloe, and carbendazim on the physiological and biochemical indices and grain yield of winter wheat. *Pestic. Biochem. Physiol.* **2010**, *98*, 151–157. [CrossRef]
42. Iqbal, A.; Jing, N.; Dong, Q.; Wang, Z.; Gui, H.; Zhang, H.; Pang, N.; Zhang, X.; Song, M. Physiological characteristics of cotton subtending leaf are associated with yield in contrasting nitrogen-efficient cotton genotypes. *Front. Plant Sci.* **2022**, *13*, 825116. [CrossRef] [PubMed]
43. Li, L.; Qi, Q.; Zhang, H.; Dong, Q.; Iqbal, A.; Gui, H.; Kayoumu, M.; Song, M.; Zhang, X.; Wang, X. Ameliorative Effects of Silicon against Salt Stress in *Gossypium hirsutum* L. *Antioxidants* **2022**, *11*, 1520. [CrossRef] [PubMed]
44. Yokoyama, S.; Hiramatsu, J.-I. A modified ninhydrin reagent using ascorbic acid instead of potassium cyanide. *J. Biosci. Bioeng.* **2003**, *95*, 204–205. [CrossRef]
45. Bradford, M.M. A rapid and sensitive method for the quantitation of microgram quantities of protein utilizing the principle of protein-dye binding. *Anal. Biochem.* **1976**, *72*, 248–254. [CrossRef]

46. Asif, I.; Dong, Q.; Wang, X.; Gui, H.; Zhang, H.; Pang, N.; Zhang, X.; Song, M. Genotypic variation in root morphology, cotton subtending leaf physiology and fiber quality against nitrogen. *J. Cotton Res.* **2021**, *4*, 32. [CrossRef]
47. Shields, R.; Burnett, W. Determination of protein-bound carbohydrate in serum by modified anthrone method. *Anal. Chem.* **1960**, *32*, 885–886. [CrossRef]
48. Iqbal, A.; Dong, Q.; Wang, Z.; Gui, H.; Niu, J.; Li, L.; Zhang, H.; Pang, N.; Zhang, X.; Song, M. N-efficient cotton genotype grown under low nitrogen shows relatively large root system, high biomass accumulation and nitrogen metabolism. *Agron. J.* **2022**, *114*, 582–600. [CrossRef]
49. Iqbal, A.; Gui, H.; Dong, Q.; Wang, X.; Zhang, H.; Zhang, X.; Song, M. Differential responses of contrasting low phosphorus tolerant cotton genotypes under low phosphorus and drought stress. *BMC Plant Biol.* **2023**, *23*, 168. [CrossRef]
50. Imtiaz, M.; Tu, S.; Xie, Z.; Han, D.; Ashraf, M.; Rizwan, M.S. Growth, V uptake, and antioxidant enzymes responses of chickpea (*Cicer arietinum* L.) genotypes under vanadium stress. *Plant Soil* **2015**, *390*, 17–27. [CrossRef]
51. Daud, M.K.; Ali, S.; Variath, M.T.; Zhu, S.J. Differential physiological, ultramorphological and metabolic responses of cotton cultivars under cadmium stress. *Chemosphere* **2013**, *93*, 2593–2602. [CrossRef] [PubMed]
52. Metwally, A.; Safronova, V.I.; Belimov, A.A.; Dietz, K.-J. Genotypic variation of the response to cadmium toxicity in *Pisum sativum* L. *J. Exp. Bot.* **2005**, *56*, 167–178. [CrossRef] [PubMed]
53. Afzal, J.; Hu, C.; Imtiaz, M.; Elyamine, A.M.; Rana, M.S.; Imran, M.; Farag, M.A. Cadmium tolerance in rice cultivars associated with antioxidant enzymes activities and Fe/Zn concentrations. *Int. J. Environ. Sci. Technol.* **2019**, *16*, 4241–4252. [CrossRef]
54. He, J.; Ren, Y.; Chen, X.; Chen, H. Protective roles of nitric oxide on seed germination and seedling growth of rice (*Oryza sativa* L.) under cadmium stress. *Ecotoxicol. Environ. Saf.* **2014**, *108*, 114–119. [CrossRef]
55. Rai, V.; Khatoon, S.; Bisht, S.S.; Mehrotra, S. Effect of cadmium on growth, ultramorphology of leaf and secondary metabolites of *Phyllanthus amarus* Schum. and Thonn. *Chemosphere* **2005**, *61*, 1644–1650. [CrossRef]
56. Chen, L.; Long, X.H.; Zhang, Z.H.; Zheng, X.T.; Rengel, Z.; Liu, Z.P. Cadmium accumulation and translocation in two Jerusalem artichoke (*Helianthus tuberosus* L.) cultivars. *Pedosphere* **2011**, *21*, 573–580. [CrossRef]
57. He, S.; Yang, X.; He, Z.; Virupax, C.B. Morphological and physiological responses of plants to cadmium toxicity: A review. *Pedosphere* **2017**, *27*, 421–438. [CrossRef]
58. Zhang, G.; Fukami, M.; Sekimoto, H. Influence of cadmium on mineral concentrations and yield components in wheat genotypes differing in Cd tolerance at seedling stage. *Field Crops Res.* **2002**, *77*, 93–98. [CrossRef]
59. Faizan, S.; Kausar, S.; Perveen, R. Varietal differences for cadmium-induced seedling mortality, foliar toxicity symptoms, plant growth, proline and nitrate reductase activity in chickpea (*Cicer arietinum* L.). *Biol. Med.* **2011**, *3*, 196–206.
60. Zhou, W.; Qiu, B. Effects of cadmium hyperaccumulation on physiological characteristics of *Sedum alfredii* Hance (Crassulaceae). *Plant Sci.* **2005**, *169*, 737–745. [CrossRef]
61. Adhikari, T.; Tel-Or, E.; Libal, Y.; Shenker, M. Effect of cadmium and iron on rice (*Oryza sativa* L.) plant in chelator-buffered nutrient solution. *J. Plant Nutr.* **2006**, *29*, 1919–1940. [CrossRef]
62. Wahid, A.; Ghani, A. Varietal differences in mungbean (*Vigna radiata*) for growth, yield, toxicity symptoms and cadmium accumulation. *Ann. Appl. Biol.* **2008**, *152*, 59–69. [CrossRef]
63. Adrees, M.; Ali, S.; Iqbal, M.; Bharwana, S.A.; Siddiqi, Z.; Farid, M.; Ali, Q.; Saeed, R.; Rizwan, M.J.E. Mannitol alleviates chromium toxicity in wheat plants in relation to growth, yield, stimulation of anti-oxidative enzymes, oxidative stress and Cr uptake in sand and soil media. *Ecotoxicol. Environ. Saf.* **2015**, *122*, 1–8. [CrossRef] [PubMed]
64. Farooq, M.A.; Ali, S.; Hameed, A.; Ishaque, W.; Mahmood, K.; Iqbal, Z. Alleviation of cadmium toxicity by silicon is related to elevated photosynthesis, antioxidant enzymes; suppressed cadmium uptake and oxidative stress in cotton. *Ecotoxicol. Environ Saf.* **2013**, *96*, 242–249. [CrossRef]
65. Saidi, I.; Ayouni, M.; Dhieb, A.; Chtourou, Y.; Chaïbi, W.; Djebali, W. Oxidative damages induced by short-term exposure to cadmium in bean plants: Protective role of salicylic acid. *S. Afr. J. Bot.* **2013**, *85*, 32–38. [CrossRef]
66. Ehsan, S.; Ali, S.; Noureen, S.; Mahmood, K.; Farid, M.; Ishaque, W.; Shakoor, M.B.; Rizwan, M. Citric acid assisted phytoremediation of cadmium by *Brassica napus* L. *Ecotoxicol. Environ. Saf.* **2014**, *106*, 164–172. [CrossRef]
67. Wang, L.-F. Physiological and molecular responses to drought stress in rubber tree (*Hevea brasiliensis* Muell. Arg.). *Plant Physiol. Biochem.* **2014**, *83*, 243–249. [CrossRef]
68. Wang, H.; Yang, Z.; Yu, Y.; Chen, S.; He, Z.; Wang, Y.; Jiang, L.; Wang, G.; Yang, C.; Liu, B. Drought enhances nitrogen uptake and assimilation in maize roots. *Agron. J.* **2017**, *109*, 39–46. [CrossRef]
69. Gonzalez-Dugo, V.; Durand, J.L.; Gastal, F.; Bariac, T.; Poincheval, J. Restricted root-to-shoot translocation and decreased sink size are responsible for limited nitrogen uptake in three grass species under water deficit. *Environ. Exp. Bot.* **2012**, *75*, 258–267 [CrossRef]
70. Sunil, B.; Talla, S.K.; Aswani, V.; Raghavendra, A.S. Optimization of photosynthesis by multiple metabolic pathways involving interorganelle interactions: Resource sharing and ROS maintenance as the bases. *Photosynth. Res.* **2013**, *117*, 61–71. [CrossRef]

1. Chang, Z.; Liu, Y.; Dong, H.; Teng, K.; Han, L.; Zhang, X. Effects of cytokinin and nitrogen on drought tolerance of creeping bentgrass. *PLoS ONE* **2016**, *11*, e0154005. [CrossRef] [PubMed]
2. Gill, S.S.; Anjum, N.A.; Gill, R.; Mahajan, M.; Tuteja, N. Abiotic stress tolerance and sustainable agriculture: A functional genomics perspective. In *Elucidation of Abiotic Stress Signaling in Plants: Functional Genomics Perspectives*; Springer: New York, NY, USA, 2015; Volume 2, pp. 439–472.
3. Cruz de Carvalho, M.H. Drought stress and reactive oxygen species: Production, scavenging and signaling. *Plant Signal. Behav.* **2008**, *3*, 156–165. [CrossRef] [PubMed]
4. Serraj, R.; Sinclair, T.R. Osmolyte accumulation: Can it really help increase crop yield under drought conditions? *Plant Cell Environ.* **2002**, *25*, 333–341. [CrossRef] [PubMed]
5. Thu Hoai, N.T.; Shim, I.S.; Kobayashi, K.; Kenji, U. Accumulation of some nitrogen compounds in response to salt stress and their relationships with salt tolerance in rice (*Oryza sativa* L.) seedlings. *Plant Growth Regul.* **2003**, *41*, 159–164. [CrossRef]
6. Kiani, S.P.; Talia, P.; Maury, P.; Grieu, P.; Heinz, R.; Perrault, A.; Nishinakamasu, V.; Hopp, E.; Gentzbittel, L.; Paniego, N. Genetic analysis of plant water status and osmotic adjustment in recombinant inbred lines of sunflower under two water treatments. *Plant Sci.* **2007**, *172*, 773–787. [CrossRef]
7. Walch-Liu, P.; Filleur, S.; Gan, Y.; Forde, B.G. Signaling mechanisms integrating root and shoot responses to changes in the nitrogen supply. *Photosynth. Res.* **2005**, *83*, 239–250. [CrossRef]
8. Fariduddin, Q.; Yusuf, M.; Chalkoo, S.; Hayat, S.; Ahmad, A. 28-homobrassinolide improves growth and photosynthesis in *Cucumis sativus* L. through an enhanced antioxidant system in the presence of chilling stress. *Photosynthetica* **2011**, *49*, 55–64. [CrossRef]
9. Özdemir, F.; Bor, M.; Demiral, T.; Türkan, İ. Effects of 24-epibrassinolide on seed germination, seedling growth, lipid peroxidation, proline content and antioxidative system of rice (*Oryza sativa* L.) under salinity stress. *Plant Growth Regul.* **2004**, *42*, 203–211. [CrossRef]

antioxidants

MDPI

Article

Genome-Wide Identification of the Sulfate Transporter Gene Family Reveals That *BolSULTR2;1* Regulates Plant Resistance to *Alternaria brassicicola* Through the Modulation of Glutathione Biosynthesis in Broccoli

Guize Wu [1,2,3,4], Yunhua Ding [2,3], Ning Li [2,3,4], Hongji Zhang [1,*] and Ning Liu [2,3,4,*]

[1] Key Laboratory of Vegetable Biology of Yunnan Province, Yunnan Agricultural University, Kunming 650201, China

[2] Beijing Vegetable Research Center, National Engineering Research Center for Vegetables, Beijing Academy of Agriculture and Forestry Sciences, Beijing 100097, China

[3] State Key Laboratory of Vegetable Biobreeding, Beijing Academy of Agriculture and Forestry Sciences, Beijing 100097, China

[4] Key Laboratory of Urban Agriculture (North China), Ministry of Agriculture and Rural Affairs, Beijing 100097, China

* Correspondence: zhanghongji111@163.com (H.Z.); liuning@nercv.org (N.L.)

Abstract: Sulfate transporters (SULTRs) are key players that regulate sulfur acquisition and distribution within plants, thereby influencing cellular redox hemostasis under pathogen attacks, such as *Alternaria brassicicola* (*Ab*). In this study, a total of 23 *BolSULTR* (*Brassica oleracea* SULTR) genes were identified from the *Brassica* genome. These *BolSULTR*s are distributed across nine chromosomes, with all collinear *BolSULTR* gene pairs undergoing purifying selections. Phylogenetic analysis reveals that the SULTR family is evolutionarily conserved among plant kingdoms. qRT-PCR analysis demonstrated that the expression of *BolSULTR*s varies across different plant organs and is modulated by hormonal signals. Furthermore, transcriptome analysis identified several *BolSULTR*s whose expression levels were depressed in *Ab*-challenged leaves in broccoli. Among them, the *BolSULTR2;1* gene emerged as a key player in the plant's response to *Ab*. Virus-induced gene silencing (VIGS) of *BolSULTR2;1*s resulted in elevated glutathione (GSH) levels and enhanced tolerance to *Ab*. Taken together, these findings underscore the role of *BolSULTR2;1* in maintaining redox homeostasis and enhancing plant disease resistance, suggesting its potential as a target for genome editing to develop broccoli varieties with improved pathogen tolerance.

Keywords: sulfate transporters; evolutionary analysis; bioinformatics; *Alternaria brassicicola*; *BolSULTR2;1*; glutathione

1. Introduction

Sulfur is one of the essential elements in all living organisms, and it plays a fundamental role in various biochemical processes in higher plants [1]. The leaves and roots of plants are essential tissues for the acquisition of external sulfur (S), playing a crucial role in the plant's overall nutrient uptake and metabolism. Plants actively absorb hydrogen sulfide (H_2S) through their leaves as a source of sulfur. The sulfur-deprived *Brassica* plants can be more easily restored to the level of plants grown in the presence of sulfate with H_2S [2]. Plant roots take up sulfur from the soil mainly in the form of sulfates (SO_4^{2-}), which is commonly mediated by several integral membrane proteins, namely sulfate transporters (SULTRs) [3,4]. Once imported into root cells, sulfate is translocated to plastids in leaves

https://doi.org/10.3390/antiox14040496

and also via SULTRs, where sulfate is converted to adenosine 5′-phosphosulfate (APS) [5]. Subsequently, APS undergoes two consecutive enzymatic reactions and is sequentially reduced to sulfite and sulfide by APS reductase (APR) and sulfite reductase (SiR), respectively. Finally, sulfide is incorporated into the carbon skeleton of O-acetylserine (OAS), and the resulting product cysteine is the first reduced sulfur donor involved in the biosynthesis of other sulfur biomolecules, such as methionine, sulfolipides, vitamins, glutathione (GSH), and glucosinolates [6–9].

It is reported that the typical SULTR proteins are generally H^+/SO_4^{2-} co-transporters with 12 transmembrane domains (TMDs), a Sulfate_transp domain in the N-terminal region, and an anti-sigma antagonist (STAS) domain in the C-terminal region [10,11]. Thus, the *SULTR* genes can be identified from the plant genome according to the sequences of conserved domains. For example, the model plant *Arabidopsis thaliana* contains 12 *SULTR* genes in its genomes [12,13]. Based on sequence similarity, the SULTR gene family can be divided into four groups. In *Arabidopsis*, Group I includes *SULTR1;1*, *SULTR1;2*, and *SULTR1;3*, known as high-affinity transporters distributed in the root surface cell layer of the roots [14,15]. However, Group II includes *SULTR2;1* and *SULTR2;2*, which are low-affinity transporters responsible for sulfate transport from root to stem. It has been accepted that SULTRs in Group II mainly participate in long-distance transport, maintaining sulfur metabolite levels between the aerial parts [16]. Group III is the largest group with preferential expression in leaves, but their exact roles remain under investigation. It is proposed that SULTR3s facilitate SO_4^{2-} uptake into chloroplasts and consequently influence downstream sulfate assimilation [17,18]. Interestingly, a yeast-based complementation experiment suggested that SULTR3;5 could directly interact with SULTR2;1, enhancing the transporter ability of SULTR2;1 under sulfur deficiency [19]. *SULTR4;1* and *SULTR4;2* belong to Group IV, and these SULTR4s are localized to the vacuolar membrane, where they actively export SO_4^{2-} from the vacuole to the cytoplasm [20]. Thus, these functionally diversified SULTRs play central roles in the absorption and distribution of sulfate within different plant organs, cell types, and subcellular compartments.

Intensive studies have demonstrated the critical roles of *SULTR* genes in plant growth and development. In *Arabidopsis*, sulfur starvation strongly activates the expression of *SULTR1;1*, whereas under normal sulfate supply, SULTR1;2 is the major transporter [14,21]. Sulfate translocation from roots to shoots is largely mediated by *microRNA395* (*miR395*) [22], which affects sulfate transport from root to stem and from old leaves to new leaves via direct suppression of *SULTR2;1*. The *Arabidopsis sultr2;1* mutant exhibits an early flowering phenotype while keeping plant biomass unchanged [23], whereas several *sultr3* mutants display stunted growth to varying degrees, resulting in a smaller thallus and shorter roots due to the reduced sulfate uptake to chloroplasts [17,24]. The SULTR4;1 and SULTR4;2 transporters are associated with sulfate transport in the vacuole, fulfilling the S requirements in young leaves and seeds [25,26]. Vacuoles isolated from callus of the *sultr4;1* and *sultr4;2* double knockout showed excess accumulation of sulfate [27]. However, our knowledge about their functions under abiotic and biotic stresses is still scarce. In rice, the *miR395-OsSULTR2s* module also regulates sulfate accumulation in leaves, resulting in broad-spectrum resistance to bacterial pathogens [28]. In the model legume *Medicago truncatula*, the majority of *MtSULTR3* genes were strongly upregulated by drought or salinity [29], implying that they might be a potential target for abiotic stresses. Indeed, the loss of *SULTR3;1* function led to lower ABA levels in *Arabidopsis* seedlings [30], further indicating that SULTRs could be the cross-talking point between sulfur metabolism and stress signaling pathways. Those findings suggest that SULTRs could help plants maintain sufficient sulfur supply, thereby protecting plants from environmental stresses as well as pathogen attacks.

Broccoli (*Brassica oleracea* L. var. *italica*) is a cruciferous vegetable that is widely cultivated worldwide owing to its nutritional value [31,32], and, in 2023, global production of broccoli (cauliflower included) reached approximately 25.9 million tons (http://faostat.fao.org/ (accessed on 15 October 2024)). Black spot disease, a necrotrophic pathogen caused by fungal *Alternaria brassicicola* (*Ab*), severely affects broccoli production. So far, fully *Ab*-resistant broccoli cultivars have not been reported yet, and, currently, all commercial varieties are highly susceptible to *Ab* [33]. Therefore, genetic manipulation is the most efficient method for the rapid improvement of broccoli resistance to *Ab* [34]. As sulfur is implicated in plant immunity, modulation of sulfur transport might benefit plant sulfur metabolism and consequently confer *Ab* resistance to broccoli.

Although the *SULTR* genes have been investigated in several plant species, their genomic distribution and biological functions in broccoli, to our knowledge, have been largely unexplored. Therefore, in this study, a genome-wide, comprehensive analysis of *SULTR* genes was performed to search for SULTRs associated with broccoli *Ab* resistance. In total, 23 *SULTR* genes were identified from broccoli and further confirmed through sequencing. The physical and chemical characteristics, genomic structures, chromosomal locations, evolutionary relationship, and expression profiles of the broccoli *SULTR* gene family were examined in detail. Furthermore, the roles of *BolSULTR2;1a* were analyzed through the virus-induced gene silencing (VIGS) approach, supporting its critical role in *Ab* resistance. Our study provides molecular information regarding the *SULTR* gene family in broccoli, providing a theoretical foundation for future genetic engineering to generate *Ab*-resistant vegetables.

2. Materials and Methods

2.1. Identification and Characterization of BolSULTRs

The newest *Brassica oleracea* cv. JZS V2.0 data were obtained from the Brassicaceae Database (BRAD) (http://brassicadb.cn (accessed on 10 December 2024)) [35]. The genome CDS sequence data were extracted and transformed into protein sequences using TBtools-II v2.154. Then, all *Arabidopsis thaliana* SULTR protein sequences were downloaded from the Arabidopsis Genome Database (TAIR) using SULTR as a keyword. These sequences were then used as seed sequences to search the *B. oleracea* protein database for the candidate BolSULTR using BLASTp with an e-value $\leq 1 \times 10^{-5}$ [36]. In addition, the *B. oleracea* protein sequences of *SULTR* were blasted by using Hidden Markov Models (HMMs) for Sulfate_transp (PF00916) and STAS (PF01740) domains downloaded from the Pfam database [37]. Domain structures were verified using NCBI-CDD and the Pfams database [38,39], yielding 23 SULTR genes. Physicochemical properties were predicted using ExPASy (http://web.expasy.org/protparam (accessed on 10 November 2024)) [40], and subcellular localization was predicted using Plant-mPLoc [41].

2.2. Phylogenetic Analysis, Classification, and Evolutionary Analysis

Protein sequences of *SULTR* genes were retrieved from TAIR and BRAD. Multiple sequence alignment of broccoli and *Arabidopsis SULTR* proteins was performed using MUSCLE in MEGA X. Phylogenetic trees were constructed through the neighbor-joining method with 1000 bootstrap replicates [42]. Based on their kinship with *Arabidopsis*, the broccoli *SULTR* genes were designated as *BolSULTR*. Tree visualization and subfamily classification were performed using iTOL (https://itol.embl.de/login.cgi (accessed on 11 December 2024)). The neighbor-joining tree was constructed based on the alignment of SULTR amino acid sequences from *Chlamydomonas reinhardtii*, *Physcomitrella patens*, *Oryza sativa*, *Solaunum tuberosum*, *Solanum lycoperiscum*, *Brassica oleracea* L. var. *italica*, and *Arabidopsis thaliana*. The percent bootstrap support for 1000 replicates is shown on each branch.

2.3. Chromosome Location and Homologous Gene Analysis

The chromosome location of *BolSULTR* genes appeared on the BRAD. The density of each gene was analyzed in the entire chromosome region. We divided the number of genes by the region size and then multiplied it by a conversion factor of 1,000,000 to obtain the number of genes per million base pairs (Mb). Moreover, gene duplication events were examined using MCScanX with default parameters [43]. The events of gene duplication in broccoli were revealed through the Advanced Circus tool from TBtools. Meanwhile, the synonymous mutation rate (Ks), the nonsynonymous mutation rate (Ka), and the ratio of nonsynonymous to synonymous mutation rates (Ka/Ks) for duplicated *BolSULTR* genes were calculated using the Ka/Ks Calculator tool in TBtools. In addition, we downloaded the genome annotation files for *Arabidopsis thaliana* and rice from the Ensembl Plants database (https://plants.ensembl.org/index.html (accessed on 10 November 2024)) to create synteny maps and analyzed the distribution of homologous genes across these different species by using the TBtools.

2.4. Analysis of Conserved Motifs and Gene Structures of the BolSULTRs

We conducted a protein motif analysis of *BolSULTR* family members utilizing the MEME tool (http://meme-suite.org/tools/meme (accessed on 11 December 2024)) with default settings. Briefly, motif sites were expected to be distributed in sequences with zero or one occurrence per sequence, and a maximum of 12 patterns were expected. Other settings were as follows: the motif E-value threshold: no limit; sites per motif: 16–23; and motif width: 15–50. The motifs were annotated in the form of amino acid logo plots. The quantitative structures of exons and introns were determined using the genome annotation file (GFF file format). TBtools was utilized to integrate and visualize phylogenetic relationships, motif patterns, and gene structures.

2.5. Promoter Cis-Acting Element

We used TBtools to extract promoter sequences of upstream 2000 bp *BolSULTR* genes, and the cis-regulatory elements were identified using PlantCAR [44]. The redundant elements were removed, and, according to its potential functions, all elements were classified. TBtools was used to visualize element distribution (Figure S3).

2.6. Protein–Protein Interaction Prediction

The BolSULTR protein interactions based on data from *Arabidopsis* orthologs were analyzed using the online tool STRING (https://string-db.org/ (accessed on 11 November 2024)). The interacting proteins were functionally characterized through Gene Ontology (GO) and Kyoto Encyclopedia of Genes and Genomes (KEGG) pathway enrichment analyses.

2.7. Plant Material and Treatment

Seeds of the susceptible cultivar '41-3' of *Brassica oleracea* L. var. *italica* (broccoli) were germinated in a soil–perlite mixture (2:1) in a plant growth chamber with a photoperiod of 16 h of light and 8 h of darkness at a temperature of 25 °C ± 1 °C and a relative humidity of approximately 65%. The 8-week-old broccoli seedlings were used for Ab infection experiments. The *A. brassicicola* wild-type strain used in this study was routinely grown and maintained on potato dextrose agar at 24 °C.

2.8. Expression Patterns of BolSULTR Genes Based on RNA-Seq Data

The concentration of the *A. brassicicola* conidial suspension was 5×10^5 spores per milliliter with 0.05% (*v*/*v*) Tween 20 was added as a surfactant, which was applied using a spray bottle directly onto each leaf. The control group received the same volume of Tween 20 as the treatment groups. Additionally, the relative humidity of the environment was increased to 80%. We selected leaves at different time points (0 days, 1 day, 3 days, 5 days, and 7 days) after treatment to perform transcriptome sequencing (RNA-Seq); each time point included three samples, with each sample consisting of four biological replicates. Additionally, we searched seven RNA-seq datasets of *Brassica* from the GSE42891 (GEO database), specifically for the root, stem, leaf, flower, silique, bud, and callus. The *BolSULTR* genes were used as a template to screen for highly homologous genes in RNA-seq datasets, and then the expression pattern of the target gene was extracted based on the FPKM value. Subsequently, log2 transformation was applied to generate a heatmap using TBtool. RNA-seq dataset information is presented in Supplementary Tables S4 and S5.

2.9. Expression of SULTR Genes in Broccoli Treated with Different Phytohormones

We used different phytohormones to treat broccoli seedling leaves. The exogenous hormones were dissolved in a small amount of ethanol and then diluted to the desired concentration for each treatment with distilled water, including 100 mg/L of Ethephon (ETH, Solarbio, 16672-87-0, Beijing, China), 1 mM/L of MeJA (Macklin, 39924-52-2, Beijing, China), and 1 mM/L of SA (Biotopped, 69-72-7, Beijing, China). Working fluids were applied to the plants through spraying and distilled water was used as a control treatment while ensuring that the concentration of anhydrous ethanol in the spray solution was the same. Additionally, 0.1% (*v*/*v*) Tween 20 was added as a surfactant. We ensured that each plant leaf was sprayed thoroughly until it dripped completely. For each treatment, we selected three samples at different time points (0 h, 4 h, 24 h), and for each sample we used 3 biological replicates. The samples were verified through qRT-PCR. These samples were then rapidly frozen in liquid nitrogen and stored at −80 °C for further experiments.

2.10. RNA Extraction and qRT-PCR Analysis

Total RNA was extracted from young broccoli leaves using the FastPure Plant Total RNA Isolation Kit (Vazyme, Nanjing, China), with modifications [45]. RNA was then treated with DNase I to remove genomic DNA in one step, and 3 μg of total RNA was used for first-strand cDNA synthesis using a cDNA synthesis kit (Transgen, Beijing, China). The resulting cDNA sequences served as templates for quantitative reverse transcription real-time PCR (qRT-PCR) using a Roche LightCycler 480 II and the LightCycler 480 SYBR Green I Master Mix (Toyobo, Shanghai, China), following the manufacturer's instructions. Fold changes were calculated from $2^{-\Delta\Delta Ct}$ values, with each assay replicated at least three times using three biological replicates (independent RNA preparations). The *Actin-2* was employed as a reference gene. The relevant primer sequences are presented in Supplemental Table S3.

2.11. Virus-Induced Gene Silencing (VIGS) of BolSULTR2;1

We used the tobacco rattle virus (TRV) vector system, consisting of pTRV1 and pTRV2 vectors, to achieve virus-induced *BolSULTR2;1* silencing (VIGS). The 315 bp sequence from CDS was subcloned into the viral vector pTRV2. The *Agrobacterium tumefaciens* strain GV3101 harboring the recombinant plasmids pTRV2:*BolSULTR2;1* and pTRV1 was used for in vitro agroinoculation through leaf injection. The *Brassica oleracea* 15-cis-phytoene desaturase (PDS) was used as a positive control, and the negative control was the pTRV2 empty vector (TRV:00). The *Agrobacterium tumefaciens* strain GV3101 carrying the constructs was cultured overnight at 28 °C and centrifuged to remove culture medium containing antibiotics when OD_{600} reached 1.5–2.0. The concentrated bacterial solutions were fully suspended with 1 mL of 10 mM magnesium chloride solution containing 200 μM of acetylsyringone and then diluted into a working solution (OD_{600} = 0.5). Then, a needleless syringe was used to infiltrate the abaxial surface of the broccoli cotyledons. Each treatment was inoculated with fifty broccoli seedlings. All of the plants were grown in the same growth environment at 25/23 °C (day/night), with a 16 h light/8 h dark cycle.

When observing a bleached leaf phenotype in seedlings infiltrated with *pTRV:PDS*, we performed a qRT-PCR analysis on the selected seedlings (pTRV:*BolSULTR2;1*). Significant downregulation of *BolSULTR2;1* in the bleached seedlings compared to negative controls indicated successful silencing. At that moment, we collected a sufficient amount of leaf tissue for further analysis. Subsequently, the silenced seedlings and control plants with comparable growth conditions were selected for inoculation with the *Ab* suspension liquid (5×10^5 conidia/mL). Determination of the stress-related physiological index contents of the leaves was performed on the fifth day after inoculation, and ImageJ software (Ver. 1.54k) was used to calculate the area of the lesions. The VIGS experiments were repeated at least three times, and each treatment was applied to more than 15 plants. The primers used for vector construction are listed in Supplemental Table S3.

2.12. Measurements of ROS-Related Physiological and Biochemical Parameters

Quantifications of reduced glutathione (GSH), superoxide dismutase (SOD), catalase (CAT), and ascorbate peroxidase (APX) were conducted using commercial kits (BC1175 for GSH, BC0175 for SOD, BC0205 for CAT, and BC0225 for APX) purchased from Solarbio Life Sciences Co., Ltd. (Beijing, China) following the manufacturer's instructions. The spectrophotometry was performed with an Agilent BioTek microplate reader (Santa Clara, CA, USA). Each assay was performed with three biological replicates. The detection of reactive oxygen species (ROS) was conducted using Coolaber DAB (3,3′-diaminobenzidine) detection kit (Beijing, China). The leaf staining procedure was conducted as previously described [46].

3. Results

3.1. Identification of SULTR Genes in Brassica Oleracea Broccoli

In this study, 23 unique *BolSULTR* members were obtained from broccoli and named according to their homology with *Arabidopsis* counterparts (Figure S1 and Table S1). Information on the gene name, loci number chromosomal location, and physicochemical characteristics of putative proteins of individual *BolSULTR* genes is listed in Table 1. Of these, the molecular weight of putative BolSULTR proteins ranged from 43.33 kDa (BolSULTR3;1c) to 108.00 kDa (BolSULTR4;2). Furthermore, the range of isoelectric points extends from a slightly acidic pI of 5.96 (BolSULTR3;1b) to a more basic pI of 9.63 (BolSULTR3;4b). Notably, except for BolSULTR3;1b protein, which was distinguished by its hydrophilic character, the other members, which exhibit positive GRAVY values, showed a generally hydrophobic nature. Consistent with transporter roles, 19 BolSULTR proteins were predicted to be localized on the cell membrane, while all BolSULTR4 members might be localized to the chloroplasts or vacuoles. Surprisingly, BolSULTR3;1b could be targeted to the nucleus, probably reflecting its distinct functions other than as a transporter. Together, the results reinforced the assumption that these BolSULTRs might functionally diverge and participate in different biochemical processes under disparate environments.

Table 1. Physicochemical properties of SULTR protein in broccoli.

Gene_Name [a]	Gene_ID [b]	Putative Proteins' Physical and Chemical Parameters [c]					Predicted Location(s) [d]
		Peptide Length (aa)	MW (kDa)	pI	Instability Index	GRAVY	
BolSULTR1.2b	BolC02g034190.2J	652	71.64	9.12	33.97	0.410	Cell membrane
BolSULTR1;1	BolC02g019620.2J	648	70.59	9.12	35.14	0.406	Cell membrane
BolSULTR1;2a	BolC06g049010.2J	655	71.84	8.81	34.81	0.401	Cell membrane
BolSULTR1;3	BolC05g020480.2J	683	74.96	9.22	34.71	0.456	Cell membrane
BolSULTR2;1a	BolC02g004100.2J	677	73.97	9.04	37.58	0.427	Cell membrane
BolSULTR2;1b	BolC09g062520.2J	686	75.39	8.86	39.45	0.414	Cell membrane
BolSULTR2;1c	BolC09g062540.2J	629	68.97	8.21	37.44	0.473	Cell membrane
BolSULTR2;1d	BolC09g062560.2J	624	68.22	8.54	36.2	0.489	Cell membrane
BolSULTR2;2	BolC06g049000.2J	471	51.29	9.43	30.44	0.586	Cell membrane
BolSULTR3;1a	BolC07g044940.2J	759	83.77	8.99	33.55	0.263	Cell membrane
BolSULTR3;1b	BolC08g036100.2J	876	97.30	5.96	40.41	−0.093	Cell membrane, nucleus
BolSULTR3;1c	BolC04g040390.2J	390	43.33	8.5	40.64	0.407	Cell membrane
BolSULTR3;2	BolC09g002080.2J	650	71.74	8.81	42.51	0.438	Cell membrane
BolSULTR3;3a	BolC07g020360.2J	656	72.22	9.1	35.68	0.461	Cell membrane
BolSULTR3;3b	BolC05g021440.2J	632	69.42	9.3	36.37	0.425	Cell membrane
BolSULTR3;4a	BolC05g050200.2J	656	71.95	9.51	40.83	0.407	Cell membrane
BolSULTR3;4b	BolC01g046840.2J	660	72.09	9.63	39.27	0.409	Cell membrane
BolSULTR3;4c	BolC03g044170.2J	653	71.60	9.54	37.55	0.44	Cell membrane
BolSULTR3;5a	BolC03g010280.2J	636	70.53	8.48	35.85	0.442	Cell membrane
BolSULTR3;5b	BolC02g009920.2J	634	70.38	8.85	32.21	0.447	Cell membrane
BolSULTR4;1a	BolC03g006540.2J	692	75.88	8.64	45.62	0.324	Chloroplast
BolSULTR4;1b	BolC09g059530.2J	693	75.89	8.34	43.5	0.288	Chloroplast
BolSULTR4;2	BolC03g041900.2J	989	108.00	6.03	39.73	0.138	Chloroplast, vacuole

[a], Name refers to systematic designation of members of the *BolSULTR* family according to the homology against *Arabidopsis*. [b], Gene accession number in BRAD database. [c], The various physical and chemical parameters calculated from the protein sequences we entered. [d], Subcellular localization was predicted using Plant-mPLoc.

3.2. Phylogenetic Analysis of SULTR Gene Family

To investigate the phylogenetic relationship between the members of the BolSULTR family, a neighbor-joining tree was constructed based on the multiple alignment of SULTR protein sequences from green algae (*Chlamydomonas reinhardtii*), bryophyte (*Physcomitrium patens*), green algae (*Chlamydomonas reinhardtii*), rice (*Oryza Stativa*), tomato (*Solanum tuberosum*), potato (*Solanum lycopersicum*), *Arabidopsis* (*Arabidopsis thaliana*), and broccoli. Among these subfamilies, subclade III stands out with 34 members, which makes it the largest group; subclade I includes 26 members, followed by subclade IV (16 SULTRs) and subclade II (11 SULTRs). Only two and four SULTR proteins were identified from the genome of green algae and bryophytes, respectively (Figure 1). The reduced number of SULTRs in lower photosynthetic organisms may reflect evolutionary adaptations to their specific environmental conditions and physiological requirements. Conversely, in higher plants, such as *Arabidopsis*, tomato, and rice, dozens of SULTRs were found in their genomes, which indicates the expansion of the SULTR gene family in higher plants.

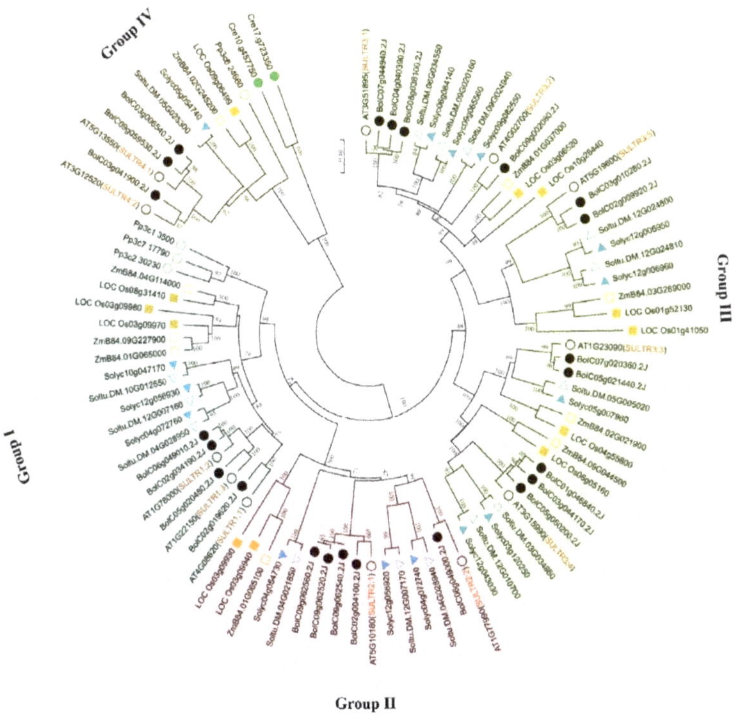

Figure 1. Phylogenetic analysis of SULTR proteins in algae, liverwort, moss, and selected flowering plants. The neighbor-joining tree was constructed based on the alignment of SULTR amino acid sequences from *Chlamydomonas reinhardtii* (green circle), *Physcomitrella patens* (empty green circle), *Solanum tuberosum* (blue triangle), *Solanum lycopersicum* (empty blue triangle), *Oryza sativa* (empty yellow square), *Zea mays* (yellow square), *Brassica oleracea* L. var. *Italica* (black circle), and *Arabidopsis thaliana* (empty black circle). The position of the root was determined from an outgroup consisting of two SULTR-like sequences from yeast (*Saccharomyces cerevisiae*). The percent bootstrap support for 1000 replicates is shown on each branch.

3.3. Conserved Domains of BolSULTRs

To better understand the structural similarity of these SULTRs, the Hidden Markov Models were used to identify the TMD, STAS, and other conserved domains of BolSULTR, and the predicted domains were visualized in ChiPlot (Figure S2). Almost all BolSULTR proteins possess the Sulfate_transp domain (PF00916) and the STAS domain (PF01740), indicating that the BolSULTRs are highly conserved in broccoli. However, SULTR2;2 only has the intact TMD region, and its STAS domain was abolished. Thus, it is plausible that the transporter ability of SULTR2;2 could be impaired due to the sequence truncations. Other domains were identified in some BolSULTRs, which might confer novel functions. For example, in addition to TMD and STAS, BolSULTR4;2 also contains the Chitin-binding and glycoside hydrolase family 19 domains in its C-terminal. Similarly, LSm7 RNA-binding and high mobility group (HMG)-box motifs were found in BolSULTR3;1a and BolSULTR3;1b, respectively (Table S1).

3.4. Analysis of Conserved Motifs and Gene Structure of BolSULTRs

A total of 12 different motifs were detected among the BolSULTR proteins, with lengths ranging from 15 to 50 amino acids. The number of motifs varied from 7 to 12 among 23

BolSULTR proteins. The motifs 1, 2, and 5, containing the Sulfate_transp motif, or motifs 7, 9, and 11, containing the STAS signature motif, were characteristic in all BolSULTR proteins. These findings are consistent with the domain analysis of the SULTR family (Figure 2B,D). Additionally, motif 8 was exclusively absent in Group IV, and several motifs were deleted in BolSULTR2;2, BolSULTR3;1c, implying the different functions of these BolSULTRs. Additionally, most *BolSULTRs* within the same subfamily generally exhibit similar genetic structures in terms of the number and length of introns and exons (Figure 2C). *BolSULTR2;1* members might arise from recent duplication events, as evidenced by the extremely high similarity in gene structure and sequences. In Group III, the number of exons ranged from 8 to 21, exhibiting a complex pattern. The *SULTR* members in Group IV had 15 to 19 exons, and *BolSULTR4;2* possessed 2 longer exons at the 3′ end due to the additional domains. Overall, the conserved motifs and complex gene structure might provide more chances of heredity and variation for *BolSULTR* genes.

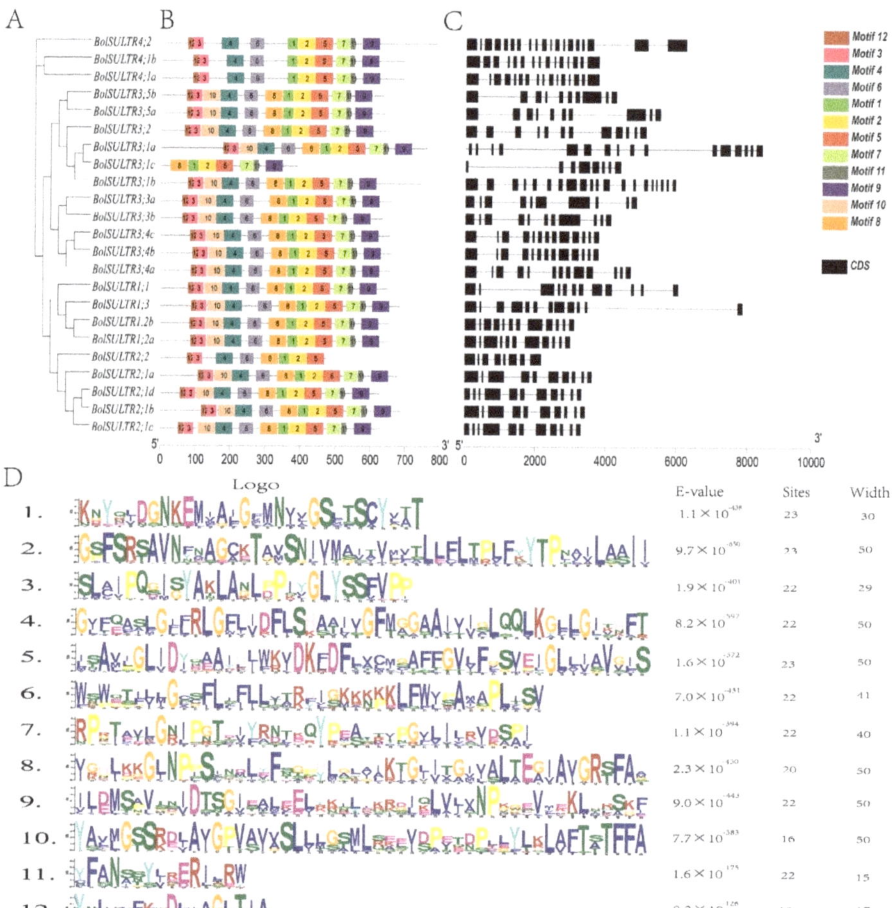

Figure 2. Conserved motifs and gene structural analyses of the *BolSULTRs*. (**A**) Creating a phylogenetic tree involved arranging 23 BolSULTR conserved motifs. (**B**) The two conserved domains are divided into 12 motifs. The specific motifs or sequence patterns are retained across different *BolSULTR* proteins due to their functional importance. (**C**) General gene structure of BolSULTRs. Black lines and black boxes indicate introns, exons, and UTR, respectively. (**D**) The functional active site or binding site corresponds to the motif's detailed sequence.

3.5. The Chromosome Localization, Gene Duplication, Ka/Ks, and Synteny Analysis

The results showed that 23 *BolSULTR* genes were unevenly distributed in nine chromosomes (Figure 3A). The majority of *BolSULTR* genes were mainly located on chromosomes C02, C03, and C09. Notably, only one *BolSULTR3* gene was found on chromosomes C01, C04, and C08, while most *BolSULTR2*s were clustered on chromosome C09. In addition, the collinearity relationships showed *BolSULTR1;1*, *BolSULTR2;2*, *BolSULTR3;2*, and *BolSULTR4;2* without associated collinear genes and a high degree of functional conservation among cruciferous species (Figure 3B). This may be due to biased gene loss in redundant gene members lacking collinear relationships [47]. Furthermore, consistent with previous observations, all collinear *BolSULTR* gene pairs had a Ka/Ks ratio of less than one, indicating that these genes underwent purifying selection. Interestingly, the genes *BolSULTR2;1a* and *BolSULTR2;1b* were involved in segmental duplication, while *BolSULTR2;1b*, *BolSULTR2;1c*, and the *BolSULTR2;1d* resulted from tandem duplication (Figure 3A and Table S2).

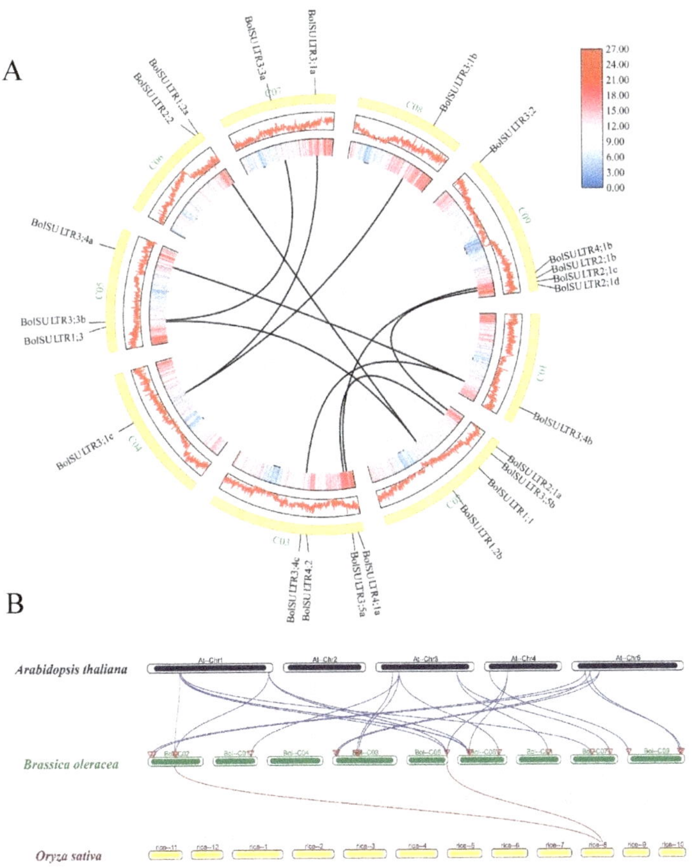

Figure 3. Synteny analysis of *BolSULTRs* in the Brassica oleracea genome and between different species. (**A**) Chromosomal locations and synteny analysis of *BolSULTRs* in broccoli. Gray lines represent all synteny blocks in the *B. oleracea* genome, while black lines indicate segmental duplication genes between *BolSULTRs*. Chromosome gene density is represented by a hot map (inner circle) and column map (middle circle), with the outer circle showing the length of chromosomes. (**B**) Cognate relationship of *BolSULTRs* in broccoli, *Arabidopsis*, and rice; colored lines emphasize the pairs of syntenic *BolSULTRs*.

3.6. Analysis of the Promoter Region of the BolSULTRs

Through PlantCare analytics, we can identify the potential regulatory regime through the cis elements in the 2 kb upstream of all identified *BolSULTR* genes. It was clear of cis-acting element distribution (Figure S3), and we drew on their functional annotations to categorize three types: plant growth and development, abiotic and biotic stresses, and phytohormone responses (Figure 4). The results of type I showed that *BolSULTR* genes were associated with plant growth and development. This is because there were abundant optical signal sensing and transduction elements distributed in the promoter region, like Box-4 (ATTAAT), G-box (TACGTG), and GT1-motif, which can regulate gene expression related to photomorphogenesis. Important cis-acting elements for the cell cycle and metal ion response were found, such as CAT-box and MRE. These elements were also related to responses to environmental changes.

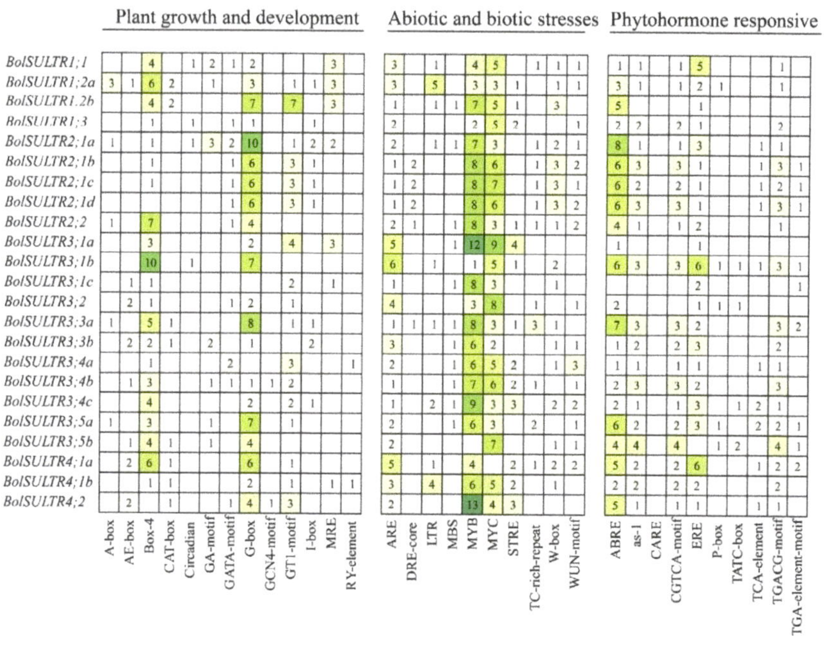

Figure 4. Analysis of cis-acting elements of 23 *BolSULTR* gene promoters. Cis-acting elements of the 23 *BolSULTR* gene promoters are grouped, including phytohormone-responsive, biotic/abiotic stress, and plant growth and development. The different colors and number of boxes indicate different promoter elements in the *BolSULTR* genes.

The abiotic and biotic stress response elements were divided into the second category, which is associated with stress responses, particularly those related to defense against pathogens and other environmental stresses. Among them, the elements responding to abiotic stress were widely distributed, including MYB (CCAAT), MYC (CACATG), ARE (anaerobic response element), MBS (drought inducibility), and LTR (low-temperature-responsive) binding sites. The second was the functional element in response to biological stress, including TC-rich-repeat, W-box (WRKY-box), WUN-motif, and STRE (Stress-Responsive Element). These elements are involved in the activation of *BolSULTR* genes that are part of the plant's response to various biotic stresses. The phytohormone response elements were grouped into the third category, which mainly included ABRE (ABA-responsive element), CGTCA-motif (MeJA-responsiveness), TGACG-motif (cis-acting regulatory element involved in MeJA-responsiveness), as-1element, and TCA-element (salicylic acid responsiveness). Additionally, ERE (Ethylene Response Element), TGA-element (auxin-responsive element), and P-box (gibberellin-responsive element) were identified in those *BolSULTR* promoters. These motifs are crucial for the regulation of genes involved in secondary metabolite production, defense compound biosynthesis, and other protective responses. Together, cis-element analyses indicate that *BolSULTRs* might be regulated by light, hormones, stress responses, and developmental cues.

3.7. BolSULTR Interaction Prediction

As interacting objects, amino acid transporter AVT1E (the amino acid/polyamine transporter 2 family), LHCB5 (Chlorophyll a-b binding protein), PSAO (Photosystem I subunit O), and PSAG (Photosystem I reaction center subunit V) were predicted to be the potential partners of BolSULTRs (Figure 5 and Table S6), which suggests that sulfur assimilation is closely associated with the antioxidant defense mechanisms of photosynthesis [48]. As expected, the BolSULTR proteins could interact with some proteins related to sulfur metabolism, including APS4 (ATP sulfurylase 4) and SIR (assimilatory sulfite reductase). This shows that BolSULTR proteins could form complexes with other proteins involved in the sulfur assimilation pathway to enhance efficiency and regulation.

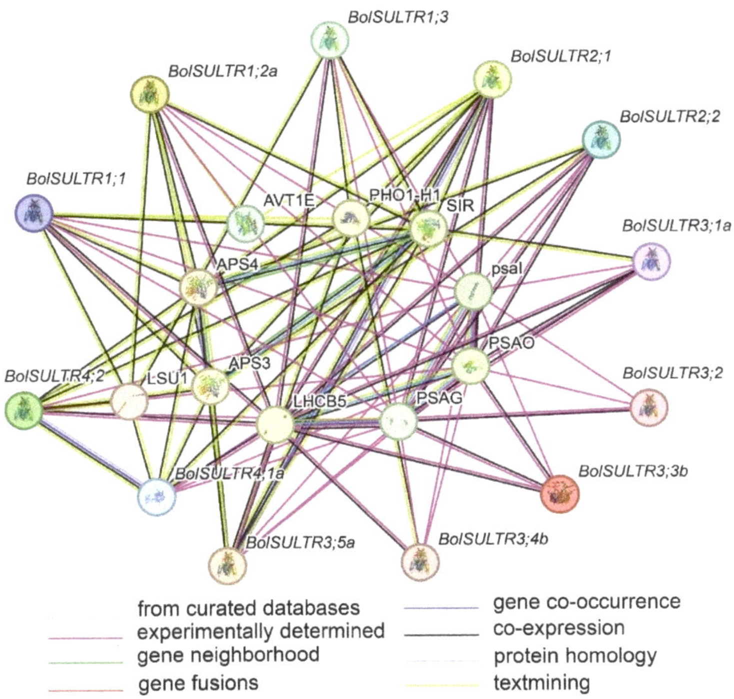

Figure 5. Interactions between BolSULTR proteins and other proteins. Line color indicates the type of interaction evidence; edges represent protein–protein associations.

3.8. Expression Patterns of the BolSULTR Genes

The expression profiles of *BolSULTR* were obtained from transcriptome analysis. Expression patterns of *BolSULTR* genes in different organs and under the *Ab* are shown in Figure 6A,B. The seven RNA-seq datasets of *Brassica* told us that *BolSULTR3;1a* and *BolSULTR4;1b* were ubiquitously expressed throughout the different organs, whereas other genes exhibited organ-preferential patterns. Specifically, *BolSULTR2;1a* was highly expressed in the stem. *BolSULTR3;1b*, *BolSULTR2;2*, and *BolSULTR3;3b* were highly expressed in flowers and the silique, and *BolSULTR3;5b* was predominantly expressed in the callus and roots. In contrast, *BolSULTR3;1c*, *BolSULTR1;3*, *BolSULTR1;1*, and *BolSULTR3;4a* displayed relatively lower expression levels across all organs. In summary, the expression of *BolSULTR* genes could be detected in diverse tissues and organs. The transcriptome datasets showed that initial gene expression levels (*BolSULTR2;1a*, *BolSULTR3;1a/b, and BolSULTR4;1b* and *BolSULTR3;3b*, *BolSULTR3;4b*, and *BolSULTR3;5b*) exhibited much higher expression levels than the other *BolSULTRs*. After the broccoli plants were challenged with *Ab*, their gene expression levels were robust biology stress responses and remained low until a correction occurred later in the course of the disease, particularly for *BolSULTR2;1a*. But, the homologous genes of *BolSULTR2;1b/c/d* did not respond. This suggests that *BolSULTR2;1a* may be important in early infections. Furthermore, the expression of *BolSULT3;1b* reached its maximum at 7 dpi, implying that it might be late-responsive genes during *Ab* infections.

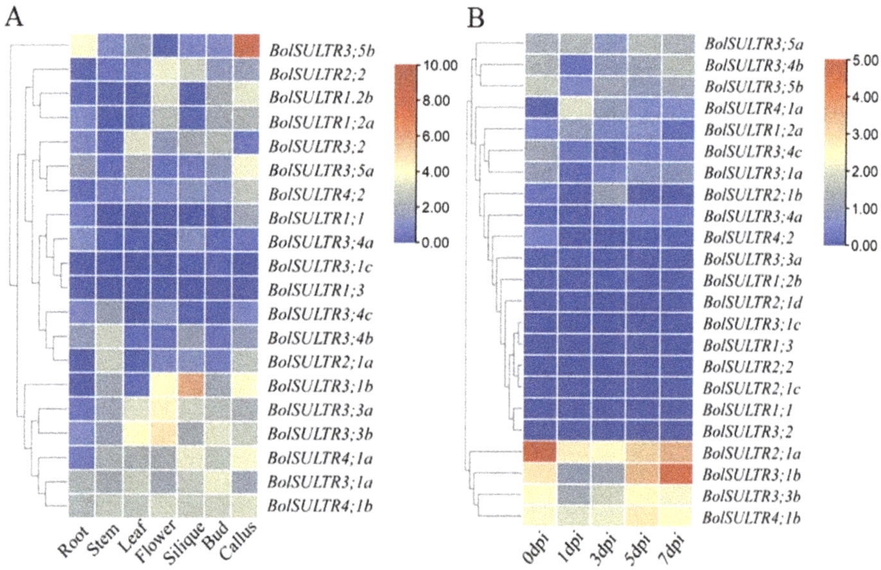

Figure 6. Expression patterns of the *BolSULTR* genes. (**A**) Different tissues' and organs' expression of *BolSULTR* genes; RNA-seq datasets belong to the GSE42891 (GEO database). (**B**) Expression of the *BolSULTR* genes under *Ab* challenges at different points in time (0, 1, 3, 5, and 7 days). The transcriptome data were averaged through log2 transformation and TBtools software to generate a heatmap. The FPKM values are from three independent biological replicates.

3.9. Expression of BolSULTR Genes Under the Phytohormone Treatments

To further investigate the potential roles of the *BolSULTR* gene family in response to the effects of different hormones on *BolSULTR* expression in broccoli, we selected and examined the relative expression levels of twelve *BolSULTR* genes after various hormone treatments for 0 h, 4 h, and 12 h using qRT-PCR (Figure 7). The results indicate that the expression of most *BolSULTR* genes was upregulated robustly under exogenous MeJA treatment conditions, suggesting that MeJA could induce plant resistance to saprophytic pathogens [49–51]. Interestingly, the *BolSULTR2;1a* gene was downregulated in response to various hormones.

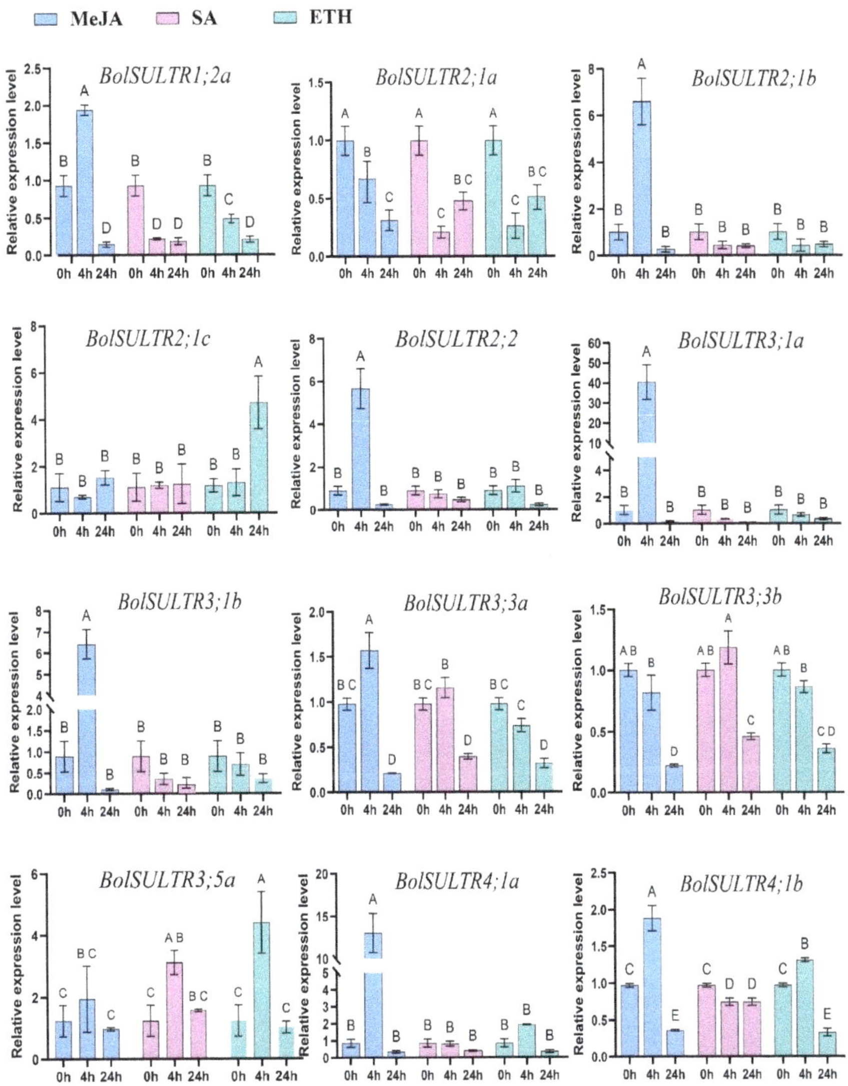

Figure 7. The expression levels of *BolSULTRs* in different hormones using qRT-PCR analysis. In total, 100 mg/L of Ethephon, 1 mM/L of MeJA, and 1 mM/L of SA were sprayed onto seedling leaves. The leaves of mature broccoli were selected at different time points (0 h, 4 h, 24 h) and three from independent biological replicates. Capital letters indicate significant differences ($p < 0.01$). Error bars represent the standard deviation.

3.10. Silencing of BolSULTR2;1 Resulted in Reduced Leaf Spot Stress

To further verify the functions of the *BolSULTR2;1a* gene in broccoli immunity, we obtained 25 independent silent plants of pTRV:*BolSULTR2;1* via the VIGS approach. The qRT-PCR analysis suggested that the genes' expression of *BolSULTR2;1* was specifically downregulated in silent plants compared with the control group, and glutathione synthetases2 (GSH2) gene expression was upregulated (Figures 8C and S5B). Then, these silent plants were subjected to phenotypic analysis and physiological parameter measurements. After 5 days of *A. brassicicola* inoculation, the silent plants exhibited local necrosis, and the areas of the necrotic spots were significantly smaller than those of the control plants.

DAB staining revealed less extensive lesions of reactive oxygen species (ROS) in the silent plant leaves (Figure 8A). As glutathione (GSH) is the major antioxidant molecule in plants, we measured the GSH (glutathione) content in the leaves' tissue surrounding the lesion. Consistent with the observed phenotypes, the GSH levels in silent plants were much higher than those of the control. And, with the development of the disease, the activity of various antioxidant enzymes in the silenced plants increased to a higher level than that of the control group (Figure 8D). Thus, these results illustrated that the repression of *BolSULTR2;1* could lead to enhanced tolerance under *Ab* infections.

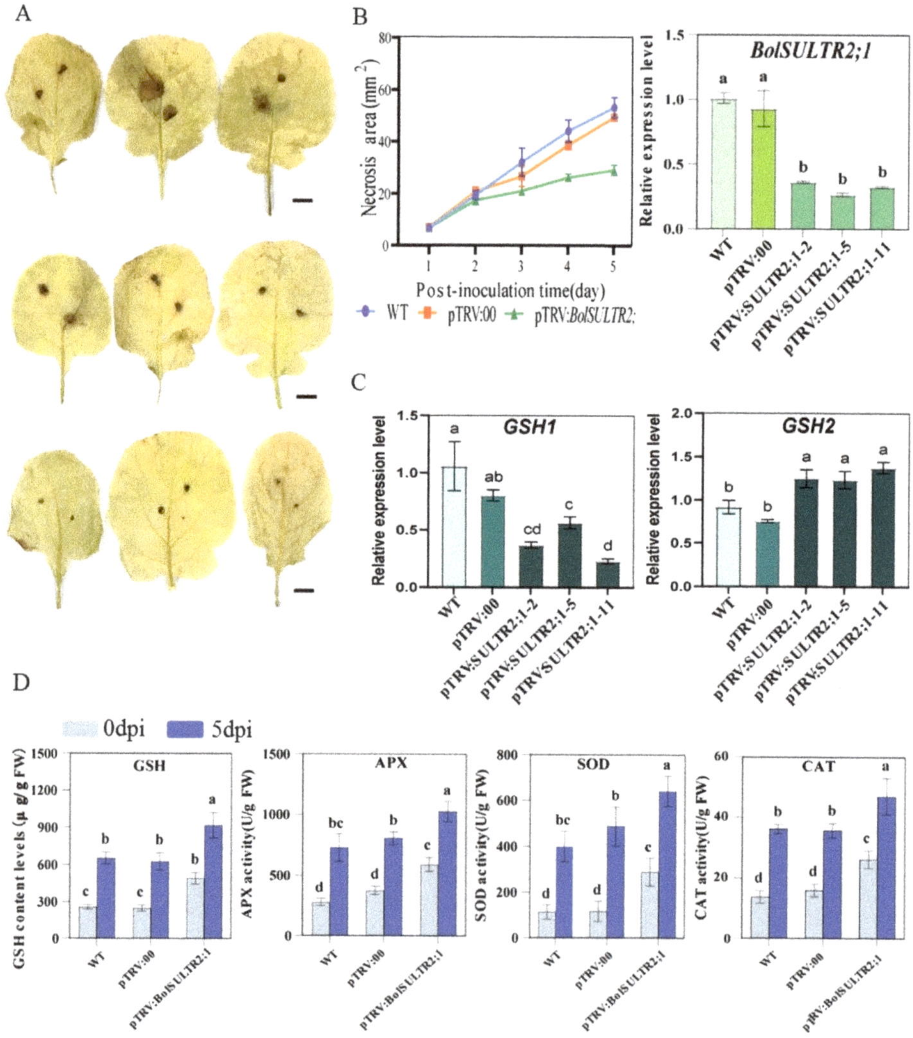

Figure 8. Physiological parameter measurements of *BolSULTR2;1*-silencing and control plants. (**A**) DAB reactive oxygen species staining in leaves. Bars = 2 cm. (**B**) Analysis of infection site area; *BolSULTR2;1* was specifically downregulated in silent plants. (**C**) Analysis of expression level of glutathione synthetases gene. (**D**) Physiological parameters of GSH, APX, SOD, and CAT with comparison between 0 days and 5 days under *Ab* stress in *BolSULTR2;1*-silencing plants. The mean values were derived from three independent biological replicates. Letters above each column indicate significant differences ($p < 0.05$ and $p < 0.01$). Error bars represent the standard deviation.

4. Discussion

4.1. The Paradox Between Genome Complexity and SULTR Copy Number in Broccoli

Plants use sulfate (SO_4^{2-}) as a primary sulfur source to produce a wide range of secondary organic metabolites that have imperative roles in plant development, metabolism, and stress responses [52]. Once sulfate is acquired from the rhizosphere, it is assimilated to biosynthesize cysteine and other sulfur-containing metabolites in the leaf chloroplasts. Thus, higher plants need to translocate the sulfate from underground to aerial organs, which is accomplished by SULTR transporters [4,53,54]. The sulfate movements between organs, cells, and organelles are strictly regulated by internal and external signals. In the last two decades, the SULTR gene family has been studied much more intensively in model plant *Arabidopsis* and, to varying degrees, in other plant species, such as rice, maize, and cabbage [12,55–57]. In this study, a total of 23 *BolSULTR* genes were identified where 2, 4, 15, 15, 9, and 12 SULTRs were found in the genomes of green algae, moss, tomato, potato, maize, and rice, respectively. The results suggested that SULTRs might be encoded by a moderate-sized gene family.

Regardless of the variable size of their genome, the number of *SULTR*s is comparable in the species investigated. Among them, *Zea mays* B84 has the largest genome of 2.1 gigabases (Gb), while only nine *ZmSULTR*s were reported in the maize genome. However, the *Arabidopsis* with the smallest genome of 130 megabases (Mb) contains 12 *SULTR*s. As a close relative of *Arabidopsis*, broccoli belongs to the Brassicaceae family. Due to several rounds of whole genome duplication (WGD) events [47,58,59], the broccoli genome (613 Mb) is nearly five times larger than its *Arabidopsis* counterpart. Interestingly, the members of *BolSULTR* only slightly doubled compared to their *Arabidopsis* counterparts because the phylogenetic analysis did identify duplication events within the *BolSULTR* gene family. It seems that the increased complexity of the genome has little effect on the expansion of the *SULTR* gene family in these plants. Accordingly, the observations might provide a plausible hypothesis for the paradox between the increased genome complexity and the relatively stable copy number of *SULTRs* in diverse species. In this scenario, although several WGD events during Brassicaceae evolution technically increase the copy number of *SULTR*s in the genome, the duplicated *SULTR*s are eventually lost through natural selection that favors the deletion of redundant copies unless the duplicated *SULTR* can acquire new functions [47,60]. Indeed, Ka/Ks ratio analysis indicated that the majority of *BolSULTR*s have undergone strongly purifying selection in their evolutionary history. However, we are unable to deny other possibilities. For instance, other plant ion transporters, instead of SULTRs, might also have the ability to absorb sulfate from the environment, which might eliminate the need for more SULTRs in these higher plants with more complex genomes. Moreover, Group III represents the largest subfamily of SULTRs according to our analysis, which is in good agreement with previous findings. Although the functional identities of BolSULTR members in Groups II and III have not been determined in broccoli, the increase in *BolSULTR* copies might contribute to the sulfur assimilations and the biosynthesis of glucosinolates and their derivatives because broccoli can accumulate glucosinolates, a group of sulfur-rich secondary metabolites abundantly found in broccoli.

4.2. Group II and III SULTRs Might Represent Flowering-Plant-Specific Transporters Associated with Long-Distance Movement of Sulfate

The SULTRs are widely distributed across the kingdom Plantae. In previous studies, two and four SULTRs were reported in *Chlamydomonas reinhardtii* and *moss Physcomitrium patens*, respectively. According to the phylogenetic trees, two green algae SULTRs were present only in Group IV, implying that *SULTR* duplication predated the divergence of unicellular and multicellular life, suggesting that they may represent the ancestral precursors of SULTR found in land plants. Likewise, four bryophyte SULTRs were found in Groups I and IV, suggesting that the SULTR family diverged into at least two groups in the common ancestor of bryophytes. It is noteworthy that all SULTR members in Groups II and III belong to flowering plants, which leads to the suggestion that they might be associated with the long-distance transport of sulfate in plants. In support of this concept, *Arabidopsis SULTR2;1* is specifically expressed in the pericycle and xylem parenchyma cells, controlling the amount of sulfate to be transported from the rhizosphere to the aboveground organs. Similarly, *Arabidopsis SULTR3;5* was found to be preferably expressed in root vasculature, and the *sultr3;5* mutant displayed a severe decrease in root to shoot sulfate movement [19]. Hence, it is not surprising that Group II and II SULTRs are absent in the non-vascular plants moss and algae. Therefore, we hypothesize that the expansion of two groups of SULTRs might have evolved in the vascular lineage as the multicellular land plants need long-distance sulfate transport from root to shoot.

4.3. BolSULTR2 Might Be Involved in Plant Defense Against Alternaria brassicicola (Ab), Presumably Through the Regulation of Glutathione Biosynthesis

It has been widely recognized that a sufficient sulfur supply is a prerequisite for plant responses to biotic stresses [29,61]. Because SULTRs are the major contributors to sulfate distributions within plants, they should theoretically impact plant immunity. In our study, the expression levels of several *BolSULTR* genes, including *BolSULTR2;1a/b*, *BolSULTR3;1a/b*, *BolSULTR3;3b*, and *BolSULTR3;4b/c*, were decreased at early stages, although some of them gradually returned to initial levels. Our results support the notion that some *BolSULTRs* might be repressed during the early stage of *Ab* attack, presumably altering the sulfate movements that favor plant immunity. Accordingly, promoter analysis of these *BolSULTRs* revealed the existence of the cis elements related to the MeJA, SA, and ethylene signaling pathways that are implicated in biotic stresses. Consistent with our findings, the expression levels of *OsSultr2;1/2;2* and *OsSultr3;1* were significantly reduced in the rice challenged by the fugal *Magnaporthe grisea* [28]. Similarly, the accumulation of *StuSULTR2* transcripts was reduced significantly when potato leaves were inoculated by the oomycetes *Phytophthora infestans* [62]. Thus, these findings indicate that plants could control sulfate transport through the regulation of SULTR genes, particularly at the early stage of pathogen infections. Similarly, we also observed that transient silencing of *BolSULTR2;1a* resulted in enhanced tolerance to *Ab* in broccoli, implying that BolSULTR2;1a might be a negative regulator of *Ab* resistance. However, stable transgenic lines are still needed to further determine how sulfur metabolism and plant resistance are regulated by *BolSULTR2;1a*. In the future, we will generate the *bolsultr2;1* knockout line through a gene editing method and investigate the distribution of sulfate in *bolsultr2;1* lines.

Glutathione (GSH) is an important sulfur compound and a key cellular redox-buffering tripeptide with irreplaceable roles in plant defense mechanisms. It has been well-established that glutathione is oxidized by reactive oxygen species (ROS) as the antioxidant barrier that ameliorates oxidative stress imposed on sensitive cellular components [63–65]. GSH acts as a non-enzymatic antioxidant that directly removes reactive oxygen species (ROS) and works in concert with enzymatic antioxidant systems, such as the ASA–GSH cycle. For example, ascorbate peroxidase (APX) relies on GSH to regenerate ascorbic acid (ASA), and elevated levels of GSH may enhance APX activity, facilitating the co-removal of hydrogen peroxide (H_2O_2) [66]. In our study, the GSH content of the *BolSULTR2;1*-silencing in the leaves was much higher in comparison to the wild-type and TRV:00 lines. As expected, elevated levels of GSH can enhance antioxidant enzymes' activity to work together to remove ROS, thereby protecting cells from oxidative stress, which provides an explanation for the *BolSULTR2;1*-silencing broccoli's increased resistance to *Ab*.

The *SULTR2.1* gene is involved in long-distance sulfate transport between leaves in *Arabidopsis*, and it is likely that *BolSULTR2;1s* might have similar capabilities in controlling sulfate movement. Therefore, our results perhaps show that upon *Ab* attack, broccoli leaves reduce the expression of *BolSULTR2;1s*, which transitorily reduces sulfate movement to non-infected counterparts. It is thus reasonable to assume that the accumulation of intracellular sulfate is conducive to elevated GSH biosynthesis during early infection, thus conferring increased tolerance to oxidative injuries imposed by *Ab* attack.

5. Conclusions

Sulfur is an essential element in plant growth resistance, and plant sulfur transporters play an essential role in the acquisition and distribution of sulfur. In this study, we identified 23 *BolSULTR* genes from the *Brassica oleracea* genome, classified and named them, studied the evolutional relationship, gene structure, and cis-regulatory elements of *BolSULTRs* in detail, and predicted the interaction proteins of the *BolSULTR* families. We characterized the expression patterns of *BolSULTRs* in different tissues and under different biological stresses. The VIGS experiment confirmed that silencing of *BolSULTR2;1* resulted in enhanced tolerance to *Ab* by regulating GSH biosynthesis. These findings provide new insights into broccoli sulfate transporters in plant disease tolerance. Furthermore, functional studies are needed to reveal the molecular mechanism of *BolSULTRs* in sulfate transport and *Ab* infections.

Supplementary Materials: The following supporting information can be downloaded at https://www.mdpi.com/article/10.3390/antiox14040496/s1, Figure S1: The 23 BolSULTR members were named according to their homology with their *Arabidopsis* counterparts; Figure S2: The conserved regions of Sulfate_transp (PF00916) or STAS (PF01740) were visualized on the ChiPlot; Figure S3: Distribution map of promoter action elements for 2 kb upstream *BolSULTR* genes; Figure S4: Highly homologous gene sequences of *BolSULTR2;1a*; Figure S5: Schematic diagram of contagious disease; Table S1: The 23 BolSULTR members were identified and named according to their homology with their Arabidopsis counterparts; Table S2: The BolSULTR gene pairs had a Ka/Ks ratio and duplication type; Table S3: The relevant primer sequences; Table S4: RNA-seq datasets of Brassica from the GSE42891 (GEO database); Table S5: RNA-seq data of *BolSULTR* genes from leaves at five time points under Alternaria brassicicola stress in broccoli susceptible cultivar '41-3'; Table S6: Browse interactions in tabular form (lists only one-way edges: A-B).

Author Contributions: Conceptualization, N.L. (Ning Liu); funding acquisition, H.Z. and N.L. (Ning Liu); investigation, G.W. and N.L. (Ning Li); resources, Y.D.; supervision, H.Z. and N.L. (Ning Liu); writing—original draft, G.W.; writing—review and editing, N.L. (Ning Liu). All authors have read and agreed to the published version of the manuscript.

Funding: This project was financially supported by the "Collaborative Innovation Program (XTCX202302)", the "Innovation and Development Program (KYCX202303)", the "Reform & Development Program (KYCX202406)", the "Cultivation of Green and High-quality Vegetable Varieties (030360007)" program of the Beijing Vegetable Research Center, and the "Young Talent Supporting Program (YCXTD00002-10)" of the Beijing Academy of Agriculture and Forestry Sciences.

Informed Consent Statement: Not applicable.

Data Availability Statement: The original contributions presented in this study are included in the article/Supplementary Materials; further inquiries can be directed to the corresponding author.

Acknowledgments: We thank to Decai Yu (Yunnan Agricultural University) for fruitful discussions on this manuscript.

Conflicts of Interest: The authors declare that they have no known competing financial interests or personal relationships that could have appeared to influence the work reported in this paper.

References

1. Zenda, T.; Liu, S.; Dong, A.; Duan, H. Revisiting Sulphur—The Once Neglected Nutrient: It's Roles in Plant Growth, Metabolism, Stress Tolerance and Crop Production. *Agriculture* **2021**, *11*, 626. [CrossRef]
2. Ausma, T.; De Kok, L.J. Atmospheric H_2S: Impact on Plant Functioning. *Front. Plant Sci.* **2019**, *10*, 743. [CrossRef]
3. Buchner, P.; Stuiver, C.E.E.; Westerman, S.; Wirtz, M.; Hell, R.d.; Hawkesford, M.J.; De Kok, L.J. Regulation of Sulfate Uptake and Expression of Sulfate Transporter Genes in *Brassica oleracea* as Affected by Atmospheric H_2S and Pedospheric Sulfate Nutrition. *Plant Physiol.* **2004**, *136*, 3396–3408. [CrossRef] [PubMed]
4. Takahashi, H.; Buchner, P.; Yoshimoto, N.; Hawkesford, M.J.; Shiu, S.-H. Evolutionary Relationships and Functional Diversity of Plant Sulfate Transporters. *Front. Plant Sci.* **2012**, *2*, 119. [CrossRef]
5. Mugford, S.G.; Yoshimoto, N.; Reichelt, M.; Wirtz, M.; Hill, L.; Mugford, S.T.; Nakazato, Y.; Noji, M.; Takahashi, H.; Kramell, R. et al. Disruption of Adenosine-5′-Phosphosulfate Kinase in *Arabidopsis* Reduces Levels of Sulfated Secondary Metabolites. *Plant Cell* **2009**, *21*, 910–927. [CrossRef] [PubMed]
6. Kopriva, S.; Calderwood, A.; Weckopp, S.C.; Koprivova, A. Plant Sulfur and Big Data. *Plant Sci.* **2015**, *241*, 1–10. [CrossRef]
7. Saito, K. Sulfur Assimilatory Metabolism. The Long and Smelling Road. *Plant Physiol.* **2004**, *136*, 2443–2450. [CrossRef]
8. Kopriva, S. Control of Sulphate Assimilation and Glutathione Synthesis: Interaction With N and C Metabolism. *J. Exp. Bot.* **2004** *55*, 1831–1842. [CrossRef]
9. Kopriva, S. Regulation of Sulfate Sssimilation in *Arabidopsis* and Beyond. *Ann. Bot.* **2006**, *97*, 479–495. [CrossRef]
10. Shibagaki, N.; Grossman, A.R. Binding of Cysteine Synthase to the STAS Domain of Sulfate Transporter and Its Regulatory Consequences. *J. Biol. Chem.* **2010**, *285*, 25094–25102. [CrossRef]
11. Wang, L.; Chen, K.; Zhou, M. Structure and Function of an *Arabidopsis thaliana* Sulfate Transporter. *Nat. Commun.* **2021**, *12*, 4455. [CrossRef] [PubMed]
12. Hawkesford, M.J. Transporter Gene Families in Plants: The Sulphate Transporter Gene Family—Redundancy or Specialization? *Physiol Plant.* **2003**, *117*, 155–163. [CrossRef]
13. Gigolashvili, T.; Kopriva, S. Transporters in plant sulfur metabolism. *Front. Plant Sci.* **2014**, *5*, 442. [CrossRef] [PubMed]
14. Yoshimoto, N.; Takahashi, H.; Smith, F.W.; Yamaya, T.; Saito, K. Two Distinct High-Affinity Sulfate Transporters with Different Inducibilities Mediate Uptake of Sulfate in *Arabidopsis* Roots. *Plant J.* **2002**, *29*, 465–473. [CrossRef]
15. Yoshimoto, N.; Inoue, E.; Saito, K.; Yamaya, T.; Takahashi, H. Phloem-Localizing Sulfate Transporter, Sultr1;3, Mediates Re-distribution of Sulfur From Source to Sink Organs in *Arabidopsis*. *Plant Physiol.* **2003**, *131*, 1511–1517. [CrossRef]
16. Takahashi, H.; Watanabe-Takahashi, A.; Smith, F.W.; Blake-Kalff, M.; Hawkesford, M.J.; Saito, K. The Roles of Three Functional Sulphate Transporters Involved in Uptake and Translocation of Sulphate in *Arabidopsis thaliana*. *Plant J.* **2000**, *23*, 171–182 [CrossRef]
17. Chen, Z.; Zhao, P.-X.; Miao, Z.-Q.; Qi, G.-F.; Wang, Z.; Yuan, Y.; Ahmad, N.; Cao, M.-J.; Hell, R.; Wirtz, M.; et al. SULTR3s Function in Chloroplast Sulfate Uptake and Affect ABA Biosynthesis and the Stress Response. *Plant Physiol.* **2019**, *180*, 593–604. [CrossRef]
18. Cao, M.J.; Wang, Z.; Wirtz, M.; Hell, R.; Oliver, D.J.; Xiang, C.B. SULTR3;1 is a chloroplast-localized sulfate transporter in Arabidopsis thaliana. *Plant J.* **2013**, *73*, 607–616. [CrossRef]
19. Kataoka, T.; Hayashi, N.; Yamaya, T.; Takahashi, H. Root-to-shoot Transport of Sulfate in *Arabidopsis*. Evidence for the Role of SULTR3;5 as a Component of Low-Affinity Sulfate Transport System in the Root Vasculature. *Plant Physiol.* **2004**, *136*, 4198–4204. [CrossRef]
20. Kataoka, T.; Watanabe-Takahashi, A.; Hayashi, N.; Ohnishi, M.; Mimura, T.; Buchner, P.; Hawkesford, M.J.; Yamaya, T.; Takahashi, H. Vacuolar Sulfate Transporters Are Essential Determinants Controlling Internal Distribution of Sulfate in *Arabidopsis*. *Plant Cell* **2004**, *16*, 2693–2704. [CrossRef]
21. Maruyama-Nakashita, A.; Inoue, E.; Watanabe-Takahashi, A.; Yamaya, T.; Takahashi, H. Transcriptome Profiling of Sulfur-Responsive Genes in *Arabidopsis* Reveals Global Effects of Sulfur Nutrition on Multiple Metabolic Pathways. *Plant Physiol.* **2003**, *132*, 597–605. [CrossRef]
22. Liang, G.; Yang, F.; Yu, D. MicroRNA395 Mediates Regulation of Sulfate Accumulation and Allocation in *Arabidopsis thaliana*. *Plant J.* **2010**, *62*, 1046–1057. [CrossRef] [PubMed]
23. Soudthedlath, K.; Nakamura, T.; Ushiwatari, T.; Fukazawa, J.; Osakabe, K.; Osakabe, Y.; Maruyama-Nakashita, A. SULTR2;1 Adjusts the Bolting Timing by Transporting Sulfate from Rosette Leaves to the Primary Stem. *Plant Cell Physiol.* **2024**, *65*, 770–780. [CrossRef] [PubMed]
24. Takahashi, H. Regulation of Sulfate Transport and Assimilation in Plants. *Int. Rev. Cell Mol. Biol.* **2010**, *281*, 129–159. [CrossRef] [PubMed]
25. Bachelet, F.; Sanchez, M.; Aimé, D.; Naudé, F.; Rossin, N.; Ourry, A.; Deulvot, C.; Le Signor, C.; Vernoud, V.; Neiers, F.; et al. The Vacuolar Sulfate Transporter PsSULTR4 is a Key Determinant of Seed Yield and Protein Composition in Pea. *Plant J.* **2024**, *119*, 2919–2936. [CrossRef]

26. Dubousset, L.; Abdallah, M.; Desfeux, A.S.; Etienne, P.; Meuriot, F.; Hawkesford, M.J.; Gombert, J.; Ségura, R.; Bataillé, M.P.; Rezé, S.; et al. Remobilization of Leaf S Compounds and Senescence in Response to Restricted Sulphate Supply During the Vegetative Stage of Oilseed Rape are Affected by Mineral N Availability. *J. Exp. Bot.* **2009**, *60*, 3239–3253. [CrossRef]
27. Liu, N.; Hu, M.; Liang, H.; Tong, J.; Xie, L.; Wang, B.; Ji, Y.; Han, B.; He, H.; Liu, M.; et al. Physiological, Transcriptomic, and Metabolic Analyses Reveal that Mild Salinity Improves the Growth, Nutrition, and Flavor properties of Hydroponic Chinese Chive (*Allium tuberosum* Rottler ex Spr). *Front. Nutr.* **2022**, *9*, 1000271. [CrossRef]
28. Yang, Z.; Hui, S.; Lv, Y.; Zhang, M.; Chen, D.; Tian, J.; Zhang, H.; Liu, H.; Cao, J.; Xie, W.; et al. miR395-Regulated Sulfate Metabolism Exploits Pathogen Sensitivity to Sulfate to Boost Immunity in Rice. *Mol. Plant* **2022**, *15*, 671–688. [CrossRef]
29. Gallardo, K.; Courty, P.E.; Le Signor, C.; Wipf, D.; Vernoud, V. Sulfate Transporters in the Plant's Response to Drought and Salinity: Regulation and Possible Functions. *Front. Plant Sci.* **2014**, *5*, 580. [CrossRef]
30. Cao, M.J.; Wang, Z.; Zhao, Q.; Mao, J.L.; Speiser, A.; Wirtz, M.; Hell, R.; Zhu, J.K.; Xiang, C.B. Sulfate Availability Affects ABA Levels and Germination Response to ABA and Salt Stress in *Arabidopsis thaliana*. *Plant J.* **2014**, *77*, 604–615. [CrossRef]
31. Abdull Razis, A.F.; Noor, N.M. Cruciferous Vegetables: Dietary Phytochemicals for Cancer Prevention. *Asian Pac. J. Cancer Prev.* **2013**, *14*, 1565–1570. [CrossRef] [PubMed]
32. Salehi, B.; Quispe, C.; Butnariu, M.; Sarac, I.; Marmouzi, I.; Kamle, M.; Tripathi, V.; Kumar, P.; Bouyahya, A.; Capanoglu, E.; et al. Phytotherapy and Food Applications from *Brassica* Genus. *Phytother. Res.* **2021**, *35*, 3590–3609. [CrossRef]
33. Zhang, Q.; Branca, F.; Li, N.; Liu, N.; Ding, Y. Identification of Black Spot Resistance in Broccoli (*Brassica oleracea* L. var. *italica*) Germplasm Resources. *Appl. Sci.* **2024**, *14*, 2883. [CrossRef]
34. Kumar, P.; Srivastava, D.K. Biotechnological Advancement in Genetic Improvement of Broccoli (*Brassica oleracea* L. var. *italica*), an Important Vegetable Crop. *Biotechnol. Lett.* **2016**, *38*, 1049–1063. [CrossRef]
35. Cai, X.; Wu, J.; Liang, J.; Lin, R.; Zhang, K.; Cheng, F.; Wang, X. Improved *Brassica oleracea* JZS Assembly Reveals Significant Changing of LTR-RT Dynamics in Different Morphotypes. *Theor. Appl. Genet.* **2020**, *133*, 3187–3199. [CrossRef] [PubMed]
36. Chen, C.; Chen, H.; Zhang, Y.; Thomas, H.R.; Frank, M.H.; He, Y.; Xia, R. TBtools: An Integrative Toolkit Developed for Interactive Analyses of Big Biological Data. *Mol. Plant* **2020**, *13*, 1194–1202. [CrossRef]
37. Mistry, J.; Chuguransky, S.; Williams, L.; Qureshi, M.; Salazar, G.A.; Sonnhammer, E.L.L.; Tosatto, S.C.E.; Paladin, L.; Raj, S.; Richardson, L.J.; et al. Pfam: The Protein Families Database in 2021. *Nucleic Acids Res.* **2021**, *49*, D412–D419. [CrossRef]
38. Marchler-Bauer, A.; Derbyshire, M.K.; Gonzales, N.R.; Lu, S.; Chitsaz, F.; Geer, L.Y.; Geer, R.C.; He, J.; Gwadz, M.; Hurwitz, D.I.; et al. CDD: NCBI's Conserved Domain Database. *Nucleic Acids Res.* **2015**, *43*, D222–D226. [CrossRef]
39. Finn, R.D.; Bateman, A.; Clements, J.; Coggill, P.; Eberhardt, R.Y.; Eddy, S.R.; Heger, A.; Hetherington, K.; Holm, L.; Mistry, J.; et al. Pfam: The Protein Families Database. *Nucleic Acids Res.* **2014**, *42*, D222–D230. [CrossRef]
40. Duvaud, S.; Gabella, C.; Lisacek, F.; Stockinger, H.; Ioannidis, V.; Durinx, C. Expasy, the Swiss Bioinformatics Resource Portal, as Designed by its Users. *Nucleic Acids Res.* **2021**, *49*, W216–W227. [CrossRef]
41. Newbigin, E.; Chou, K.-C.; Shen, H.-B. Plant-mPLoc: A Top-Down Strategy to Augment the Power for Predicting Plant Protein Subcellular Localization. *PLoS ONE* **2010**, *5*, e11335. [CrossRef]
42. Kumar, S.; Stecher, G.; Li, M.; Knyaz, C.; Tamura, K.; Battistuzzi, F.U. MEGA X: Molecular Evolutionary Genetics Analysis across Computing Platforms. *Mol. Biol. Evol.* **2018**, *35*, 1547–1549. [CrossRef] [PubMed]
43. Wang, Y.; Tang, H.; DeBarry, J.D.; Tan, X.; Li, J.; Wang, X.; Lee, T.-h.; Jin, H.; Marler, B.; Guo, H.; et al. MCScanX: A Toolkit for Detection and Evolutionary Analysis of Gene Synteny and Collinearity. *Nucleic Acids Res.* **2012**, *40*, e49. [CrossRef]
44. Lescot, M.; Déhais, P.; Thijs, G.; Marchal, K.; Moreau, Y.; Van de Peer, Y.; Rouzé, P.; Rombauts, S. PlantCARE, a Database of Plant Cis-Acting Regulatory Elements and a Portal to Tools for in Silico Analysis of Promoter Sequences. *Nucleic Acids Res.* **2002**, *30*, 325–327. [CrossRef] [PubMed]
45. Li, N.; Xie, L.; Hu, M.; Tong, J.; Wang, B.; Ji, Y.; Chen, J.; Liang, H.; Liu, W.; Liu, M.; et al. Effects of Blanching Cultivation on the Chemical Composition and Nutritional Quality of Chinese Chive. *Food Chem.* **2025**, *464*, 141824. [CrossRef]
46. Lai, M.; Cheng, Z.; Xiao, L.; Klosterman, S.J.; Wang, Y. The bZip Transcription Factor VdMRTF1 is a Negative Regulator of Melanin Biosynthesis and Virulence in *Verticillium dahliae*. *Microbiol. Spectr.* **2022**, *10*, e02581-21. [CrossRef]
47. Schnable, J.C.; Wang, X.; Pires, J.C.; Freeling, M. Escape from Preferential Retention Following Repeated Whole Genome Duplications in Plants. *Front. Plant Sci.* **2012**, *3*, 94. [CrossRef]
48. Rahman, M.M.; Ghosh, P.K.; Akter, M.; Al Noor, M.M.; Rahman, M.A.; Keya, S.S.; Roni, M.S.; Biswas, A.; Bulle, M. Green Vanguards: Harnessing the Power of Plant Antioxidants, Signal Catalysts, and Genetic Engineering to Combat Reactive Oxygen Species under Multiple Abiotic Stresses. *Plant Stress* **2024**, *13*, 100547. [CrossRef]
49. Roychowdhury, R.; Hada, A.; Biswas, S.; Mishra, S.; Prusty, M.R.; Das, S.P.; Ray, S.; Kumar, A.; Sarker, U. Jasmonic Acid (JA) in Plant Immune Response: Unravelling Complex Molecular Mechanisms and Networking of Defence Signalling Against Pathogens. *J. Plant Growth Regul.* **2024**, *44*, 89–114. [CrossRef]

50. Uji, Y.; Kashihara, K.; Kiyama, H.; Mochizuki, S.; Akimitsu, K.; Gomi, K. Jasmonic Acid-induced VQ-motif-Containing Protein OsVQ13 Influences the OsWRKY45 Signaling Pathway and Grain Size by Associating with OsMPK6 in Rice. *Int. J. Mol. Sci.* **2019**, *20*, 2917. [CrossRef]
51. Wang, Z.; Tan, X.; Zhang, Z.; Gu, S.; Li, G.; Shi, H. Defense to Sclerotinia Sclerotiorum in Oilseed Rape is Associated with the Sequential Activations of Salicylic Acid Signaling and Jasmonic Acid Signaling. *Plant Sci.* **2012**, *184*, 75–82. [CrossRef]
52. Capaldi, F.R.; Gratão, P.L.; Reis, A.R.; Lima, L.W.; Azevedo, R.A. Sulfur Metabolism and Stress Defense Responses in Plants. *Trop Plant Biol.* **2015**, *8*, 60–73. [CrossRef]
53. Leustek, T.; Saito, K. Sulfate Transport and Assimilation in Plants. *Plant Physiol.* **1999**, *120*, 637–644. [CrossRef] [PubMed]
54. Kopriva, S.; Takahashi, H. Sulfate Transport Systems in Plants: Functional Diversity and Molecular Mechanisms Underlying Regulatory Coordination. *J. Exp. Bot.* **2019**, *70*, 4075–4087. [CrossRef]
55. Kumar, S.; Asif, M.H.; Chakrabarty, D.; Tripathi, R.D.; Trivedi, P.K. Differential Expression and Alternative Splicing of Rice Sulphate Transporter Family Members Regulate Sulphur Status during Plant Growth, Development and Stress Conditions. *Funct Integr. Genom.* **2011**, *11*, 259–273. [CrossRef]
56. Huang, Q.; Wang, M.; Xia, Z. The SULTR Gene Family in Maize (*Zea mays* L.): Gene Cloning and Expression Analyses under Sulfate Starvation and Abiotic Stress. *J. Plant Physiol.* **2018**, *220*, 24–33. [CrossRef]
57. Yang, X.; Liao, X.; Yu, L.; Rao, S.; Chen, Q.; Zhu, Z.; Cong, X.; Zhang, W.; Ye, J.; Cheng, S.; et al. Combined Metabolome and Transcriptome Analysis Reveal the Mechanism of Selenate Influence on the Growth and Quality of Cabbage (*Brassica oleracea* var *capitata* L.). *Food Res. Int.* **2022**, *156*, 111135. [CrossRef]
58. Clark, J.W.; Donoghue, P.C.J. Whole-Genome Duplication and Plant Macroevolution. *Trends Plant Sci.* **2018**, *23*, 933–945. [CrossRef]
59. Liang, Z.; Schnable, J.C. Functional Divergence between Subgenomes and Gene Pairs after Whole Genome Duplications. *Mol Plant* **2018**, *11*, 388–397. [CrossRef]
60. De Smet, R.; Adams, K.L.; Vandepoele, K.; Van Montagu, M.C.E.; Maere, S.; Van de Peer, Y. Convergent Gene Loss Following Gene and Genome Duplications Creates Single-Copy Families in Flowering Plants. *Proc. Natl. Acad. Sci. USA* **2013**, *110*, 2898–2903. [CrossRef]
61. Kruse, C.; Haas, F.H.; Jost, R.; Reiser, B.; Reichelt, M.; Wirtz, M.; Gershenzon, J.; Schnug, E.; Hell, R. Improved Sulfur Nutrition Provides the Basis for Enhanced Production of Sulfur-Containing Defense Compounds in *Arabidopsis thaliana* upon Inoculation with *Alternaria brassicicola*. *J. Plant Physiol.* **2012**, *169*, 740–743. [CrossRef]
62. Vatansever, R.; Koc, I.; Ozyigit, I.I.; Sen, U.; Uras, M.E.; Anjum, N.A.; Pereira, E.; Filiz, E. Genome-wide Identification and Expression Analysis of Sulfate Transporter (SULTR) Genes in Potato (*Solanum tuberosum* L.). *Planta* **2016**, *244*, 1167–1183. [CrossRef] [PubMed]
63. Salih, H.; Bai, W.; Liang, Y.; Yang, R.; Zhao, M.; Muhammd, S.M.; Zhang, D.; Li, X. ROS Scavenging Enzyme-encoding Genes Play Important Roles in the Desert Moss Syntrichia Caninervis Response to Extreme Cold and Desiccation Stresses. *Int. J. Biol. Macromol.* **2024**, *254*, 127778. [CrossRef] [PubMed]
64. Ito, T.; Ohkama-Ohtsu, N.; Marsolais, F. Degradation of Glutathione and Glutathione Conjugates in Plants. *J. Exp. Bot.* **2023**, *74*, 3313–3327. [CrossRef] [PubMed]
65. Singh, D. Juggling with Reactive Oxygen Species and Antioxidant Defense System—A Coping Mechanism under Salt Stress *Plant Stress* **2022**, *5*, 100093. [CrossRef]
66. Li, S. Novel Insight into Functions of Ascorbate Peroxidase in Higher Plants: More than a Simple Antioxidant Enzyme. *Redox Biol.* **2023**, *64*, 102789. [CrossRef]

antioxidants

MDPI

Article

Greater Biomass Production Under Elevated CO_2 Is Attributed to Physiological Optimality, Trade-Offs in Nutrient Allocation, and Oxidative Defense in Drought-Stressed Mulberry

Songmei Shi [1,2], Huakang Li [1], Xinju Wang [1], Ziran Wang [1], Junqiang Xu [1], Xinhua He [2,3,4,5,*] and Zheng'an Yang [1,*]

1 Key Laboratory of Vegetable Biology in Yunnan, College of Landscape and Horticulture, Yunnan Agricultural University, Kunming 650201, China; shismei2022@ynau.edu.cn (S.S.); 2023210206@stu.ynau.edu.cn (H.L.); 2022210207@stu.ynau.edu.cn (X.W.); wangziran@cau.edu.cn (Z.W.); xujunqiang101@163.com (J.X.)

2 Centre of Excellence for Soil Biology, College of Resources and Environment, Southwest University, Chongqing 400716, China

3 College of Forestry, Sichuan Agricultural University, Chengdu 611130, China

4 School of Biological Sciences, University of Western Australia, Perth, WA 6009, Australia

5 Department of Land, Air and Water Resources, University of California at Davis, Davis, CA 95616, USA

* Correspondence: xinhua.he@uwa.edu.au (X.H.); yangzhengan@ynau.edu.cn (Z.Y.)

Abstract: Mulberry (*Morus alba* L.), a species of significant ecological and economic importance, is widely cultivated for sericulture, soil conservation, and environmental restoration. Despite its remarkable resilience to environmental stresses, the combined impact of elevated CO_2 (eCO_2) and drought stress on aboveground–root–soil interactions remains poorly understood, particularly in the context of global climate change. Here, we investigated the effects of eCO_2 and drought on physiological leaf and root indicators, nutrient absorption and allocation, and soil properties in mulberry seedlings. Mulberry seedlings were grown in environmentally auto-controlled growth chambers under ambient CO_2 (420/470 ppm, day/night) or eCO_2 (710/760 ppm) and well-watered (75–85% soil relative water content, RWC), moderate-drought (55–65% RWC), or severe-drought (35–45% RWC) conditions. Results showed that both above- and below-ground plant biomass production were significantly promoted by eCO_2, particularly by 36% and 15% under severe drought, respectively. This could be attributed to several factors. Firstly, eCO_2 improved leaf photosynthesis by 25–37% and water use efficiency by 104–163% under drought stresses while reducing negative effects of drought on the effective quantum yield of PSII photochemistry and the photochemical quenching coefficient. Secondly, eCO_2 significantly decreased proline accumulation while increasing soluble sugar contents, as well as peroxidase and superoxide dismutase activities, in both leaves and roots under drought stress. Lastly, eCO_2 promoted soil sucrase, urease, and phosphatase activities, as well as plant nitrogen, phosphorus and potassium uptake while facilitating their allocation into roots under drought stress. These findings demonstrate that eCO_2 enhanced the drought tolerance of mulberry plants through improvements in photosystem II efficiency, water use efficiency, antioxidative defense capacity, and nutrient uptake and allocation, providing critical insights for sustainable mulberry plantation management under future climate change scenarios.

Keywords: antioxidant mechanisms; chlorophyll fluorescence; leaf photosynthesis; nutrient uptake; oxidative stress; plant biomass

1. Introduction

Ambient CO_2 (aCO_2) concentrations have increased from 280 ppm at the beginning of the industrial revolution to the current ~420 ppm, primarily due to anthropogenic activities such as extensive utilization of fossil fuels, deforestation, etc. According to the IPCC (2018), aCO_2 could potentially reach to 550 ppm or higher by the end of this century, with significant implications for global climate patterns and plant responses [1]. Elevated CO_2 (eCO_2) contributes to temperature increases through the greenhouse effect, inducing spatially heterogeneous precipitation frequencies and patterns on a global scale, eventually leading to regional droughts under future climate change [2,3]. Additionally, eCO_2 is commonly practiced in controlled environments such as greenhouses to enhance plant growth and productivity, making it relevant for both natural and agricultural systems. Drought stress typically affects various cellular and physiological processes, such as protein degradation, changes in membrane fluidity, and the inhibition of leaf photosynthesis, ultimately resulting in a decrease in crop yield by 25–30% and deteriorated quality [4–6]. However, it is worth noting that the growth-stimulating effect of eCO_2 can partially offset the impacts caused by drought stress, particularly in C_3 species [7–9]. Therefore, it is crucial to consider the abovementioned effects when assessing the overall impact of climate change on plant growth.

To comprehend the potential mechanism of agricultural production in response to eCO_2 and drought stress, it is imperative to investigate alterations in the physiological processes and biochemical components of plants that are sensitive to both eCO_2 and drought stress, such as leaf photosynthesis, chlorophyll fluorescence, soluble sugars, proline, and antioxidative ability [9–12]. Plant photosynthesis is a drought-sensitive process, and water deficit usually leads to a significant decline in leaf photosynthesis [5,10,13]. The decrease in leaf photosynthesis under water deficit may be associated with reduced stomatal conductance [14], nitrogen (N) and chlorophyll contents [15], and 1,5-bisphophate carboxylase/oxygenase (Rubisco) activity [16]. It has been observed that drought stress decreases the effective quantum yield of PSII photochemistry in wheat while also causing increased accumulation of proline and soluble sugar contents in barley and buckwheat [17,18]. Additionally, antioxidant enzymes such as superoxide dismutase (SOD), peroxidase (POD), and catalase (CAT) were found to be highly activated in soybean under drought stress in order to adapt and scavenge the free radicals induced by drought stress, accompanied by increased accumulation of malondialdehyde (MDA) [19,20]. Therefore, drought stress limits crop growth and grain yield by reducing aboveground biomass and modifying the shoot-to-root ratio through the development of deeper roots capable of acquiring more water [21].

The increase in atmospheric CO_2 concentration, the primary C source for plants, exerts significant CO_2 fertilization effects on crops [22–24] because the Rubisco enzyme in C_3 plants is not saturated at the current atmospheric CO_2 concentration [25,26]; thus, eCO_2 can stimulate leaf photosynthesis and enhance plant growth and grain yield by facilitating C_3 the uptake of more CO_2 molecules by plants [27]. Previous studies have revealed that eCO_2 can mitigate the adverse effects of drought stress on plant growth and physiological functioning by increasing the intercellular CO_2 concentration (Ci), improving water use efficiency (WUE) while reducing transpiration rates [28–31]. A meta-analysis showed that eCO_2 significantly increases both the net photosynthetic rate (37.2–46.3%) and WUE (84.2–88.9%) [9] while decreasing evapotranspiration by ~10% [32]. The increased WUE under eCO_2 implies that plants may exhibit tolerance to future high CO_2 concentrations under drought conditions [8,33]. Moreover, enhanced photosynthesis under eCO_2 leads to an accumulation of carbohydrates under drought stress, resulting in increased cell osmotic potential and promoting osmotic adjustment [34]. Concurrently, the formation of reactive oxygen species (ROS) was also found to be mitigated under eCO_2 due to increased electron consumption in C fixation processes [7,35]. However, the capacity of the CO_2 fertilization effect to alleviate drought stress may vary not only among different plant species [36,37] but also depending on the levels of water deficit [30,38]. For instance, some studies have reported that eCO_2 can effectively mitigate drought stress by enhancing WUE and photosynthesis in certain species, such as coffee [21] and soybean [19,33]. In contrast, other investigations have revealed that eCO_2 did not significantly improve drought tolerance or even exacerbated water stress in species like wheat and sorghum, particularly under severe drought conditions [39,40]. These inconsistent findings may be attributed to differences in species-specific traits, experimental conditions, and the severity of the applied drought stress. For example, C_4 plants, which have a different photosynthetic pathway, may not benefit as much from eCO_2 as C_3 plants under drought stress [41,42]. Thus, there is currently no consistency regarding the specific physiological mechanisms by which the eCO_2 concentration can affect plant growth and crop yield under drought stress.

In addition to its impact on plant physiology, drought also significantly affects the uptake, accumulation, and distribution of soil nutrients (particularly nitrogen (N), phosphorus (P), and potassium (K)). A recent meta-analysis revealed that drought stress generally decreases nutrient absorption from soil and leads to decreased N and P concentrations in plant tissue [43–45]. In contrast, eCO_2 can alter the shoot-to-root ratio by stimulating deeper root growth, thereby enabling better water and nutrients uptake [9,30]. In general, plants under eCO_2 store more C in their above- and below-ground biomass compared to those under aCO_2 [35], leading to changes in their contents of N, P, and K and other nutrients in both plants and soil [46]. Meanwhile, eCO_2 increases fluxes of organic compounds into soil due to a higher rate of plant litterfall and rhizodeposition, resulting in changes in soil organic carbon (SOC) content [47], the availability of soil nutrients [44], composition and activities of the microbial community [48,49], and activities of soil enzymes involved in C and N cycling [47,50]. Such changes in soil are predicted to result in stronger but uncertain feedback at the ecosystem level to eCO_2. However, most studies have explored the individual effects of focusing on either eCO_2 or drought stress on a plant–soil–microbe system, with limited studies addressing their combined effects on plant nutrient distribution and accumulation, soil nutrient availability, and enzyme activity.

Mulberry (*Morus alba* L.), a rapidly growing multipurpose cash plant, is extensively cultivated in southwest China, accounting for approximately 70% of China's total plantations (~758,200 hectares) due to its significant economic value. However, this region is currently experiencing reduced precipitation and increased aridity, leading to a scarcity of freshwater resources for agricultural production [51]. These climatic changes pose a significant threat to mulberry cultivation, especially during prolonged drought periods [10,52]. Despite its sensitivity to extreme drought, mulberry exhibits remarkable adaptability to moderate environmental stressors, making it an ideal model for studying climate change adaptation [53,54]. Its high plasticity, deep root system, and efficient water use mechanisms enable it to thrive in arid and semi-arid regions, positioning it as a critical crop for sustainable agriculture in water-limited environments [55,56]. Additionally, mulberry leaves are the primary food source for silkworms, which are integral to the sericulture industry, a major economic driver in many developing regions [57]. The plant's ability to produce high levels of biomass under moderate stress conditions, coupled with its role in carbon sequestration, further underscores its importance in mitigating climate change impacts [58]. Although studies have reported the morphological, physiological, and biochemical responses of mulberry plants to drought stress [10,52,59,60], most of these studies primarily focused on different mulberry cultivars and/or altered soil–water gradients without considering the potential effect of CO_2. We previously observed a significant 33–40% increase in leaf biomass production of mulberry when the CO_2 concentration was elevated from

410 to 710 ppm [61,62]. Leaf photosynthesis and WUE were enhanced by 550–800 ppm eCO_2 in various mulberry cultivars, including Selection-13, Kanva-2, 'Qinglong' mulberry, Gui-sang-you 62, Qiangsang-1, and Nongsang-14, under well-watered conditions [61–65]. However, understanding regarding how eCO_2 affects drought-induced changes in photosynthesis, PSII function, antioxidant defense mechanisms, and nutrient acquisition and partitioning, as well as soil enzyme activities. By understanding how mulberry responds to combined drought and eCO_2 conditions, we can develop strategies to enhance its resilience and productivity, thereby supporting both economic and environmental sustainability. Here, we hypothesized that eCO_2 might have a positive impact on mulberry performance under drought conditions. Specifically, the objectives were to investigate (1) the combined effects of eCO_2 and drought stress on mulberry biomass accumulation; (2) the correlations between drought tolerance, physiological traits, and nutrient acquisition under eCO_2; and (3) the potential mechanisms through which eCO_2 could mitigate negative effects induced by drought stress via alterations in photosynthetic characteristics, antioxidant defense systems, nutrient uptake/distribution, and soil enzyme activities. Therefore, the following physiological traits of mulberry were investigated: leaf gas exchange, PSII efficiency, CAT, POD, SOD, MAD, and contents of proline and soluble sugar at both the leaf and root levels. Meanwhile, the uptake and partitioning of N, P, and K, along with soil enzyme activity. The anticipated findings will contribute to the development of effective water management strategies by elucidating the physiological responses of plants to future drought stress induced by elevated atmospheric CO_2 levels.

2. Materials and Methods

2.1. Growth Chamber Experiments

This experiment was conducted in 12 environmentally controlled plant growth chambers that were located at the National Monitoring Base for Purple Soil Fertility and Fertilizer Efficiency (29°48′ N, 106°24′ E, 266.3 m above the sea level) on the campus of Southwest University, Chongqing, China. The rectangular structure of each chamber measures $1.5 \times 1.0 \times 2.5$ m and is supported by a steel frame suspended 50 cm above a cement floor. The bottom floor of the growth chamber consists of polyvinyl chloride panels, while the four side walls and top roof are lined with by tempered glass (10 mm thick) with transparency allowing approximately 90% of natural sunlight to pass through. Humidity and temperature inside and outside the growth chamber were automatically maintained within ±0.5 °C air temperature and ±5% humidity using electronic control systems. When the humidity inside a chamber was higher than that outside, the inside air was pumped out using pump controlled by a mini-computer and filtered with solid anhydrous calcium chloride. The temperature was automatically maintained at 0.5 °C variation between inside and outside using an air conditioner controlled by a mini-computer. To maintain a targeted CO_2 concentration (±30 ppm) inside each chamber, 90% pure CO_2 gas was injected using a solenoid valve controlled by a mini-computer [42].

The experiment was conducted using a split-plot design consisting two factors (CO_2 and water or drought), with CO_2 as the main factor and drought as the sub-factor. Based on previously observed daytime and night-time atmospheric CO_2 concentrations at the study site, two different CO_2 concentrations were applied: ambient CO_2 (aCO_2, 420 ppm daytime/470 ppm nighttime) for six growth chambers and eCO_2 (710 ppm daytime/760 ppm nighttime) for other six growth chambers. Daytime was defined as 07:00 a.m. to 19:00 p.m., while nighttime was defined 19:00 p.m. to 07:00 a.m. The experimental designs and the procedures are illustrated in Figure 1.

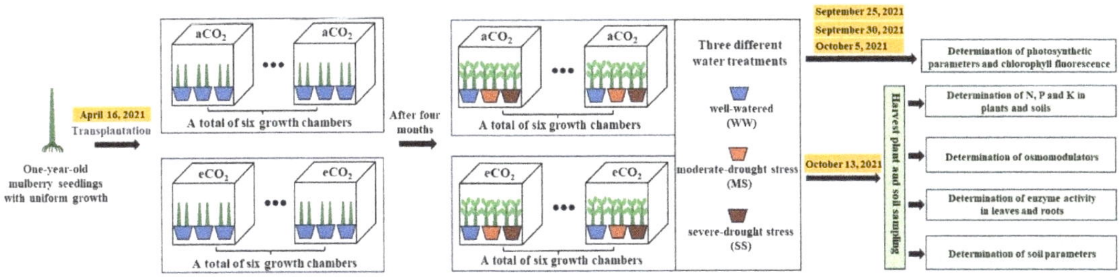

Figure 1. A schematic diagram showing the experimental designs and procedures. aCO_2: ambient CO_2 (420/470 ppm, day/night); eCO_2: elevated CO_2 (710/760 ppm).

One-year-old, uniformly growing mulberry seedlings (*Morus multicaulis* Perr. QiangSang-1) were used as experimental materials. Two seedlings were planted in a plastic pot (32 cm in diameter and 20 cm in height). The pots were filled with 11 kg air-dried soil (Eutric Regosol, FAO Soil Classification System). The soil was sieved (2 mm) and had a pH of 6.8, organic carbon content of 7.03 g kg^{-1}, available N content of 31.21 mg kg^{-1}, available P content of 11.47 mg kg^{-1}, and available K content of 127 mg kg^{-1}. Three pots were placed into each growth chamber. Mulberry seedlings were transplanted on 16 April 2021. During the first four months, all of the seedlings were regularly watered two or three times per week with 400 mL of tap water for each pot so that soil relative water content (RWC) was maintained at 75–85%. Meanwhile, 2.2 g N, 0.77 g P_2O_5, and 1.1 g K_2O were applied per pot to meet the plants' nutrient requirements.

After the four-month experimental period, three pots from each CO_2 chamber were subjected to three different water treatments: well-watered (WW), moderate drought stress (MS), and severe drought stress (SS). Under the well-watered condition, plants were consistently irrigated as described above to maintain an RWC of 75–85%. Plants experiencing moderate drought stress received irrigation twice weekly with 200 mL of water per pot in order to maintain an RWC of 55–65%. Plants exposed to severe stress were irrigated twice weekly with 100 mL of water per pot to maintain an RWC of 35–45%. Soil volumetric water contents were measured using a soil moisture sensor (JXBS-3001, Qingdao, China), then converted to gravimetric water contents. Soil relative water content, which was calculated as the ratio of gravimetric water to saturated moisture content. These three water treatments were performed in each chamber; thus, there were six replicated chambers and six replicate pots for each CO_2 concentration treatment. Additionally, pots were relocated among different chambers once fortnightly to minimize any potential effect of environmental variations across different growth chambers.

2.2. Determination of Photosynthetic Parameters

Measurements of photosynthetic parameters were conducted prior to harvest between 9:00 and 11:00 a.m. on sunny days, specifically on 25, 30 September and 5 October 2021 (data averaged for these three days). The third fully expanded leaf from the top was selected for determination of the net photosynthetic rate (Pn), stomatal conductance (gs), transpiration rate (E), and intercellular CO_2 concentration (Ci) using a Li-6800 portable photosynthesis system (LICOR, Lincoln, NE, USA). Throughout the measurement, the saturated photosynthetically active radiation was maintained at 1600 μmolm^{-2} s^{-1} using an integrated LED light source (Model 6800-01A, LI-COR Inc.) within the leaf chamber (Model 6800-02A Large Leaf Chamber, LI-COR Inc.). This light intensity was selected based on preliminary light-response curve experiments, which confirmed that it was sufficient to achieve light-saturated photosynthetic rates in the studied mulberry leaves [64]. The LED source provides uniform light distribution and stable intensity, ensuring accurate and reproducible measurements. Meanwhile, the leaf temperature in the leaf chamber was set at 25 ± 1 °C. For plants growing under eCO_2 and aCO_2, the leaf cavity's CO_2 concentration was maintained at 700 ppm and 420 ppm, respectively. The vapor-pressure deficit at the leaf surface was consistently maintained at 1.2 ± 0.2 kPa. Water use efficiency (WUE) was calculated as the ratio of Pn to E. The intercellular CO_2 and ambient CO_2 concentrations were used to the calculate Ci/Ca ratio.

2.3. Determination of Chlorophyll Fluorescence

The chlorophyll fluorescence was also determined using the Li-6800 fluorescence leaf chamber (LI-COR, Lincoln, USA) connected to an Li-6800 portable photosynthesis system on the same leaves used for photosynthetic leaf gas exchange measurements on 25 and 30 September and 5 October 2021 (data averaged for these three days). To ensure that all photosystem-II (PS-II) reaction centers were open, the leaves were wrapped with tin foil and darkened for 30 min. The following parameters were measured: initial fluorescence yield (F0) and maximum fluorescence yield (Fm) in the dark-adapted stage, minimum fluorescence yield (F0′) and maximum fluorescence (Fm′) in the light-adapted stage, and steady-stage fluorescence yield (Fs). Subsequently, the effective quantum yield of PSII photochemistry (Φ_{PSII}), maximal quantum yield of PSII photochemistry (Fv/Fm), actual photosynthetic efficiency of PSII under illumination (Fv′/Fm′), photochemical quenching coefficient (qP), non-photochemical quenching coefficient (NPQ), and PSII electron transport rate (ETR) were calculated using the aforementioned measured parameters. The calculation formulas are expressed as follows: $\Phi_{PSII} = (Fm' - Fs)/Fm'$, $NPQ = (Fm - Fm')/Fm'$, $qP = (Fm' - Fs)/(Fm' - F0')$, and $ETR = 0.8 \times \Phi_{PSII} \times PPFD$ [66].

2.4. Plant and Soil Sampling

The plants were harvested on 13 October 2021, and the leaves used to measure photosynthetic parameters were collected, immediately frozen in liquid nitrogen, and subsequently stored in a −80 °C freezer for determination of enzyme activity and osmomodulators. The plant tissues were divided into leaves, stems, and roots. The fresh roots were thoroughly washed with tap water, followed by rinsing with deionized water. A portion of the fresh roots was immediately frozen in liquid nitrogen after collection and stored at −80 °C until analysis of enzyme activity and osmomodulators. The remaining portion of the roots, leaves, and stems was dried at 105 °C for 30 min initially, then further dried at 75 °C for ≥ 48 h until reaching a constant weight to determine the biomass and N, P, K, etc. Soil samples from each growth pot were collected after thorough mixing and removal of debris and fine roots and subsequently air-dried (2 mm sieved) to determine their chemical properties and enzyme activities.

2.5. Determination of N, P, and K in Plants and Soils

The dried leaf, stem, and roots of each sample were finely ground into a powder before being subjected to digestion with 98% sulfuric acid and 30% hydrogen peroxide. Tissue N, P, and K concentrations were determined by the trace Kjeldahl method, vanadium molybdate yellow colorimetry, and flame photometry [67], respectively. Soil available N (AN) was measured using the micro-diffusion technique after alkaline hydrolysis. Soil available P (AP) was extracted with 0.5 M $NaHCO_3$, followed by measurement using the Mo-Sb anti spectrophotometric method. Soil available K (AK) was extracted with 1.0 M ammonium acetate, then determined by flame photometry [67].

2.6. Determination of Osmomodulators

The soluble sugars were determined using anthrone colorimetry [57]. Briefly, 0.5 g of fresh samples was extracted with 10 mL H_2O in boiling water for 20 min. The extraction was mixed with 0.5 mL of anthrone ethyl acetate and 5 mL of sulfuric acid, followed by incubation in boiling water for 10 min. The absorbance was measured spectrophotometrically at a wavelength of 620 nm.

The determination of proline was conducted using the sulfosalicylic acid method [68]. Briefly, 0.5 g of fresh sample was extracted with 5 mL of 3% sulfosalicylic acid in boiling water for 10 min. Then, 2 mL of filtrate was mixed with 2 mL glacial acetic acid and 2 mL of acid ninhydrin. The mixture was then subjected to a boiling water bath for 30 min. After cooling, 4 mL toluene was added into the mixture. The absorbance was measured at a wavelength of 520 nm.

MDA was determined using the thiobarbituric acid (TAB) method [68]. Briefly, 0.5 g of fresh sample was homogenized with 5 mL of 5% TAB solution. Then, 2 mL of the extracted solution was mixed with 2 mL of 0.67% TAB solution and boiled for 15 min. The absorbance of the incubated supernatant was spectrophotometrically measured at wavelengths of 450 nm, 532 nm, and 600 nm.

2.7. Determination of Enzyme Activity in Leaves and Roots

The activity of POD was determined using the guaiacol method and expressed in terms of enzyme units per gram fresh weight (U g^{-1} FW) [68]. Briefly, 0.5 g of fresh sample was extracted with 10 mL of 0.1 mM phosphate buffer (PBS, pH 6.0), followed by centrifugation at 4000 rpm for 15 min. The supernatant was collected in a 100 mL volumetric flask. Subsequently, the reaction mixture contained 1 mL of enzyme extraction, 2.5 mL guaiacol (0.1 M), and 0.5 mL 30% H_2O_2. The change in absorbance at 470 nm, measured as 1.0 per min, was considered one unit of POD activity.

The activity of SOD was determined using the nitrogen blue tetrazolium (NBT) assay method [68]. Briefly, 0.5 g of fresh sample was extracted with 10 mL of 0.05 mM PBS (pH 7.8), then centrifuged at 10,000 rpm for 10 min. The supernatant was collected in a 100 mL volumetric flask. Subsequently, the reaction mixture contained 0.3 mL of enzyme extraction, 1.5 mL of 0.05 mM PBS, 0.3 mL of 130 mM Met solution, 0.3 mL 750 μM NBT solution, 0.3 mL 100 μM EDTA-Na_2 solution, and 0.3 mL 20 μM riboflavin. The reaction mixture was exposed to fluorescence under 4000 luxes for 20 min; then, the absorbance was measured at a wavelength of 560 nm. The SOD activity was defined as the amount of enzyme required to inhibit 50% of NBT reduction per g fresh weight (U g^{-1} FW).

The activity of CAT was determined by potassium permanganate titration [57]. Briefly, 0.5 g of fresh sample was extracted with 10 mL of 0.2 mM PBS (pH 7.0), then centrifuged at 4000 rpm for 10 min. The supernatant was collected in a 100 mL volumetric flask. Subsequently, 5 mL of the extraction was added to 2.5 mL of 0.1 M H_2O_2 and placed in a 30 °C water bath for 10 min; then 2.5 mL of 10% sulfuric acid was added, and the mixture was titrated with 0.1 M $KMnO_4$ until a pink color was achieved, followed by 30 s without color change. CAT activity is expressed as mg of hydrogen peroxide decomposed per minute per g fresh weight of sample.

2.8. Determination of Soil Enzyme Activity

The soil invertase activity was determined using the 3,5-dinitrosalicylic acid colorimetry method [69]. Briefly, 1.0 g of air-dried soil was incubated with a mixture containing 0.2 of ml toluene, 3 mL of 8% (*w*/*v*) sucrose, and 1 mL of PBS (pH 5.5) at 37 °C for 24 h. After incubation, 1 mL extraction with 3 mL 3,5-dinitrosalicylic acid was incubated in boiling water for 5 min. Subsequently, the absorbance was measured at a wavelength of 508 nm. Results are expressed as mg of glucose released per g dry soil per h.

The soil urease activity was determined using sodium phenol–sodium hypochlorite colorimetry [69]. Briefly, 1.0 g of air-dried soil was incubated with 0.2 mL of toluene, 2 mL of 10% urea solution, and 4 mL of citrate buffer (pH 6.7) at 37 °C for 24 h. After incubation, the reaction mixture contained 1 mL enzyme extraction, 4 mL sodium phenol solution, and 3 mL sodium hypochlorite solution. The absorbance of the reaction solution was measured at a wavelength of 578 nm. Results are expressed as mg of NH_4^+–N released per g soil per h.

The neutral phosphatase activity was determined using disodium benzene phosphate colorimetry [69]. Briefly, 1.0 g of air-dried soil was incubated with 0.2 mL of toluene and 5 mL of 0.5% disodium phenyl phosphate solution for 24 h at 37 °C. After incubation, 1 mL extraction with 0.5 mL of citrate buffer (pH 7.0), 1 mL of 8% potassium ferrocyanide (w/v), and 1 mL of 2% 4-aminoantipyrine were thoroughly mixed. Subsequently, the absorbance of the reaction solution was measured at a wavelength of 510 nm. Results are expressed as mg of phenol per g soil per 4 h.

2.9. Statistical Analysis

Statistical analysis was performed using SPSS 19.0 software (SPSS Inc., Chicago, IL, USA). Results (means ± SE, n = 6) between treatments were analyzed by two-way analysis of variance (ANOVA), and significant differences were compared with the Duncan multiple range test at $p < 0.05$ unless mentioned otherwise. Structural equation modeling (SEM) was employed in the Amos 21.0 software package to investigate potential causative relationships among measured factors (Smallwaters Corporation, Chicago, IL, USA). Figures were generated using Origin Pro 2018 software (OriginLab Corp., Northampton, MA, USA).

3. Results

3.1. Effects of eCO_2 and Drought on Photosynthetic Parameters of Leaves of Mulberry Plants

Compared to well-watered conditions, drought stress significantly decreased the net photosynthetic rates, stomatal conductance, transpiration rates, intercellular CO_2 concentration, and Ci/Ca ratio of mulberry plants, regardless of CO_2 concentrations (Figure 2). However, eCO_2 significantly mitigated these effects. Under eCO_2, the net photosynthetic rates increased by 27%, 25%, and 37% (p = 0.001, Figure 2A), while stomatal conductance decreased by 38%, 41%, and 48% (p = 0.001, Figure 2B) and leaf transpiration rates decreased by 26%, 51%, and 48% (p = 0.001, Figure 2C) under well-watered conditions, moderate drought, and severe drought stress, respectively. As a result, leaf WUE was dramatically enhanced by 96%, 104%, and 163%, respectively (p = 0.001, Figure 2E), under the same conditions. Meanwhile, mulberry plants under eCO_2 exhibited higher Ci values than those grown under aCO_2, regardless of soil water status (Figure 2D). Nevertheless, the Ci/Ca ratio was significantly lower under eCO_2 than under aCO_2 under both well-watered and moderate-drought-stress conditions (Figure 2F). Two-way ANOVA revealed significant interaction effects between eCO_2 and drought on leaf WUE (p = 0.020), Ci (p = 0.004) and Ci/Ca ratio (p = 0.001) but not on the net photosynthetic rate (p = 0.37), stomatal conductance (p = 0.78), or transpiration rate (p = 0.94) (Figure 2).

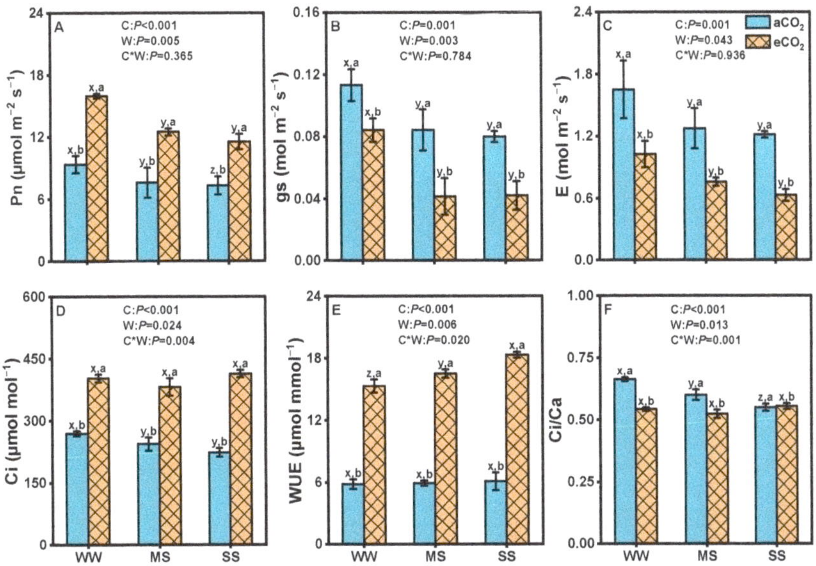

Figure 2. Net photosynthetic rate (Pn, **A**), stomatal conductance (gs, **B**), transpiration rate (E, **C**), intercellular CO_2 (Ci, **D**), leaf water use efficiency (WUE, **E**), and the ratio of Ci to ambient CO_2 (Ci/Ca, **F**) of the fully expanded upper leaf of 18-month-old mulberry plants under ambient CO_2 (aCO_2, 420/470 ppm, day/night), elevated CO_2 (eCO_2, 710/760 ppm), and varying water regimes (well-watered, WW; moderate drought stress, MS; severe drought stress, SS). Data (means ± SE, $n = 6$) followed by different letters indicate significant differences between CO_2 treatments for the same water regime (a, b) and among water treatments for the same CO_2 concentration (x, y, z) at $p < 0.05$. Statistical comparisons (two-way ANOVA) between water regimes and CO_2 treatments, as well as their interaction (water × CO_2), are presented for each variable.

3.2. Effects of eCO$_2$ and Drought on Leaf Chlorophyll Fluorescence of Mulberry Plants

Compared to well-watered conditions, drought stress significantly decreased Φ_{PSII} and ETR under eCO_2 and aCO_2, as well as qP under aCO_2, while increasing NPQ, regardless of CO_2 concentrations (Figure 3). The qP was 21.9% lower in eCO_2 than in aCO_2 under well-watered conditions but 33.3% higher under severe drought (Figure 3C). Similarly, eCO_2 also decreased Φ_{PSII} by 11.1% under both well-watered and moderate-drought conditions (Figure 3E) and decreased the ETR by 27.0% and 12.9% under well-watered and severe-drought conditions, respectively (Figure 3F). Conversely, eCO_2 significantly increased NPQ by 7.1%, 4.1%, and 13.4% under well-watered, moderate-drought, and severe-drought conditions, respectively (Figure 3D). Neither Fv/Fm nor Fv′/Fm′ was affected by eCO_2 or drought stress ($p > 0.05$, Figure 3A,B). Two-way ANOVA revealed significant interaction effects between eCO_2 and drought on qP ($p = 0.001$), NPQ ($p < 0.05$), Φ_{PSII} ($p < 0.05$) and ETR ($p = 0.001$) but not on Fv/Fm ($p > 0.285$) or Fv′/Fm′ ($p > 0.5$, Figure 3).

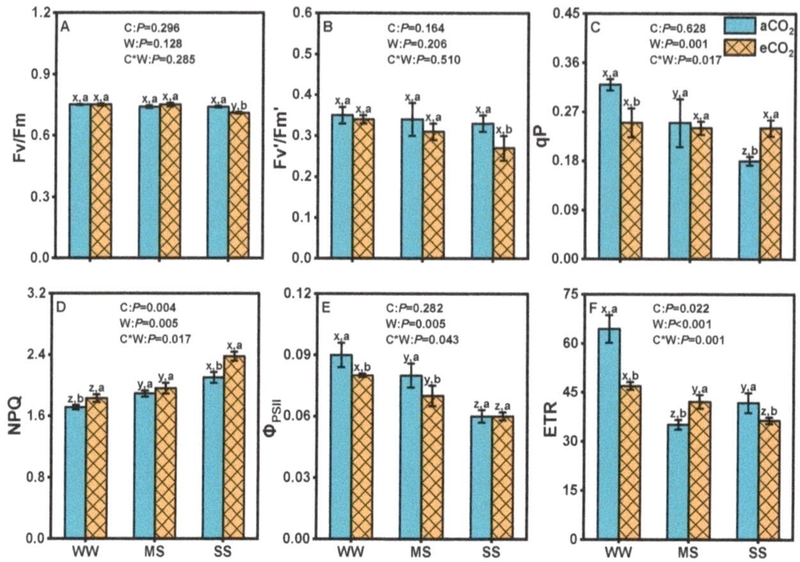

Figure 3. Fv/Fm (**A**), Fv′/Fm′ (**B**), qP (**C**), NPQ (**D**), Φ_{PSII} (**E**) and ETR (**F**) of the fully expanded upper leaf of mulberry plants under ambient CO_2 (aCO_2, 420/470 ppm, day/night), elevated CO_2 (eCO_2, 710/760 ppm), and varying water regimes (well-watered, WW; moderate drought stress, MS; severe drought stress, SS). Data (means ± SE, n = 6) followed by different letters indicate significant differences between CO_2 treatments for the same water regime (a, b) and among water treatments for the same CO_2 concentration (x, y, z) at $p < 0.05$. Statistical comparisons (two-way ANOVA) between water regimes and CO_2 treatments, as well as their interaction (water × CO_2), are presented for each variable.

3.3. Effects of eCO_2 and Drought on Osmotic Regulators of Mulberry Plants

In comparison to well-watered conditions, leaf MDA concentration was substantially increased by 25.0% and 38.1% under moderate and severe stresses at aCO_2 ($p < 0.05$, Figure 4A), while no significant changes were observed ateCO_2 ($p > 0.05$, Figure 4A). Drought stress significantly increased root MDA, irrespective of CO_2 concentrations ($p < 0.001$, Figure 4D). In contrast, eCO_2 decreased leaf MDA by 12.4% and 13.0% under moderate and severe stresses, respectively, and root MDA by 10.2% and 13.2% under well-watered and severe stress conditions, respectively ($p < 0.05$, Figure 4D). Similarly, eCO_2 significantly reduced proline concentrations in leaves by 11.2%, 13.3%, and 26.7% and in roots by 9.4%, 21.2%, and 22.0% under well-watered, moderate stress, and severe stress conditions, respectively. However, both leaf and root proline concentrations were obviously increased with declining soil water under aCO_2 ($p < 0.05$, Figure 4B,E). Conversely, eCO_2 significantly enhanced soluble sugar concentrations in leaves by 27.1%, 15.0%, and 6.7% and in roots by 43.9%, 71.8%, and 62.3% under well-watered, moderate stress, and severe stress conditions, respectively ($p < 0.05$, Figure 4C,E). Two-way ANOVA results showed significant interaction effects between eCO_2 and drought on leaf soluble sugar ($p = 0.03$) and leaf malondialdehyde ($p = 0.02$) but not on root soluble sugar ($p = 0.33$), root malondialdehyde ($p = 0.13$), leaf proline ($p = 0.23$), or root proline ($p = 0.23$, Figure 4).

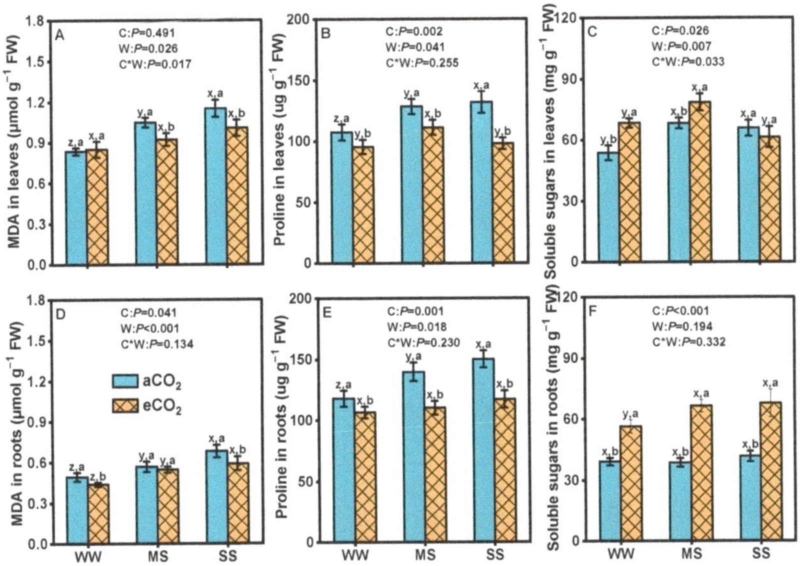

Figure 4. Malondialdehyde (MDA, **A**,**D**), proline (**B**,**E**), and soluble sugar (**C**,**F**) in leaves and roots of mulberry plants under ambient CO_2 (aCO_2, 420/470 ppm, day/night), elevated CO_2 (eCO_2, 710/760 ppm), and varying water regimes (well-watered, WW; moderate drought stress, MS; severe drought stress, SS). Data (means ± SE, $n = 6$) followed by different letters indicate significant differences between CO_2 treatments for the same water regime (a, b) and among water treatments for the same CO_2 concentration (x, y, z) at $p < 0.05$. Statistical comparisons (two-way ANOVA) between water regimes and CO_2 treatments, as well as their interaction (water × CO_2), are presented for each variable.

3.4. Effects of eCO_2 and Drought on Antioxidant Enzyme Activities in Mulberry Plants

The activities of CAT, POD, and SOD in both leaves and roots consistently increased as soil water declined under both aCO_2 and eCO_2 (Figure 5). Elevated CO_2 significantly increased CAT activity in leaves by 18.3% under well-watered conditions, and in roots by 44.5%, 188.2%, and 100.4% under well-watered, moderate-drought, and severe-drought conditions, respectively ($p < 0.05$, Figure 5A,D). Similarly, POD activity increased in leaves by 15.5% under severe drought, and in roots by 20.6%, 12.4%, and 10.3% under well-watered, moderate-drought, and severe-drought conditions, respectively ($p < 0.05$, Figure 5E). SOD activity exhibited a similar pattern, with higher activity under eCO_2 compared to under aCO_2 under well-watered and moderate-drought conditions, but no significant changes were observed under severe-drought ($p < 0.05$, Figure 5C,F). Two-way ANOVA results showed a significant interaction effect between eCO_2 and drought on CAT activity in leaves ($p < 0.001$) and roots ($p = 0.003$), POD activity in roots ($p = 0.037$), and SOD activity in leaves ($p = 0.050$) and roots ($p < 0.001$) but not on POD activity in leaves ($p = 0.354$) (Figure 5).

3.5. Effects of eCO_2 and Drought on Tissue N, P, and K in Mulberry Plants

Compared to well-watered conditions, drought stress significantly decreased the stem N concentration under both aCO_2 and eCO_2, as well as the root N concentration under aCO_2, but increased the leaf N concentration under aCO_2 (Table 1). Under eCO_2, leaf N decreased by 6.1%, 10.1%, and 13.0% ($p = 0.001$) and stem N decreased by 10.5%, 15.9%, and 2.2% ($p = 0.042$) under well-watered, moderate-drought, and severe-drought conditions, respectively (Table 1), while root N remained unaffected ($p > 0.05$, Table 1).

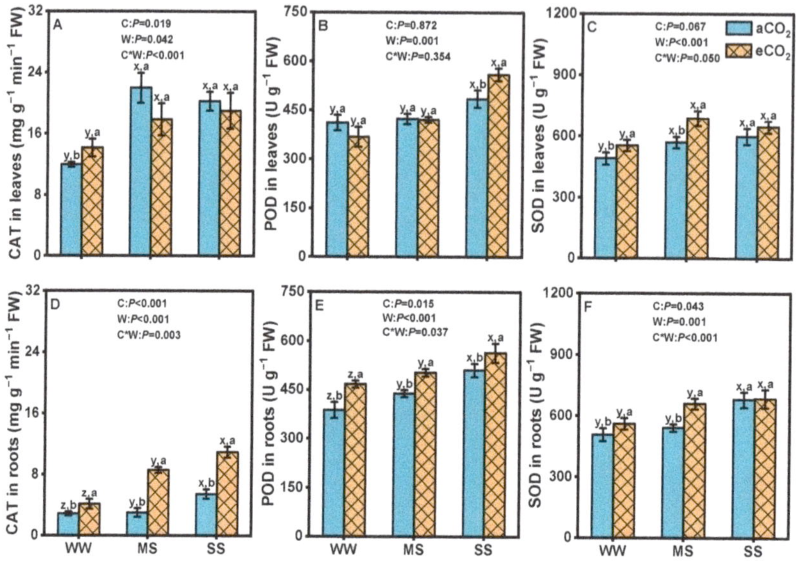

Figure 5. Catalase (CAT: **A**,**D**), peroxidase (POD: **B**,**E**), and superoxide dismutase (SOD: **C**,**F**) activities in leaves and roots of 18-month-old mulberry plants under ambient CO_2 (aCO_2, 420/470 ppm, day/night), elevated CO_2 (eCO_2, 710/760 ppm), and varying water regimes (well-watered, WW; moderate drought stress, MS; severe drought stress, SS). Data (means ± SE, n = 6) followed by different letters indicate significant differences between CO_2 treatments for the same water regime (a, b) and among water treatments for the same CO_2 concentration (x, y, z) at $p < 0.05$. Statistical comparisons (two-way ANOVA) between water regimes and CO_2 treatments, as well as their interaction (water × CO_2), are presented for each variable.

Table 1. Variations in tissue nitrogen (N), phosphorus (P), and potassium (K) concentrations of mulberry under ambient (aCO_2) and elevated CO_2 (eCO_2) under well-watered, moderate-drought, and severe-drought conditions.

Variable	*p*-Value			CO_2 Treatment	Water Treatment			Moderate Water Deficit Effect (%)	Severe Water Deficit Effect (%)
	CO_2	Water	C*W		Well-Watered	Moderate Drought	Severe Drought		
Leaf N concentration (mg g^{-1})	0.001	0.055	0.276	aCO_2 eCO_2 CO_2 effect (%)	34.5 ± 0.6 y,a 32.4 ± 0.7 x,b −6.1	35.0 ± 1.1 y,a 31.4 ± 0.6 x,b −10.1	38.0 ± 0.1 x,a 33.0 ± 1.4 x,b −13.0	+1.3 −3.2	+10.1 +1.8
Stem N concentration (mg g^{-1})	0.042	0.001	0.159	aCO_2 eCO_2 CO_2 effect (%)	16.2 ± 0.4 x,a 14.5 ± 0.4 x,b −10.5	13.3 ± 0.5 y,a 11.3 ± 0.5 y,b −15.9	13.4 ± 0.8 y,a 13.11 ± 0.8 x,a −2.2	−17.9 −21.7	−17.2 −9.6
Root N concentration (mg g^{-1})	0.308	0.526	0.045	aCO_2 eCO_2 CO_2 effect (%)	16.5 ± 0.6 x,a 13.9 ± 0.5 x,b −15.9	14.0 ± 0.9 y,a 15.0 ± 0.9 x,a +7.6	14.6 ± 0.3 y,a 14.5 ± 0.3 x,a −0.9	−15.6 +8.3	−11.2 +4.8
Leaf P concentration (mg g^{-1})	0.007	0.136	0.283	aCO_2 eCO_2 CO_2 effect (%)	1.30 ± 0.06 x,a 1.08 ± 0.08 y,b −16.7	1.18 ± 0.04 y,a 1.10 ± 0.03 y,b −6.6	1.28 ± 0.04 x,a 1.20 ± 0.03 x,b −5.8	−9.1 +3.2	−0.9 +8.2
Stem P concentration (mg g^{-1})	0.012	0.010	0.226	aCO_2 eCO_2 CO_2 effect (%)	0.99 ± 0.04 x,a 0.85 ± 0.04 x,b −14.5	0.84 ± 0.02 y,a 0.72 ± 0.04 y,b −14.2	0.86 ± 0.04 y,a 0.85 ± 0.06 x,a −0.9	−15.1 −14.7	−13.1 −0.8
Root P concentration (mg g^{-1})	0.044	0.039	0.005	aCO_2 eCO_2 CO_2 effect (%)	0.95 ± 0.04 x,a 0.70 ± 0.03 y,b −26.6	0.74 ± 0.05 y,a 0.84 ± 0.06 x,a +12.6	0.95 ± 0.06 x,a 0.87 ± 0.02 x,b −8.4	−22.0 +20.8	−0.4 +26.1
Leaf K concentration (mg g^{-1})	0.294	0.003	0.271	aCO_2 eCO_2 CO_2 effect (%)	20.0 ± 0.4 y,a 18.8 ± 0.3 z,b −6.1	19.2 ± 0.8 y,a 19.7 ± 0.3 y,a +2.7	21.6 ± 0.3 x,a 21.15 ± 0. 6 x,a −2.7	−3.7 +4.8	+8.4 +10.2

Table 1. *Cont.*

Variable	p-Value			CO_2 Treatment	Water Treatment			Moderate Water Deficit Effect (%)	Severe Water Deficit Effect (%)
	CO_2	Water	C*W		Well-Watered	Moderate Drought	Severe Drought		
Stem K concentration (mg g^{-1})	0.677	0.106	0.785	aCO_2	7.6 ± 0.6 x,a	6.0 ± 0.2 y,a	7.2 ± 0.7 x,a	−21.1	−5.6
				eCO_2	7.2 ± 0.8 x,a	6.3 ± 0.1 x,a	6.8 ± 0.6 x,a	−12.7	−6.0
				CO_2 effect (%)	−5.3	+4.5	−5.8		
Root K concentration (mg g^{-1})	0.802	0.150	0.108	aCO_2	8.3 ± 0.6 x,a	6.5 ± 0.7 y,a	8.2 ± 0.1 x,a	−22.2	−1.1
				eCO_2	7.5 ± 0.4 x,a	7.3 ± 0.5 x,a	7.6 ± 0.5 x,a	−2.0	+1.8
				CO_2 effect (%)	−10.0	13.1	−7.2		

Data (means ± SE, n = 6) followed by different letters indicate significant differences between CO_2 treatments for the same water regime (a, b) and among water treatments for the same CO_2 concentration (x, y, z) at $p < 0.05$. Statistical comparisons (two-way ANOVA) between water regimes and CO_2 treatments, as well as their interaction (water × CO_2), are presented for each variable.

Similarly, P concentrations in leaves, stems, and roots were consistently lower under eCO_2 than under aCO_2 across all water treatments (all $p < 0.05$, Table 1), except for stem P under severe drought and root P under moderate drought, which showed no significant differences. Compared to the well-watered treatment, moderate drought and severe drought reduced stem P by 15.1% and 13.1%, respectively, under aCO_2, and by 14.7% and 0.8%, respectively, under eCO_2. Conversely, the root P concentration increased significantly under moderate drought and severe drought under eCO_2 but not under aCO_2 ($p = 0.044$; Table 1).

In addition, only the leaf K concentration exhibited a consistent increase with declining soil water, regardless of CO_2 conditions ($p = 0.003$; Table 1). In contrast, eCO_2 had no significant effect on K concentrations in leaves, stems, or roots (all $p > 0.05$, Table 1). Two-way ANOVA revealed significant interactions between eCO_2 and drought for both N and P concentrations in roots ($p < 0.05$; Table 1).

3.6. *Effects of eCO_2 and Drought on Tissue Accumulations and Allocation of N, P, and K in Mulberry Plants*

In general, drought stress significantly decreased the accumulation of N, P, and K in the leaves, stems, roots, and total plant under both aCO_2 and eCO_2, except for leaf P accumulation (Figure 6). Severe drought resulted in reductions in N, P, and K accumulation in the total plant of 26.7%, 31.0%, and 24.3% under aCO_2 and of 20.7%, 13.5%, and 14.4% under eCO_2, respectively. Nevertheless, eCO_2 significantly increased N, P, and K accumulation in leaves, stems, roots, and the total plant subjected to well-watered, moderate-drought, and severe-drought conditions. Compared to aCO_2, eCO_2 significantly increased total-plant N accumulation by 13.3%, 18.8%, and 22.6% ($p = 0.007$, Figure 6D) and total-plant K accumulation by 16.1%, 36.0%, and 31.3% under well-watered, moderate-drought, and severe-drought conditions, respectively ($p < 0.001$, Figure 6L). Similarly, eCO_2 also increased total-plant P accumulation by 30.7% and 25.9% under moderate drought and severe drought, respectively ($p = 0.016$), but had no significant effect under well-watered conditions. Two-way ANOVA revealed significant interaction effects between eCO_2 and drought on stem N and stem and root P accumulation ($p < 0.05$, Figure 6). Furthermore, severe drought significantly enhanced the allocations of N, P, and K to the leaves but decreased their allocations to the stems and roots of mulberry, regardless of CO_2 concentration. Conversely, eCO_2 decreased N, P, and K allocations to the leaves and stems but increased their allocations to roots under well-watered and severe-drought conditions (Figure 7).

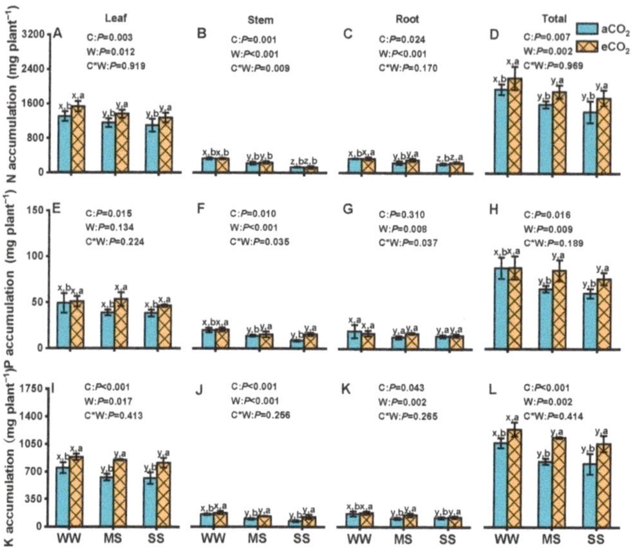

Figure 6. Accumulation of nitrogen (N: **A–D**), phosphorus (P: **E–H**), and potassium (K: **I–L**) in leaves, stems, roots, and in total plants in 18-month-old mulberry under ambient CO_2 (aCO_2, 420/470 ppm, day/night), elevated CO_2 (eCO_2, 710/760 ppm), and varying water regimes (well-watered, WW; moderate drought stress, MS; severe drought stress, SS). Data (means ± SE, n = 6) followed by different letters indicate significant differences between CO_2 treatments for the same water regime (a, b) and among water treatments for the same CO_2 concentration (x, y, z) at $p < 0.05$. Statistical comparisons (two-way ANOVA) between water regimes and CO_2 treatments, as well as their interaction (water × CO_2), are presented for each variable.

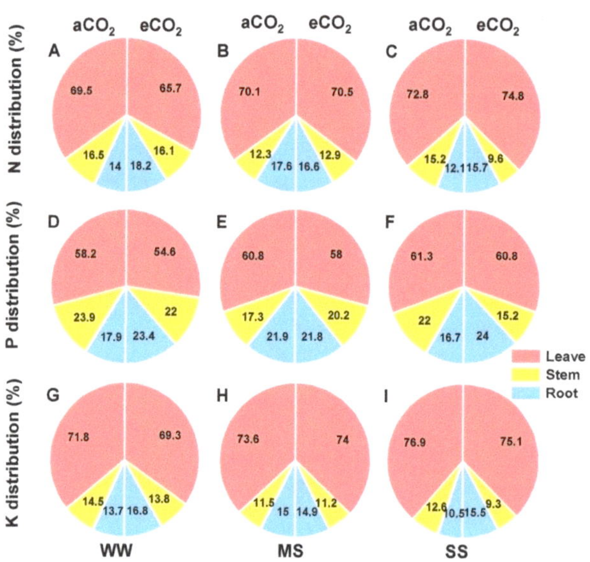

Figure 7. Leaf, stem, and root distribution of nitrogen (N: **A–C**), phosphorus (P: **D–F**), and potassium (K: **G–I**) in mulberry plants under ambient CO_2 (aCO_2, 420/470 ppm, day/night), elevated CO_2 (eCO_2, 710/760 ppm), and varying water regimes (well-watered, WW; moderate drought stress, MS; severe drought stress, SS).

3.7. Effects of eCO_2 and Drought on Soil Nutrients and Soil Enzyme Activities

The response of soil AN, AP, and AK to decreasing soil water status exhibited an increasing trend under both aCO_2 and eCO_2 (Figure 8). Soil AN showed an asymmetric response to CO_2 concentration along the soil water gradient, where it was significantly lower under eCO_2 than under aCO_2 under the well-watered condition but higher under eCO_2 than under aCO_2 under moderate-drought and severe-drought conditions ($p = 0.018$, Figure 8A). Soil AP and AK were significantly lower under eCO_2 than under aCO_2 along the soil water gradient ($p < 0.05$, Figure 8B,C).

Moreover, regardless of soil water status, eCO_2 generally significantly increased the activities of soil sucrase and phosphatase, although these enzyme activities obviously decreased with declining soil water content under both aCO_2 and eCO_2 ($p < 0.05$, Figure 8D,F). In contrast to the observed trends in soil sucrase and phosphatase activities, urease activity exhibited a different pattern: it decreased along the soil water gradient under aCO_2 but slightly increased under moderate-drought conditions under eCO_2 ($p < 0.05$, Figure 8E). Two-way ANOVA showed significant interaction effects between eCO_2 and drought on soil AN ($p = 0.023$), AP ($p = 0.027$), and sucrase ($p < 0.001$, Figure 8).

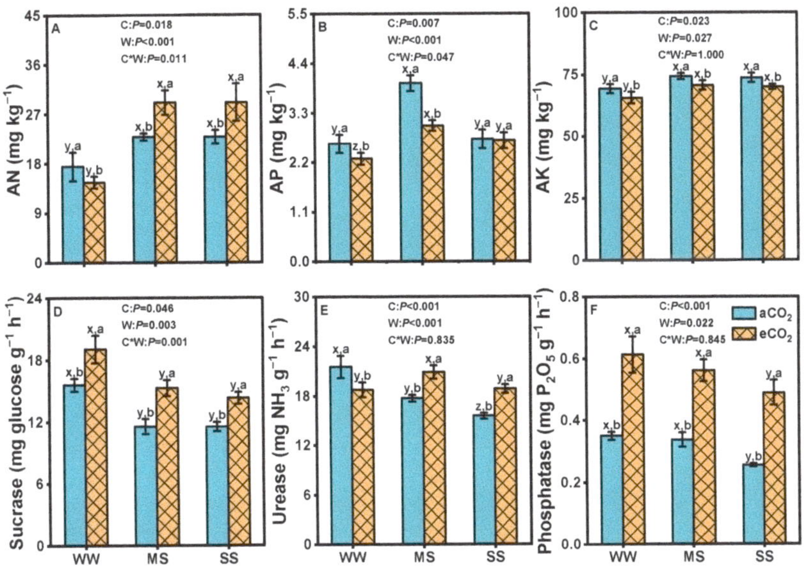

Figure 8. Available nitrogen (AN, **A**), available phosphorus (AP, **B**), available potassium (AK, **C**) in soil, as well as soil sucrase (**D**), urease (**E**), and phosphatase (**F**) activities of mulberry plants, under ambient CO_2 (aCO_2, 420/470 ppm, day/night), elevated CO_2 (eCO_2, 710/760 ppm), and varying water regimes (well-watered, WW; moderate drought stress, MS; severe drought stress, SS). Data (means $\pm$ SE, $n = 6$) followed by different letters indicate significant differences between CO_2 treatments for the same water regime (a, b) and among water treatments for the same CO_2 concentration (x, y, z) at $p < 0.05$. Statistical comparisons (two-way ANOVA) between water regimes and CO_2 treatments, as well as their interaction (water $\times$ CO_2), are presented for each variable.

3.8. Effects of eCO_2 and Drought on Biomass Production in Mulberry Plants

Compared to the well-watered treatment, drought stress significantly decreased the biomass production in leaves, stems, roots, and total plants, regardless of CO_2 concentration. Moderate drought and severe drought resulted 14.7% and 31.8% reductions in total biomass production, respectively, under aCO_2 and 11.3% and 22.8% reductions, respectively, under eCO_2. Conversely, a very strong effect of CO_2 fertilization on mulberry growth was observed under both well-watered and drought stress conditions, with the effect more pronounced as soil water status declined. Specifically, eCO_2 increased the leaf, stem, root, and total plant biomass production by 26.1%, 20.1%, 20.4%, and 23.1%, respectively, under well-watered conditions; by 32.0%, 29.4%, 18.8%, and 28.0%, respectively, under moderate drought; and by 35.7%, 83.7%, 15.3%, and 39.2% (all $p \leq 0.01$), respectively, under severe drought (Table 2). Additionally, eCO_2 significantly decreased the root/shoot ratio under severe drought. Two-way ANOVA showed a significant interaction effect between eCO_2 and drought on stem biomass production only ($p = 0.037$, Table 2).

Table 2. Variations in plant tissue biomass production of mulberry plants under ambient CO_2 (aCO_2); elevated CO_2 (eCO_2); and well-watered, moderate-drought, and severe-drought conditions.

Variable	*p*-Value			CO_2 Treatment	Water Treatment			Moderate Water Deficit Effect (%)	Severe Water Deficit Effect (%)
	CO_2	Water	C*W		Well-Watered	Moderate Drought	Severe Drought		
Leaf biomass (g plant^{-1})	0.002	0.000	0.981	ACO_2	37.5 ± 1.8 x,b	32.9 ± 1.0 y,b	28.6 ± 2.9 z,b	−12.4	−23.8
				ECO_2	47.3 ± 0.4 x,a	43.4 ± 0.5 y,a	38.8 ± 2.7 z,a	−8.3	−18.0
				CO_2 effect (%)	+26.1	+32.0	+35.7		
Stem biomass (g plant^{-1})	0.000	0.000	0.037	ACO_2	20.3 ± 0.9 x,b	16.5 ± 0.9 y,b	10.0 ± 0.6 z,b	−18.9	−50.6
				ECO_2	24.4 ± 1.4 x,a	21.3 ± 0.8 y,a	18.4 ± 0.9 z,a	−12.6	−24.4
				CO_2 effect (%)	+20.1	+29.4	+83.7		
Root biomass (g plant^{-1})	0.000	0.001	0.574	ACO_2	19.8 ± 0.9 x,b	16.9 ± 0.8 y,b	14.3 ± 1.2 z,b	−14.8	−27.8
				ECO_2	23.9 ± 0.4 x,a	20.1 ± 0.8 y,a	16.5 ± 0.8 z,a	−16.0	−30.9
				CO_2 effect (%)	+20.4	+18.8	+15.3		
Total biomass (g plant^{-1})	0.000	0.000	0.876	ACO_2	77.7 ± 3.4 x,b	66.2 ± 1.3 y,b	53.0 ± 3.8 z,b	−14.7	−31.8
				ECO_2	95.6 ± 2.1 x,a	84.8 ± 1.1 y,a	73.8 ± 14.2 z,a	−11.3	−22.8
				CO_2 effect (%)	+23.1	+28.0	+39.2		
Root : shoot ratio	0.027	0.164	0.080	ACO_2	0.34 ± 0.01 x,a	0.34 ± 0.01 x,a	0.37 ± 0.04 x,a	−0.3	+9.4
				ECO_2	0.33 ± 0.01 x,a	0.31 ± 0.02 x,b	0.28 ± 0.02 x,b	−6.8	−13.1
				CO_2 effect (%)	−2.9	−9.2	−22.9		

Data (means ± SE, $n = 6$) followed by different letters indicate significant differences between CO_2 treatments for the same water regime (a, b) and among water treatments for the same CO_2 concentration (x, y, z) at $p < 0.05$. Statistical comparisons (two-way ANOVA) between water regimes and CO_2 treatments, as well as their interaction (water × CO_2), are presented for each variable.

3.9. Multivariate Analyses

To explore the potential mechanisms underlying the impact of eCO_2 on plant response to drought stress, particularly in terms of changes in biomass production, we employed structural equation modeling (SEM) to analyze the relationships between eCO_2 concentration, drought treatments, gas exchange, chlorophyll fluorescence, osmotic regulators, antioxidant enzyme activities, plant nutrient uptake, soil properties, and biomass production. The SEM analysis showed that drought had significant direct, positive effects on osmotic substances and soil nutrients while exerting significant direct, negative effects on photosynthesis, total biomass production, and soil enzyme activities. In contrast, eCO_2 had significant direct, positive effects on photosynthesis, total biomass, and soil enzyme activities and significant direct, negative effects on chlorophyll fluorescence; osmotic substances; and plant N, P, and K (Figure 9).

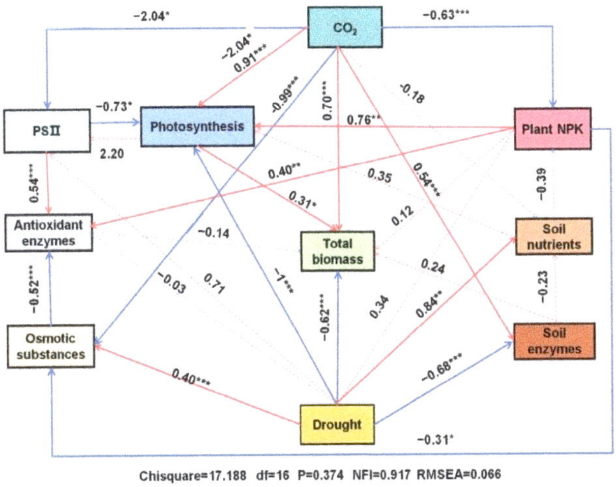

Figure 9. Structural equation model (SEM) showing relationships among CO_2 concentration; soil water regime; photosynthetic parameters; PSII efficiency; antioxidant enzyme activities; osmotic substance concentrations; N, P, and K concentrations; soil nutrients; soil enzyme activities; and total biomass production of mulberry plants. Numbers above arrows are path coefficients. Bold and dashed lines indicate significant ($p < 0.05$) and non-significant ($p > 0.05$) paths, respectively. *, ** and *** significant at $p \leq 0.05$, $p \leq 0.01$ and $p \leq 0.001$, respectively.

4. Discussion

Mulberry plants growing in the majority of southwest China frequently experience a moderate drought due to a progressive reduction in soil water availability [10,70,71]. Drought stress has been widely reported to significantly reduce both photosynthesis and biomass accumulation in mulberry plants [72], in agreement with our study findings (Figure 2A, Table 2). In addition, mulberry production faces future challenges due to increases in , drought frequency, and atmospheric CO_2 concentration. This study expands upon previous knowledge regarding how mulberry plants can cope with water scarcity and discusses the dynamics through which eCO_2 mitigates the effects of prolonged drought stress by enhancing CO_2 uptake and biomass production, altering nutrient uptake and partitioning while decreasing oxidative pressure.

4.1. eCO_2 Alleviates Adverse Effects of Drought on Plant Biomass Production and Photosynthesis

Under drought conditions, 710 ppm eCO_2 significantly stimulated the above- and below-ground biomass of mulberry, resulting in a substantial mitigation effect on mulberry productivity (Table 2). This finding agrees with previous studies conducted on coffee plants, *Picea abies*, *Pinus halepensis*, and *Populus tremula* under 700–867 ppm eCO_2 in open-top chambers or free-air CO_2 enrichment experiments [7,30,73,74]. The increase in plant biomass production due to eCO_2 can be primarily attributed to biochemical and photochemical processes such as photosynthesis and respiration [75,76]. Specifically, plants grown under eCO_2 typically exhibit a 30–60% greater leaf photosynthetic rate due to increased Rubisco carboxylation while competitively suppressing plant photorespiration and dark respiration [41,77]. In this study, eCO_2 directly stimulated photosynthesis in mulberry through increased CO_2 uptake (Figure 2A). A considerably greater intercellular CO_2 concentration was observed under eCO_2 (Figure 2D), indicating that the carboxylation rate of Rubisco was constrained by CO_2 diffusion into the carboxylation sites under aCO_2. Consequently, eCO_2 favored the carboxylation reaction of Rubisco over its oxygenation reaction (Figure 5), thereby enhancing carbon assimilation by increasing Ci for Rubisco carboxylation. These results are consistent with findings reported by Kelly et al. (2019) [78] and Fan et al. (2020) [38]. The enhanced photosynthetic rates under eCO_2 directly contribute to higher carbon assimilation, which is a key driver of biomass accumulation and overall productivity in mulberry. This is particularly evident in the significant increase in above- and below-ground biomass observed in our study (Table 2). The greater carbon fixation under eCO_2 provides more resources for growth and development, enabling mulberry to maintain productivity, even under drought stress. Additionally, the increased carbon assimilation not only supports biomass production but also enhances the plant's ability to cope with drought stress. By maintaining higher photosynthetic rates, mulberry can accumulate more carbohydrates, which are essential for osmotic adjustment and energy metabolism under water-limited conditions [79]. This explains the mitigation of drought stress observed in our study.

Despite the significant impact of eCO_2 on leaf photosynthesis, its effects may be partially mitigated by drought stress. This is evidenced by a greater decline in net photosynthetic rate observed in plants experiencing moderate drought and severe drought compared to well-watered plants (Figure 2A). Such a reduction in the net photosynthetic rate under drought stress could be attributed to both diffusive limitations (such as gs and Ci, as shown in Figure 2B,D) and non-diffusive limitations related to biochemical processes [80]. These observations were supported by decreased leaf N, P, and K levels and their correlations with leaf photosynthesis in mulberry plants (Figure 9). Despite the partial mitigation of photosynthetic rates under drought stress, the overall enhancement of carbon assimilation under eCO_2 still plays a critical role in sustaining mulberry growth and productivity. The ability of mulberry to maintain higher photosynthetic rates under combined eCO_2 and drought conditions highlights its potential as a resilient crop for sustainable agriculture in regions affected by climate change.

However, it is important to critically assess potential limitations and trade-offs associated with long-term exposure to eCO_2. One potential limitation is the acclimation response, where plants may exhibit reduced photosynthetic efficiency over time due to downregulation of Rubisco and other photosynthetic enzymes [62,81]. This acclimation response could diminish the initial benefits of eCO_2 in terms of carbon assimilation and growth, particularly under prolonged exposure [82]. Future studies should investigate whether mulberry exhibits similar acclimation responses under long-term eCO_2 exposure.

4.2. *eCO_2 Mitigated the Negative Effects of Drought by Improving Water Use Efficiency*

Stomatal responses to eCO_2 exhibit considerable variability and are influenced by environmental conditions. In this study, lower stomatal conductance and transpiration rates were observed under eCO_2 compared to aCO_2 under both well-watered conditions and drought stresses (Figure 2B,C). These findings are consistent with previous studies on various herbaceous and woody species [30,83,84]. The reduction in stomatal conductance under eCO_2 can be an adaptive mechanism for water conservation while maintaining high levels of photosynthesis, which is associated with decreased stomatal density and an altered stomatal aperture [8]. The Ci/Ca ratio reflects the relationship between stomatal conductance and photosynthetic capacity. Our results showed that Ci/Ca values decreased with declining soil water status under aCO_2 but remained unchanged under eCO_2 (Figure 2F), indicating that stomata were sensitive to drought stress under aCO_2 but exhibited reduced sensitivity or physiological adaptation to eCO_2. These results indicate that stomatal conductance is not the limiting factor for the increase in the net photosynthetic rate under eCO_2. Instead, eCO_2 compensates for the decrease in CO_2 caused by stomatal closure during drought stress, thereby alleviating the adverse effects of drought stress on plants [85,86]. Plants typically minimize water loss by reducing their leaf transpiration rate as a key adaptive strategy under short-term drought stress [87]. Our results demonstrate that eCO_2 dramatically improved leaf WUE in mulberry plants subjected to well-watered, moderate-drought, and severe-drought conditions (Figure 2A,C,E). The enhancement in WUE under eCO_2 can be attributed to the combined effects of reduced stomatal conductance and increased photosynthetic efficiency. Elevated CO_2 concentrations reduce the stomatal aperture, thereby minimizing water loss via transpiration while ensuring sufficient CO_2 uptake for photosynthesis. This stomatal regulation is mediated by alterations in guard-cell turgor pressure and the signaling pathways involving abscisic acid and other phytohormones [31]. Furthermore, eCO_2 enhances the carboxylation efficiency of Rubisco, allowing plants to achieve higher photosynthetic rates with lower stomatal conductance [26]. These physiological adaptations enable mulberry to optimize water use and sustain growth under drought stress, highlighting the critical roles of stomatal behavior and photosynthetic efficiency in improving WUE.

4.3. *eCO_2 Mitigated the Negative Effects of Drought by Improving Photosystem II Efficiency*

Chlorophyll fluorescence was employed to assess the extent of damage to the photosynthetic apparatus—in particular, to photosystem II (PSII). Under drought stress, impairment of PSII reaction centers impedes primary photochemistry, thereby affecting the process of photosynthetic electron transport. In this study, we observed a reduction in PSII efficiency in mulberry plants due to drought stress, as evidenced by decreased Φ_{PSII}, ETR, and qP values (Figure 3). Importantly, the detrimental effects of severe drought on Φ_{PSII}, ETR, and qP were more pronounced under aCO_2 compared to under eCO_2 (Φ_{PSII}, 33% and 25% lower than the control; ETR, 35% and 13% lower; qP, 44% and 4% lower; Figure 3). There was a significant interactive effect between drought stresses and eCO_2 on Φ_{PSII}, ETR, and qP ($p < 0.05$, Figure 3), which partially mitigated the damage caused by drought stress. Similar findings have been reported with respect to cucumber [88], coffee plants [30], and wheat [89]. The reduction in PSII efficiency under drought stress directly impacts the photosynthetic electron transport chain, leading to lower energy conversion efficiency and reduced carbon assimilation. This, in turn, limits biomass accumulation and overall productivity in mulberry [90,91]. However, the partial mitigation of PSII efficiency under eCO_2, as indicated by higher ΦPSII, ETR, and qP values, suggests that mulberry can maintain better photochemical efficiency and energy utilization under combined eCO_2 and drought conditions. This enhanced photochemical efficiency likely contributes to the observed increase in biomass and productivity under eCO_2 (Table 2) [41]. Furthermore, a slight decrease in Fv/Fm under eCO_2+SS indicates that mulberry plants exposed to drought conditions may not have undergone photoinhibition. Therefore, mulberry plants activate non-photochemical quenching and reduced drought-induced PSII efficiency under eCO_2 (Figure 3). The activation of NPQ under eCO_2 and drought stress is a protective mechanism that dissipates excess light energy as heat, thereby preventing damage to the photosynthetic apparatus [92]. This mechanism allows mulberry to maintain higher photosynthetic activity and carbon assimilation levels under stress conditions, which are critical for sustaining growth and productivity [93]. The ability of mulberry to optimize energy utilization and protect PSII under combined eCO_2 and drought stress highlights its potential as a resilient crop for cultivation in water-limited environments [94].

4.4. *eCO_2 Mitigated the Negative Effects of Drought by Boosting Mulberry's Defenses*

Drought stress significantly reduces the efficiency of the photosynthetic electron transport chain, leading to a decline in photosynthetic efficiency and an increased electron flow towards molecular oxygen and production of superoxide radicals [95]. The accumulation of ROS is detrimental to biomolecules such as nucleic acids, proteins, chloroplast pigments, etc. [96,97]. In response to drought-induced oxidative stress, plants develop a complex antioxidant system to regulate cellular redox status. CAT, POD, and SOD serve as the first line of defense against reactive oxygen species. In this study, the activities of CAT, POD, and SOD were significantly increased in both leaves and roots under drought stress, which is consistent with previous results [10,98]. Furthermore, eCO_2 further improved the activities of CAT, POD, and SOD in leaves and/or roots under drought stress conditions (Figure 5), in line with other studies [12,76]. The impact was more pronounced for SOD under moderate drought and for POD under severe drought, indicating distinct responses in antioxidant metabolism. The mitigation of ROS formation under eCO_2 can be attributed to increased electron consumption in C fixation, reducing the diversion of electrons towards ROS-generating pathways such as photorespiration or the Mehler reaction [99,100]. Enhanced carbon assimilation under eCO_2 provides more resources for the synthesis of antioxidant enzymes and osmoprotectants, thereby mitigating oxidative damage and supporting cellular functions under drought stress. Soluble sugars, acting as osmolyte and metabolite signaling molecules, play a crucial role in modulating plant responses to stress [82]. The elevated concentration of soluble sugars under eCO_2 observed in this study (Figure 4C,D) indicates enhanced C assimilation, which improves turgor pressure through osmotic adjustment. This finding is consistent with a previous study showing that soluble sugars (such as glucose and sugar alcohols) contributed to increased root biomass in *Sorghum bicolor* and alleviated drought stress under eCO_2 [101]. The accumulation of soluble sugars under eCO_2 not only supports osmotic adjustment but also provides energy for growth and stress responses [79]. Proline serves as both a stress marker and an osmoprotectant, implying that plants may accumulate more proline under drought stress or less proline when experiencing reduced stress levels [102]. Our data demonstrate that eCO_2 decreased proline content in both leaves and roots under drought stress (Figure 4A,B). Hence, the reduced accumulation of proline in plants exposed to eCO_2 suggests that excessive proline buildup is not necessary to withstand drought stress [103]. The reduced proline accumulation under eCO_2 indicates that mulberry experiences less severe drought stress due to improved water use efficiency and antioxidant defense. This allows the plant to allocate more resources towards growth and productivity rather than stress mitigation [104]. These changes in antioxidant metabolites and osmolytes clearly indicate that mulberry plants modify their biochemical system to counteract the detrimental effects of drought stress under eCO_2.

4.5. eCO_2 Mitigated the Negative Effects of Drought by Regulating NPK Partitioning

Drought stress can inhibit plant growth by reducing nutrient uptake and affecting the relocation of N, P, and K in plants [105,106]. A meta-analysis revealed that drought stress results in decreased levels of N and P in plant tissues, as well as reduced nutrient absorption from soil [39]. In this study, the concentrations of N, P, and K in stems and roots were significantly lower under drought stress compared to well-watered conditions under aCO_2, and a greater decrease was observed under moderate drought than under severe drought (Table 2). The impacts of drought on N, P, and K concentrations were more pronounced in roots and stems compared to leaves, indicating that soil drought negatively affected the nutrient uptake capacity of roots and stems. The decline in nutrient levels could be attributed to decreased photosynthesis and transpiration rates under drought conditions, resulting in reduced nutrient uptake (Figures 2 and 8). This coincided with an increase in the allocation of N, P, and K to leaves and a decrease in their allocations to stems and/or roots (Figure 7). However, these findings are inconsistent with studies suggesting drought may accelerate N transfer from green leaves to sink tissues such as roots or grains [107]. The increased allocation of N, P, and K to leaves observed in this study could be a mechanism of drought avoidance through the synthesis and accumulation of osmoprotectants and antioxidant enzymes, which are predominantly found in higher concentrations in leaves than in roots (Figures 5 and 6), as further supported by a significantly positive correlation between plant N, P, and K concentrations and antioxidant enzyme activities (Figure 9).

Fortunately, the detrimental impacts of drought on the concentrations of N, P, and K in stems and roots were mitigated under eCO_2 compared to under aCO_2, and even drought stress increased root P concentrations under eCO_2 (Table 1). Previous studies have reported a decline in plant-tissue N due to CO_2 enrichment [46,61,62,108], which subsequently affects other components involved in N metabolism, such as proline levels [109]. Similar results were observed in this study (Table 1 and Figure 4B,E). Under both well-watered and severe-drought conditions, eCO_2 led to decreases in N and P contents in mulberry tissues, but the reduction was less pronounced under severe drought than under well-watered conditions (Table 1). This phenomenon may be attributed to the nutrient dilution effect, where increased biomass production under eCO_2 leads to lower concentrations of essential nutrients such as N and P in plant tissues [110]. This nutrient dilution could have implications for the nutritional quality of mulberry leaves, which are critical for silkworm feeding and sericulture productivity. Future research should explore strategies to mitigate nutrient dilution effects, such as optimized fertilization practices or genetic improvements in nutrient use efficiency.

Furthermore, eCO_2 induced allocations of N, P, and K to root tissues under drought stresses (Figure 8). This adaptation to eCO_2 facilitates greater allocation to the expanded root system rather than the photosynthetic apparatus of green tissues, optimizing energy costs [111]. Thus, eCO_2 and drought stress may increase nutrient use efficiency by regulating nutrient reallocation [112].

4.6. eCO_2 Mitigated the Negative Effects of Drought by Improving Nutrient Uptake and Stimulating Soil Enzyme Activities

Most studies on eCO_2 and/or drought have primarily focused on plant characteristics, with limited research conducted on the feedback between plant nutrient uptake and soil nutrient supply. It is widely acknowledged that soil nutrient reserves play a crucial role in determining plant growth. However, the impacts of eCO_2 and/or drought on the soil nutrient pool and plant requirements remain unpredictable. Our results demonstrate that plants exhibited reduced uptake of N, P, and K when subjected to decreased soil moisture (Figure 6), resulting in higher contents of available forms of these nutrients in soil (Figure 8). Reduced nutrient uptake during drought may be attributed to factors such as diminished mineralization-mediated nutrient supply, as well as reduced nutrient diffusion and mass flow in soil [113]. Additionally, altered dynamics of root nutrient uptake induced by drought can also contribute to decreased acquisition of essential elements by plants [114]. In contrast, our results indicate that eCO_2 significantly enhanced N, P, and K accumulation in mulberry plants under both well-watered and drought-stressed conditions (Figure 6), leading to decreased levels of AN, AP, and AK in soil, except for AN under drought stress (Figure 8). These findings suggest that eCO_2 ameliorated the adverse impacts of drought by improving nutrient uptake. Consistent with our findings, previous studies have displayed similar effects; for example, 700 ppm eCO_2 upregulated gene expression related to nutrient uptake proteins, enhancing the rates of N and P assimilation [110].

The activities of soil enzymes indicate microbial functions in decomposing organic matter and mobilizing nutrients [48]. In this study, we observed significant decreases in sucrase, urease, and phosphatase activities under drought conditions (Figure 8D–F), consistent with previous studies that have reported negative correlations between soil moisture and enzymatic activities [45,51,115]. The reduction in enzyme activities could also be attributed to the low microbial biomass and physiology [116] or decreased enzyme production and turnover rates [117]. Drought-induced inhibition of plant growth and reductions in rhizodeposition result in decreased substrate availability. However, our study revealed significantly higher activities of soil sucrase, urease, and phosphatase under eCO_2 than under aCO_2 as soil water content declined (Figure 8D–F), indicating a positive correlation between eCO_2 levels and soil enzyme activities (Figure 9). Similar increases in enzyme activities have been reported in herbaceous and woody soils under 550–700 ppm eCO_2 [47,49,118]. A two-year FACE study conducted in a semi-arid grassland at the United States Department of Agriculture's Agricultural Research Service High Plains Grasslands Research Station demonstrated that 600 ppm eCO_2 had a positive effect on soil C, N, and P cycling by increasing the abundance of microbial functional genes, which were closely correlated with the increase in soil moisture [119]. The increase in soil enzyme activities is directly related to the plant root exudates [120]. Therefore, eCO_2 could accelerate C turnover in the rhizosphere and stimulated soil enzyme activities, thereby influencing the microbial biomass and microbial activity that was not tested in this study, which might ultimately have potential to ameliorate the adverse impacts of drought.

However, the long-term sustainability of eCO_2 effects on mulberry growth and productivity remains uncertain. While eCO_2 can enhance nutrient uptake and soil enzyme activities in the short term, prolonged exposure to elevated CO_2 may alter soil nutrient dynamics and microbial activity, potentially affecting nutrient availability and plant health over time [121]. For example, increased carbon inputs under eCO_2 may accelerate soil organic matter decomposition, leading to changes in nutrient cycling and microbial community composition [122]. Future studies should investigate the long-term impacts of eCO_2 on mulberry growth, nutrient cycling, and ecosystem interactions to fully understand its potential benefits and limitations.

5. Conclusions

We conducted a comprehensive study on the responses of mulberry plants to drought under eCO_2. As shown in Figure 10, eCO_2 promotes mulberry's photosynthesis and WUE under drought. It also mitigates the adverse effects of drought on ETR and ΦPSII, thereby decreasing damage to photosystems and ROS accumulation in mulberry under drought stress. Moreover, eCO_2 enhanced the tolerance of mulberry to drought stress by increasing activities of antioxidant enzymes such as SOD, POD, and CAT, and the contents of osmolytes such as proline, malondialdehyde, and soluble sugars in both leaves and roots. Furthermore, eCO_2 stimulated the activities soil enzymes under such as soil sucrase, urease, and phosphatase, under drought, improving soil fertility. This stimulation facilitated nutrient uptake of soil N, P, and K, leading to their increased accumulation in mulberry tissues while promoting their allocation to an expanded root system under drought. The enhanced photosynthetic capacity, reduced oxidative stress, and increased nutrient absorption induced by eCO_2 can effectively mitigate the detrimental impact of drought on mulberry growth by augmenting both above- and below-ground biomass production. Our findings provide crucial insights to guide water management strategies in mulberry plantations while increasing leaf growth and productivity through the regulation of N, P, and K nutrition under future conditions of elevated CO_2 and drought.

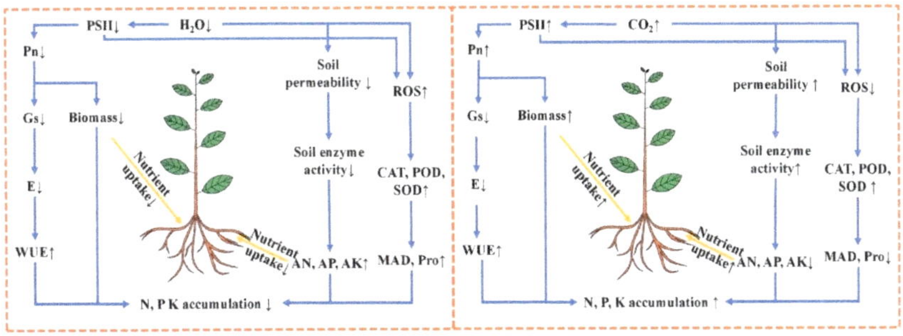

Figure 10. Comparative schematic model showing how mulberry plants can respond to drought and elevated CO_2 concentrations. PSII, photosystem-II; Pn, net photosynthetic rate; gs, stomatal conductance; E, transpiration rate; WUE, water use efficiency in leaves; N, nitrogen; P, phosphorus; K, potassium; AN, available nitrogen; AP, available phosphorus; AK, available potassium; ROS, reactive oxygen species; CAT, catalase; POD, peroxidase; SOD, superoxide dismutase; MDA, malondialdehyde; Pro, proline. An upward arrow represented an increase, whereas a downward arrow signified a reduction.

Author Contributions: S.S.: Conceptualization, Formal analysis, Methodology, Validation, Project administration, Writing—original draft, and Writing—review and editing. H.L.: Data curation, Formal analysis, Methodology, and Writing—review and editing. X.W.: Data curation, Formal analysis, Methodology, and Writing—review and editing. Z.W.: Formal analysis, Methodology, Visualization, and Writing—review and editing. J.X.: Data curation, Formal analysis, Methodology, and Writing—review and editing. X.H.: Conceptualization, Supervision, Validation, Writing—original draft, and Writing—review and editing. Z.Y.: Conceptualization, Funding acquisition, Supervision, Writing—original draft, and Writing—review and editing. All authors approved the manuscript submission. All authors have read and agreed to the published version of the manuscript.

Funding: This study was jointly supported by the National Youth Natural Science Foundation of China (31800612), the National Natural Science Foundation of China (42267056), and the High-level Scientific Research Foundation of Yunnan Agricultural University (A2032022004).

Data Availability Statement: Data is contained within the article.

Conflicts of Interest: The authors declare no conflicts of interest.

Abbreviations

The following abbreviations are used in this manuscript:

aCO_2	Ambient CO_2
eCO_2	Elevated CO_2
Ci	Intercellular CO_2 concentration
N	Nitrogen
P	Phosphorus
K	Potassium
WW	Well-watered
MS	Moderate drought stress
SS	Severe drought stress
WUE	Water use efficiency
ROS	Reactive oxygen species
CAT	Catalase
POD	Peroxidase
SOD	Superoxide dismutase
MAD	Malondialdehyde
TAB	Thiobarbituric acid

NBT	Nitrogen blue tetrazolium
EDTA-Na_2	Ethylenediaminetetraacetic acid disodium salt
$KMnO_4$	Potassium permanganate
SOC	Soil organic carbon
AN	Soil available nitrogen
AP	Soil available phosphorus
AK	Soil available potassium
F0	Initial fluorescence yield under the dark-adapted stage
Fm	Maximum fluorescence yield under the dark-adapted stage
F0′	Minimum fluorescence yield under the light-adapted stage
Fm′	Maximum fluorescence under the light-adapted stage
ΦPSII	Effective quantum yield of PSII photochemistry
Fv/Fm	Maximal quantum yield of PSII photochemistry
Fv′/Fm′	Actual photosynthetic efficiency of PSII under illumination
qP	Photochemical quenching coefficient
NPQ	Non-photochemical quenching coefficient
ETR	PSII electron transport rate
SEM	Structural equation modeling

References

1. IPCC. *Summary for Policymakers-Global Warming of 1.5 °C. An IPCC Special Peport*; Cambridge University Press: Cambridge, UK New York, NY, USA, 2018; pp. 3–24. [CrossRef]
2. Schewe, J.; Heinke, J.; Gerten, D.; Haddeland, I.; Arnell, N.W.; Clark, D.B.; Dankers, R.; Eisner, S.; Fekete, B.M.; Colón-González, F.J.; et al. Multimodel assessment of water scarcity under climate change. *Proc. Natl. Acad. Sci. USA* **2014**, *111*, 3245–3250. [CrossRef] [PubMed]
3. Zhang, W.; Zhou, T.; Wu, P. Anthropogenic amplification of precipitation variability over the past century. *Science* **2024**, *385*, 427–432. [CrossRef] [PubMed]
4. Cohen, I.; Zandalinas, S.I.; Huck, C.; Fritschi, F.B.; Mittler, R. Meta-analysis of drought and heat stress combination impact on crop yield and yield components. *Physiol. Plant.* **2020**, *171*, 66–76. [CrossRef]
5. Patel, J.; Mishra, A. Plant aquaporins alleviate drought tolerance in plants by modulating cellular biochemistry, root-architecture, and photosynthesis. *Physiol. Plant.* **2021**, *172*, 1030–1044. [CrossRef]
6. Liu, Q.; Peng, C.; Schneider, R.; Cyr, D.; McDowell, N.G.; Kneeshaw, D. Drought-induced increase in tree mortality and corresponding decrease in the carbon sink capacity of Canada's boreal forests from 1970 to 2020. *Glob. Change Biol.* **2023**, *29*, 2274–2285. [CrossRef] [PubMed]
7. Birami, B.; Nägele, T.; Gattmann, M.; Preisler, Y.; Gast, A.; Arneth, A.; Ruehr, N.K. Hot drought reduces the effects of elevated CO_2 on tree water-use efficiency and carbon metabolism. *New Phytol.* **2020**, *226*, 1607–1621. [CrossRef]
8. Zhang, D.; Li, A.; Lam, S.K.; Li, P.; Zong, Y.; Gao, Z.; Hao, X. Increased carbon uptake under elevated CO_2 concentration enhances water-use efficiency of C_4 broomcorn millet under drought. *Agric. Water Manag.* **2021**, *245*, 106631. [CrossRef]
9. Wang, Z.; Wang, C.; Liu, S. Elevated CO_2 alleviates adverse effects of drought on plant water relations and photosynthesis: A global meta-analysis. *J. Ecol.* **2022**, *110*, 2836–2849. [CrossRef]
10. Zhai, Z.; Fang, Y.; Cheng, J.; Tian, Y.; Liu, L.; Cao, X. Intrinsic morphology and spatial distribution of non-structural carbohydrates contribute to drought resistance of two mulberry cultivars. *Plant Biol.* **2023**, *25*, 771–784. [CrossRef]
11. Roy, S.; Kapoor, R.; Mathur, P. Revisiting changes in growth, physiology and stress responses of plants under the effect of enhanced CO_2 and temperature. *Plant Cell Physiol.* **2023**, *65*, 4–19. [CrossRef]
12. Zhang, X.; Duan, Y.; Xing, Q.; Duan, R.; Shen, J.; Zong, Y.; Zhang, D.; Shi, X.; Li, P.; Hao, X. Elevated CO_2 concentration enhances drought tolerance by mitigating oxidative stress and enhancing carbon assimilation in foxtail millet (*Setaria italica*). *J. Agron. Crop. Sci.* **2024**, *210*, e12778. [CrossRef]
13. Tankari, M.; Wang, C.; Ma, H.; Li, X.; Li, L.; Soothar, R.K.; Cui, N.; Zaman-Allah, M.; Hao, W.; Liu, F.; et al. Drought priming improved water status, photosynthesis and water productivity of cowpea during post-anthesis drought stress. *Agric. Water. Manag.* **2021**, *245*, 106565. [CrossRef]
14. Martin-StPaul, N.; Delzon, S.; Cochard, H. Plant resistance to drought depends on timely stomatal closure. *Ecol. Lett.* **2017**, *20*, 1437–1447. [CrossRef]
15. Ding, L.; Lu, Z.; Gao, L.; Guo, S.; Shen, Q. Is Nitrogen a key determinant of water transport and photosynthesis in higher plants upon drought stress? *Front. Plant Sci.* **2018**, *9*, 1143. [CrossRef]

6. Bota, J.; Medrano, H.; Flexas, J. Is photosynthesis limited by decreased Rubisco activity and RuBP content under progressive water stress? *New Phytol.* **2004**, *162*, 671–681. [CrossRef]
7. Filek, M.; Łabanowska, M.; Kościelniak, J.; Biesaga-Kościelniak, J.; Kurdziel, M.; Szarejko, I.; Hartikainen, H. Characterization of barley leaf tolerance to drought stress by chlorophyll fluorescence and electron paramagnetic resonance studies. *J. Agron. Crop. Sci.* **2014**, *201*, 228–240. [CrossRef]
8. Yuan, H.; Wang, Q.; Qi, A.; Li, S.; Hu, Y.; Hu, Z.; Guo, L.; Liang, C.; Li, W.; Liu, C.; et al. Morphological, physiological, and photosynthetic differences of Tartary buckwheat induced by post-anthesis drought. *Plants* **2024**, *13*, 2161. [CrossRef] [PubMed]
9. Li, B.; Feng, Y.; Zong, Y.; Zhang, D.; Hao, X.; Li, P. Elevated CO_2-induced changes in photosynthesis, antioxidant enzymes and signal transduction enzyme of soybean under drought stress. *Plant Physiol. Biochem.* **2020**, *154*, 105–114. [CrossRef]
0. Zhang, L.; Xu, Q.; Yong, X.; Wu, M.; Jiang, B.; Jia, Y.; Ma, J.; Mou, L.; Tang, S.; Pan, Y. Effects of water deficit on two cultivars of Hibiscus mutabilis: A comprehensive study on morphological, physiological, and metabolic responses. *Plant Physiol. Bioc.* **2024**, *217*, 109269. [CrossRef]
1. Avila, R.T.; de Almeida, W.L.; Costa, L.C.; Machado, K.L.; Barbosa, M.L.; de Souza, R.P.; Martino, P.B.; Juárez, M.A.; Marçal, D.M.; Martins, S.C.; et al. Elevated air [CO_2] improves photosynthetic performance and alters biomass accumulation and partitioning in drought-stressed coffee plants. *Environ. Exp. Bot.* **2020**, *177*, 104137. [CrossRef]
2. Haverd, V.; Smith, B.; Canadell, J.G.; Cuntz, M.; Mikaloff-Fletcher, S.; Farquhar, G.; Woodgate, W.; Briggs, P.R.; Trudinger, C.M. Higher than expected CO_2 fertilization inferred from leaf to global observations. *Glob. Change Biol.* **2020**, *26*, 2390–2402. [CrossRef]
3. Ainsworth, E.A.; Long, S.P. 30 years of free-air carbon dioxide enrichment (FACE): What have we learned about future crop productivity and its potential for adaptation? *Glob. Change Biol.* **2020**, *27*, 27–49. [CrossRef]
4. Song, L.; Tao, Y.; van Groenigen, K.J.; Chang, S.X.; Peñuelas, J.; Zhang, J.; You, L.; Cai, C.; Wang, S.; Jiang, Y.; et al. Rising atmospheric carbon dioxide concentrations increase gaps of rice yields between low- and middle-to-high-income countries. *Nat. Food* **2024**, *5*, 754–763. [CrossRef]
5. Busch, F.A.; Sage, R.F. The sensitivity of photosynthesis to O_2 and CO_2 concentration identifies strong Rubisco control above the thermal optimum. *New Phytol.* **2016**, *213*, 1036–1051. [CrossRef] [PubMed]
6. Ancín, M.; Gámez, A.L.; Jauregui, I.; Galmes, J.; E Sharwood, R.; Erice, G.; A Ainsworth, E.; Tissue, D.T.; Sanz-Sáez, A.; Aranjuelo, I. Does the response of Rubisco and photosynthesis to elevated [CO_2] change with unfavourable environmental conditions? *J. Exp. Bot.* **2024**, *75*, 7351–7364. [CrossRef] [PubMed]
7. Wang, W.; Cai, C.; He, J.; Gu, J.; Zhu, G.; Zhang, W.; Zhu, J.; Liu, G. Yield, dry matter distribution and photosynthetic characteristics of rice under elevated CO_2 and increased temperature conditions. *Field Crop. Res.* **2020**, *248*, 107605. [CrossRef]
8. Robredo, A.; Pérez-López, U.; de la Maza, H.S.; González-Moro, B.; Lacuesta, M.; Mena-Petite, A.; Muñoz-Rueda, A. Elevated CO_2 alleviates the impact of drought on barley improving water status by lowering stomatal conductance and delaying its effects on photosynthesis. *Environ. Exp. Bot.* **2007**, *59*, 252–263. [CrossRef]
9. Pazzagli, P.T.; Weiner, J.; Liu, F. Effects of CO_2 elevation and irrigation regimes on leaf gas exchange, plant water relations, and water use efficiency of two tomato cultivars. *Agric. Water Manag.* **2016**, *169*, 26–33. [CrossRef]
0. Avila, R.T.; Cardoso, A.A.; de Almeida, W.L.; Costa, L.C.; Machado, K.L.; Barbosa, M.L.; de Souza, R.P.; Oliveira, L.A.; Batista, D.S.; Martins, S.C.; et al. Coffee plants respond to drought and elevated [CO_2] through changes in stomatal function, plant hydraulic conductance, and aquaporin expression. *Environ. Exp. Bot.* **2020**, *177*, 104148. [CrossRef]
31. Li, S.; Li, X.; Wei, Z.; Liu, F. ABA-mediated modulation of elevated CO_2 on stomatal response to drought. *Curr. Opin. Plant Biol.* **2020**, *56*, 174–180. [CrossRef]
32. Baker, J.T.; Allen, L.H. Rice growth, yield and photosynthetic responses to elevated atmospheric carbon dioxide concentration and drought. *J. Crop. Improv.* **2005**, *13*, 7–30. [CrossRef]
33. Li, A.; Lv, D.; Zhang, Y.; Zhang, D.; Zong, Y.; Shi, X.; Li, P.; Hao, X. Elevated CO_2 concentration enhances drought resistance of soybean by regulating cell structure, cuticular wax synthesis, photosynthesis, and oxidative stress response. *Plant Physiol. Biochem.* **2023**, *206*, 108266. [CrossRef]
34. Ulfat, A.; Shokat, S.; Li, X.; Fang, L.; Großkinsky, D.K.; Majid, S.A.; Roitsch, T.; Liu, F. Elevated carbon dioxide alleviates the negative impact of drought on wheat by modulating plant metabolism and physiology. *Agric. Water Manag.* **2021**, *250*, 106804. [CrossRef]
35. Liu, W.; Liu, L.; Yan, R.; Gao, J.; Wu, S.; Liu, Y. A comprehensive meta-analysis of the impacts of intensified drought and elevated CO_2 on forage growth. *J. Environ. Manag.* **2022**, *327*, 116885. [CrossRef] [PubMed]
36. Miranda-Apodaca, J.; Pérez-López, U.; Lacuesta, M.; Mena-Petite, A.; Muñoz-Rueda, A. The interaction between drought and elevated CO 2 in water relations in two grassland species is species-specific. *J. Plant Physiol.* **2018**, *220*, 193–202. [CrossRef]
37. Wei, Z.; Abdelhakim, L.O.A.; Fang, L.; Peng, X.; Liu, J.; Liu, F. Elevated CO_2 effect on the response of stomatal control and water use efficiency in amaranth and maize plants to progressive drought stress. *Agric. Water Manag.* **2022**, *266*, 107609. [CrossRef]

38. Fan, X.D.; Cao, X.; Zhou, H.R.; Hao, L.H.; Dong, W.; He, C.L.; Xu, M.; Wu, H.X.; Wang, L.S.; Chang, Z.J. Carbon dioxide fertilization effect on plant growth under soil water stress associates with changes in stomatal traits, leaf photosynthesis, and foliar nitrogen of bell pepper (*Capsicum annuum* L.). *Environ. Exp. Bot.* **2020**, *179*, 104203. [CrossRef]
39. Wu, D.-X.; Wang, G.-X.; Bai, Y.-F.; Liao, J.-X. Effects of elevated CO_2 concentration on growth, water use, yield and grain quality of wheat under two soil water levels. *Agric. Ecosyst. Environ.* **2004**, *104*, 493–507. [CrossRef]
40. Ottman, M.J.; Kimball, B.A.; Pinter, P.J.; Wall, G.W.; Vanderlip, R.L.; Leavitt, S.W.; Lamorte, R.L.; Matthias, A.D.; Brooks, T.J. Elevated CO_2 increases sorghum biomass under drought conditions. *New Phytol.* **2001**, *150*, 261–273. [CrossRef]
41. Ainsworth, E.A.; Long, S.P. What have we learned from 15 years of free-air CO_2 enrichment (FACE)? A meta-analytic review of the responses of photosynthesis, canopy properties and plant production to rising CO_2. *New Phytol.* **2005**, *165*, 351–372. [CrossRef]
42. Leakey, A.D.B.; Ainsworth, E.A.; Bernacchi, C.J.; Rogers, A.; Long, S.P.; Ort, D.R. Elevated CO_2 effects on plant carbon, nitrogen and water relations: Six important lessons from FACE. *J. Exp. Bot.* **2009**, *60*, 2859–2876. [CrossRef] [PubMed]
43. He, M.; Dijkstra, F.A. Drought effect on plant nitrogen and phosphorus: A meta-analysis. *New Phytol.* **2014**, *204*, 924–931. [CrossRef] [PubMed]
44. Zhao, Q.; Guo, J.; Shu, M.; Wang, P.; Hu, S. Impacts of drought and nitrogen enrichment on leaf nutrient resorption and root nutrient allocation in four Tibetan plant species. *Sci. Total. Environ.* **2020**, *723*, 138106. [CrossRef] [PubMed]
45. Xiao, L.; Min, X.; Liu, G.; Li, P.; Xue, S. Effect of plant–plant interactions and drought stress on the response of soil nutrient contents, enzyme activities and microbial metabolic limitations. *Appl. Soil Ecol.* **2022**, *181*, 104666. [CrossRef]
46. Shi, S.; Luo, X.; Wen, M.; Dong, X.; Sharifi, S.; Xie, D.; He, X. *Funneliformis mosseae* improve growth and nutrient accumulation in wheat by facilitating soil nutrient uptake under elevated CO_2 at daytime, not nighttime. *J. Fungi* **2021**, *7*, 458. [CrossRef]
47. Singh, A.K.; Rai, A.; Kushwaha, M.; Chauhan, P.S.; Pandey, V.; Singh, N. Tree growth rate regulate the influence of elevated CO_2 on soil biochemical responses under tropical condition. *J. Environ. Manag.* **2019**, *231*, 1211–1221. [CrossRef]
48. Kuzyakov, Y.; Horwath, W.R.; Dorodnikov, M.; Blagodatskaya, E. Review and synthesis of the effects of elevated atmospheric CO_2 on soil processes: No changes in pools, but increased fluxes and accelerated cycles. *Soil Biol. Biochem.* **2019**, *128*, 66–78. [CrossRef]
49. Kumar, U.; Nayak, A.; Sahoo, S.; Kumar, A.; Kaviraj, M.; Shahid, M. Combined effects of elevated CO_2, N fertilizer and water deficit stress on diazotrophic community in sub-humid tropical paddy soiltropical paddy soil. *Appl. Soil Ecol.* **2020**, *155*, 103682. [CrossRef]
50. Guenet, B.; Lenhart, K.; Leloup, J.; Giusti-Miller, S.; Pouteau, V.; Mora, P.; Nunan, N.; Abbadie, L. The impact of long-term CO_2 enrichment and moisture levels on soil microbial community structure and enzyme activities. *Geoderma* **2012**, *170*, 331–336. [CrossRef]
51. Wang, Y.; Wang, Y.; Chen, Y.; Chen, H.; Li, X.; Ding, Z.; Han, X.; Tang, X. Spatial and temporal characteristics of drought events in southwest China over the past 120 years. *Remote. Sens.* **2023**, *15*, 3008. [CrossRef]
52. Reddy, K.S.; Sekhar, K.M.; Sreeharsha, R.V.; Reddy, A.R. Hydraulic dynamics and photosynthetic performance facilitate rapid screening of field grown mulberry (*Morus* spp.) genotypes for drought tolerance. *Environ. Exp. Bot.* **2019**, *157*, 320–330. [CrossRef]
53. Kaushal, R.; Kumar, A.; Mandal, D.; Tomar, J.; Jinger, D.; Islam, S.; Panwar, P.; Jayaprakash, J.; Uthappa, A.; Singhal, V. et al. Mulberry based agroforestry system and canopy management practices to combat soil erosion and enhancing carbon sequestration in degraded lands of Himalayan foothills. *Environ. Sustain. Indic.* **2024**, *24*, 100467. [CrossRef]
54. Manzoor, S.; Qayoom, K. Environmental importance of mulberry: A review. *J. Exp. Agric. Int.* **2024**, *46*, 95–105. [CrossRef]
55. Sarkar, T.; Doss, S.G.; Sivaprasad, V.; Teotia, R.S. *Stress Tolerant Traits in Mulberry (Morus spp.) Resilient to Climate Change, Mulberry: Genetic Improvement in Context of Climate Change*; CRC Press: Boca Raton, FL, USA, 2021; pp. 165–179. [CrossRef]
56. Sasmita, N.; Purba, J.H.; Yuniti, I.G.A.D. Adaptation of *Morus alba* and *Morus cathayana* plants in a different climate and environment conditions in Indonesia. *Biodiversitas J. Biol. Divers.* **2019**, *20*, 544–554. [CrossRef]
57. Ghosh, A.; Gangopadhyay, D.; Chowdhury, T. Economical and environmental importance of mulberry: A review. *Int. J. Plant Environ.* **2017**, *3*, 67–72. [CrossRef]
58. Bajwa, G.A.; Umair, M.; Nawab, Y.; Rizwan, Z. Morphometry of leaf and shoot variables to assess aboveground biomass structure and carbon sequestration by different varieties of white mulberry (*Morus alba* L.). *J. For. Res.* **2021**, *32*, 2291–2300. [CrossRef]
59. Durgadevi, R.; Vijayalakshmi, D. Mulberry with increased stomatal frequency regulates gas exchange traits for improved drought tolerance. *Plant Physiol. Rep.* **2020**, *25*, 24–32. [CrossRef]
60. Sun, Z.; Yin, Y.; Zhu, W.; Zhou, Y. Morphological, physiological, and biochemical composition of mulberry (*Morus* spp.) under drought stress. *Forests* **2023**, *14*, 949. [CrossRef]
61. Shi, S.; Qiu, Y.; Wen, M.; Xu, X.; Dong, X.; Xu, C.; He, X. Daytime, not nighttime, elevated atmospheric carbon dioxide exposure improves plant growth and leaf quality of mulberry (*Morus alba* L.) seedlings. *Front. Plant Sci.* **2021**, *11*, 609031. [CrossRef]
62. Shi, S.; Xu, X.; Dong, X.; Xu, C.; Qiu, Y.; He, X. Photosynthetic acclimation and growth responses to elevated CO_2 associate with leaf nitrogen and phosphorus concentrations in mulberry (*Morus multicaulis* Perr.). *Forests* **2021**, *12*, 660. [CrossRef]
63. Sekhar, K.M.; Rachapudi, V.S.; Mudalkar, S.; Reddy, A.R. Persistent stimulation of photosynthesis in short rotation coppice mulberry under elevated CO_2 atmosphere. *J. Photochem. Photobiol. B Biol.* **2014**, *137*, 21–30. [CrossRef]

14. Sekhar, K.M.; Sreeharsha, R.V.; Reddy, A.R. Differential responses in photosynthesis, growth and biomass yields in two mulberry genotypes grown under elevated CO 2 atmosphere. *J. Photochem. Photobiol. B Biol.* **2015**, *151*, 172–179. [CrossRef] [PubMed]
15. Liu, X.; Zhang, H.; Wang, J.; Wu, X.; Ma, S.; Xu, Z.; Zhou, T.; Xu, N.; Tang, X.; An, B. Increased CO_2 concentrations increasing water use efficiency and improvement PSII function of mulberry seedling leaves under drought stress. *J. Plant Interact.* **2019**, *14*, 213–223. [CrossRef]
16. Kramer, D.M.; Johnson, G.; Kiirats, O.; Edwards, G.E. New fluorescence parameters for the determination of Q_A redox state and excitation energy fluxes. *Photosynth. Res.* **2004**, *79*, 209–218. [CrossRef]
17. Yang, J.; Wang, C.; Dai, H. *Soil Agrochemical Analysis and Environmental Monitoring Techniques*; Chinese Dadi Press: Beijing, China, 2008; pp. 18–64.
18. Li, H.S. *The Experiment Principle and Technique on Plant Physiology and Biochemistry*; Higher Education Press: Beijing, China, 2000; pp. 78–102.
19. Guan, S.Y.; Zhang, D.; Zhang, Z. *Soil Enzyme and Its Research Methods*; Agricultural Press: Beijing, China, 1986; pp. 274–340.
20. Huang, X.; Liu, Y.; Li, J.; Xiong, X.; Chen, Y.; Yin, X.; Feng, D. The response of mulberry trees after seedling hardening to summer drought in the hydro-fluctuation belt of Three Gorges Reservoir Areas. *Environ. Sci. Pollut. Res.* **2012**, *20*, 7103–7111. [CrossRef]
21. Deng, J.; Zhang, Y.; Yin, Y.; Zhu, X.; Zhu, W.; Zhou, Y. Comparison of soil bacterial community and functional characteristics following afforestation in the semi-arid areas. *PeerJ* **2019**, *7*, e7141. [CrossRef] [PubMed]
22. Reddy, K.S.; Sekhar, K.M.; Reddy, A.R. Genotypic variation in tolerance to drought stress is highly coordinated with hydraulic conductivity–photosynthesis interplay and aquaporin expression in field-grown mulberry (Morus spp.). *Tree Physiol.* **2017**, *37*, 926–937. [CrossRef]
23. Lauriks, F.; Salomón, R.L.; De Roo, L.; Sobrino-Plata, J.; Rodríguez-García, A.; Steppe, K. Limited mitigating effects of elevated CO_2 in young aspen trees to face drought stress. *Environ. Exp. Bot.* **2022**, *201*, 104942. [CrossRef]
24. Ofori-Amanfo, K.K.; Klem, K.; Veselá, B.; Holub, P.; Agyei, T.; Juráň, S.; Grace, J.; Marek, M.V.; Urban, O. The effect of elevated CO_2 on photosynthesis is modulated by nitrogen supply and reduced water availability in *Picea abies*. *Tree Physiol.* **2023**, *43*, 925–937. [CrossRef]
25. Ayub, G.; Zaragoza-Castells, J.; Griffin, K.L.; Atkin, O.K. Leaf respiration in darkness and in the light under pre-industrial, current and elevated atmospheric CO_2 concentrations. *Plant Sci.* **2014**, *226*, 120–130. [CrossRef]
26. Kumar, A.; Nayak, A.; Das, B.; Panigrahi, N.; Dasgupta, P.; Mohanty, S.; Kumar, U.; Panneerselvam, P.; Pathak, H. Effects of water deficit stress on agronomic and physiological responses of rice and greenhouse gas emission from rice soil under elevated atmospheric CO_2. *Sci. Total Environ.* **2018**, *650*, 2032–2050. [CrossRef] [PubMed]
27. Liberloo, M.; Tulva, I.; Raïm, O.; Kull, O.; Ceulemans, R. Photosynthetic stimulation under long-term CO_2 enrichment and fertilization is sustained across a closed *Populus* canopy profile (EUROFACE). *New Phytol.* **2006**, *173*, 537–549. [CrossRef]
28. Kelly, J.W.G.; Duursma, R.A.; Atwell, B.J.; Tissue, D.T.; Medlyn, B.E. Drought × CO_2 interactions in trees: A test of the low-intercellular CO_2 concentration (C_i) mechanism. *New Phytol.* **2015**, *209*, 1600–1612. [CrossRef] [PubMed]
29. Regier, N.; Streb, S.; Cocozza, C.; Schaub, M.; Cherubini, P.; Zeeman, S.C.; Frey, B. Drought tolerance of two black poplar (*Populus nigra* L.) clones: Contribution of carbohydrates and oxidative stress defence. *Plant Cell Environ.* **2009**, *32*, 1724–1736. [CrossRef]
30. Semedo, J.N.; Rodrigues, A.P.; Lidon, F.C.; Pais, I.P.; Marques, I.; Gouveia, D.; Armengaud, J.; Silva, M.J.; Martins, S.; Semedo, M.C.; et al. Intrinsic non-stomatal resilience to drought of the photosynthetic apparatus in *Coffea* spp. is strengthened by elevated air [CO_2]. *Tree Physiol.* **2020**, *41*, 708–727. [CrossRef]
31. Halpern, M.; Bar-Tal, A.; Lugassi, N.; Egbaria, A.; Granot, D.; Yermiyahu, U. The role of nitrogen in photosynthetic acclimation to elevated [CO_2] in tomatoes. *Plant Soil* **2018**, *434*, 397–411. [CrossRef]
32. Sanches, R.F.E.; Centeno, D.d.C.; Braga, M.R.; da Silva, E.A. Impact of high atmospheric CO_2 concentrations on the seasonality of water-related processes, gas exchange, and carbohydrate metabolism in coffee trees under field conditions. *Clim. Change* **2020**, *162*, 1231–1248. [CrossRef]
33. Zheng, Y.; He, C.; Guo, L.; Hao, L.; Cheng, D.; Li, F.; Peng, Z.; Xu, M. Soil water status triggers CO_2 fertilization effect on the growth of winter wheat (*Triticum aestivum*). *Agric. For. Meteorol.* **2020**, *291*, 108097. [CrossRef]
34. Yousefvand, P.; Pilehvar, B.; Nasrolahi, A.H. Morphological, physiological, and biochemical responses of Pistacia atlantica seedlings to elevated CO_2 concentration and drought stress. *Eur. J. For. Res.* **2023**, *142*, 657–670. [CrossRef]
35. Woodward, F.I. Stomatal numbers are sensitive to increases in CO_2 from pre-industrial levels. *Nature* **1987**, *327*, 617–618. [CrossRef]
36. Morison, J. Stomatal response to increased CO_2 concentration. *J. Exp. Bot.* **1998**, *49*, 443–452. [CrossRef]
37. Correia, B.; Hancock, R.D.; Amaral, J.; Gomez-Cadenas, A.; Valledor, L.; Pinto, G. Combined drought and heat activates protective responses in eucalyptus globulus that are not activated when subjected to drought or heat stress alone. *Front. Plant Sci.* **2018**, *9*, 819. [CrossRef] [PubMed]
38. Li, Q.; Liu, B.; Wu, Y.; Zou, Z. Interactive effects of drought stresses and elevated CO_2 concentration on photochemistry efficiency of cucumber seedlings. *J. Integr. Plant Biol.* **2008**, *50*, 1307–1317. [CrossRef] [PubMed]

89. Sommer, S.G.; Han, E.; Li, X.; Rosenqvist, E.; Liu, F. The chlorophyll fluorescence parameter F_v/F_m correlates with loss of grain yield after severe drought in three wheat genotypes grown at two CO_2 concentrations. *Plants* **2023**, *12*, 436. [CrossRef]
90. Flexas, J.; Medrano, H. Drought-inhibition of photosynthesis in C3 plants: Stomatal and non-stomatal limitations revisited. *Ann. Bot.* **2002**, *89*, 183–189. [CrossRef]
91. Chaves, M.M.; Pereira, J.S.; Marôco, J.; Rodrigues, M.L.; Ricardo, C.P.P.; Osório, M.L.; Carvalho, I.; Faria, T.; Pinheiro, C. How plants cope with water stress in the field? photosynthesis and growth. *Ann. Bot.* **2002**, *89*, 907–916. [CrossRef]
92. Muller, P.; Li, X.-P.; Niyogi, K.K. Non-photochemical quenching. A response to excess light energy. *Plant Physiol.* **2001**, *125*, 1558–1566. [CrossRef]
93. Murchie, E.H.; Niyogi, K.K. Manipulation of photoprotection to improve plant photosynthesis. *Plant Physiol.* **2011**, *155*, 86–92. [CrossRef]
94. Long, S.P.; Ainsworth, E.A.; Leakey, A.D.B.; Nosberger, J.; Ort, D.R. Food for thought: Lower-than-expected crop yield stimulation with rising CO_2 concentrations. *Science* **2006**, *312*, 1918–1921. [CrossRef]
95. Wujeska, A.; Bossinger, G.; Tausz, M. Responses of foliar antioxidative and photoprotective defence systems of trees to drought: A meta-analysis. *Tree Physiol.* **2013**, *33*, 1018–1029. [CrossRef]
96. Hasanuzzaman, M.; Bhuyan, M.B.; Zulfiqar, F.; Raza, A.; Mohsin, S.M.; Al Mahmud, J.; Fujita, M.; Fotopoulos, V. Reactive oxygen species and antioxidant defense in plants under abiotic stress: Revisiting the crucial role of a universal defense regulator. *Antioxidants* **2020**, *9*, 681. [CrossRef] [PubMed]
97. Ozturk, M.; Unal, B.T.; García-Caparrós, P.; Khursheed, A.; Gul, A.; Hasanuzzaman, M. Osmoregulation and its actions during the drought stress in plants. *Physiol. Plant.* **2020**, *172*, 1321–1335. [CrossRef] [PubMed]
98. Gayathri, T.; Mogili, T.; Krishnan, R.; Doss, S.G.; Sarkar, T.; Teotia, R.S.; Tewary, P. Physio-biochemical evaluation and in situ localization of reactive oxygen species (ROS) in mulberry genotypes under optimal growth conditions. *Acta Physiol. Plant* **2022**, *44*, 141. [CrossRef]
99. Asada, K. The water-water cycle in chloroplasts: Scavenging of active oxygens and dissipation of excess photons. *Annu. Rev. Plant Biol.* **1999**, *50*, 601–639. [CrossRef]
100. Silva, E.N.; Ferreira-Silva, S.L.; Fontenele, A.d.V.; Ribeiro, R.V.; Viégas, R.A.; Silveira, J.A.G. Photosynthetic changes and protective mechanisms against oxidative damage subjected to isolated and combined drought and heat stresses in *Jatropha curcas* plants. *J. Plant Physiol.* **2010**, *167*, 1157–1164. [CrossRef]
101. Rodrigues, T.d.S.; Arge, L.W.P.; Guedes, F.A.d.F.; Travassos-Lins, J.; de Souza, A.P.; Cocuron, J.; Buckeridge, M.S.; Grossi-De-Sá, M.F.; Alves-Ferreira, M. Elevated CO_2 increases biomass of *Sorghum bicolor* green prop roots under drought conditions via soluble sugar accumulation and photosynthetic activity. *Physiol. Plant.* **2023**, *175*, e13984. [CrossRef]
102. Yamada, M.; Morishita, H.; Urano, K.; Shiozaki, N.; Yamaguchi-Shinozaki, K.; Shinozaki, K.; Yoshiba, Y. Effects of free proline accumulation in petunias under drought stress. *J. Exp. Bot.* **2005**, *56*, 1975–1981. [CrossRef]
103. Nader, A.A.; Hauka, F.I.A.; Afify, A.H.; El-Sawah, A.M. Drought-tolerant bacteria and arbuscular mycorrhizal fungi mitigate the detrimental effects of drought stress induced by withholding irrigation at critical growth stages of soybean (*Glycine max*, L.) *Microorganisms* **2024**, *12*, 1123. [CrossRef]
104. Verbruggen, N.; Hermans, C. Proline accumulation in plants: A review. *Amino Acids* **2008**, *35*, 753–759. [CrossRef]
105. Sardans, J.; Peñuuelas, J.; Estiarte, M.; Prieto, P. Warming and drought alter C and N concentration, allocation and accumulation in a Mediterranean shrubland. *Glob. Change Biol.* **2008**, *14*, 2304–2316. [CrossRef]
106. Xu, S.; Sardans, J.; Zhang, J.; Peñuelas, J. Variations in foliar carbon:nitrogen and nitrogen:phosphorus ratios under global change: A meta-analysis of experimental field studies. *Sci. Rep.* **2020**, *10*, 12156. [CrossRef]
107. Xu, Z.; Zhou, G.; Wang, Y. Combined effects of elevated CO_2 and soil drought on carbon and nitrogen allocation of the desert shrub Caragana intermedia. *Plant Soil* **2007**, *301*, 87–97. [CrossRef]
108. Wujeska-Klause, A.; Crous, K.Y.; Ghannoum, O.; Ellsworth, D.S. Lower photorespiration in elevated CO_2 reduces leaf N concentrations in mature *Eucalyptus* trees in the field. *Glob. Change Biol.* **2019**, *25*, 1282–1295. [CrossRef] [PubMed]
109. Bista, D.R.; Heckathorn, S.A.; Jayawardena, D.M.; Boldt, J.K. Effect of drought and carbon dioxide on nutrient uptake and levels of nutrient-uptake proteins in roots of barley. *Am. J. Bot.* **2020**, *107*, 1401–1409. [CrossRef]
110. Loladze, I.; Korea, R.O. Hidden shift of the ionome of plants exposed to elevated CO_2 depletes minerals at the base of human nutrition. *eLife* **2014**, *3*, e02245. [CrossRef] [PubMed]
111. Franklin, O. Optimal nitrogen allocation controls tree responses to elevated CO_2. *New Phytol.* **2007**, *174*, 811–822. [CrossRef]
112. Sekhar, K.M.; Kota, V.R.; Reddy, T.P.; Rao, K.V.; Reddy, A.R. Amelioration of plant responses to drought under elevated CO_2 by rejuvenating photosynthesis and nitrogen use efficiency: Implications for future climate-resilient crops. *Photosynth. Res.* **2020**, *150*, 21–40. [CrossRef]
113. Lambers, H.; Chapin, F.S.; Pons, T.L. *Plant Physiological Ecology*; Springer: New York, NY, USA, 2008.
114. Bassirirad, H. Kinetics of nutrient uptake by roots: Responses to global change. *New Phytol.* **2000**, *147*, 155–169. [CrossRef]

15. Alster, C.J.; German, D.P.; Lu, Y.; Allison, S.D. Microbial enzymatic responses to drought and to nitrogen addition in a southern California grassland. *Soil Biol. Biochem.* **2013**, *64*, 68–79. [CrossRef]
16. Gao, W.; Reed, S.C.; Munson, S.M.; Rui, Y.; Fan, W.; Zheng, Z.; Li, L.; Che, R.; Xue, K.; Du, J.; et al. Responses of soil extracellular enzyme activities and bacterial community composition to seasonal stages of drought in a semiarid grassland. *Geoderma* **2021**, *401*, 115327. [CrossRef]
17. Ahmed, M.A.; Sanaullah, M.; Blagodatskaya, E.; Mason-Jones, K.; Jawad, H.; Kuzyakov, Y.; Dippold, M.A. Soil microorganisms exhibit enzymatic and priming response to root mucilage under drought. *Soil Biol. Biochem.* **2018**, *116*, 410–418. [CrossRef]
18. Wang, Y.; Yan, D.; Wang, J.; Ding, Y.; Song, X. Effects of elevated CO_2 and drought on plant physiology, soil carbon and soil enzyme activities. *Pedosphere* **2017**, *27*, 846–855. [CrossRef]
19. Yu, H.; Deng, Y.; He, Z.; Van Nostrand, J.D.; Wang, S.; Jin, D.; Wang, A.; Wu, L.; Wang, D.; Tai, X.; et al. Elevated CO_2 and warming altered grassland microbial communities in soil top-layers. *Front. Microbiol.* **2018**, *9*, 1790. [CrossRef]
20. Bhattacharyya, P.; Roy, K.S.; Neogi, S.; Manna, M.C.; Adhya, T.K.; Rao, K.S.; Nayak, A.K. Influence of elevated carbon dioxide and temperature on belowground carbon allocation and enzyme activities in tropical flooded soil planted with rice. *Environ. Monit. Assess.* **2013**, *185*, 8659–8671. [CrossRef] [PubMed]
21. van Groenigen, K.J.; Qi, X.; Osenberg, C.W.; Luo, Y.; Hungate, B.A. Faster decomposition under increased atmospheric CO_2 limits soil carbon storage. *Science* **2014**, *344*, 508–509. [CrossRef]
22. Phillips, R.P.; Finzi, A.C.; Bernhardt, E.S. Enhanced root exudation induces microbial feedbacks to N cycling in a pine forest under long-term CO_2 fumigation. *Ecol. Lett.* **2010**, *14*, 187–194. [CrossRef]

antioxidants

MDPI

Review

Harnessing Antioxidants for Abiotic Stress Management: Mechanistic Insights and Prospects for Sustainable Agriculture

Fasih Ullah Haider [1], Tianhao Liu [1], Luis Carlos Ramos Aguila [2], Babar Shahzad [3], Habiba [4], Peng Zhang [1,*] and Xiangnan Li [1,*]

1 Key Laboratory of Black Soil Conservation and Utilization, Northeast Institute of Geography and Agroecology, Chinese Academy of Sciences, Changchun 130102, China; haider281@iga.ac.cn (F.U.H.); liutianhao@iga.ac.cn (T.L.)
2 State Key Laboratory of Biocontrol, School of Ecology, Sun Yat-sen University, Shenzhen 518000, China; carlos3@mail.sysu.edu.cn
3 School of Agriculture, Food and Wine, Adelaide University, Adelaide, SA 5064, Australia; babar.shahzad@adelaide.edu.au
4 Department of Biological Science, Science Hall-3401, Lehman College-CUNY, Bronx, NY 10468, USA; fnu.habiba79@login.cuny.edu
* Correspondence: zhangpeng06@iga.ac.cn (P.Z.); lixiangnan@iga.ac.cn (X.L.)

Abstract

Abiotic stresses disrupt redox homeostasis and reduce crop productivity. Antioxidant networks support resilience by limiting excess reactive oxygen species (ROS) and maintaining redox signalling for stress perception, gene expression, and metabolic reprogramming. We summarize advances (2000–2025) in ROS generation, detoxification mechanisms, and signalling across organelles, including chloroplasts, mitochondria, peroxisomes, and the apoplast. This includes compartmentalized enzymes—superoxide dismutase (SOD), catalase (CAT), ascorbate peroxidase (APX), glutathione peroxidase (GPX), and glutathione reductase (GR)—as well as the peroxiredoxin–thioredoxin system and non-enzymatic buffers like ascorbate, glutathione, tocopherols, carotenoids, and flavonoids. We uniquely synthesize these findings in a compartment-resolved "redox rheostat" model, linking ROS concentration–time windows (signaling vs. damage) to antioxidant network design (kinetic tiers, compartmentation, and trade-offs) and identifying intervention points for breeding, genome editing, and field-scale priming. We emphasize constraints, such as NADPH supply and antioxidant recycling capacity, that lead to context-dependent outcomes. We evaluate omics, transgenic strategies, genome editing (CRISPR and Cas systems), exogenous applications, and plant–microbe associations. This synthesis clarifies how antioxidant systems protect photosynthetic and respiratory machinery while supporting signalling, thus outlining routes to climate-resilient, yield-stable crops across varied environments and stresses.

Keywords: abiotic stress; reactive oxygen species; redox signalling; ascorbate–glutathione cycle; peroxiredoxin–thioredoxin; NADPH; genome editing; crop resilience

1. Introduction

Agriculture is the cornerstone of global food security, yet it increasingly faces diverse climatic threats [1]. Rapid shifts in climate patterns, rising temperatures, altered precipitation, intensifying droughts, soil salinization, and recurrent heatwaves represent some of the toughest challenges to crop stability and productivity worldwide [2]. Quantitative data underscore these effects: abiotic stresses cause yield losses in major crops such as

https://doi.org/10.3390/antiox15030337

wheat (*Triticum aestivum* L.), rice (*Oryza sativa* L.), and maize (*Zea mays* L.), with magnitude varying by stress type, severity, timing, genotype, and environment [3,4]. Empirical models show that warming results in about 3–7% yield loss per 1 °C in major cereals, with disproportionate losses in vulnerable tropical and subtropical regions [5,6]. By mid-century, many projections indicate that combined climatic stresses could greatly reduce cereal productivity without adaptation, increasing pressure on global food security as populations grow [7]. Thus, it is essential to broaden our understanding of plant physiological and molecular responses to abiotic stress and boost crop resilience for sustainable agriculture [8–10].

A common feature among different abiotic stressors is their strong disruption of cellular redox homeostasis [1–4]. Under normal conditions, plants naturally produce reactive oxygen species (ROS). These act as metabolic switches within organelles such as chloroplasts, mitochondria, and peroxisomes [11]. ROS, which include superoxide radicals ($O_2^{\bullet-}$), hydrogen peroxide (H_2O_2), hydroxyl radicals (−OH), and singlet oxygen ($^1O_2$), usually exist at low basal levels. They serve important roles as secondary messengers in processes like development, hormone signalling, stomatal regulation, and pathogen defence [12,13]. When abiotic stress occurs, this balance is disrupted by excess ROS production [8]. For example, drought stress can rapidly increase hydrogen peroxide levels in leaf mesophyll cells, whereas saline exposure triggers oxidative bursts primarily in roots [12]. Though ROS are critical for signalling, excessive amounts lead to oxidative damage. This affects membrane lipids, protein complexes (including those for photosynthesis), and nucleic acids [13]. Thus, ROS are both vital messengers and potential cytotoxins. Their regulation is crucial for plant survival under stress [14,15].

Plants fight oxidative damage through a complex antioxidant defence network that includes both enzymatic and non-enzymatic components. This network maintains redox balance and supports adaptive signalling [4,6,16]. Enzymatic antioxidants include superoxide dismutase (SOD), catalase (CAT), and ascorbate peroxidase (APX). These enzymes detoxify ROS efficiently and specifically [17]. For instance, SOD converts superoxide radicals into H_2O_2 at a very rapid rate. This allows enzymes like CAT to further break down H_2O_2 into harmless water and oxygen, preventing the formation of dangerous hydroxyl radicals [8,13]. Alongside these are non-enzymatic antioxidants. These include ascorbic acid (vitamin C), glutathione (GSH), carotenoids, tocopherols, and various phenolic and flavonoid compounds. They help buffer ROS levels and support redox-sensitive signaling pathways critical for stress acclimation [18–20]. Genetic studies show the importance of antioxidants. Rice plants that overexpress chloroplast-targeted SOD or APX show better stress tolerance. In contrast, GSH biosynthesis mutants are more susceptible to oxidative stress [21]. Thus, antioxidants are dynamic regulators that shape cellular signalling, stress memory, and flexibility during adaptation to abiotic stress [22].

Despite progress in recent decades, we still lack a full understanding of how antioxidant systems integrate with broader cellular stress networks. Crosstalk between ROS signalling and phytohormonal pathways, such as abscisic acid (ABA), salicylic acid (SA), jasmonates, and ethylene, is only partly understood, especially with respect to timing and feedback. Moreover, while applying antioxidants like APX or GSH externally can boost tolerance in controlled settings, their effectiveness in the field is unpredictable. Results vary across studies and depend on the form of application, including foliar, seed, and soil treatments [1,20]. Additionally, maintaining chronically high antioxidant activity has metabolic costs. This leads to trade-offs between stress tolerance and achieving optimal growth or yield [16,17]. Taken together, these challenges show the need for research that combines mechanistic detail with practical applications. Such approaches must link molecular biology to agronomy to realize the promise of antioxidants for crop improvement. In this review, we focus on causal evidence, genetic changes, targeted interventions, and yield

https://doi.org/10.3390/antiox15030337

validation, and note when claims rely mainly on short-term or inconsistent field results. Antioxidant-based strategies succeed only if they balance ROS signalling and prevent ongoing oxidative overload. This lesson should guide breeding and management decisions.

Abiotic stress tolerance depends on more than just antioxidant abundance. Rather, it is governed by the way these networks are organized across cell compartments. These networks act as a quantitative rheostat, keeping ROS within beneficial signalling ranges and preventing oxidative damage. This review addresses four main questions: (i) where and when ROS are produced under stress (Section 3); (ii) how compartmentalized antioxidant systems buffer redox signals (Section 4); (iii) how this regulation adapts under drought, salinity, temperature extremes, and heavy metals (Section 5); and (iv) which interventions can realistically adjust this rheostat in crops (Sections 6 and 7), considering trade-offs, feasibility, and field constraints. Importantly, we assess antioxidant strategies in light of agronomic results, yield, stability, and quality, focusing on performance under combined or sequential stresses as the real test for field utility. By linking biological mechanisms to practical actions, our framework highlights recurring challenges, such as limited NADPH supply, insufficient recycling, and trade-offs between growth and defence. These issues explain why many antioxidant modifications do not translate to field success. Therefore, we discuss omics, transgenic, CRISPR, and field interventions only when they directly affect the ROS–antioxidant network and show real improvements in stress tolerance. Topics not central to redox regulation are not covered in detail.

2. Research Methodology

A structured literature search and evidence synthesis were conducted to examine antioxidant function in plants under abiotic stress. We searched databases (Web of Science, Scopus, PubMed, SpringerLink, ScienceDirect, and Google Scholar) for studies published from January 2000 to July 2025. Boolean combinations of terms covered abiotic stress, ROS/redox processes, antioxidant systems, and major stress classes (drought, salinity, temperature extremes, and heavy metals). We applied filters for translational approaches (transgenics, CRISPR/Cas, nanoparticles, and agronomic interventions). To address field relevance, we prioritized studies that reported agronomic outcomes (yield components, yield stability, or crop quality traits). We included evidence from combined or sequential stress designs (e.g., drought–heat, salinity–nutrient limitation) and multi-environment validation where available. Studies were screened for relevance and methodological quality, prioritizing recent advances (2015–2025) while retaining seminal work. The final synthesis is organized around four themes: ROS dynamics under stress; compartmentalized antioxidant network logic; stress-specific reparameterization; and intervention classes, including their feasibility, trade-offs, and field relevance.

3. Compartment-Specific ROS Dynamics Under Abiotic Stress (What Changes vs. Normal and Why It Matters)

Abiotic stresses such as drought, salinity, temperature extremes, and heavy metal toxicity disrupt plant metabolism. They quickly disturb cellular redox homeostasis and change ROS production, accumulation, and signalling [1,3,7,23], [Figure 1]. Under normal conditions, ROS are produced continuously as by-products of aerobic metabolism. During stress, ROS can build up faster than the cell can detoxify them [24]. ROS serve as signalling messengers, but can also cause oxidative damage. How cells respond depends on the type of ROS, where they are produced, their amount, and how long they persist [20,24]. Understanding stress tolerance requires a detailed, compartment-specific view of ROS dynamics, rather than assuming more ROS is always harmful.

https://doi.org/10.3390/antiox15030337

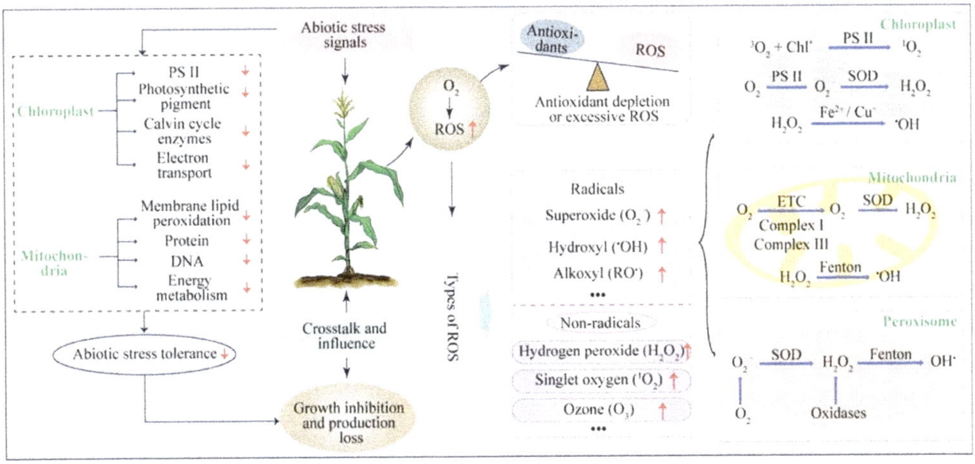

Figure 1. Reactive oxygen species (ROS) balance shifts from controlled signalling (normal) to harmful stress (abiotic stress). Under normal conditions, cells naturally produce ROS. Antioxidants keep these levels in check. This balance allows ROS to send helpful signals. Under abiotic stress, the balance breaks down. Antioxidants run low, and/or too much ROS builds up (red ↑). This disruption spreads as chloroplasts, mitochondria, and peroxisomes interact with one another. Chloroplasts (excited chlorophyll/PSII) make singlet oxygen ($^1O_2$) and superoxide radicals ($O_2^{\bullet-}$). Superoxide dismutase (SOD) changes $O_2^{\bullet-}$ into hydrogen peroxide (H_2O_2). Iron or copper helps make $^{\bullet}OH$. Mitochondria mainly make $O_2^{\bullet-}$ via their electron transport chains (ETCs). SOD then converts $O_2^{\bullet-}$ to H_2O_2 and generates more $^{\bullet}OH$. In peroxisomes, oxidases generate H_2O_2, leading to more $^{\bullet}OH$. These changes between cell parts slow growth and lower yield. Stress-linked ROS are radicals (like $O_2^{\bullet-}$, $^{\bullet}OH$, $RO^{\bullet}$) and non-radicals (like H_2O_2, $^1O_2$, O_3); all rise under stress (red ↑). The red arrows indicate that stress reduces photosynthesis and respiration. This increases cell injury and lowers overall stress tolerance. Abbreviations: ROS, reactive oxygen species; SOD, superoxide dismutase; PSII, photosystem II; ETC, electron transport chain.

3.1. Sources of ROS in Plant Cells

Plant cells contain multiple ROS-generating sites that are interconnected through redox and Ca^{2+} signalling networks [6–8]. ROS produced in one compartment can influence dynamics in other compartments, illustrating an integrated cellular redox system rather than isolated organelle events [11–13].

3.1.1. Chloroplasts: The Photosynthetic Hub of ROS

Chloroplasts are major sources of ROS under abiotic stress. Photosynthetic electron transport is highly sensitive to disruptions in CO_2 assimilation and energy balance [25,26]. Drought, salinity, and heat often limit CO_2 availability and carbon fixation. This leads to over-reduction in the electron transport chain and promotes $O_2^{\bullet-}$ formation at PSI (Mehler-type reactions) [10–12], [Figure 1]. Stromal SOD rapidly converts $O_2^{\bullet-}$ to H_2O_2. This links superoxide formation to downstream peroxide signalling and detoxification [27]. Excess excitation energy generates $^1O_2$ at PSII. $^1O_2$ acts locally due to its high reactivity and short lifetime [28]. H_2O_2 is more stable and diffusible. This enables chloroplast-to-nucleus retrograde signalling and transcriptional reprogramming during stress acclimation [29]. At physiological levels, chloroplast-derived ROS participate in photoprotective regulation (such as NPQ and antenna adjustments). Sustained ROS accumulation promotes photoinhibition, lipid peroxidation, and cell death [30]. Chloroplast ROS homeostasis provides a sensitive readout of the photosynthetic "redox budget" during stress.

https://doi.org/10.3390/antiox15030337

3.1.2. Mitochondria: Balancing Respiration and ROS

Mitochondria generate ROS mainly by electron leakage from complexes I and III of the respiratory electron transport chain during oxidative phosphorylation [Figure 1]. Under steady-state conditions, leakage is limited. Drought, chilling, and heavy metals slow downstream electron flow and increase electron transfer to O_2, raising $O_2^{\bullet-}$ formation [31]. Mn-SOD converts mitochondrial $O_2^{\bullet-}$ to H_2O_2 in the matrix, and peroxidases detoxify it to support ROS homeostasis [32]. The alternative oxidase (AOX) pathway is a key adaptation. It relieves over-reduction in the ETC and reduces ROS generation, but decreases ATP yield. This supports survival under stress [33]. Mitochondrial H_2O_2 also functions in retrograde signalling and influences nuclear gene expression programmes. These programmes adjust antioxidant capacity and increase stress acclimation [34–36].

3.1.3. Peroxisomes: Hotspots of Hydrogen Peroxide

Peroxisomes are prominent H_2O_2-producing organelles due to oxidative pathways. These include photorespiration (via glycolate oxidase), fatty acid β-oxidation, and purine catabolism [37], [Figure 1]. Photorespiration increases strongly under conditions that lower internal CO_2, such as drought, high light, or salinity. This makes peroxisomes major contributors to cellular peroxide load and redox signalling [38]. H_2O_2 can diffuse and participate in inter-organelle communication. Thus, peroxisome-derived H_2O_2 can influence chloroplast and mitochondrial redox states and integrate stress signalling across compartments [39].

3.1.4. Plasma Membrane NADPH Oxidases (RBOHs): Deliberate ROS Producers

Plasma membrane NADPH oxidases (RBOHs) generate ROS deliberately as part of signal transduction, unlike organelle-derived "leak" ROS [40]. RBOHs transfer electrons from cytosolic NADPH to apoplastic oxygen, producing $O_2^{\bullet-}$, which is converted to H_2O_2 and then triggers downstream signalling. Their activity is tightly controlled by Ca^{2+} binding, phosphorylation, and interactions with regulatory proteins. This enables rapid, localized ROS production in response to stress cues [41]. For example, ABA-induced RBOH activation in guard cells promotes apoplastic H_2O_2 accumulation and Ca^{2+} influx. This contributes to stomatal closure [38]. In roots, RBOH-driven ROS bursts regulate ion transport and mediate stress-induced root architectural changes. They can also propagate as systemic ROS waves that coordinate whole-plant acclimation with Ca^{2+} signalling [30–32].

3.1.5. Integration Across Organelles: A Redox Communication Network

ROS production is coordinated across compartments through amplification loops and signalling crosstalk. It is not confined to isolated sites [24]. Chloroplast ROS can influence mitochondrial ROS generation, a process called "ROS-induced ROS release." This reinforces stress signalling. RBOH-mediated apoplastic ROS and Ca^{2+} waves support long-distance communication from local stress sites to distant tissues [38]. Peroxisomal H_2O_2 can connect photorespiratory flux with organellar redox status. This helps coordinate stress responses at the systems level [25]. The compartment of origin and the timing of ROS are decisive for cell fate. This balance enables plants to signal for acclimation while limiting oxidative damage.

3.2. ROS Homeostasis: The Fine Balance

ROS embody a central biological paradox: at controlled levels, they serve as indispensable messengers. When they exceed cellular buffering capacity, however, they become damaging oxidants [10–12]. This balance is determined by three interacting variables: ROS amplitude, subcellular distribution, and duration. Together, these determine whether

https://doi.org/10.3390/antiox15030337

ROS acts as a signal or a stressor [39]. Plants maintain ROS homeostasis via coordinated enzymatic and non-enzymatic antioxidants, redox-sensitive transcriptional regulation, and crosstalk with hormonal pathways [18–20]. Here, we use these control variables to formalize a compartment-resolved "redox rheostat." This framework distinguishes signalling transients from damaging overload, providing a unifying logic for interpreting antioxidant responses across stresses.

3.2.1. Beneficial Thresholds and the Signaling Role of ROS

At low-to-moderate levels (often in the low micromolar range for H_2O_2, ROS act as second messengers. Their biological specificity comes from their limited lifetime and spatial confinement [40]. In guard cells, for example, transient H_2O_2 elevations activate Ca^{2+}-dependent kinase cascades. These cascades drive stomatal closure during drought responses [1,5,9,10]. In photosynthetic tissues, $^1O_2$ generated at PSII is short-lived and locally restricted. Yet, it can initiate retrograde signalling pathways that induce nuclear protective programmes and acclimation [15,20]. Similarly, localized apoplastic H_2O_2 gradients contribute to root growth and cell-wall remodelling. These instances show that ROS convey spatial and temporal information, not functioning solely as metabolic by-products [23,25].

3.2.2. Harmful Thresholds: The Switch to Oxidative Stress

When ROS accumulation becomes sustained and exceeds antioxidant capacity, the signalling regime collapses. For example, high cytosolic H_2O_2 in the tens to hundreds of micromolar range drives direct cellular injury [32,36]. The $^{\bullet}OH$, produced via Fenton chemistry, is particularly destructive. It reacts with DNA, proteins, and membrane lipids at the site of formation, causing irreversible oxidative damage [15,20]. Downstream consequences include lipid peroxidation, accumulation of cytotoxic aldehydes (e.g., MDA), and impairment of protein function. If ROS remain elevated, they activate programmed cell death and accelerate senescence [23,34,41]. Thus, ROS intensity and persistence act as a cell-fate switch. Transient, buffered pulses support acclimation, while sustained overload promotes cytotoxicity and tissue decline [34].

3.2.3. Crosstalk Between ROS and Hormonal Networks

ROS homeostasis is tightly integrated with hormone signalling. Hormones shape ROS production and help interpret ROS signals in a context-dependent manner [34–36]. ABA promotes RBOH-dependent apoplastic ROS generation in guard cells, amplifying Ca^{2+} signalling and enabling rapid stomatal closure. ROS can also enhance ABA sensitivity, reinforcing drought responses [15,38]. Salicylic acid increases ROS accumulation by modulating ROS production and suppressing detoxification pathways, such as CAT, thereby amplifying defence signalling and priming loops [8,23]. Jasmonates frequently balance these responses by inducing antioxidant enzymes that keep ROS within a signalling-competent window during wounding or salinity responses [42]. Ethylene links ROS dynamics to development and ageing. Transient ROS can support adaptive growth responses, while chronic ROS contributes to senescence [43]. Hormone–ROS coupling encodes both the amplitude and meaning of ROS signals.

3.2.4. ROS as Systemic Signals: ROS Waves and Whole-Plant Communication

ROS signalling does not stay confined to local production sites. RBOH-driven ROS waves can propagate through tissues and along vasculature, often coupled to Ca^{2+} waves. This enables rapid long-distance communication of stress status across the plant [20,44]. Such systemic signalling lets unstressed organs pre-activate protective programmes, anticipating stress spreading through the plant [40]. Antioxidant systems do not eliminate

https://doi.org/10.3390/antiox15030337

these waves. Instead, they shape signal amplitude and prevent escalation into uncontrolled oxidative damage, preserving communication while keeping redox safety [25].

Overall, ROS behaviour under abiotic stress is a balance between adaptive signalling and oxidative injury. This balance is quantitative and compartment-dependent. When ROS are produced as transient, spatially restricted pulses within buffering capacity, they coordinate gene expression, metabolic adjustments, and developmental acclimation. But when ROS production becomes sustained or widespread, they drive lipid peroxidation, protein dysfunction, and cell death. The main determinants of this equilibrium are organelle electron-transport kinetics, antioxidant recycling capacity, hormone–ROS integration, and systemic signalling. These factors shape whether plants acclimate or succumb under stress. They represent central leverage points for developing climate-resilient crops.

4. Antioxidant Systems in Plants

4.1. Enzymatic Antioxidants

Plants rely on a sophisticated enzymatic defence system to maintain ROS homeostasis, with each enzyme playing a distinct yet interconnected role [1,15,20]. By detoxifying ROS and integrating into redox signalling pathways, these enzymes ensure that ROS are channelled into adaptive functions rather than unchecked damage [23,32,38], [Figure 2]. This interconnectedness lays the groundwork for understanding how individual enzymes operate within the broader network.

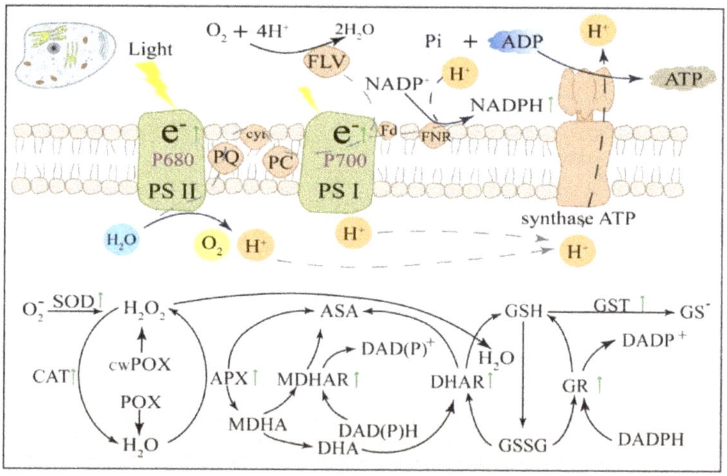

Figure 2. Chloroplast electron transport links stress-induced ROS (reactive oxygen species) to detoxification networks that maintain redox balance. In non-stressed chloroplasts, light-driven electron flow moves from PSII (P680) to PSI (P700) through PQ, cyt, and PC. This process helps form NADPH via FNR and generates a proton gradient for ATP synthesis. During environmental stress, excess electrons leak to O_2, increasing ROS like $O_2^{\bullet-}$ and H_2O_2, which can damage the photosystems if not controlled. Chloroplasts manage ROS through antioxidant systems. SOD converts $O_2^{\bullet-}$ to H_2O_2, which is then broken down by CAT, peroxidases (POX; cwPOX), or the ascorbate–glutathione cycle (APX, MDHAR, DHAR, GR). GST helps buffer redox changes and aids detoxification. Green ↑ indicates increased activity of protective antioxidant systems. Abbreviations: PSII/PSI, photosystems II/I; PQ, plastoquinone; cyt, cytochrome complex; PC, plastocyanin; Fd, ferredoxin; FNR, ferredoxin–NADP$^+$ reductase; FLV, flavodiiron proteins; APX, ascorbate peroxidase; MDHAR, monodehydroascorbate reductase; DHAR, dehydroascorbate reductase; GR, glutathione reductase; GST, glutathione S-transferase; ASA, ascorbate; GSH/GSSG, reduced/oxidized glutathione.

https://doi.org/10.3390/antiox15030337

4.1.1. Superoxide Dismutase (SOD)

Superoxide dismutase (SOD) is the primary enzymatic barrier against $O_2^{\bullet-}$. It serves as the first defence against uncontrolled radical accumulation under stress [44]. $O_2^{\bullet-}$ is generated mainly by photosynthetic and respiratory electron transport. This radical is highly reactive and can damage Fe–S clusters in metabolic enzymes, disrupt carbon fixation, and promote the formation of more toxic ROS through redox cycling [5,23,45]. SOD rapidly converts $O_2^{\bullet-}$ into H_2O_2 and molecular oxygen according to $2O_2^{\bullet-} + 2H^+ \rightarrow H_2O_2 + O_2$. It operates at near diffusion-limited rates (~109 M^{-1} s^{-1}) [44]. This reaction is protective, stopping local superoxide damage and forming H_2O_2, which is more stable and can be detoxified or used in signalling [46,47]. Plants express several SOD isoforms, placing them in key ROS-producing sites for "interception at the source." Cu/Zn-SOD is usually in the cytosol and chloroplasts [Figure 2]. It buffers $O_2^{\bullet-}$ produced during photosynthetic stress and helps maintain photosystem performance and carbon assimilation steady when CO_2 fixation declines [46]. Mn-SOD is found in the mitochondrial matrix. It protects respiratory complexes during stresses that increase electron leakage, such as drought, chilling, or heavy metal exposure [10]. Fe-SOD, present in the chloroplasts of species such as maize and *Arabidopsis*, maintains chloroplast redox balance during stress that involves photorespiratory and photochemical ROS [23]. Compartment-specific deployment is essential because $O_2^{\bullet-}$ is short-lived and poorly diffusible. Effective scavenging must occur near its generation site [45]. Experimental evidence shows SOD is quantitatively important for stress tolerance. Salinity stress strongly induces mitochondrial Mn-SOD activity, which coincides with reduced $O_2^{\bullet-}$ accumulation and improved respiratory performance [44]. Transgenic tobacco (*Nicotiana tabacum* L.) lines overexpressing chloroplast Cu/Zn-SOD show better photosynthesis and less photoinhibition under high light [25]. *Arabidopsis* mutants lacking Fe-SOD have more lipid peroxidation under drought, highlighting this isoform's role in protecting membranes and photosynthetic integrity during stress [22,37]. SOD also shapes redox signalling. It channels $O_2^{\bullet-}$ into H_2O_2, allowing downstream processing by catalase, peroxidases, and the ascorbate–glutathione cycle. This enzyme also supports organelle-to-nucleus communication, which is needed for acclimation [46–48]. So, SOD acts not only as a detoxification enzyme but also as a gatekeeper. It turns a reactive radical into a regulated peroxide signal in compartment-specific antioxidant networks.

4.1.2. Catalase (CAT)

Catalase is a major enzyme that removes H_2O_2 and maintains redox equilibrium under basal and stress conditions [44]. CAT catalyzes the breakdown of H_2O_2 to water and oxygen ($2H_2O_2 \rightarrow 2H_2O + O_2$) without using reductants, making it an efficient route for peroxide detoxification [49]. Predominantly found in peroxisomes, CAT manages the high H_2O_2 flux produced during photorespiration and oxidase-dependent pathways [23]. Without rapid CAT activity, peroxisomal peroxide can exceed the capacity of lower-capacity peroxidases, increasing the risk of oxidative injury [38]. CAT's high turnover rate (kcat ~10^7 s^{-1}) makes it effective for detoxification, though its substrate affinity is low (Km in the millimolar range) [50]. Thus, CAT acts as a high-capacity, low-affinity "safety valve" during peroxide surges. APX and peroxiredoxins are more efficient at lower (micromolar) H_2O_2 levels needed for signalling [51]. These combined properties establish a hierarchy: peroxidases fine-tune H_2O_2 levels at the cellular level, while CAT responds to sharp increases in H_2O_2 during stress. CAT activity often increases under abiotic stress, supporting oxidative protection. For example, drought-stressed maize shows a substantial rise in CAT activity, reflecting greater H_2O_2 detoxification [52]. In salt-stressed wheat, CAT induction correlates with reduced lipid peroxidation and better chlorophyll retention, supporting photosynthesis [46]. Besides detoxification, hormonal regulation influences CAT's role in

https://doi.org/10.3390/antiox15030337

redox signalling. Salicylic acid (SA) can transiently inhibit CAT, allowing localized H_2O_2 buildup to initiate defence gene expression and priming [23,48,53]. Mechanistically, CAT acts as both a peroxide scavenger and a signalling gatekeeper: it prevents harmful H_2O_2 overload but permits controlled, localized H_2O_2 signals as needed [49–51,54]. This dual function places CAT at the centre of peroxide homeostasis and stress adaptation in plants.

4.1.3. Ascorbate Peroxidase (APX)

Ascorbate peroxidase (APX) is a high-affinity peroxide scavenger. It fine-tunes H_2O_2 levels at low, signalling-relevant concentrations. In this role, APX complements CAT, which is optimized for bulk peroxide removal [5,15,34]. Specifically, APX reduces H_2O_2 to water (H_2O_2 + Asc → $2H_2O$ + DHA) using ascorbate (Asc) as an electron donor. This enzyme operates efficiently in the micromolar range due to its relatively high substrate affinity ($K_m \approx$ 20–50 μM for H_2O_2) [55]. Thanks to this kinetic property, APX can remove excess H_2O_2 without abolishing ROS-dependent signalling. Consequently, APX helps maintain redox homeostasis during stress acclimation [27,53]. Notably, APX comprises multiple isoforms tailored to distinct cellular compartments, which aligns with the need for localized peroxide control [56]. For instance, cytosolic APX buffers H_2O_2 near signalling hubs and membranes. Similarly, chloroplast isoforms (stromal and thylakoid-bound APX) protect the photosynthetic apparatus from light-driven oxidative bursts [57]. In addition, peroxisomal APX complements catalase by processing micromolar levels of H_2O_2, while CAT handles higher-flux peroxide surges [58]. Mitochondrial APX also contributes by protecting respiratory function, detoxifying peroxide during oxidative phosphorylation, especially when electron flow is perturbed by stress [27]. Supporting the functional importance of APX, genetic and physiological studies have shown that Arabidopsis mutants lacking thylakoid APX exhibit increased lipid peroxidation and impaired PSII performance under high light. This evidence highlights APX's role in photoprotection [43]. Conversely, APX overexpression in crop systems has been associated with reduced H_2O_2 accumulation, improved membrane integrity, and enhanced stress survival under salinity or drought [21,59]. Mechanistically, APX operates within the ascorbate–glutathione (AsA–GSH) cycle. The DHA produced by APX is recycled back to ascorbate by DHAR, using reduced glutathione (GSH) as a cofactor. This process links peroxide detoxification to cellular redox buffering and NADPH-dependent recycling [23,59]. Through this coupling, APX acts as both a detoxification enzyme and a signal-preserving modulator, maintaining H_2O_2 within a physiological window that supports acclimation while preventing escalation into oxidative damage [54,56,57].

4.1.4. Glutathione Peroxidase (GPX)

Glutathione peroxidases (GPXs) detoxify H_2O_2 and, importantly, organic hydroperoxides (ROOH). They play a key role in limiting lipid hydroperoxides that accumulate in membranes under stress [44]. Plant GPXs are usually selenium-independent cysteine-based enzymes. This differs from many animal selenoprotein GPXs. They are distributed across chloroplasts, mitochondria, and the cytosol. These are sites where ROS generation intersects with peroxidation-prone lipids [55,60,61]. GPX uses reduced glutathione as an electron donor to reduce ROOH to the corresponding alcohol (ROH) (ROOH + 2GSH → ROH + GSSG + H_2O). This interrupts lipid peroxidation chain reactions. Without this, the integrity of the thylakoid and mitochondrial membranes could be compromised, impairing photosynthesis and respiration [55,61]. Stress-responsive induction of GPX supports a protective role for membrane function. Basal activities are typically measurable in healthy tissues and often increase under drought, salinity, or heavy-metal exposure. This is consistent with inducible antioxidant defence [62,63]. In transgenic rice, GPX overexpression has been

https://doi.org/10.3390/antiox15030337

associated with reduced lipid peroxidation (lower MDA), improved membrane stability, and better maintenance of chlorophyll under salinity stress [64]. Similarly, increased GPX activity in *Arabidopsis* has been linked to reduced electrolyte leakage under cadmium (Cd) stress. This is consistent with improved membrane protection [65]. Functionally, GPX activity is integrated with glutathione recycling. The GSSG formed during peroxide reduction is converted back to GSH by glutathione reductase (GR) in an NADPH-dependent reaction. This couples the GPX function to the broader cellular redox economy and sustains turnover during stress [65,66]. Within the antioxidant hierarchy, GPXs complement APX and CAT by targeting peroxide species that directly threaten membranes, especially lipid hydroperoxides. They help preserve photosynthetic efficiency, mitochondrial energy metabolism, and overall cellular viability under abiotic stress [60,61]. This membrane-focused specialization makes GPX a critical component of oxidative stress tolerance, especially where lipid peroxidation is a dominant damage pathway.

4.1.5. Glutathione Reductase (GR)

Glutathione reductase (GR) is a central enzyme in redox homeostasis. It regenerates reduced glutathione (GSH) from oxidized glutathione (GSSG) using NADPH (GSSG + NADPH $\rightarrow$ 2GSH + $NADP^+$) [66,67]. This reaction sustains the AsA–GSH cycle. It enables continuous peroxide detoxification via APX, GPX, and associated recycling enzymes such as DHAR [68]. Without efficient GR-mediated recycling, GSH becomes depleted. GSSG accumulates, leading to a collapse of peroxide-scavenging capacity. The intracellular redox environment destabilizes under stress [69]. GR is present in multiple compartments. These include chloroplasts, mitochondria, cytosol, and peroxisomes. This distribution aligns with the widespread production of ROS and the demand for antioxidants. In chloroplasts, GR supports ascorbate turnover. It protects photosystems under high light and photooxidative stress [70]. Mitochondrial GR helps buffer ROS generated when electron transport is constrained. It preserves respiratory efficiency under adverse conditions [44]. In peroxisomes, GR aids detoxification of photorespiration-derived H_2O_2. It works with peroxidase systems [23]. GR's stress responsiveness further highlights its role. GR activity often increases under abiotic stress. This maintains GSH pools and a high GSH/GSSG ratio. This ratio strongly associates with reduced lipid peroxidation and improved photosynthetic performance in stressed plants [15,71,72]. Mechanistically, GR links antioxidant function to cellular energy metabolism. It does this by coupling glutathione recycling to NADPH availability from photosynthetic reactions, the oxidative pentose phosphate pathway, and mitochondrial metabolism [73]. GR does not directly remove ROS. Instead, it acts as the redox "recycler." GR determines the throughput and resilience of glutathione-dependent detoxification pathways during prolonged stress exposure [74].

4.1.6. Peroxiredoxins and Thioredoxins

Peroxiredoxins (Prxs) and thioredoxins (Trxs) form an interconnected thiol-based redox system. This system combines peroxide detoxification with redox regulation and signalling [75]. Prxs are cysteine-dependent peroxidases that reduce H_2O_2 and organic hydroperoxides. In some contexts, they also reduce reactive nitrogen species. This reduction occurs via a conserved catalytic cysteine, which is oxidized during peroxide reduction and then restored by Trx via disulfide reduction [75,76]. Prxs typically exhibit high affinity for H_2O_2, often in the low micromolar range. They are well-suited to buffer signalling-range peroxide concentrations below the CAT efficient operating range. This supports fine control of basal redox status across compartments [77]. Prxs are widely distributed in chloroplasts, mitochondria, peroxisomes, and the cytosol, enabling localized regulation of ROS production and interpretation [5,38]. Beyond detoxification, Prxs can function as redox sensors and stress-responsive switches. Under stronger oxidative conditions, Prxs can become overoxidized at the catalytic cysteine. This transiently reduces peroxidase activity and, in some cases, promotes chaperone-like functions that support proteostasis during stress [78]. Trxs complement these roles as broad protein disulfide reductases. They relay reducing equivalents from NADPH (via thioredoxin reductase) to diverse targets [79]. In chloroplasts, Trx activity is also linked to light-driven electron flow through the ferredoxin–thioredoxin system. This couple's photosynthetic energy status regulates the Calvin–Benson cycle enzymes and redox-sensitive metabolic steps [75]. This integration helps synchronize carbon metabolism, antioxidant capacity, and stress acclimation [23,34]. Genetic and physiological evidence support the importance of the Prx–Trx module in stress tolerance and photosynthetic stability. Perturbation of chloroplast Prx systems can increase peroxide accumulation and sensitize plants to high-light stress and photoinhibition. Enhanced Trx capacity has been associated with improved maintenance of photosynthesis and stress resilience in multiple systems [76–78,80]. Collectively, Prxs act as peroxide buffers and sensors. Trxs translate redox changes into reversible protein regulation, connecting ROS control to metabolic reprogramming during stress [81].

Collectively, enzymatic antioxidants function as a hierarchical, coupled network. SOD converts $O_2^{\bullet -}$ to H_2O_2. Catalase removes high-flux peroxide. APX and GPX, together with the AsA–GSH cycle, regulate micromolar H_2O_2 and hydroperoxides. This regulation is relevant to signalling and membrane protection. GR sustains glutathione-dependent turnover through NADPH-driven recycling. The Prx–Trx module links peroxide buffering to redox regulation of metabolism and stress signalling. Through these interlinked tiers, plants maintain ROS within adaptive bounds. This preserves signalling while preventing oxidative damage across diverse abiotic stress conditions.

4.2. Non-Enzymatic Antioxidants

While enzymatic antioxidants provide highly regulated, compartmentalized ROS detoxification, plants also rely on diverse non-enzymatic antioxidants. These act as rapid scavengers, redox buffers, and essential cofactors [1,20]. Their metabolites help prevent ROS from exceeding damaging thresholds while preserving signalling functions. They integrate tightly with enzymatic pathways such as APX/GPX systems and thiol-based redox circuits, forming overlapping layers of redox control [23,34], [Figure 2].

4.2.1. Ascorbic Acid (Vitamin C)

Ascorbic acid (Asc) is the dominant water-soluble antioxidant in plant cells. It serves as a central hub of redox metabolism [44]. Asc often accumulates in chloroplasts to high millimolar levels, typically 1–20 mM, and to higher concentrations in stress-acclimated tissues. This supports both direct ROS scavenging and turnover of the antioxidant network [81]. Asc directly quenches multiple ROS species, including highly reactive radicals. It is the preferred electron donor for ascorbate peroxidases (APXs) in H_2O_2 reduction, helping link non-enzymatic buffering to enzymatic peroxide control [82]. Beyond detoxification, Asc is a cofactor for violaxanthin de-epoxidase in the xanthophyll cycle. It supports non-photochemical quenching (NPQ) and photoprotection under excess light [83,84]. The function of Asc depends on pool size and rapid recycling. After donating electrons, Asc forms monodehydroascorbate (MDHA). MDHA can be reduced back to Asc by MDHAR using NAD(P)H or can disproportionate to dehydroascorbate (DHA) [82,85]. DHA is then reduced to Asc by DHAR in a GSH-dependent reaction. This integrates Asc turnover into the broader AsA–GSH cycle, linking Asc status to glutathione pools, GR activity, and NADPH supply [86]. This recycling capacity enables sustained APX turnover and stabilizes redox homeostasis across compartments, including chloroplasts, cytosol, mitochondria,

https://doi.org/10.3390/antiox15030337

and the apoplast [87]. Physiological and genetic evidence show Asc's importance for stress resilience and photosynthetic stability. Asc-deficient mutants exhibit increased lipid peroxidation and accelerated photoinhibition under salinity or high light, indicating compromised ROS buffering and impaired NPQ [88,89]. Enhancing Asc recycling, for example, by DHAR overexpression, can sustain larger Asc pools and improve tolerance to drought and salinity. This supports the idea that recycling throughput is a key determinant of stress performance, not just Asc abundance [82]. In summary, Asc is both a direct antioxidant and an enabling cofactor. It coordinates photoprotection, peroxide detoxification, and redox signalling in response to abiotic stress.

4.2.2. Glutathione (GSH)

Glutathione (GSH; γ-glutamyl-cysteinyl-glycine) is the most abundant non-protein thiol in plants and acts as a central "redox currency." It supports detoxification, redox buffering, and signalling [90,91]. GSH typically occurs at millimolar concentrations in most tissues. Levels can rise further in chloroplasts, especially during stress acclimation [90]. Functionally, GSH controls ROS through several coupled pathways. It donates electrons for GPXs and glutathione S-transferases (GSTs), which reduce H_2O_2 and ROOH and limit lipid peroxidation in membranes [90–94]. GSH is also essential for the AsA–GSH cycle. In this cycle, DHAR uses GSH to regenerate Asc from DHA. This process sustains APX-dependent H_2O_2 detoxification and keeps peroxide levels within ranges compatible with signalling [95,96]. The ratio of reduced to oxidized glutathione (GSH/GSSG) sensitively reflects cellular redox status. This ratio often shifts under stress, indicating oxidative pressure and triggering antioxidant gene expression [94,96]. Stress often increases total GSH and induces GR, which reduces GSSG back to GSH using NADPH. These responses are frequently associated with improved redox buffering and reduced oxidative damage [97]. Genetic studies highlight GSH's centrality. Mutants with impaired GSH biosynthesis show heightened sensitivity to oxidative and metal stress. Elevated GSH biosynthetic capacity commonly improves tolerance to salinity, ozone, and heavy metals [98–100]. Beyond buffering, GSH also participates in redox signalling through reversible S-glutathionylation. This process modulates protein activity and stability during stress responses [94,101]. In summary, GSH integrates detoxification, NADPH-dependent recycling, and signalling regulation, making it a core determinant of redox resilience under abiotic stress.

4.2.3. Tocopherols (Vitamin E)

Tocopherols are lipophilic antioxidants enriched in plant membranes. The most abundant and biologically active isoform in many tissues is α-tocopherol [102]. Tocopherols are especially concentrated in chloroplast thylakoid membranes. In these locations, they protect polyunsaturated fatty acids from lipid peroxidation initiated by ROS (notably $^1O_2$) and lipid peroxyl radicals formed during oxidative stress [102–104]. Tocopherols terminate peroxidation chain reactions by donating a hydrogen atom to lipid radicals. This produces a relatively stable tocopheroxyl radical, which can be recycled back to α-tocopherol by ascorbate and other reductants. This process functionally links tocopherol defence to the broader cellular antioxidant network [105,106]. Stress-responsive induction and mutant phenotypes demonstrate the importance of tocopherols for membrane stability and photoprotection. In *Arabidopsis*, α-tocopherol levels increase during drought, high light, and chilling, supporting its protective role during oxidative stress [107]. In contrast, tocopherol-deficient mutants (e.g., *vte1*) show pronounced lipid peroxidation, chlorosis, and accelerated senescence under high-light stress. This demonstrates that tocopherols are needed to restrain membrane oxidative cascades under high excitation pressure [105,108]. Tocopherols limit lipid peroxide amplification and stabilize membrane properties, pre-

https://doi.org/10.3390/antiox15030337

serving photosystem function and sustaining photosynthetic performance under abiotic stress [104,107,109]. Thus, tocopherols act as specialized membrane-phase antioxidants. They complement soluble redox buffers by protecting thylakoid integrity during stress.

4.2.4. Carotenoids

Carotenoids are isoprenoid pigments serving two critical roles: as light-harvesting cofactors and as chloroplast antioxidants, making them central to photoprotection under abiotic stress [110]. By quenching triplet chlorophyll, they prevent ROS formation, and they also dissipate excess excitation energy that would otherwise produce $^1O_2$. When oxidative load rises, carotenoids can chemically react with ROS [110–112]. β-carotene, located near reaction centres, is particularly important for suppressing $^1O_2$ formation; it intercepts excited chlorophyll states, thereby protecting photosystem cores from oxidative damage [113,114]. In addition, carotenoids drive the xanthophyll cycle, a system that regulates non-photochemical quenching (NPQ) through rapid interconversion of violaxanthin, antheraxanthin, and zeaxanthin under fluctuating light and stress [115,116]. Accumulation of zeaxanthin is strongly associated with increased NPQ and reduced photoinhibition during high light or drought-related excitation pressure [117,118]. Genetic evidence further supports these functions. For example, mutants impaired in zeaxanthin formation (such as *Arabidopsis npq1*, deficient in violaxanthin de-epoxidase) exhibit enhanced photoinhibition and oxidative damage under high-light stress, which demonstrates the necessity of carotenoid-dependent energy dissipation for redox stability [119,120]. If oxidative stress exceeds buffering capacity, carotenoid oxidation can produce apocarotenoids, which act as signalling molecules linking chloroplast redox state to downstream acclimation programmes [104,121]. In sum, carotenoids coordinate ROS prevention, energy dissipation, and stress signalling, thus sustaining photosynthetic efficiency and chloroplast integrity under abiotic stress.

4.2.5. Flavonoids, Phenolics, and Secondary Metabolites

Flavonoids and phenolic compounds comprise a diverse class of secondary metabolites that contribute to stress tolerance by combining direct antioxidant activity with photoprotection and modulation of signalling [5,15]. These compounds accumulate primarily in vacuoles, epidermal tissues, and the apoplast, positioning them at cellular interfaces where ROS exposure is high (e.g., under UV, drought, salinity, and oxidative bursts) [34]. Their polyhydroxylated structures enable them to donate electrons or hydrogen atoms to reactive species such as $O_2^{\bullet-}$ and $^{\bullet}OH$, forming resonance-stabilized radicals and thereby limiting oxidative chain reactions, particularly when enzymatic systems become saturated [122,123]. A significant additional function is UV screening and photoprotection. Flavonols such as quercetin and kaempferol absorb UV-B and reduce photooxidative excitation in underlying tissues, indirectly lowering chloroplast ROS formation and helping preserve photosynthetic capacity under high irradiance [124,125]. Phenolic acids (e.g., ferulic and caffeic acids) can further contribute by reinforcing cell walls through cross-linking and lignification-related chemistry, affecting both oxidative diffusion dynamics and stress resilience at the tissue level [123]. Genetic and physiological evidence support their protective role: mutants impaired in flavonoid biosynthesis (e.g., *Arabidopsis tt4*) exhibit elevated ROS levels and reduced tolerance to drought and UV stress, consistent with a loss of this peripheral antioxidant buffer [126]. Beyond scavenging, flavonoids can modulate hormone-linked growth programmes, for example, by affecting auxin transport, thereby integrating redox status with developmental adjustments under stress [122,127]. Thus, flavonoids and phenolics function as inducible, spatially targeted redox buffers that extend antioxidant protection beyond core organelles and couple oxidative cues to adaptive growth and defence responses.

https://doi.org/10.3390/antiox15030337

4.3. Compartment-Specific Antioxidant Networks

ROS are generated in most plant compartments, but their effects depend on local production rates, diffusion, and strength of compartment-specific antioxidants [128]. To balance ROS's dual role as toxic and signalling molecules, plants use spatially organized antioxidants tailored to each organelle's metabolic needs [129]. Chloroplasts are especially vulnerable because photosynthesis produces $O_2^{\bullet-}$ and $^1O_2$ at PSI/PSII, particularly under excess light or CO_2 shortage [130]. To cope, chloroplastic defences are both redundant and layered: SOD isoforms quickly convert $O_2^{\bullet-}$ to H_2O_2. Stromal and thylakoid APXs keep H_2O_2 at signalling-friendly micromolar levels [131]. High ascorbate and glutathione pools support the AsA–GSH cycle and constant recycling via GR [15]. Membrane-phase antioxidants like tocopherols and carotenoids reduce lipid peroxidation and quench $^1O_2$ in thylakoids [34]. Genetic data show this compartmental need: *Arabidopsis tapx* mutants have higher lipid peroxidation under high light, while tocopherol-deficient vte1 lines display chlorosis and faster senescence, highlighting the vital role of chloroplast antioxidants in photoprotection and photosynthetic stability [61,71].

Mitochondria generate ROS mainly at complexes I and III of the ETC. ROS leakage rises when electron flow is limited by drought, chilling, or heavy metals [88]. Mitochondrial Mn-SOD limits $O_2^{\bullet-}$ buildup, while APX/GPX and thiol systems detoxify H_2O_2 and regulate redox-sensitive metabolism needed for respiration [35,132]. Stress-induced increases in these defences often lower oxidative damage and improve stress tolerance, supporting mitochondria's central role in linking redox balance to energy metabolism [81]. Peroxisomes also act as high-flux peroxide organelles, making large amounts of H_2O_2 during photorespiration and other oxidase reactions [25]. Their antioxidant setup focuses on capacity: catalase rapidly clears H_2O_2 surges, while peroxisomal APX and Prxs fine-tune lower-level peroxide signals and support redox communication between organelles [132]. Peroxisomal antioxidant activity often increases under stresses that increase photorespiration, suggesting an adjustment to peroxide load [39].

The cytosol acts as an integration hub. It buffers ROS from organelles and maintains redox balance using soluble antioxidants and enzymes such as Cu/Zn-SOD, APX, and large ascorbate/glutathione pools [39,97]. Cytosolic scavenging shapes signalling precision by controlling H_2O_2 influx from the apoplast, partly by diffusion through aquaporins [133]. The apoplast, by contrast, often produces ROS intentionally via RBOHs during stress sensing and signalling. These temporary oxidative pulses are limited by apoplastic peroxidases and redox-active phenolics or flavonoids, which prevent excess ROS yet keep signal integrity [134–136]. Together, these compartment-specific antioxidant systems form a connected redox network: chloroplasts protect from light stress, mitochondria manage respiration, peroxisomes handle peroxide surges, the cytosol buffers ROS traffic, and the apoplast enables ROS-based signals [137,138]. This setup allows plants to balance their protective and signalling roles as conditions change.

5. Antioxidant-Mediated Tolerance to Specific Abiotic Stresses

Plants face a wide range of abiotic stresses, non-living environmental factors such as drought, salinity, and extreme temperatures, that disrupt cellular redox homeostasis, or the balance between antioxidants and reactive molecules in cells, by triggering excessive production of ROS [20,23]. ROS can cause cellular damage but also serve as crucial signalling molecules [1,38]. To counterbalance ROS overaccumulation, plants rely on sophisticated antioxidant systems. These systems mitigate oxidative stress, a harmful condition resulting from an imbalance between ROS and antioxidants, and facilitate acclimation [15,34]. Antioxidant defences operate through distinct yet often interconnected mechanisms. They

https://doi.org/10.3390/antiox15030337

are tailored to modulate ROS levels and maintain cellular equilibrium under diverse stress conditions (Table 1).

Table 1. Enzymatic antioxidant responses contributing to oxidative stress mitigation in plants exposed to diverse environmental stresses.

Type of Stress	Plant Name	Experimental Conditions	Key Findings	References
Drought stress (20% PEG)	*Arabidopsis*	Controlled growth chamber conditions	*AgAPX1*, an ascorbate peroxidase gene from celery, is upregulated under drought stress and encodes a protein optimally active at pH 7.0 and 55 °C. Transgenic *Arabidopsis* overexpressing *AgAPX1* display increased *AsA* content (~1.4-fold), enhanced total antioxidant capacity (+21–29%), and improved drought tolerance.	[139]
Drought stress (15% PEG)	Soybean	Hydroponic experiment	SOD activity increased notably under severe drought by about 15%, while CAT and APX contents were elevated during drought stress, reflecting enhanced enzymatic antioxidant defense during water limitation	[140]
Drought stress (20% PEG)	Rapeseed	Hydroponic experiment	MDHAR and DHAR activities decreased under drought stress, while GSH levels increased by 31–26%, respectively. APX and AsA contents rose at lower stress levels, with GR remaining stable initially but increasing by 30% at drought stress.	[141]
Drought stress (20% PEG)	Rapeseed	Hydroponic experiment	Drought significantly increased oxidative stress (MDA + 82%, H_2O_2 + 131%), while proline pretreatment lowered them by up to 34% and 29%. Drought also boosted antioxidant enzymes, but osmolytes modulated responses differently (e.g., proline: MDHAR +15%, GR −12%; glycine betaine and trehalose varied). Overall, treatments improved glutathione redox status (↑GSH, ↓GSSG by 35–47%) and strengthened methylglyoxal detoxification via higher glyoxalase activity (up to +53%).	[142]
Drought stress (30% field capacity)	Tampala	Controlled growth chamber conditions	Key antioxidant enzymes such as SOD, CAT, APX, and GR are upregulated, helping to detoxify harmful ROS like superoxide radicals and hydrogen peroxide. Non-enzymatic antioxidants, including ascorbic acid and glutathione, also accumulate to maintain redox balance	[143]
Salinity stress (99 mM NaCl)	Tomato	Controlled growth chamber conditions	*FeSOD* overexpression strengthened antioxidant defences in tomato roots, stabilizing ROS levels and maintaining ordered microtubule structure under NaCl stress.	[144]

https://doi.org/10.3390/antiox15030337

Table 1. *Cont.*

Type of Stress	Plant Name	Experimental Conditions	Key Findings	References
Salinity stress (300 mM NaCl)	Sweet potato	Petri dishes experiment	Overexpression of the *swpa4* peroxidase gene and CAT in sweet potato enhanced class III peroxidase activity (3–13×), significantly improving oxidative stress tolerance under H_2O_2 and salinity conditions	[145]
Salinity stress (100 mM NaCl)	*Arabidopsis*	Controlled growth chamber conditions	Overexpression of *AtGPXL5* (*glutathione peroxidase-like 5*) enhanced salt tolerance in *Arabidopsis* by maintaining lower H_2O_2 and MDA levels, elevating GSH content, and sustaining a more negative redox potential (≈−251 mV).	[146]
Salinity stress (200 mM NaCl)	Mustard	Controlled growth chamber conditions	Overexpressing cytosolic APX (*AtApx1*) in mustard improved salt tolerance (200 mM NaCl) by boosting ROS scavenging: H_2O_2 and MDA dropped ~2.8- and ~2.5-fold, while APX and GPX activities rose ~1.9- and ~1.5-fold. This strengthened the ascorbate–glutathione cycle, maintained redox balance, and reduced oxidative damage.	[147]
Salinity stress (100 mM NaCl)	Tomato	Hydroponic experiment	*SlMDHAR* expression, activity, and S-nitrosylation increase under salt stress, boosting ROS detoxification and redox balance. Transgenic plants overexpressing *SlMDHAR* show up to 48% higher germination, 2.3-fold longer roots, reduced ROS, and elevated antioxidant activities (SOD, CAT, POD, DHAR, APX, GR), enhancing salt tolerance effectively.	[148]
Salinity stress (400 mM NaCl)	Sugar beet and *Arabidopsis*	Controlled growth chamber conditions	*BvM14-MDHAR* from sugar beet is upregulated under salt stress, enhancing ascorbic acid regeneration and maintaining redox homeostasis. Overexpression in *Arabidopsis* improves salt tolerance by increasing MDHAR and DHAR activities, chlorophyll content, root growth, and AsA/DHA ratio, while reducing membrane damage.	[149]
Salinity stress (150 mM NaCl)	Kiwifruit	Controlled growth chamber conditions	*Actinidia chinensis* cytosolic APXs (*AcAPX1*/*AcAPX2*) are salt-induced antioxidants; when overexpressed in Arabidopsis, they improve salinity tolerance by raising APX activity (+31–39%), lowering H_2O_2 (−23 to −37%), and increasing total ascorbate and glutathione to strengthen redox homeostasis.	[150]

https://doi.org/10.3390/antiox15030337

Table 1. *Cont.*

Type of Stress	Plant Name	Experimental Conditions	Key Findings	References
Salinity stress (150 mM NaCl)	Broad beans (Hassawi-3 and ILB-4347)	Controlled growth chamber conditions	Oxidative stress indicators (H_2O_2, MDA, electrolyte leakage) rose 60–120%, but their accumulation was 25–30% lower in ILB-4347 than in Hassawi-3. Antioxidant activities—SOD, CAT, GR, and AsA—increased by 25–85%, with ILB-4347 exhibiting 1.5–2 times higher enzyme levels than Hassawi-3.	[151]
Salinity stress (100 mM NaCl)	Wheat (Suntop and Sunmate)	Hydroponic experiment	Under salinity, the sensitive cv. Sunmate showed the highest MDA (greater oxidative damage), whereas tolerant lines (Suntop, CM72) had 1.5–2.2× higher SOD, POD, CAT, APX, and GR with earlier leaf peaks, improving ROS scavenging and cutting lipid peroxidation ~40–50% to maintain redox stability.	[152]
Salinity stress (8.55 mM NaCl)	Tomato	Controlled growth chamber conditions	AsA levels decreased, along with reductions in both reduced and oxidized glutathione (GSH and GSSG), leading to a lowered GSH/GSSG ratio. Concurrently, APX and GR activities increased by 28% and 14%, respectively.	[153]
Salinity stress (200 mM NaCl)	Common bean	Controlled growth chamber conditions	Under salt stress, key antioxidant enzymes rise strongly in tolerant genotypes (CAT up to +400%, APX +600–700%, GR +60%, with higher SOD), reducing lipid peroxidation (MDA −44–56%). Non-enzymatic antioxidants also increase (ascorbate +26–33%, flavonoids +47–70%), strengthening overall defense.	[154]
Heat stress (40 °C temperature for 5 h)	*Arabidopsis*	Controlled growth chamber conditions	Overexpression of *BcAPX* genes from *Brassica campestris* in *Arabidopsis* enhanced heat tolerance through increased APX activity and reduced oxidative damage. Specific transgenic lines (e.g., *BcAPX1-3, 2-1, 3-5*) showed 2–4× higher APX activity and lower H_2O_2 and MDA levels at 40 °C, indicating improved ROS detoxification and heat resilience.	[155]
Heat stress (40 °C temperature for 48 h)	Mung bean	Controlled growth chamber conditions	MDHAR and DHAR enzyme activities decline under stress, whereas APX and GR activities increase, contributing effectively to ROS detoxification and enhancing stress tolerance.	[156]
Heat stress (42 °C temperature for 12 h)	Tomato	Controlled growth chamber conditions	Under heat stress, tomato seedlings showed oxidative injury with MDA and H_2O_2 increasing by 48% and 60%, respectively. Antioxidant enzymes acted as the main defence, where SOD, POD, CAT, and APX activities rose by 11%, 47%, 75%, and 26% compared with control plants.	[157]

https://doi.org/10.3390/antiox15030337

Table 1. *Cont.*

Type of Stress	Plant Name	Experimental Conditions	Key Findings	References
Heat stress (45 °C temperature for 12 h)	Rice	Controlled growth chamber conditions	Under 45 °C heat stress, rice seedlings showed severe oxidative damage, with MDA and electrolyte leakage rising >60%. This was mitigated by increased SOD and POD activity (~1.4–1.6-fold), which reduced H_2O_2 and O_2^- by 30–35%. The upregulation of *OsCATB*, *Fe-SOD*, and *OsAXP1* genes confirmed enhanced antioxidant defence.	[158]
Cold stress (4 °C temperature for 48 h)	Potato	Controlled growth chamber conditions	Overexpression of *StSOD1* in potato enhanced antioxidant defence under cold stress. SOD activity in overexpressing lines rose 1.38-fold versus non-transgenic plants, while RNAi lines showed reduced activity. Increased SOD correlated with lower MDA content, and elevated POD (~1.24-fold) and CAT (~1.37-fold) activities.	[159]
Cold stress (4 °C temperature for 48 h)	Hardy orange	Controlled growth chamber conditions	Overexpression of *CsPIF8* enhanced cold tolerance in orange and transgenic tomato by activating SOD-mediated antioxidant defence. *CsPIF8* bound the E-box of the *CsSOD* promoter, elevating SOD expression and activity and reducing ROS and MDA accumulation.	[160]
Cold stress (4 °C temperature for 192 h)	*Arabidopsis*	Controlled growth chamber conditions	Overexpression of *AtSOD* and *CmSOD* in *Arabidopsis thaliana* markedly enhanced cold tolerance through stronger antioxidant defence. Transgenic lines showed 2–3 fold higher SOD activity and reduced superoxide accumulation, leading to lower oxidative damage after 4 °C exposure. Cold-responsive genes *AtCBF2*, *AtRD29A*, and *AtRD29B* were upregulated, with *AtRD29A* also induced by ABA, suggesting ABA-linked regulation.	[161]
Cold stress (4 °C temperature for 24 h)	Hardy orange	Controlled growth chamber conditions	Overexpression of *PtrbHLH* in *orange* enhanced cold tolerance by improving ROS detoxification. Transgenic lines showed lower ROS, electrolyte leakage, and MDA levels, along with higher CAT, POD, and SOD activities.	[162]
Cold stress (5 °C temperature for 24 h)	Rice	Controlled growth chamber conditions	Overexpression of wheat CAT in rice enhances cold tolerance by efficiently detoxifying H_2O_2, resulting in markedly higher CAT activity—especially under cold stress—while SOD and APX remain unchanged, confirming strengthened antioxidant defence and improved stress resilience.	[163]

https://doi.org/10.3390/antiox15030337

Table 1. *Cont.*

Type of Stress	Plant Name	Experimental Conditions	Key Findings	References
Heavy metal stress (100 μM Cd)	Tobacco	Controlled growth chamber conditions	Overexpression of *BjCAT3* in tobacco improves Cd tolerance by enhancing CAT-driven ROS detoxification, leading to higher CAT activity, reduced H_2O_2 and MDA accumulation, and increased SOD and POD activities under Cd stress.	[164]
Heavy metal stress (200 μM Al)	*Arabidopsis*	Hydroponic experiment	Overexpression of *AtGR1* in *Arabidopsis thaliana* improved aluminum tolerance by boosting glutathione reductase activity, maintaining higher *GSH* levels and *GSH/GSSG* ratio, and reducing ROS and RCS accumulation.	[165]
Heavy metal stress (100 μM Pb)	Cotton	Controlled growth chamber conditions	Lead toxicity induces ROS that damage plant cellular components. In response, plants upregulate antioxidant enzymes (SOD, CAT, GPX, APX) to detoxify ROS, lowering oxidative damage indicators (MDA, H_2O_2, electrolyte leakage) and preserving cellular function under heavy-metal stress.	[166]
Heavy metal stress (300 ppm Pb)	Mustard	Controlled growth chamber conditions	Lead toxicity in plants triggers oxidative stress by increasing ROS, causing damage to cellular components. Antioxidant enzymes such as APX and GR are upregulated, increasing their activities by up to 57.84% and 51.97%, respectively, helping to regulate ROS and maintain redox balance. Elevated oxidative markers like H_2O_2 (up to 303.34%) and MDA (up to 34.82%) impair plant development, but enhanced antioxidant activity under heavy metal stress.	[167]
Heavy metal stress (600 ppm Cd)	Mustard	Controlled growth chamber conditions	Cd stress elevates H_2O_2 (up to 203%), MDA (41.3%), and electrolyte leakage (468%), indicating severe oxidative damage; antioxidant enzymes respond markedly, with CAT activity recovering and increasing up to 17.2% with treatment, GR rising up to 151%, and APX up to 192%, collectively protecting cellular function under Cd toxicity.	[168]
Heavy metal stress (1 mM Pb)	Wheat	Hydroponic experiment	Pb stress increased MDA (up to 179%) and H_2O_2 (95%) while suppressing antioxidant enzymes and ascorbate–GSH pools; GSH supplementation restored CAT and GPX, enhanced MDHAR (53%), DHAR (55%), GR (126%), SOD, and glyoxalase I/II (39% and 48%), reduced methylglyoxal (27%), improved leaf water content (13%), and lowered proline accumulation (24%).	[169]

https://doi.org/10.3390/antiox15030337

5.1. Drought Stress

Drought is the most severe abiotic stress affecting global agriculture. It causes yield losses of 40–70% in key cereals such as maize, wheat, and rice during severe events [139,140]. Water deficit induces stomatal closure, reducing CO_2 assimilation by 50–70%. This disrupts the balance between light absorption and carbon fixation [140]. The imbalance increases electron leakage from photosystems I and II. This electron leakage generates $O_2^{\cdot-}$ via the Mehler reaction at rates of up to 2–3% of the total electron flux [141]. Mitochondrial electron transport is similarly over-reduced, releasing superoxide mainly from complexes I and III [142]. Concurrently, peroxisomes produce substantial H_2O_2 during intensified photorespiration. In drought-stressed mesophyll cells, concentrations can exceed 100 μM [143]. These ROS act as signalling molecules to trigger defence pathways but also serve as toxicants, causing lipid peroxidation, protein carbonylation, and DNA damage [170]. A central defence mechanism involves a robust network of antioxidant enzymes to detoxify ROS. For instance, SODs convert $O_2^{\cdot-}$ to H_2O_2. This prevents formation of –OH [28]. In drought-stressed wheat, SOD activity increases 2.5-fold. This increase correlates with a 45% reduction in MDA, a lipid peroxidation marker [59]. Peroxisomal CAT detoxifies accumulated H_2O_2. This results in a threefold increase in activity in drought-treated maize [110]. Similarly, APXs in chloroplasts, mitochondria, and cytosol precisely regulate H_2O_2 at signalling levels [138]. *Arabidopsis apx1* mutants accumulate 60% more H_2O_2 under drought. They suffer earlier PSII photoinhibition and show a 20% survival decline. In contrast, APX-overexpressing lines maintain Fv/Fm ratios 25% higher, indicating protective effects [64,139]. The AsA–GSH cycle acts as a pivotal redox amplifier, buffering ROS fluxes [171]. Drought elevates the GSH/GSSG from ~4:1 to nearly 9:1 in soybean, sustaining APX activity [140,171]. Wheat exhibits a 3.8-fold increase in GR during prolonged drought. This change ensures efficient GSH recycling [59]. Metabolomic profiling in barley (*Hordeum vulgare* L.) reveals a 2.5-fold rise in ascorbate and a threefold increase in GSH pools. This confirms synergistic antioxidant engagement [97].

Non-enzymatic antioxidants, such as tocopherols, are also key regulators of drought stress in maize. They preserve thylakoid membrane integrity and enhance zeaxanthin up to tenfold via the xanthophyll cycle. This boosts non-photochemical quenching, allowing excess energy to dissipate [128,135,170]. Compatible solutes–proline, glycine betaine, and soluble sugars–are accumulated. These solutes modulate osmotic adjustment by maintaining turgor and stabilizing cellular structures [19,27]. In maize during drought, proline rises from ~3 to 15 μmol g^{-1} fresh weight. This increase is accompanied by a twofold increase in APX and a 35% decrease in ROS leakage [171]. Beyond osmoprotection, proline directly quenches singlet oxygen. It also upregulates antioxidant genes through redox-sensitive transcription factors [172]. Glycine betaine stabilizes Rubisco activase and thylakoid membranes, thereby indirectly mitigating ROS production in photosynthesis [173]. Drought elicits ABA-mediated ROS bursts in guard cells via RBOHs at the signaling hub [62]. ABA induces apoplastic H_2O_2 accumulation (10-20 μM) within minutes. This activates Ca^{2+}-dependent kinases and promotes stomatal closure [174]. Antioxidants buffer these transient ROS elevations, preventing oxidative damage and preserving signal fidelity. Systems-level analyses identify APX2, 2-Cys peroxiredoxins, and GR as core nodes integrating ROS detoxification with ABA- and DREB-driven transcriptional programmes [175]. Proteomic studies reveal that drought induces reversible cysteine oxidation in over 80 proteins, including Calvin–Benson and glycolytic enzymes. These proteins are subsequently reactivated through thioredoxin-dependent reduction [176]. These redox switches exemplify the intertwined regulation of metabolism and stress signaling. Overall, drought-responsive antioxidant systems do more than detoxify. They orchestrate a nuanced balance between signalling and oxidative defence. By safeguarding the photosynthetic apparatus, stabilizing

https://doi.org/10.3390/antiox15030337

membranes, coordinating osmotic balance, and modulating hormone crosstalk, antioxidants create a biochemical and molecular framework. This framework underpins crop resilience under water-limiting conditions.

5.2. Salinity Stress

Soil salinization affects approximately 20% of irrigated croplands and is also increasing in areas of rainfed agriculture, a problem that is expected to worsen with climate change [54,177]. High sodium (Na^+) and chloride (Cl^-) concentrations pose two primary challenges for plants: osmotic stress, which limits water uptake and mimics drought, and ionic toxicity, where Na^+ and Cl^- displace essential nutrients such as potassium, calcium, and magnesium [100,144]. This disrupts ion balance, damages membranes, impairs enzyme activity, and reduces photosynthetic efficiency [145]. Mechanistically, salinity stress limits CO_2 assimilation and increases photorespiration, leading to excess excitation energy directed toward oxygen, producing O_2-$^-$ in chloroplasts [146]. Additionally, Na^+ influx depolarises membranes which triggers potassium efflux and activates RBOHs, leading to localized ROS surges that damage lipids, proteins, and DNA [147–151]. For example, rice seedlings exposed to 150 mM NaCl show a ~60% increase in MDA and a ~40% decrease in chlorophyll content, indicating membrane damage and a decline in photosynthetic activity [145].

In salt-tolerant barley, SOD and APX activities increase up to 2.3-fold and 1.9-fold, respectively, reducing H_2O_2 accumulation by 45% compared to sensitive genotypes [154]. In rice, APX overexpression improves seedling survival under salinity by 35%, stabilizing membranes and limiting ROS [178]. The glutathione system maintains redox homeostasis, as GR and GPX regulate the GSH/GSSG ratio, which decreases from ~12:1 to 3–4:1 during salt stress [152,153]. Arabidopsis overexpressing GR sustains higher ratios (~6:1), delaying senescence and preserving photosystem II efficiency [151,152]. Non-enzymatic antioxidants, such as ascorbate and glutathione, increase 2–3-fold in tolerant genotypes driving the ascorbate–glutathione cycle and recycling antioxidant capacity [179,180]. Ion and redox signalling cross-talk is essential in salinity acclimation [146]. Na^+ influx disrupts K^+ homeostasis and fuels ROS production, creating a feedback loop that modulates ROS waves, regulating ion transporters (e.g., *SOS1*, *NHX*) and osmoprotectant synthesis, such as proline and glycine betaine [181,182]. Application of ascorbate in salt-stressed wheat has been shown to improve K^+/Na^+ balance, lowering MDA by 35%, and raise relative water content by ~20%, enhancing metabolic function and osmotic adjustment [183]. Overall, the synchronized induction of antioxidant enzymes and metabolites delineates the threshold between salt-induced damage and adequate acclimation.

Soil salinization affects around 20% of irrigated croplands. It is rising in rainfed areas, a problem expected to worsen with climate change [54,177]. High sodium (Na^+) and chloride (Cl^-) concentrations pose two main challenges for plants: osmotic stress and ionic toxicity. Osmotic stress limits water uptake and mimics drought. Ionic toxicity occurs when Na^+ and Cl^- displace potassium (K^+), calcium (Ca^{2+}), and magnesium (Mg^{2+}) [100,144]. This disrupts ion balance, damages membranes, impairs enzyme activity, and reduces photosynthetic efficiency [145]. Salinity stress also limits CO_2 assimilation and increases photorespiration. As a result, excess excitation energy is diverted toward oxygen, producing $O_2^{\cdot-}$ in chloroplasts [146]. Additionally, Na^+ influx depolarises membranes, triggers potassium efflux, and activates RBOHs. This process leads to localized ROS surges that damage lipids, proteins, and DNA [147–151]. For example, rice seedlings exposed to 150 mM NaCl show a ~60% increase in MDA and a ~40% decrease in chlorophyll content. These indicate membrane damage and reduced photosynthetic activity [145]. In salt-tolerant barley, SOD and APX activities increase up to 2.3-fold and 1.9-fold, respectively.

https://doi.org/10.3390/antiox15030337

As a result, this reduces H_2O_2 accumulation by 45% compared to sensitive genotypes [154]. In rice, APX overexpression improves seedling survival under salinity by 35%. It stabilizes membranes and limits ROS [178]. The glutathione system also maintains redox homeostasis. GR and GPX regulate the GSH/GSSG ratio, which drops from ~12:1 to 3–4:1 during salt stress [152,153]. Interestingly, *Arabidopsis* overexpressing GR maintains higher ratios (~6:1), which delays senescence and preserves photosystem II efficiency [151,152]. Non-enzymatic antioxidants, such as ascorbate and glutathione, increase 2–3-fold in tolerant genotypes. This drives the ascorbate–glutathione cycle, recycling antioxidant capacity [179]. Ion and redox signalling cross-talk is essential in salinity acclimation [146]. Na^+ influx disrupts K^+ homeostasis and fuels ROS production, creating a feedback loop. This modulates ROS waves and regulates ion transporters (e.g., SOS1, NHX). It also drives the synthesis of osmoprotectants, such as proline and glycine betaine [181,182]. Applying ascorbate to salt-stressed wheat improves K^+/Na^+ balance, lowers MDA by 35%, and raises relative water content by ~20%. These changes enhance metabolic function and osmotic adjustment [183]. Synchronized induction of antioxidant enzymes and metabolites marks the threshold between salt-induced damage and effective acclimation.

5.3. Heat Stress

Rising global temperatures pose a severe challenge to agricultural productivity by disrupting plant physiological, biochemical, and molecular functions [184]. Heat stress accelerates ROS production, leading to protein denaturation and destabilization of membranes and enzymes [26]. In chloroplasts, this excess thermal energy impairs the photosynthetic electron transport chain, increasing electron leakage to oxygen and generating $O_2^{\cdot-}$ and H_2O_2. H_2O_2 levels can increase fivefold in heat-stressed leaves, while mitochondrial ROS output may double due to reduced respiratory efficiency [155,156,184,185]. Heat-induced increases in membrane lipid fluidity promote lipid peroxidation, with MDA levels rising 40–70% in wheat, maize, and rice, illustrating extensive oxidative damage [155,156,183–185].

To counteract these effects, plants activate coordinated antioxidant defences [186]. Enzymes such as SOD, APX, and CAT detoxify ROS in chloroplasts, mitochondria, and the cytosol to maintain photosystem integrity [187]. In rice seedlings at 42 °C, Cu/Zn-SOD activity rises 2.3-fold while APX and CAT increase two- to threefold, reducing MDA by half [188]. Overexpression of APX in tomato preserves Rubisco activase and photosynthetic electron transport, highlighting its critical role in heat tolerance [189]. Heat stress also unfolds and aggregates proteins, but antioxidants act synergistically with heat shock proteins (HSPs) to stabilize protein structure and maintain enzymatic function [155,156]. The AsA–GSH cycle buffers redox homeostasis, maintaining high GSH/GSSG ratios to support ATP synthase and Calvin–Benson cycle enzymes even under thermal stress [190]. Thioredoxins and peroxiredoxins reverse heat-induced thiol oxidation, as evidenced by *Arabidopsis* 2-Cys Prx mutants that exhibit elevated ROS and rapid photosystem II photoinhibition under heat stress [30,191]. Membrane preservation is critical, as lipid-phase antioxidants such as tocopherols and carotenoids quench singlet oxygen and lipid radicals [191–193]. Heat stress induces a two- to threefold increase in α-tocopherol in thylakoid membranes, preserving fluidity and stability [192]. Additionally, β-carotene and zeaxanthin generated via the xanthophyll cycle dissipate excess thermal energy, thereby protecting photosynthetic pigments [42]. *Arabidopsis* mutants deficient in α-tocopherol exhibit accelerated senescence and lipid peroxidation under heat stress, underscoring the protective role of α-tocopherol in plants [193].

Heat-induced ROS also act as signalling molecules activating heat shock transcription factors (HSFs), especially *HSFA2*, which induces both HSPs and antioxidant en-

https://doi.org/10.3390/antiox15030337

zymes [194,195]. Hormonal crosstalk involving abscisic acid and ethylene further modulates osmotic regulation, stomatal closure, and water retention, thereby alleviating oxidative damage [196]. Transient H_2O_2 increases (~10 μM) directly upregulating SOD and APX, and establishing a feedback loop that primes antioxidant defences during and after heat episodes [197]. Field and multi-omics studies have consistently shown that a strong antioxidant capacity is correlated with heat tolerance and yield stability [198]. Wheat varieties with high CAT and APX activities suffer less than 15% yield loss under 38 °C compared to over 40% in sensitive types [157,199]. Heat-tolerant rice varieties accumulate up to 12 mM glutathione and exhibit 30% less electrolyte leakage, reinforcing the importance of redox balance in thermal resilience [188]. Overall, integrating enzymatic and non-enzymatic antioxidants with redox-regulated chaperones and signalling networks provides the basis for plant adaptation to rising temperatures, a trait crucial for food security under climate change.

5.4. Cold Stress

Low or chilling temperatures pose a severe metabolic challenge, especially for plants from warm climates. They disrupt membrane fluidity, impair enzyme function, and increase ROS levels, thereby exacerbating cellular damage [158]. Sensitive crops, such as rice, tomatoes, and maize, may suffer yield losses of up to 50–80% after cold snaps [158–161,200,201]. Cold reduces membrane fluidity by inducing a gel-phase state in lipid bilayers. This impairs ion transport and destabilizes the chloroplast and mitochondrial electron transport chains. The result is electron over-reduction and ROS surges [162,163]. For example, maize seedlings at 5 °C show a 3.5-fold increase in chloroplastic superoxide. Barley and tomato exhibit 40–70% rises in MDA, indicating lipid peroxidation and membrane injury [158]. Genotypes with higher unsaturation of membrane lipids, notably linolenic acid, maintain fluidity and reduce ROS leakage during cold stress [160].

Cold triggers the rapid induction of enzymatic and non-enzymatic antioxidants, such as SOD, which converts superoxide to H_2O_2 [162]. This is then detoxified by APX and CAT across cellular compartments [200]. In wheat and rice, SOD activity doubles within 24 h of chilling, whereas in barley, CAT activity increases from approximately 120 to 350 μmol H_2O_2 decomposed per minute per gram of protein, resulting in a 60% reduction in H_2O_2 levels [158]. Thylakoid-bound APX protects photosystem II by regulating H_2O_2 concentrations; *Arabidopsis* overexpressing tAPX retain 35% higher photosynthetic efficiency and 40% less lipid peroxidation under cold conditions [202–204]. The ascorbate–glutathione cycle balances redox status by regenerating antioxidants; in rapeseed at 4 °C, glutathione pools increase nearly threefold, raising the GSH/GSSG ratio from 3:1 to 8:1 and supporting APX activity [15,38]. Deficiencies in GR or DHAR lead to cold sensitivity, chlorosis, and ROS accumulation [23]. Cold acts as both a stress and a signal, with membrane rigidification activating calcium influx, triggering MAP kinase cascades, and activating the ICE1-CBF-COR transcriptional pathway [205]. H_2O_2 pulses (10–20 μM) activate CBF transcription factors, promoting the accumulation of osmoprotectants, such as proline and sugars, which stabilize membranes [158]. Exogenous ascorbate upregulates cold-responsive genes and reduces electrolyte leakage by 30%, improving cold tolerance [206]. However, excess ROS above 100 μM induces oxidative damage, disrupting thylakoid membranes and photosynthetic enzymes, underscoring the importance of redox homeostasis for cold acclimation [207].

Cold stress also disrupts mitochondrial function and carbon metabolism [200]. Mitochondrial alternative oxidase serves as an electron sink that limits ROS generation, while thioredoxins and peroxiredoxins maintain key enzymes in reduced, active states [28,207]. Cold-hardy varieties express up to 4 times more thioredoxins, enabling more flexible

https://doi.org/10.3390/antiox15030337

metabolic regulation [160]. Lipid-soluble antioxidants like tocopherols and carotenoids protect membranes by quenching singlet oxygen and lipid radicals; α-tocopherol increases two- to threefold within 48 h of chilling in Arabidopsis, reducing photobleaching and preserving photosynthesis [203,204]. In summary, antioxidant-mediated cold tolerance integrates membrane stabilization, rapid antioxidant deployment, and cross-talk between ROS and hormone signalling, enabling plants to mitigate oxidative bursts and maintain electron transport. This coordinated defence ensures survival in extreme cold and supports acclimation, which is critical for cultivation in temperate and high-altitude regions.

5.5. Heavy Metal Stress

Heavy metals (HMs), including cadmium (Cd), chromium (Cr), nickel (Ni), copper (Cu), and vanadium (V), pose a major abiotic threat. They disrupt metabolic processes and redox homeostasis, often driving excessive ROS formation ($O_2^{\bullet-}$, H_2O_2, $^{\bullet}OH$). This leads to oxidative damage to lipids, proteins, and nucleic acids [15,38,208]. Across diverse species, HM exposure commonly elevates H_2O_2 and lipid peroxidation (MDA), while reducing photosynthetic efficiency and growth. For example, Cd stress increases oxidative markers and impairs photosynthesis in Arabidopsis and cotton (*Gossypium hirsutum* L.). Cr and Ni can also compromise PSII performance and carbon assimilation [209–212]. To counter ROS surges, plants activate a coordinated antioxidant network. SOD converts $O_2^{\bullet-}$ to H_2O_2 and CAT or APX remove peroxide, supported by GR-driven recycling within the AsA–GSH cycle [15,28,208,211]. Enhanced APX and GR activity can stabilize chloroplast redox status and improve tolerance under Cd or Cu stress. This is consistent with the requirement for sustained antioxidant turnover—rather than single-enzyme changes [213]. Non-enzymatic antioxidants (Asc, GSH, tocopherols, flavonoids, phenolics) also buffer ROS and limit chain reactions. GSH plays a dual role as a substrate for phytochelatin synthesis, linking redox buffering to its capacity for metal detoxification [214,215]. A key feature of heavy metal tolerance is chelation and sequestration, which reduces free cytosolic metal ions and limits Fenton-type ROS amplification. Glutathione-derived phytochelatins bind metal ions to form HM–PC complexes. These complexes are transported into vacuoles via ABC transporters, lowering reactive metal availability in sensitive compartments [209,216,217]. Metallothioneins (MTs), rich in cysteine residues, provide binding capacity for metals such as Cd, Cu, and Zn and contribute to ionic homeostasis [218]. Increased PC production under Cd stress reduces free Cd^{2+} and partially recovers photosynthetic function, for example, in Brassica juncea L. Coordinated induction of MTs and GSH-dependent enzymes is linked with improved tolerance to Ni and Cr in crop systems [205,219,220]. Hormonal and redox signalling networks also shape these antioxidant and chelation responses. SA, JA, NO, and melatonin modulate antioxidant gene expression and redox capacity. Exogenous treatments (such as melatonin, SA, or silicon) have been reported to enhance activities of core antioxidant enzymes and associated protective pathways under metal stress. These treatments often reduce oxidative injury and support photosynthetic performance [205,207,221–224]. ROS–NO interactions, including redox enzyme modifications such as S-nitrosylation, can affect enzyme stability and activity. This further reinforces plant defence outputs [220,222]. In summary, heavy-metal tolerance is an integration of several defences. Antioxidant systems buffer ROS, chelators restrict metal reactivity and compartmentalize ions, and signalling networks coordinate these defences across tissues and organelles. Together, these systems enable plants to maintain metabolism and growth in contaminated environments [215,220,222].

https://doi.org/10.3390/antiox15030337

6. Genetic and Biotechnological Levers That Tune Antioxidant Network Balance

Sections 3–5 show that abiotic stress (environmental factors such as drought or extreme temperatures) disrupts the balance of ROS, and the effect depends on the cell compartment (specific locations within the cell, such as the chloroplast or mitochondrion). Under normal conditions, ROS are molecules that work in controlled signalling. But stress can lead to harmful ROS buildup if antioxidant defences (cellular systems that neutralize ROS) are overwhelmed. ROS metabolism and defence often overlap and support each other, so a single-gene fix rarely improves stress tolerance [225–227]. Section 6 outlines genetic and biotechnological ways to adjust antioxidant networks: (i) limiting ROS production by managing relevant enzymes and pathways, (ii) boosting detoxification by improving key enzymes (superoxide dismutase, ascorbate peroxidase/peroxiredoxin, catalase: SOD–APX/Prx–CAT), (iii) improving antioxidant recycling through the ascorbate–glutathione (AsA–GSH) and Trx-dependent systems, and (iv) targeting specific cell compartments and maintaining ROS signalling. These strategies are checked against stress responses and, where possible, crop yield data (Table 2). Traditional breeding has made some progress, but newer tools such as omics (comprehensive studies of cell molecules), transgenics (inserted genes), and genome editing (precise genetic changes) enable more targeted improvements across the whole network [Figure 3].

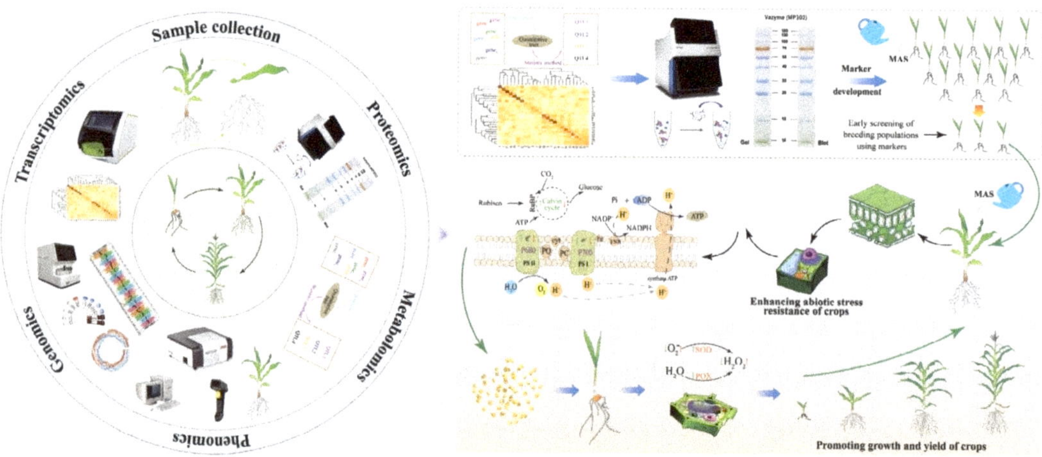

Figure 3. Integrative pipeline connecting stress-induced redox perturbation to deployable tolerance via omics, breeding, and antioxidant interventions. Left: Researchers collect samples from different developmental stages and contrasting genotypes under normal and abiotic-stress conditions. This enables multi-layer profiling using genomics, transcriptomics, proteomics, metabolomics, and phenomics. These approaches help discover candidate loci and biomarkers linked to redox control and stress tolerance. Right (top): Identified loci are translated into marker development. These markers are deployed through marker-assisted selection (MAS) for early screening of breeding populations. Right (middle): Omics-guided targeting of photosynthetic electron transport and related redox pathways links mechanistic antioxidant capacity to improved stress resilience. Right (bottom): Antioxidant-based interventions, such as priming or exogenous antioxidants, modulate the ROS and antioxidant balance. These changes are visualized as improved growth and yield stability under stress. Green arrows (↑/↓) indicate enhancement or inhibition after adding antioxidants. Red indicators (↑/↓) mark increases or decreases in response to abiotic stress relative to the normal state. Abbreviations: MAS, marker-assisted selection; QTL, quantitative trait locus; ROS, reactive oxygen species; SOD, superoxide dismutase; POX, peroxidase.

https://doi.org/10.3390/antiox15030337

Table 2. Genes used to improve abiotic stress tolerance in different crop plants.

Type of Stress	Crops	Techniques	Key Findings	References
Drought	Rice	CRISPR/Cas9	This study used eleven rice genes—*OsPDS*, *OsPMS3*, *OsEPSPS*, *OsDERF1*, *OsMSH1*, *OsMYB5*, *OsMYB1*, *OsROC5*, *OsSPP*, and *OsYSA*—to assess CRISPR/Cas9 editing efficiency and stress response. *OsDERF1*, *OsMYB1*, and *OsMYB5* regulate drought, salinity, and oxidative stress tolerance; *OsMSH1* maintains genomic stability; and *OsROC5* supports adaptive growth. *OsPDS*, *OsEPSPS*, and *OsYSA* acted as marker genes confirming successful editing.	[228]
Drought	Rice	CRISPR/Cas9	CRISPR/Cas9 knockout of *SRL1* and *SRL2* enhanced rice drought tolerance by boosting antioxidant defense. Mutants showed 45% higher survival, lower MDA, and 1.5–2× higher SOD and CAT activity than wild type. Proteomic data confirmed enrichment of antioxidant and stress-response proteins, indicating improved oxidative stress protection and drought resilience.	[229]
Drought	Rice	CRISPR/Cas9	*OsAAA-1* and *OsAAA-2* genes in rice negatively regulate drought tolerance; their overexpression reduces tolerance, while CRISPR-Cas9 knockout or RNAi silencing significantly enhances drought tolerance and grain yield without adverse growth effects.	[230]
Drought	Rice	CRISPR/Cas9	CRISPR/Cas9-induced mutations in *OsERA1* enhance drought tolerance through accelerated stomatal closure and increased ABA sensitivity without affecting leaf growth. *OsERA1* acts as a key negative regulator in ABA-mediated drought response, indicating its potential as a target for genetic improvement of drought tolerance and antioxidant activity in rice	[231]
Drought	Rice	CRISPR/Cas9	*OsPYL9* is an ABA receptor in rice that regulates drought stress tolerance by enhancing ABA accumulation and antioxidant enzyme activities (SOD, POD, CAT), and reducing MDA levels. CRISPR/Cas9-edited *OsPYL9* mutants show altered protein expression involved in circadian rhythms and abiotic stress responses, including proteins like *GIGANTEA* and pseudo-response regulators, which are central to ABA-activated signaling and ROS detoxification.	[232]
Drought	Rice	CRISPR/Cas9	*OsNAC14* is a drought-inducible transcription factor highly expressed during meiosis and in response to drought, salinity, ABA, and low temperature. Its overexpression enhances drought tolerance by activating genes involved in stress response, defense, and metabolism, notably directly upregulating *OsRAD51A1*, a key DNA repair gene.	[233]

https://doi.org/10.3390/antiox15030337

Table 2. *Cont.*

Type of Stress	Crops	Techniques	Key Findings	References
Drought	Rice	CRISPR/Cas9	*OsPUB67* interacts with E2 enzymes and E3 ligases such as *OsRZFP34* and *OsDIS1* to form complexes that modulate drought response. It regulates ABA-dependent stress genes and enhances antioxidant activity, proline accumulation, and membrane stability during drought stress.	[234]
Drought	Rice	TALEN	*OsDERF1*, targeted by TALEN editing, is associated with enhanced drought resistance and plays a key role in regulating plant stress response pathways, including the modulation of antioxidant activity to improve abiotic stress tolerance.	[235]
Drought	Rice	ZFNs	A non-coding safe harbor locus on chromosome 3 was identified in rice using ZFN-mediated transformation, enabling stable integration of stress-regulatory genes such as *DREB*, *NAC*, *WRKY*, *SOD*, *CAT*, and *APX*.	[236]
Drought	Wheat	CRISPR/Cas9	Using CRISPR/Cas9, targeted editing of *TaDREB2* and *TaERF3* in wheat protoplasts confirmed their roles in drought and oxidative stress regulation. Both genes were efficiently and specifically mutated, and their upregulation under dehydration indicates key functions in enhancing stress tolerance and antioxidant defense in wheat.	[237]
Drought	Maize	CRISPR/Cas9	Maize *ARGOS8* negatively regulates ethylene sensitivity. Its elevated expression via CRISPR-Cas9 enhances drought tolerance without affecting yield under normal conditions.	[238]
Drought	Tomato	CRISPR/Cas9	*AITR* transcription factors (*AITR1*, *AITR2*, *AITR3*, *AITR4*, *AITR5*, and *AITR6*) negatively regulate ABA responses. CRISPR/Cas9 knockout mutants (*aitr256*, *aitr1256*, *aitr23456*, and *aitr123456*) show reduced ABA sensitivity and enhanced drought tolerance without growth or pathogen resistance trade-offs. Altered expression of ABA signaling genes *PYL4*, *PYL5*, *PYL6*, *HAI1*, and *ABF3* underlies improved stress resilience.	[239]
Salinity	Rice	CRISPR/Cas9	*OsbHLH024* acts as a negative regulator of salt tolerance. It reduces oxidative damage, indicated by lower ROS and MDA levels, and promotes balanced antioxidant enzyme activity. Molecularly, it upregulates *OsHKT1;3*, *OsHAK7*, and *OsSOS1*, enhancing Na+ exclusion and K+ retention to maintain redox and ionic balance, while downregulating *OsLEA3*.	[240]
Salinity	Rice	CRISPR/Cas9	CRISPR/Cas9-induced mutations in *OsRR22* enhance salinity resistance by modulating cytokinin signaling, which improves ion regulation and antioxidant defense. Mutant lines (*WPB106-cas-1* and *WPB106-cas-2*) maintain better redox balance and reduced oxidative damage under salt stress, showing that loss of *OsRR22* activates stress-responsive pathways for improved tolerance.	[241]

https://doi.org/10.3390/antiox15030337

Table 2. *Cont.*

Type of Stress	Crops	Techniques	Key Findings	References
Salinity	Rice	CRISPR/Cas9	CRISPR/Cas9 mutants (*OsRR9* and *OsRR10*) showed improved tolerance, lower oxidative stress, and better Na^+/K^+ balance. Transcriptomic data revealed upregulation of ion transporters and antioxidant-related genes, indicating that *OsRR9* and *OsRR10* suppress stress-responsive pathways controlling ion homeostasis and redox balance under salinity.	[242]
Salinity	Rice	CRISPR/Cas9	The *OsNAC041* gene positively regulates salt stress tolerance in rice. Its expression increases under salinity, and CRISPR/Cas9 mutants lacking *OsNAC041* show higher ROS and MDA levels with reduced SOD, POD, and CAT activities, indicating impaired antioxidant defense. Transcriptome analysis revealed disrupted MAPK signaling, peroxisome function, and photosynthesis-related genes.	[243]
Salinity	Rice	CRISPR/Cas9	*OsSPL10* is a nuclear transcription factor that negatively regulates salt tolerance and positively controls trichome formation in rice. Knockout mutants show enhanced salt tolerance and glabrous leaves, while overexpression increases trichome density but reduces salt tolerance.	[244]
Salinity	Rice	S-adenosylmethionine decarboxylase	The *SAMDC* gene regulates stress and antioxidant activity by enhancing polyamine biosynthesis under stress conditions. Its ABA-inducible expression under NaCl stress increases spermidine and spermine levels, which act as antioxidants that stabilize cellular structures and scavenge reactive oxygen species, thereby protecting plants from oxidative stress.	[245]
Salinity	Barley	CRISPR/Cas9	The *HvITPK1* gene, along with other ITPK genes, regulates abiotic stress response and antioxidant defense via inositol phosphate metabolism. Its expression is induced by ABA and salt, and mutants exhibit altered stress sensitivity. The gene's ABA-responsive promoter suggests it is transcriptionally regulated by oxidative and osmotic stress, positioning *HvITPK1* as a key regulator of stress signaling and antioxidant pathways.	[246]
Salinity	Barley	Overexpression by using 35S promoter	The gene *AVP1* enhances salinity stress tolerance by increasing shoot biomass and grain yield in transgenic barley under saline conditions without changing leaf sodium levels. *AVP1* promotes early growth and likely supports antioxidant activity by maintaining cellular ion balance and pH, helping plants manage oxidative stress during salinity.	[247]
Salinity	Soybean	CRISPR/Cas9	*GmAITR* genes in soybean are ABA-induced transcription repressors that regulate ABA signaling and stress responses. CRISPR/Cas9 mutants of *GmAITRs* show increased ABA sensitivity and enhanced salinity tolerance, with improved germination, growth, and yield under salt stress.	[248]

Table 2. *Cont.*

Type of Stress	Crops	Techniques	Key Findings	References
Salinity	Tomato	CRISPR/Cas9	The *SlHyPRP1* gene negatively regulates salt stress in tomato. CRISPR/Cas9-mediated precise deletion of its functional domains improved salinity tolerance by enhancing germination, growth, and survival under salt stress.	[249]
Cold stress	Rice	CRISPR/Cas9	CRISPR/Cas9 editing of *OsMYB30* enhanced cold stress tolerance and survival, while combined edits with *OsPIN5b* and *GS3* improved overall stress resilience. These gene edits help rice better manage abiotic stress, likely supporting antioxidant defences and improving tolerance to cold stress.	[250]
Cold stress	Rice	CRISPR/Cas9	The gene *OsAnn3* is crucial for rice cold tolerance, with increased expression under cold stress. Its knockout reduces survival and increases membrane damage, indicating a role in stress protection. *OsAnn3* likely helps stabilize membranes and supports antioxidant defences during cold stress.	[251]
Cold stress	Rice	CRISPR/Cas9	*OsPRP1* enhances cold tolerance in rice by regulating antioxidant enzymes and stress-related metabolites. Its knockout reduces survival, antioxidant activity, proline, chlorophyll, ABA, and ascorbic acid levels, leading to increased cold sensitivity.	[252]
Cold stress	*Arabidopsis*	CRISPR/Cas9	The *SPCP2* gene enhances plant stress tolerance by regulating senescence and antioxidant responses. Its expression under salt, drought, and hormonal stress supports cellular protection, helping maintain antioxidant balance and improve tolerance to abiotic stress.	[253]
Heat stress	Tomato	CRISPR/Cas9	The *SlAGL6* is a key regulator of heat tolerance, enabling fruit set under high-temperature stress. While not directly linked to antioxidant activity, future studies could investigate its role in stress resilience, potentially involving oxidative damage response.	[254]
Heat stress	Rice	Overexpression by using the Maize Ubi1 promoter	Overexpression of *Hsp101* enhances thermotolerance by maintaining protein homeostasis and protecting cells from heat-induced oxidative damage. Its activity supports the antioxidant defense network, reducing cellular stress and improving survival during severe heat stress.	[255]
Heat stress	Rice	Overexpression by using the Maize Ubi1 promoter	The *OsRab7* improves heat tolerance in rice by boosting antioxidant enzyme activities (CAT, SOD, APX, POD) and reducing oxidative damage markers (H_2O_2, MDA). It also upregulates key ROS-scavenging genes (*OsCATA*, *OsCATB*, *OsAPX2*, *OsSOD-Cu/Zn*) and other stress-responsive genes, enhancing osmotic adjustment and overall antioxidant defense under stress.	[256]
Heat stress	Rice	Promoter of Rca-a from *Oryza meridionalis*	The heat-stable *Rca* gene from *Oryza australiensis* maintains Rubisco function at high temperatures, preventing protein denaturation and indirectly reducing oxidative damage, thereby enhancing thermotolerance.	[257]

https://doi.org/10.3390/antiox15030337

6.1. Transgenic Overexpression of Antioxidant Genes

Transgenic overexpression of antioxidant enzymes has been widely used to enhance tolerance to abiotic stresses by limiting ROS-driven damage to lipids, proteins, and DNA [258]. However, the effectiveness of this approach depends strongly on the gene selected, its subcellular targeting, and the specific stress context, given that ROS production is spatially and temporally structured [259–261]. Among these enzymes, SODs provide an upstream defence by converting $O_2^{\bullet-}$ to H_2O_2, thereby reducing radical pressure at photosynthetic and respiratory electron transport sites [226,262]. Reflecting this, chloroplast-targeted Cu/Zn-SOD overexpression in tobacco improved photosynthetic performance and reduced photoinhibition under high light, while mitochondrial Mn-SOD in rice improved salinity tolerance by limiting respiratory ROS leakage and supporting energy metabolism [184,261]. SOD overexpression has also been associated with improved drought-related performance and symbiotic traits in legumes, underscoring its value as a buffer at ROS-generating "hotspots" [263]. Because SOD increases H_2O_2 formation, downstream peroxide-scavenging capacity becomes particularly important. In this context, APX operates effectively at signalling-range peroxide concentrations and can improve stress tolerance when appropriately expressed and targeted [15]. For example, cytosolic APX overexpression in rice improved salt-stress survival with reduced H_2O_2 accumulation, and thylakoid-associated APX in *Arabidopsis* reduced lipid peroxidation under high light, supporting chloroplast protection [179]. Notably, co-overexpression of SOD and APX in tobacco can yield additive protection, illustrating the advantage of coordinating consecutive steps rather than enhancing a single enzyme in isolation [148,262]. Furthermore, catalase provides high-capacity H_2O_2 removal, particularly relevant in peroxisomes during photorespiration, and CAT overexpression has been linked to improved drought- or salinity-related performance in crops such as maize and cotton, including reduced membrane leakage and better chlorophyll retention [25,43,120,264].

In addition to the enzymes described above, glutathione-dependent enzymes reinforce whole-cell redox buffering by sustaining recycling throughput. For example, GR overexpression can maintain a more reduced glutathione pool (a higher GSH/GSSG ratio) under stress, supporting continued APX/GPX turnover and redox stability, whereas GPX overexpression can reduce lipid peroxidation (e.g., lower MDA) and protect membranes under salinity [147,265]. Taken together, these examples show that antioxidant engineering can improve stress performance when it increases system throughput (detoxification plus recycling) rather than creating an unbalanced bottleneck. Several mechanistic principles emerge across studies. First, compartment targeting matters: expression in chloroplasts or mitochondria often yields stronger benefits than cytosolic expression because ROS must be intercepted near their sites of origin. Second, balanced pathway engineering is essential: elevating SOD without matching APX/CAT capacity can increase H_2O_2 accumulation and shift stress responses toward toxicity rather than acclimation [266]. Third, constitutive overexpression can impose growth penalties under non-stress conditions due to metabolic costs and disruption of ROS-dependent developmental signalling, motivating the use of inducible promoters and stress-responsive designs [267]. Illustrative successes, such as improved cold-related performance in potato with Cu/Zn-SOD, enhanced drought-related root function in chickpea with APX, and delayed senescence/extended photosynthesis in maize with GR, support the potential of antioxidant transgenics when these design constraints are respected [268–271]. Overall, the field is moving toward multi-gene, compartment-targeted, and conditionally regulated strategies that coordinate ROS metabolism across compartments while preserving productivity.

https://doi.org/10.3390/antiox15030337

6.2. CRISPR/Cas-Mediated Genome Editing to Enhance Antioxidant Capacity

CRISPR/Cas genome editing enables precise modification of endogenous antioxidant and redox-regulatory pathways. This approach offers a complementary route to classical transgenics for improving stress tolerance [228–230]. Edits can be introduced at native loci and, in many cases, without stable foreign DNA. CRISPR can adjust antioxidant network behaviour [231]. It helps reduce problems such as metabolic burden, mislocalization, and disruption of ROS-dependent signalling, which are associated with constitutive overexpression [232–234]. This precision is useful for engineering stress resilience while preserving yield. ROS detoxification must remain coordinated with development and signalling. CRISPR strategies for enhancing antioxidant activity generally fall into three categories. Knockout of negative regulators removes inhibitory constraints on ROS detoxification. For example, disrupting ROS-repressor genes in rice under salinity increased APX and SOD activity ($\approx$1.8-fold), reduced lipid peroxidation (~35%), and improved seedling survival (~20%) [235,236]. In tomato, editing targets affecting ascorbate metabolism increased intracellular ascorbate pools and APX activity. This was accompanied by improved drought-related performance, including photosynthetic stability and fruit setting under water limitation [237–239]. These results suggest that relieving regulatory bottlenecks can shift redox balance toward controlled ROS buffering. However, benefits remain context- and target-dependent.

Cis-regulatory (promoter) editing provides a route to inducible or tissue-specific activation of antioxidant genes. This reduces the costs of continuous expression [240,241]. Coordinated upregulation of SOD/CAT/APX through promoter activation has been reported to reduce H_2O_2 accumulation under drought-like conditions [242]. Stress-responsive alleles remain largely silent under non-stress conditions and activate in response to salinity. This illustrates an attractive design principle: deploy detoxification capacity only when ROS input rises, minimising growth penalties [243–245]. Allele replacement or optimization can improve enzyme stability or kinetics under stress, such as heat-sensitive APX variants. This maintains detoxification when proteins are destabilized or when turnover becomes limiting [272,273]. A further frontier is regulatory rewiring by inserting hormone-responsive elements, such as ABA- or SA-responsive motifs. This more tightly couples antioxidant activation to endogenous stress signalling and preserves signalling fidelity [244,249]. Overall, CRISPR enables movement from "more antioxidants" toward network tuning. It removes bottlenecks, builds conditional control, and optimizes enzyme performance. These designs must consider system constraints such as NADPH supply, AsA/GSH recycling capacity, and compartment cross-talk [252–257].

6.3. Marker-Assisted Breeding for Antioxidant Traits

Although genome engineering has expanded experimental control over redox pathways, conventional breeding remains essential. This is especially true when regulatory constraints or market acceptance limit the adoption of engineered cultivars [228,231]. Marker-assisted selection (MAS) offers a practical way to exploit natural allelic variation in antioxidant enzymes and redox-buffering pathways. This approach improves stress resilience while operating within standard breeding frameworks [272,273]. QTL mapping and association studies have identified loci linked to antioxidant activities and metabolite pools. These discoveries enable breeders to track antioxidant-related traits without relying solely on slow or expensive phenotyping [274]. In rice, QTLs associated with increased SOD and APX activity under salinity have been reported. These explain a substantial fraction of phenotypic variation. Tolerant lines carrying these loci show stronger chloroplast ROS control and reduced membrane lipid peroxidation compared to susceptible backgrounds [275,276]. In wheat, drought-associated loci affecting GR activity and glu-

https://doi.org/10.3390/antiox15030337

tathione redox homeostasis have been mapped. Introgression of these loci has been linked to improved photosynthesis and grain filling under water deficit—traits that connect redox buffering directly to yield-relevant performance [277,278]. Similar patterns are reported in other crops. Lines enriched for antioxidant buffering capacity, such as CAT/GPX activity or higher ascorbate pools, often show improved chlorophyll retention ('stay-green') and better maintenance of productivity under drought or heat stress [279,280]. A consistent lesson is that antioxidant-associated effects are usually polygenic and network-based. Pyramiding multiple loci, for example, combining QTLs linked to both SOD and APX activity, can provide stronger reductions in oxidative damage markers and more robust biomass performance than single loci. This reinforces the idea that resilience derives from coordinated antioxidant capacity rather than from isolated enzymes [281,282]. Looking forward, MAS will be most powerful when integrated with genomic selection, high-throughput phenotyping, and multi-environment trials. This integration will enable the selection of redox-associated alleles that remain beneficial across variable field regimes [283].

6.4. Omics-Driven Approaches

A complete understanding of antioxidant function cannot rely on a single enzyme or metabolite. This is because ROS metabolism is dynamic, interconnected, and strongly context-dependent [284]. Omics technologies, including transcriptomics, proteomics, and metabolomics, now provide system-level views of redox regulation. These technologies offer increasing spatial, temporal, and quantitative resolution [285]. Transcriptome profiling (especially RNA-seq) consistently shows broad remodelling of antioxidant networks during abiotic stress. For example, salt-stressed soybean induces large sets of redox-related genes spanning SOD, APX, GST, and GR families. This induction often shows marked tissue specificity [286,287]. Single-cell transcriptomics further refines this picture by revealing cell-type specialization. For example, some cells show preferential induction of peroxidase/peroxiredoxin modules in guard cells, while others show glutathione/catalase-linked programs in mesophyll. This pattern is consistent with compartment- and function-specific ROS control strategies [288]. Time-course studies add a crucial kinetic dimension. These studies often show an early induction of core enzymatic detoxification (e.g., SOD/APX). A later activation of secondary metabolism and the production of protective compounds usually follows. These findings imply a staged deployment of antioxidant defences, rather than a uniform response [289,290].

Proteomics complements transcript data by capturing post-transcriptional and post-translational regulation, which directly controls enzyme activity and signalling [291]. Redox proteomics has identified many stress-responsive thiol-modified proteins. These include Calvin–Benson cycle enzymes, photosystem components, and regulatory kinases. This supports the view that reversible cysteine oxidation acts as a redox "switchboard." Such a switchboard connects ROS dynamics to metabolic reprogramming [292,293]. Thioredoxin- and peroxiredoxin-based systems counterbalance these modifications. This process links detoxification capacity to regulation of photosynthesis and stress signalling [75]. Metabolomics directly quantifies the sizes of the antioxidant pool and its redox states. This provides stress-specific biochemical signatures [287]. Drought and related stresses commonly shift ascorbate and glutathione pools and their redox ratios. These changes determine the throughput of the AsA–GSH cycle. High light can strongly increase xanthophyll-cycle components such as zeaxanthin, which support NPQ and photoprotection [61,294,295]. Together, these datasets emphasize that stress tolerance depends not only on component abundance. It also relies on turnover, redox state, and coordination across pathways.

https://doi.org/10.3390/antiox15030337

The main value of omics lies in integrating datasets. Multi-omics network analyses can identify regulatory hubs and limiting steps ("bottlenecks") in ROS detoxification capacity and reductant recycling. This process prioritizes mechanistically grounded targets for engineering or breeding [286–288]. Integrated studies in Arabidopsis, for example, have highlighted nodes such as APX2 and 2-Cys peroxiredoxins. These nodes are central connectors between ROS buffering and heat tolerance [296]. Beyond mechanisms, omics-derived signatures (transcript modules and metabolite markers) are increasingly used as biomarkers. These markers stratify germplasm and predict stress performance. They support translation into breeding pipelines, especially when combined with multi-environment validation [296]. In a translational context, omics-driven discovery complements intervention strategies. It provides the system-level constraints needed for rational design. Transgenics provide proof of concept that ROS damage can be reduced through antioxidant manipulation. CRISPR enables more precise tuning through promoter editing, allele optimization, and multiplex rewiring [228–230]. Marker-assisted and genomic approaches can then exploit natural diversity. These approaches deploy antioxidant-associated alleles into elite backgrounds across agroecological contexts [272]. Moving forward, the most robust route is likely an integrated framework. In this framework, omics identifies hubs and trade-offs. Engineering provides targeted perturbations. Breeding, plus field phenotyping, validates whether the altered redox regulation delivers consistent benefits to yield and stability under realistic, often combined-stress conditions [297–301].

7. Field-Facing Antioxidant Interventions Linked to Redox Mechanisms (Evidence and Limits)

This section focuses on field-facing interventions, including foliar antioxidant application, soil amendments that influence redox-active compounds, and priming treatments that modulate redox homeostasis during abiotic stress. We build on the compartment-resolved framework from Sections 3–5 and the genetic or biotechnological approaches outlined in Section 6. We discuss antioxidant-related practices only when there is a plausible or demonstrated mechanistic link to ROS control. We evaluate the evidence by considering limits that affect real-world performance, such as dose, timing, formulation or delivery, developmental stage, and genotype-by-environment interactions. Because manipulating antioxidants can preserve or suppress adaptive signalling, we also highlight situations in which exogenous inputs or priming strategies yield inconsistent or non-transferable results outside controlled environments.

7.1. Exogenous Antioxidant Application

The application of antioxidants directly is a fast and straightforward way to increase the redox buffering capacity of plants during exposure to abiotic stress [302]. These interventions act through two complementary mechanisms. First, they acutely increase the local antioxidant pool to quench ROS produced under stress insults before they damage cellular macromolecules. Second, they prime endogenous defense systems, thus prolonging protection beyond the treatment period [303,304]. Foliar sprays remain the most commonly applied method. They provide direct exposure to photosynthetic organs, where ROS production is highest [305,306]. Water-soluble antioxidants, such as Asc and GSH, penetrate leaf tissues through stomata and cuticular pores and scavenge H_2O_2, $O_2^{\bullet-}$, and $^{\bullet}OH$ [55,62–64]. Ascorbate also functions as an electron donor for chloroplastic APX, reinforcing enzymatic detoxification [65]. Lipid-soluble antioxidants such as tocopherols stabilize thylakoid membranes and photosynthetic pigments. They also modulate apoplastic H_2O_2 signalling that coordinates stomatal closure and defence gene activation [116]. Data available confirm that there is a relevant physiological advantage when the right

https://doi.org/10.3390/antiox15030337

concentration and timing of formulation are used. For example, foliar spraying with 2 mM ascorbate in drought-stressed wheat increased photosynthetic rate by 25% and grain yield by 15% [55]. GSH sprays decreased H_2O_2 accumulation by 30% under Cd stress in rice [307]. Such applications are best before or at the time of early stress to achieve maximum acclimation benefit. This timing helps prevent an oxidative burst [308]. Effective foliar ascorbate concentrations generally range from 50–200 mg L^{-1}, though millimolar levels are sometimes tested [96,97]. Over-application can induce osmotic stress or economic inefficiency. This emphasizes the need for calibration [115]. The absorption can be enhanced by using adjuvants. However, they must be used with caution due to their potential for phytotoxicity [174]. Seed priming is an economical approach to establishing antioxidant "memory" before germination. Pre-treatment of seeds with antioxidants or osmoprotectants induces a low-level oxidative signaling. This signaling up-regulates defense enzymes, such as SOD, APX, CAT, and GR, as well as the synthesis of osmolytes [309]. This metabolic priming accelerates germination and provides a lasting memory of protection against subsequent stress [310,311]. In maize, priming seeds with ascorbate or proline enhanced antioxidant enzyme activity by 1.5–3-fold in seedlings exposed to drought. It also increased vigor and establishment [312]. The antioxidant concentration and duration must be optimized for each crop species. Seed priming methods are readily transferable across different cropping systems [313]. Post-priming drying also needs to be well controlled to avoid contamination by microorganisms.

Application mediated by nanoparticles is an exciting frontier in sustained antioxidant protection [314]. Modifying SOD and CAT mimetics with engineered nanomaterials, such as cerium oxide nanoparticles (CeO_2-NPs), results in catalytic "nanozymes." These nanozymes shuttle between Ce^{3+} and Ce^{4+}, mimicking SOD and CAT activities [315]. This redox-cycling is sustained without exhaustion. It allows continued ROS detoxification and, consequently, long-term protection. A 50 mg L^{-1} concentration of CeO_2-NPs in tomato and cotton reduced ~40% ROS leakage, maintained chlorophyll content, and increased fruit or biomass yield under salinity and drought conditions [316,317]. Nanoparticles offer additional benefits over conventional antioxidants. These include sustained release, increased stability, and tissue-targeting specificity through surface modification or carrier matrices [6,192]. Nevertheless, the use of nanomaterials requires careful evaluation of their environmental risks [167]. Their persistence in soil and effects on the soil microbial community, nutrient cycling, and non-target organisms are still poorly understood [315]. Effective but non-disruptive concentrations should be defined through long-term, crop-specific trials [32,192]. A comprehensive, locally specific strategy holds the greatest potential for field resistance [196]. Otherwise, foliar antioxidant sprays provide quick protection during acute stress, such as heat spikes or short-term water deficits [318]. Seed priming is a strategy to promote early vigor and stress tolerance. Nanoparticle formulations can maintain redox homeostasis over extended growth periods [319]. When combined, such strategies (for example, antioxidant priming and foliar application at key stages of plant development) typically result in additive or synergistic effects [320]. In the future, dose–response optimization in practical field conditions is essential to optimize both application procedures and cost-effectiveness [321]. Concomitant research on the long-term ecological impacts, especially those of nanomaterials, is necessary to balance efficacy with sustainability. When properly applied, exogenous antioxidant applications can connect mechanistic redox biology with the practical requirements of crop management. Such applications provide immediate tools to stabilize yields by minimizing oxidative stress in an increasingly unstable climate.

https://doi.org/10.3390/antiox15030337

7.2. Plant Growth-Promoting Rhizobacteria (PGPR) and Microbial Elicitors

Soil microorganisms play a key role in regulating plant redox biology. Their influence on oxidative balance goes beyond nutrient cycling and hormone interactions [322]. Plant growth-promoting rhizobacteria (PGPR), such as *Pseudomonas fluorescens* L., *Bacillus subtilis* L., and *Azospirillum brasilense* L., prime the antioxidant defense system and boost abiotic stress tolerance [323]. These microbes colonize the rhizosphere and root cortex, where they form a biochemical interface that helps control primary metabolism and maintain redox homeostasis [324]. The stress relief from PGPR comes from redox-dependent signaling. PGPR release phytohormones (including IAA, cytokinins, and gibberellins) and volatile organic compounds that adjust stomatal conductance, stimulate osmolyte synthesis, and activate antioxidant genes [325]. Under conditions such as drought or salinity, their inoculation consistently increases antioxidant enzyme activity [326]. For example, *Pseudomonas putida* L. inoculated into maize increased APX and GR activities by about 2-fold and reduced MDA by 50%, thereby improving membrane stability [327]. *Bacillus subtilis* L. in salt-stressed wheat increased SOD and CAT activities by 80–120%, which helped maintain photosynthetic efficiency and Fv/Fm [328]. PGPR also boosts non-enzymatic antioxidants. In tomato and rice, *Pseudomonas fluorescens* L. inoculation doubled the ascorbate and glutathione pools, supporting the ascorbate–glutathione cycle and promoting the synthesis of tocopherols, phenolic acids, and flavonoids [329,330]. This wide-reaching metabolic reprogramming shows that microbial priming extends beyond enzyme activation and creates broad redox optimization [325].

Mechanistically, induced systemic tolerance (IST) originates from microbeplant signaling [331]. Recognition of microbial molecules, lipopolysaccharides, flagellin peptides, or siderophores, by plant pattern-recognition receptors (PRRs) triggers transient ROS bursts via RBOHs, activating MAPK cascades and transcription of antioxidant genes [331,332]. This early signalling establishes a preemptive defense state that accelerates detoxification when stress occurs. PGPR also reshapes root architecture and ion balance, indirectly alleviating oxidative stress [331]. In saline soils, *Bacillus amyloliquefaciens* L. improved the K^+/Na^+ ratio by 30% and doubled APX and GR activity, sustaining photosynthesis and biomass. Enhanced nitrogen assimilation and organic acid exudation further contribute to redox stability [333]. Beyond living bacteria, microbial elicitors such as chitosan, β-glucans, and fungal metabolites mimic microbe-associated molecular patterns (MAMPs) to activate defense signaling [334]. These compounds induce systemic resistance through controlled ROS bursts and hormone crosstalk involving jasmonic acid and ethylene [335,336]. For example, foliar chitosan (100 mg L^{-1}) in salt-stressed soybean increased CAT and SOD activity by 75–90% and phenolic content by 2.5-fold, lowering H_2O_2 by ~40% [337]. Similarly, β-glucan treatment in rice enhanced APX and peroxidase activity, stabilized chlorophyll, and reduced ion leakage under drought [338]. Such elicitor-triggered oxidative bursts serve as priming cues that strengthen antioxidant responses without causing cellular damage [334–336].

Using PGPR and microbial elicitors provides an environmentally friendly approach to redox resilience [334]. These biological tools strengthen antioxidant networks, which reduce stress-induced ROS. This also protects photosynthetic and growth processes [328]. Their effects last through both vegetative and reproductive growth, unlike the short-term effects of chemical sprays [334]. Future research should identify specific redox signaling patterns in strains, develop compatible microbial mixtures for different crops and soils, and combine inoculation with molecular breeding and precision farming [338]. However, persistence in the field, strong colonization, and environmental changes remain major challenges [339]. Even with these challenges, microbial priming moves the approach from

https://doi.org/10.3390/antiox15030337

reactive protection to proactive redox management. It forms a science-based plan for sustainable stress resilience in crops.

7.3. Biofortification Strategies

Biofortification (plant nutritional quality improvement) is a sustainable approach that increases oxidative stress adaptation and enhances plant nutritive value [340]. Unlike transient antioxidant sprays, it creates heritable and stable modifications to redox-active metabolites through genetic engineering, genome editing, and conventional breeding [341]. High pools of ascorbate, glutathione, tocopherols, and carotenoids detoxify ROS during stress and act as metabolic buffers [116,155,193]. Ascorbate biofortification focuses on enhancing flux through the Smirnoff–Wheeler and D-galacturonate reductase (GalUR) pathways [193]. Overexpression of GalUR in tomato increased fruit ascorbate 2.5–3-fold, improving stress tolerance and post-harvest life [342]. Overexpression of GDP-L-galactose phosphorylase (GGP) in rice and lettuce enhanced ascorbate content by 60–150% and protected chlorophyll from salinity-induced destruction [343]. These enhancements are reflected in the AsA–GSH cycle, which requires continued NADPH turnover and H_2O_2 detoxification via APX [72,76]. Glutathione biofortification targets key enzymes of the cysteine and γ-glutamyl pathways [65]. Overexpressing γ-glutamylcysteine synthetase (GSH1) or GR in *Arabidopsis* and rice increased glutathione pools 2–3-fold, raising the GSH/GSSG ratio and reducing membrane lipid damage by 40–50% [340]. These plants showed better tolerance to Cd, ozone, and salinity, while larger glutathione pools also aided recycling of ascorbate and tocopherol [73]. Carotenoid pathway engineering is a typical example [124]. Golden Rice, which contains both the *psy* and *crtI* genes, accumulated 30–35 $\mu g\ g^{-1}$ of endosperm β-carotene, a precursor of vitamin A and a strong scavenger of singlet oxygen [341]. Similar approaches in maize and cassava (*Manihot esculenta* L.) resulted in twofold to fourfold increases in carotenoid levels, enhancing stress tolerance and yield [125,341]. Overexpression of γ-TMT increased tocopherol levels by 35–45% in wheat and soybeans, reducing seed aging and improving membrane stability during drought and heat stress [342]. α-tocopherol inhibits lipid oxidation and, with ascorbate, stabilizes redox states within plastids [343].

Recent innovations in CRISPR/Cas technology make targeted, DNA-free fine-tuning of antioxidant pathways possible [246]. For example, editing the GGP and GalUR promoters in tomato and rice has led to 30–70% increases in ascorbate levels. This method does not require transgene insertion [344]. Additionally, regulating the expression of repressors such as *SlAPX4* or *VTC2* can also increase endogenous antioxidant capacity. These plants show improved heat and drought stress tolerance [345]. Transferring stress-responsive cis-elements into promoters of antioxidants allows inducible regulation. This strategy conserves metabolic resources during optimal conditions but enables rapid antioxidant production during stress [231]. In summary, biofortification creates a win-win outcome. It enhances plant stress tolerance and improves human nutrition by raising levels of vitamin C and E, β-carotene, and phenolics [335]. Fresh market tomato lines high in ascorbate show 25% greater yield stability under drought and 30% firmer post-harvest fruit. These findings suggest a link between improved antioxidant metabolism and better agronomic and nutritional performance [346,347]. Combining biofortification with microbial priming, exogenous application, and CRISPR-based regulation gives plants a multifaceted oxidative stress defense mechanism [246,346]. The challenge is to refine metabolic fluxes and remove growth–defense trade-offs. Customizing antioxidant increases to crop physiology and environment is needed. Advances in systems biology, metabolic modeling, and breeding technologies will support the development of climate-resilient, nutrient-rich crops. These crops can sustain productivity and human health under increasing environmental stress.

https://doi.org/10.3390/antiox15030337

7.4. Translating Antioxidant Modulation into Yield and Crop Quality Under Field-Relevant Stress Regimes

Many studies quantify antioxidants using biochemical proxies such as ROS, MDA, enzyme activities, and electrolyte leakage. However, these markers often fail to predict harvest outcomes. To enable translation, antioxidant interventions must be directly linked to yield components. These include maintaining photosynthetic duration, optimizing assimilate partitioning, improving flowering and fruit set, enhancing grain filling, and increasing harvest index. Evaluation should also focus on quality traits, nutrient density, firmness, shelf life, and seed longevity, not just seedling survival under stress. The evidence summarized above demonstrates this linkage in selected systems. For instance, tomato lines with elevated ascorbate show more stable yields under drought and firmer post-harvest fruit. Antioxidant capacity tracks with heat-stress tolerance in cereals as well. However, robust deployment needs both benefits under stress and penalties under non-stress conditions to be reported. This is due to potential trade-offs between growth and defense. Performance testing across varied environments and combined or sequential stresses is necessary, since ROS regimes and resource constraints often interact non-additively. Future antioxidant engineering and management should make yield and quality primary endpoints. Use redox measurements to explain and predict these outcomes.

8. Challenges and Knowledge Gaps

Despite progress in identifying antioxidant components and stress phenotypes, translating findings from controlled studies into the field remains inconsistent. Plant redox regulation is dynamic and compartmentalized. It also includes feedback loops. Changing one part can trigger compensatory responses or alter ROS signalling. The key is not just boosting "antioxidant capacity" but adjusting the redox balance in a predictable way. This means keeping ROS at helpful levels while avoiding long-term oxidative damage if environments change. Measurement is the first challenge. ROS and redox states are highly localized and brief. For example, H_2O_2 acts for seconds to minutes over micrometre distances. Hydroxyl radicals react where they are formed. Without compartment-specific quantitative data, studies rely on endpoint markers to infer ROS patterns. This hinders comparisons across species, tissues, or stress intensities. While genetically encoded probes, redox reporters, and multi-omics approaches are advancing, high-resolution mapping of ROS and antioxidant turnover in intact tissues is still limited. This is especially true in the field.

Another barrier is network nonlinearity and coupling. Increasing one enzyme's activity can change the ROS pattern, not just decrease stress. For example, boosting superoxide dismutase can raise H_2O_2 levels. If peroxidase and reductant recycling do not keep up, defense signals or senescence may change. There is also a metabolic and developmental cost. Sustained antioxidant increases draw on carbon, nitrogen, and reductants. Excessive ROS removal can suppress essential developmental or stress signals. These factors explain why constant overexpression can improve tolerance in labs but lower growth or yield under non-stress conditions. Regulating antioxidants with stress-responsive, real-time systems is more promising. Field conditions create extra complexity. Stresses rarely occur alone; drought may combine with heat, salinity with poor nutrients, and temperature swings with oxidative bursts. These combinations mean ROS and antioxidant demands are not additive. Single-stress experiments cannot predict these effects. A key gap is understanding how the redox balance is reset during combined or sequential stresses. It is still unclear which intervention points are helpful or harmful under these conditions. Closing this gap needs factorial or sequential stress tests and compartment-specific redox tracking. Yield and quality measures are needed across reproductive development.

https://doi.org/10.3390/antiox15030337

Environmental differences and genotype-by-environment interactions still limit translation. Redox processes differ with genotype, stage, tissue, microclimate, and stress timing. When and where you sample can change outcomes. To address this, use integrated phenotyping pipelines. Methods include non-destructive imaging, remote sensing, and, when possible, in situ redox reporting. Validate in multiple environments and use predictive models. These models should link ROS flows, antioxidants, reductant supply, and growth costs. This lets you assess interventions as whole-system designs rather than just single-gene changes.

9. Future Perspectives

The next frontier in plant redox biology is to convert mechanistic insight into robust, yield-stable field interventions. This will require moving beyond static "more antioxidant" strategies. Future work should focus on dynamic control. Regulatory circuits should activate detoxification and recycling only when ROS exceed physiological thresholds. At the same time, signalling needed for acclimation and development must be preserved. Synthetic biology tools, stress-responsive promoters, tunable regulatory modules, and compartment-targeted designs provide practical ways to implement this control in crops. Genome engineering is also shifting away from coarse overexpression. The field now uses precision tuning, including multiplex CRISPR, base editing, and prime editing. These approaches generate allelic series that adjust enzyme kinetics, localization, and redox-sensitive residues without adding foreign DNA. Target selection must consider system constraints, NADPH supply, glutathione/ascorbate recycling, and organelle cross-talk. This helps avoid stress tolerance that reduces yield under normal conditions. Building predictive measurement and modelling infrastructure is equally important. Integrating transcriptomics, proteomics, metabolomics, and redox flux with digital phenotyping and data-driven models enables clear, testable design rules for shifting the redox rheostat under changing field conditions. Microbiome-based strategies offer a complementary, potentially low-input approach. They can stabilize redox buffering via hormone–ROS crosstalk and metabolite support. However, success depends on colonization stability, soil context, host genotype, and formulation methods. Translation ultimately requires integrated breeding pipelines. These must combine natural diversity, genomic prediction, targeted editing, and multi-environment validation. Only in this way can resilience traits deliver consistent benefits without hidden yield costs.

10. Conclusions

This review advances a unifying interpretation of antioxidant function in abiotic stress. Outcomes depend less on absolute antioxidant abundance than on compartment-resolved network architecture, operating under the constraints of reductant supply and recycling throughput. Framed as a quantitative redox rheostat, the same ROS species support acclimation when restricted to specific concentration–time windows; conversely, they cause injury if fluxes are sustained or misregulated in space. Thus, inconsistent results from antioxidant manipulation arise when interventions ignore kinetics, compartmentation, and system-level coupling. A practical path forward relies on measurement and models. Priority needs include quantitative, in vivo ROS and redox mapping, identifying recycling and NADPH bottlenecks across tissues and stress phases, and mechanistically understanding inter-organelle coupling. Translational strategies are most likely to succeed with dynamic, conditional control, such as stress-responsive regulation, coordinated multi-node tuning, and genotype-by-environment validation, rather than with single-component overexpression. Following these principles, antioxidant biology can shift from reactive damage control to predictive design, enabling crop resilience strategies that work across real agricultural environments.

https://doi.org/10.3390/antiox15030337

Author Contributions: Conceptualization, F.U.H., B.S., and X.L.; methodology, T.L.; software, T.L., and H.; validation, L.C.R.A., formal analysis, resources, P.Z., and X.L.; writing—original draft preparation, F.U.H., and L.C.R.A.; writing—review and editing, B.S., P.Z., H., and X.L.; funding acquisition, P.Z. and X.L. All authors have read and agreed to the published version of the manuscript.

Funding: The Science and Technology Development Program of Jilin Province (YDZJ202401498ZYTS), National Natural Science Foundation of China (32372228), Strategic Priority Research Program of the Chinese Academy of Sciences (XDA28020400).

Institutional Review Board Statement: Not applicable.

Informed Consent Statement: Not applicable.

Data Availability Statement: No new data were created or analyzed in this study. Data sharing does not apply to this article.

Acknowledgments: Language editing and manuscript preparation were assisted by AI-based tools (e.g., Grammarly) for linguistic refinement only. The research design, interpretations, and conclusions are solely those of the authors. All authors approved the final manuscript and take full responsibility for its content.

Conflicts of Interest: The authors declare no conflicts of interest.

References

1. Fathi, A.; Ghadirnezhad Shiade, S.R.; Saleem, A.; Shohani, F.; Fazeli, A.; Riaz, A.; Zulfiqar, U.; Shabaan, M.; Ahmed, I.; Rahimi, M Reactive Oxygen Species (ROS) and Antioxidant Systems in Enhancing Plant Resilience against Abiotic Stress. *Int. J. Agron.* **2025**, *2025*, 8834883. [CrossRef]
2. Manne, H.K.; Kumari, N.; Malik, K.; Avtar, R.; Nain, S.; Duhan, J.; Islam, S.; Rani, B.; Jattan, M.; Vaishnavi, K.; et al. Unraveling Plant Resilience: Breakthroughs in Abiotic Stress Responses and Defense Mechanisms. In *Role of Antioxidants in Abiotic Stress Management*; Elsevier: Amsterdam, The Netherlands, 2025; pp. 1–17. [CrossRef]
3. Hasanuzzaman, M.; Hossain, M.A.; Teixeira da Silva, J.A.; Fujita, M. Plant Responses and Tolerance to Abiotic Oxidative Stress: Antioxidant Defense Is a Key Factor. In *Crop Stress and Its Management: Perspectives and Strategies*; Bandi, V., Shanker, A.K., Shanker, C., Mandapaka, M., Eds.; Springer: Berlin/Heidelberg, Germany, 2012; pp. 261–316.
4. Singh, A.; Kumar, A.; Yadav, S.; Singh, I.K. Reactive Oxygen Species-Mediated Signaling during Abiotic Stress. *Plant Gene* **2019**, *18*, 100173. [CrossRef]
5. Mittler, R. ROS Are Good. *Trends Plant Sci.* **2017**, *22*, 11–19. [CrossRef] [PubMed]
6. Maqsood, M.F.; Shahbaz, M.; Khalid, F.; Rasheed, Y.; Asif, K.; Naz, N.; Zulfiqar, U.; Zulfiqar, F.; Moosa, A.; Alamer, K.H.; et al. Biogenic Nanoparticles Application in Agriculture for ROS Mitigation and Abiotic Stress Tolerance: A Review. *Plant Stress* **2023**, *10*, 100281. [CrossRef]
7. Dumanović, J.; Nepovimova, E.; Natić, M.; Kuča, K.; Jaćević, V. The Significance of Reactive Oxygen Species and Antioxidant Defense System in Plants: A Concise Overview. *Front. Plant Sci.* **2021**, *11*, 552969. [CrossRef]
8. Mittler, R.; Zandalinas, S.I.; Fichman, Y.; Van Breusegem, F. Reactive Oxygen Species Signalling in Plant Stress Responses. *Nat. Rev. Mol. Cell Biol.* **2022**, *23*, 663–679. [CrossRef]
9. Das, K.; Roychoudhury, A. Reactive Oxygen Species (ROS) and Response of Antioxidants as ROS-Scavengers during Environmental Stress in Plants. *Front. Environ. Sci.* **2014**, *2*, 53. [CrossRef]
10. Nadarajah, K.K. ROS Homeostasis in Abiotic Stress Tolerance in Plants. *Int. J. Mol. Sci.* **2020**, *21*, 5208. [CrossRef]
11. Sharma, P.; Jha, A.B.; Dubey, R.S.; Pessarakli, M. Reactive Oxygen Species, Oxidative Damage, and Antioxidative Defense Mechanism in Plants under Stressful Conditions. *J. Bot.* **2012**, *2012*, 217037. [CrossRef]
12. Krieger-Liszkay, A.; Fufezan, C.; Trebst, A. Singlet Oxygen Production in Photosystem II and Related Protection Mechanism. *Photosynth. Res.* **2008**, *98*, 551–564. [CrossRef]
13. Bhattacharjee, S. Reactive Oxygen Species and Oxidative Burst: Roles in Stress, Senescence and Signal Transduction in Plant. *Curr. Sci.* **2005**, *89*, 1113–1121. Available online: https://www.jstor.org/stable/24110963 (accessed on 24 February 2026).
14. Hackbarth, S.; Schlothauer, J.; Preuß, A.; Röder, B. New Insights to Primary Photodynamic Effects–Singlet Oxygen Kinetics in Living Cells. *J. Photochem. Photobiol. B Biol.* **2010**, *98*, 173–179. [CrossRef] [PubMed]
15. Khan, M.; Ali, S.; Al Azzawi, T.N.I.; Saqib, S.; Ullah, F.; Ayaz, A.; Zaman, W. The Key Roles of ROS and RNS as a Signaling Molecule in Plant–Microbe Interactions. *Antioxidants* **2023**, *12*, 268. [CrossRef] [PubMed]

https://doi.org/10.3390/antiox15030337

16. Ying, Z.; Fu, S.; Yang, Y. Signaling and Scavenging: Unraveling the Complex Network of Antioxidant Enzyme Regulation in Plant Cold Adaptation. *Plant Stress* **2025**, *16*, 100833. [CrossRef]
17. Hasanuzzaman, M.; Bhuyan, M.H.M.B.; Zulfiqar, F.; Raza, A.; Mohsin, S.M.; Mahmud, J.A.; Fujita, M.; Fotopoulos, V. Reactive Oxygen Species and Antioxidant Defense in Plants under Abiotic Stress: Revisiting the Crucial Role of a Universal Defense Regulator. *Antioxidants* **2020**, *9*, 681. [CrossRef]
18. Kasote, D.M.; Katyare, S.S.; Hegde, M.V.; Bae, H. Significance of Antioxidant Potential of Plants and Its Relevance to Therapeutic Applications. *Int. J. Biol. Sci.* **2015**, *11*, 982–991. [CrossRef]
19. Rudenko, N.N.; Vetoshkina, D.V.; Marenkova, T.V.; Borisova-Mubarakshina, M.M. Antioxidants of Non-Enzymatic Nature: Their Function in Higher Plant Cells and the Ways of Boosting Their Biosynthesis. *Antioxidants* **2023**, *12*. [CrossRef]
20. Rao, M.J.; Duan, M.; Ikram, M.; Zheng, B. ROS Regulation and Antioxidant Responses in Plants under Air Pollution: Molecular Signaling, Metabolic Adaptation, and Biotechnological Solutions. *Antioxidants* **2025**, *14*, 907. [CrossRef]
21. Ermakov, A.; Bobrovskikh, A.; Zubairova, U.; Konstantinov, D.; Doroshkov, A. Stress-Induced Changes in the Expression of Antioxidant System Genes for Rice (*Oryza sativa* L.) and Bread Wheat (*Triticum aestivum* L.). *PeerJ* **2019**, *7*, e7791. [CrossRef]
22. Shafi, A.; Chauhan, R.; Gill, T.; Swarnkar, M.K.; Sreenivasulu, Y.; Kumar, S.; Kumar, N.; Shankar, R.; Ahuja, P.S.; Singh, A.K. Expression of SOD and APX Genes Positively Regulates Secondary Cell Wall Biosynthesis and Promotes Plant Growth and Yield in *Arabidopsis* under Salt Stress. *Plant Mol. Biol.* **2015**, *87*, 615–631. [CrossRef]
23. Hasanuzzaman, M.; Bhuyan, M.H.M.B.; Parvin, K.; Bhuiyan, T.F.; Anee, T.I.; Nahar, K.; Hossen, M.S.; Zulfiqar, F.; Alam, M.M.; Fujita, M. Regulation of ROS Metabolism in Plants under Environmental Stress: A Review of Recent Experimental Evidence. *Int. J. Mol. Sci.* **2020**, *21*, 8695. [CrossRef]
24. Ulhassan, Z.; Gill, R.A.; Huang, H.; Ali, S.; Mwamba, T.M.; Ali, B.; Huang, Q.; Hamid, Y.; Khan, A.R.; Wang, J.; et al. Selenium Mitigates the Chromium Toxicity in Brassica napus L. by Ameliorating Nutrients Uptake, Amino Acids Metabolism and Antioxidant Defense System. *Plant Physiol. Biochem.* **2019**, *145*, 142–152. [CrossRef] [PubMed]
25. Sandalio, L.M.; Rodríguez-Serrano, M.; Romero-Puertas, M.C.; del Río, L.A. Role of Peroxisomes as a Source of Reactive Oxygen Species (ROS) Signaling Molecules. In *Peroxisomes and Their Key Role in Cellular Signaling and Metabolism;Subcellular Biochemistry*; del Río, L., Ed.; Springer: Dordrecht, The Netherlands, 2013; Volume 69, pp. 231–255. [CrossRef]
26. Sachdev, S.; Ansari, S.A.; Ansari, M.I.; Fujita, M.; Hasanuzzaman, M. Abiotic Stress and Reactive Oxygen Species: Generation, Signaling, and Defense Mechanisms. *Antioxidants* **2021**, *10*, 277. [CrossRef] [PubMed]
27. Pilon, M.; Ravet, K.; Tapken, W. The Biogenesis and Physiological Function of Chloroplast Superoxide Dismutases. *Biochim. Biophys. Acta Bioenerg.* **2011**, *1807*, 989–998. [CrossRef]
28. Pospíšil, P. Production of Reactive Oxygen Species by Photosystem II as a Response to Light and Temperature Stress. *Front. Plant Sci.* **2016**, *7*, 1950. [CrossRef] [PubMed]
29. Bardaweel, S.K.; Gul, M.; Alzweiri, M.; Ishaqat, A.; AlSalamat, H.A.; Bashatwah, R.M. Reactive Oxygen Species: The Dual Role in Physiological and Pathological Conditions of the Human Body. *Eurasian J. Med.* **2018**, *50*, 193–201. [CrossRef]
30. Dayem, A.A.; Hossain, M.K.; Lee, S.B.; Kim, K.; Saha, S.K.; Yang, M.; Choi, H.Y.; Cho, G. The Role of Reactive Oxygen Species (ROS) in the Biological Activities of Metallic Nanoparticles. *Int. J. Mol. Sci.* **2017**, *18*, 120. [CrossRef]
31. Ugalde, J.M.; Fuchs, P.; Nietzel, T.; Cutolo, E.A.; Homagk, M.; Vothknecht, U.C.; Holuigue, L.; Schwarzländer, M.; Müller-Schüssele, S.J.; Meyer, A.J. Chloroplast-Derived Photo-Oxidative Stress Causes Changes in H_2O_2 and EGSH in Other Subcellular Compartments. *Plant Physiol.* **2021**, *186*, 125–141. [CrossRef]
32. Mailloux, R.J.; Jin, X.; Willmore, W.G. Redox Regulation of Mitochondrial Function with Emphasis on Cysteine Oxidation Reactions. *Redox Biol.* **2014**, *2*, 123–139. [CrossRef]
33. Wang, Y.; Berkowitz, O.; Selinski, J.; Xu, Y.; Hartmann, A.; Whelan, J. Stress Responsive Mitochondrial Proteins in *Arabidopsis* thaliana. *Free Radic. Biol. Med.* **2018**, *122*, 28–39. [CrossRef]
34. Cortassa, S.; O'Rourke, B.; Aon, M.A. Redox-Optimized ROS Balance and the Relationship between Mitochondrial Respiration and ROS. *Biochim. Biophys. Acta Bioenerg.* **2014**, *1837*, 287–295. [CrossRef] [PubMed]
35. Huang, S.; Van Aken, O.; Schwarzländer, M.; Belt, K.; Millar, A.H. The Roles of Mitochondrial Reactive Oxygen Species in Cellular Signaling and Stress Response in Plants. *Plant Physiol.* **2016**, *171*, 1551–1559. [CrossRef] [PubMed]
36. He, H.; Van Breusegem, F.; Mhamdi, A. Redox-Dependent Control of Nuclear Transcription in Plants. *J. Exp. Bot.* **2018**, *69*, 3359–3372. [CrossRef] [PubMed]
37. Corpas, F.J.; González-Gordo, S.; Palma, J.M. Plant Peroxisomes: A Factory of Reactive Species. *Front. Plant Sci.* **2020**, *11*, 853. [CrossRef] [PubMed]
38. Del Río, L.A. ROS Generation in Peroxisomes and Its Role in Cell Signaling. *Plant Cell Physiol.* **2016**, *57*, 1364–1376. [CrossRef]
39. Kataya, A.R.A.; Reumann, S. *Arabidopsis* Glutathione Reductase 1 Is Dually Targeted to Peroxisomes and the Cytosol. *Plant Signal. Behav.* **2010**, *5*, 171–173. [CrossRef]
40. Schieber, M.; Chandel, N.S. ROS Function in Redox Signaling and Oxidative Stress. *Curr. Biol.* **2014**, *24*, R453–R462. [CrossRef]

https://doi.org/10.3390/antiox15030337

41. Majeed, J.A.; Bibi, S.; Mahmood, A.; Ali, L.; Safdar, M.E.; Seleiman, M.F.; Abidin, Z.U.; Alhammad, B.A.; Asghar, M.A. Optimizing Tomato (*Solanum lycopersicum*) Yield under Salt Stress: The Physiological and Biochemical Effects of Foliar Thiourea Application. *Plants* **2024**, *13*, 3318. [CrossRef]
42. Fichman, Y.; Zandalinas, S.I.; Mittler, R. Jasmonic Acid and Salicylic Acid Modulate Systemic Reactive Oxygen Species Signaling during Stress Responses. *Plant Physiol.* **2022**, *191*, 862–867. [CrossRef]
43. Li, J.; Xu, H.; Liu, C.; Zhang, W.; Lu, T. Ethylene Inhibits Root Elongation during Alkaline Stress through AUXIN1 and Associated Changes in Auxin Accumulation. *Plant Physiol.* **2015**, *168*, 1777–1791. [CrossRef]
44. Sultan, M.; Batool, I.; Bano, H.; Zafar, Z.U.; Ayyaz, A. Multifaceted Roles in Plant Abiotic Stress Responses and Tolerance Mechanisms. In *Role of Antioxidants in Abiotic Stress Management*; Elsevier: Amsterdam, The Netherlands, 2024; pp. 21–33. [CrossRef]
45. Dvořák, P.; Krasylenko, Y.; Ovečka, M.; Basheer, J.; Zapletalová, V.; Šamaj, J.; Takáč, T. In Vivo Light-Sheet Microscopy Resolves Localisation Patterns of FSD1, a Superoxide Dismutase with Function in Root Development and Osmoprotection. *Plant Cell Environ.* **2021**, *44*, 68–87. [CrossRef]
46. Ali, M.A.; Fahad, S.; Haider, I.; Ahmed, N.; Ahmad, S.; Hussain, S.; Arshad, M. Oxidative Stress and Antioxidant Defense in Plants Exposed to Metal/Metalloid Toxicity. In *Antioxidants and Antioxidant Enzymes in Higher Plants*; Springer: Cham, Switzerland, 2018; pp. 353–370. [CrossRef]
47. Gheisary, E.; Kazemeini, S.A.; Samadi, M.; Alinia, M.; Mastinu, A. Unraveling the protective roles of silicon and thiourea on salinity stress tolerance of *Triticum aestivum* L. cultivars: From antioxidant defense to photosynthetic recovery. *Soil Sci. Plant Nutr.* **2025**, *25*, 10683–10697. [CrossRef]
48. Mushtaq, Z.; Faizan, S.; Gulzar, B.; Hakeem, K.R. Inoculation of Rhizobium Alleviates Salinity Stress through Modulation of Growth Characteristics, Physiological and Biochemical Attributes, Stomatal Activities and Antioxidant Defence in *Cicer arietinum* L. *J. Plant Growth Regul.* **2021**, *40*, 2148–2163. [CrossRef]
49. Leung, D.W.M. Studies of Catalase in Plants under Abiotic Stress. In *Antioxidants and Antioxidant Enzymes in Higher Plants*; Springer: Cham, Switzerland, 2018; pp. 25–44. [CrossRef]
50. Tyagi, S.; Singh, K.; Upadhyay, S.K. Molecular Characterization Revealed the Role of Catalases under Abiotic and Arsenic Stress in Bread Wheat (*Triticum aestivum* L.). *J. Hazard. Mater.* **2021**, *403*, 123585. [CrossRef] [PubMed]
51. Riseh, R.S.; Fathi, F.; Vatankhah, M.; Kennedy, J.F. Catalase-Associated Immune Responses in Plant–Microbe Interactions: A Review. *Int. J. Biol. Macromol.* **2024**, *280*, 135859. [CrossRef]
52. Mishra, N.; Jiang, C.; Chen, L.; Paul, A.; Chatterjee, A.; Shen, G. Achieving Abiotic Stress Tolerance in Plants through Antioxidative Defense Mechanisms. *Front. Plant Sci.* **2023**, *14*, 1110622. [CrossRef]
53. Sarker, U.; Oba, S. Catalase, Superoxide Dismutase and Ascorbate–Glutathione Cycle Enzymes Confer Drought Tolerance of Amaranthus tricolor. *Sci. Rep.* **2018**, *8*, 1640. [CrossRef]
54. Gill, S.S.; Singh, L.P.; Gill, R.; Tuteja, N. Generation and Scavenging of Reactive Oxygen Species in Plants under Stress. In *Plant Adaptation Strategies in Changing Environment*; Springer: Singapore, 2018; pp. 49–70. [CrossRef]
55. Singh, Y.J.; Grewal, S.K.; Gill, R.K. Role of Glutathione in Methylglyoxal Detoxification Pathway during Yellow Mosaic Virus (YMV) Infection in Black Gram (*Vigna mungo* (L.) Hepper). *Physiol. Mol. Plant Pathol.* **2020**, *111*, 101513. [CrossRef]
56. Pandey, S.; Fartyal, D.; Agarwal, A.; Shukla, T.; James, D.; Kaul, T.; Negi, Y.K.; Arora, S.; Reddy, M.K. Abiotic Stress Tolerance in Plants: Myriad Roles of Ascorbate Peroxidase. *Front. Plant Sci.* **2017**, *8*, 581. [CrossRef]
57. Li, S. Novel Insight into Functions of Ascorbate Peroxidase in Higher Plants: More than a Simple Antioxidant Enzyme. *Redox Biol.* **2023**, *64*, 102789. [CrossRef]
58. Kuo, E.Y.; Cai, M.; Lee, T. Ascorbate Peroxidase 4 Plays a Role in the Tolerance of Chlamydomonas reinhardtii to Photo-Oxidative Stress. *Sci. Rep.* **2020**, *10*, 70247. [CrossRef]
59. El-Shazoly, R.M.; Othman, A.A.; Zaheer, M.S.; Al-Hossainy, A.F.; Abdel-Wahab, D.A. Zinc Oxide Seed Priming Enhances Drought Tolerance in Wheat Seedlings by Improving Antioxidant Activity and Osmoprotection. *Sci. Rep.* **2025**, *15*, 86824. [CrossRef]
60. Zhou, Y.; Li, J.; Wang, J.; Yang, W.; Yang, Y. Identification and Characterization of the Glutathione Peroxidase (GPX) Gene Family in Watermelon and Its Expression under Various Abiotic Stresses. *Agronomy* **2018**, *8*, 206. [CrossRef]
61. Zhai, Z.; Zhao, L.; Yin, J.; Chen, M.; Wang, Y.; Li, C.; Xu, S.; Ma, Z. Two Wheat Glutathione Peroxidase Genes Whose Products Are Located in Chloroplasts Improve Salt and H_2O_2 Tolerances in *Arabidopsis*. *PLoS ONE* **2013**, *8*, e73989. [CrossRef] [PubMed]
62. Paiva, A.L.S.; Passaia, G.; Jardim-Messeder, D.; Nogueira, F.C.; Domont, G.B.; Margis-Pinheiro, M. The Mitochondrial Isoform Glutathione Peroxidase 3 (OsGPX3) Is Involved in ABA Responses in Rice Plants. *J. Proteom.* **2021**, *232*, 104029. [CrossRef]
63. Navrot, N.; Collin, V.; Gualberto, J.; Gelhaye, E.; Hirasawa, M.; Rey, P.; Knaff, D.B.; Issakidis, E.; Jacquot, J.; Rouhier, N. Plant Glutathione Peroxidases Are Functional Peroxiredoxins Distributed in Several Subcellular Compartments and Regulated during Biotic and Abiotic Stresses. *Plant Physiol.* **2006**, *142*, 1364–1379. [CrossRef]
64. Miao, Y.; Lv, D.; Wang, P.; Wang, X.; Chen, J.; Miao, C.; Song, C. An *Arabidopsis* Glutathione Peroxidase Functions as Both a Redox Transducer and a Scavenger in Abscisic Acid and Drought Stress Responses. *Plant Cell* **2006**, *18*, 2749–2766. [CrossRef]

65. Kuźniak, E.; Skłodowska, M. Compartment-Specific Role of the Ascorbate–Glutathione Cycle in the Response of Tomato Leaf Cells to Botrytis cinerea Infection. *J. Exp. Bot.* **2005**, *56*, 921–933. [CrossRef]
66. Herbette, S.; Labrouhe, D.T.D.; Drevet, J.R.; Roeckel-Drevet, P. Transgenic Tomatoes Showing Higher Glutathione Peroxidase Antioxidant Activity Are More Resistant to an Abiotic Stress but More Susceptible to Biotic Stresses. *Plant Sci.* **2011**, *180*, 548–553. [CrossRef]
67. Hasanuzzaman, M.; Nahar, K.; Anee, T.I.; Fujita, M. Glutathione in Plants: Biosynthesis and Physiological Role in Environmental Stress Tolerance. *Physiol. Mol. Biol. Plants* **2017**, *23*, 249–272. [CrossRef]
68. Smith, I.K.; Vierheller, T.L.; Thorne, C.A. Properties and Functions of Glutathione Reductase in Plants. *Physiol. Plant.* **1989**, *77*, 449–456. [CrossRef]
69. Harshavardhan, V.T.; Wu, T.M.; Hong, C.Y. Glutathione Reductase and Abiotic Stress Tolerance in Plants. In *Glutathione in Plant Growth, Development, and Stress Tolerance*; Springer: Cham, Switzerland, 2017; pp. 203–225. [CrossRef]
70. Ding, S.; Jiang, R.; Lu, Q.; Wen, X.; Lu, C. Glutathione Reductase 2 Maintains the Function of Photosystem II in *Arabidopsis* under Excess Light. *Biochim. Biophys. Acta Bioenerg.* **2016**, *1857*, 665–677. [CrossRef] [PubMed]
71. Marty, L.; Bausewein, D.; Müller, C.; Khan Bangash, S.A.; Moseler, A.; Schwarzländer, M.; Müller-Schüssele, S.J.; Zechmann, B.; Riondet, C.; Balk, J.; et al. *Arabidopsis* Glutathione Reductase 2 Is Indispensable in Plastids, while Mitochondrial Glutathione Is Safeguarded by Additional Reduction and Transport Systems. *New Phytol.* **2019**, *224*, 1569–1584. [CrossRef] [PubMed]
72. Rai, G.K.; Kumar, P.; Choudhary, S.M.; Singh, H.; Adab, K.; Kosser, R.; Magotra, I.; Kumar, R.R.; Singh, M.; Sharma, R.; et al. Antioxidant Potential of Glutathione and Crosstalk with Phytohormones in Enhancing Abiotic Stress Tolerance in Crop Plants. *Plants* **2023**, *12*, 1133. [CrossRef]
73. Gul, N.; Ahmad, P.; Wani, T.A.; Tyagi, A.; Aslam, S. Glutathione Improves Low-Temperature Stress Tolerance in Pusa Sheetal Cultivar of *Solanum lycopersicum*. *Sci. Rep.* **2022**, *12*, 16440. [CrossRef]
74. Mhamdi, A.; Hager, J.; Chaouch, S.; Queval, G.; Han, Y.; Taconnat, L.; Saindrenan, P.; Gouia, H.; Renou, J.-P.; Noctor, G. *Arabidopsis* Glutathione Reductase1 Plays a Crucial Role in Leaf Responses to Intracellular Hydrogen Peroxide and in Ensuring Appropriate Gene Expression through Both Salicylic Acid and Jasmonic Acid Signaling Pathways. *Plant Physiol.* **2010**, *153*, 1144–1160. [CrossRef]
75. Delgado-Requerey, V.; Cejudo, F.J.; González, C. The Functional Relationship between NADPH Thioredoxin Reductase C, 2-Cys Peroxiredoxins, and m-Type Thioredoxins in the Regulation of Calvin–Benson Cycle and Malate-Valve Enzymes in *Arabidopsis*. *Antioxidants* **2023**, *12*, 1041. [CrossRef]
76. De Brasi-Velasco, S.; Aroca, A.; Romero, L.C.; Gotor, C.; Sevilla, F.; Jiménez, A. New Role for Thioredoxins in Plants: Implication of TRXo1 in Protein Depersulfidation. *Redox Biol.* **2025**, *82*, 103627. [CrossRef]
77. Sevilla, F.; Camejo, D.; Calderón, A.; Lázaro, J.J.; Jiménez, A. The Thioredoxin/Peroxiredoxin/Sulfiredoxin System: Current Overview on Its Redox Function in Plants and Regulation by Reactive Oxygen and Nitrogen Species. *J. Exp. Bot.* **2015**, *66*, 2945–2955. [CrossRef]
78. Martí, M.C.; Jiménez, A.; Sevilla, F. Thioredoxin Network in Plant Mitochondria: Cysteine S-Posttranslational Modifications and Stress Conditions. *Front. Plant Sci.* **2020**, *11*, 571288. [CrossRef]
79. Ojeda, V.; Pérez-Ruiz, J.M.; Cejudo, F.J. 2-Cys Peroxiredoxins Participate in the Oxidation of Chloroplast Enzymes in the Dark. *Mol. Plant* **2018**, *11*, 1377–1388. [CrossRef] [PubMed]
80. Jiménez-López, J.; Casatejada, A.; Gálvez-Ramírez, A.; Pérez-Ruiz, J.M.; Cejudo, F.J. Functional Relationship of Atypical Thioredoxins with NADPH-Thioredoxin Reductase C and 2-Cys Peroxiredoxins in *Arabidopsis* Chloroplasts. *J. Exp. Bot.* **2024**, *76*, eraf283. [CrossRef] [PubMed]
81. Rouhier, N.; Jacquot, J.-P. Plant Peroxiredoxins: Alternative Hydroperoxide Scavenging Enzymes. *Photosynth. Res.* **2002**, *74*, 259–268. [CrossRef] [PubMed]
82. Bilska, K.; Wojciechowska, N.; Alipour, S.; Kalemba, E.M. Ascorbic Acid—The Little-Known Antioxidant in Woody Plants. *Antioxidants* **2019**, *8*, 645. [CrossRef]
83. Xiao, M.; Li, Z.; Zhu, L.; Wang, J.; Zhang, B.; Zheng, F.; Zhao, B.; Zhang, H.; Wang, Y.; Zhang, Z. The Multiple Roles of Ascorbate in the Abiotic Stress Response of Plants: Antioxidant, Cofactor, and Regulator. *Front. Plant Sci.* **2021**, *12*, 598173. [CrossRef]
84. Venkatesh, J.; Park, S.W. Role of L-Ascorbate in Alleviating Abiotic Stresses in Crop Plants. *Bot. Stud.* **2014**, *55*, 38. [CrossRef]
85. Zechmann, B. Compartment-Specific Importance of Ascorbate during Environmental Stress in Plants. *Antioxid. Redox Signal.* **2018**, *29*, 1488–1501. [CrossRef]
86. Yeh, H.; Lin, T.; Chen, C.; Cheng, T.; Chang, H.; Lee, T. Monodehydroascorbate Reductase Plays a Role in the Tolerance of *Chlamydomonas reinhardtii* to Photooxidative Stress. *Plant Cell Physiol.* **2019**, *60*, 2167–2179. [CrossRef]
87. Yan, Y.; Pan, C.; Du, Y.; Li, D.; Liu, W. Exogenous Salicylic Acid Regulates Reactive Oxygen Species Metabolism and Ascorbate–Glutathione Cycle in *Nitraria tangutorum* Bobr. under Salinity Stress. *Physiol. Mol. Biol. Plants* **2018**, *24*, 577–589. [CrossRef]
88. Tóth, D.; Kovács, L.; Neupert, J.; Tóth, S.Z. Ascorbate Deficiency Does Not Limit Nonphotochemical Quenching in *Chlamydomonas reinhardtii*. *Plant Physiol.* **2020**, *182*, 597–611. [CrossRef]

https://doi.org/10.3390/antiox15030337

89. Toledo, M.E.A.; Ueda, Y.; Imahori, Y.; Ayaki, M. L-Ascorbic Acid Metabolism in Spinach (*Spinacia oleracea* L.) during Postharvest Storage in Light and Dark. *Postharvest Biol. Technol.* **2003**, *28*, 47–57. [CrossRef]
90. Golan, T.; Niyogi, K.K. Ascorbate-Deficient Mutants of *Arabidopsis* Grow in High Light Despite Chronic Photooxidative Stress. *Plant Physiol.* **2004**, *134*, 1163–1172. [CrossRef] [PubMed]
91. Wang, Y.; Cao, S.; Sui, X.; Wang, J.; Geng, Y.; Gao, F.; Zhou, Y. Genome-Wide Characterization, Evolution, and Expression Analysis of the Ascorbate Peroxidase and Glutathione Peroxidase Gene Families in Response to Cold and Osmotic Stress in Ammopiptanthus nanus. *J. Plant Growth Regul.* **2023**, *42*, 502–522. [CrossRef]
92. Aslam, S.; Gul, N.; Mir, M.A.; Asgher, M.; Abulfaraj, A.A.; Qari, S. Role of Jasmonates, Calcium, and Glutathione in Plants to Combat Abiotic Stresses through Precise Signaling Cascade. *Front. Plant Sci.* **2021**, *12*, 668029. [CrossRef] [PubMed]
93. Santos, C.; Santos, A.S.; Lopes, S.; Pirovani, C.P.; Micheli, F. The Family of Glutathione Peroxidase Proteins and Their Role against Biotic Stress in Plants: A Systematic Review. *Front. Plant Sci.* **2025**, *16*, 1425880. [CrossRef]
94. Bohle, F.; Klaus, A.; Ingelfinger, J.; Tegethof, H.; Safari, N.; Schwarzländer, M.; Hochholdinger, F.; Hahn, M.; Meyer, A.J.; Acosta, I.F.; et al. Contrasting Cytosolic Glutathione Redox Dynamics under Abiotic and Biotic Stress in Barley as Revealed by the Biosensor Grx1–roGFP2. *J. Exp. Bot.* **2024**, *75*, 2299–2312. [CrossRef]
95. Noctor, G.; Mhamdi, A.; Chaouch, S.; Han, Y.; Neukermans, J.; Marquez-Garcia, B.; Queval, G.; Foyer, C.H. Glutathione in Plants: An Integrated Overview. *Plant Cell Environ.* **2012**, *35*, 454–484. [CrossRef]
96. Meyer, A.J. The Integration of Glutathione Homeostasis and Redox Signaling. *J. Plant Physiol.* **2008**, *165*, 1390–1403. [CrossRef]
97. Mittra, P.K.; Rahman, M.A.; Roy, S.K.; Kwon, S.; Yun, S.H.; Kun, C.; Zhou, M.; Katsube-Tanaka, T.; Shiraiwa, T.; Woo, S. Deciphering Proteomic Mechanisms Explaining the Role of Glutathione as an Aid in Improving Plant Fitness and Tolerance against Cadmium Toxicity in *Brassica napus* L. *J. Hazard. Mater.* **2024**, *471*, 134262. [CrossRef]
98. Koramutla, M.K.; Negi, M.; Ayele, B.T. Roles of Glutathione in Mediating Abscisic Acid Signaling and Its Regulation of Seed Dormancy and Drought Tolerance. *Genes* **2021**, *12*, 1620. [CrossRef]
99. Noctor, G.; Cohen, M.; Trémulot, L.; Van Breusegem, F.; Mhamdi, A. Glutathione: A Key Modulator of Plant Defence and Metabolism through Multiple Mechanisms. *J. Exp. Bot.* **2024**, *75*, 4549–4572. [CrossRef] [PubMed]
100. Faizan, M.; Alam, P.; Rajput, V.D.; Kaur, K.; Faraz, A.; Minkina, T.; Ahmed, S.M.; Rajpal, V.R.; Hayat, S. Potential Role of Tocopherol in Protecting Crop Plants against Abiotic Stresses. *Physiol. Mol. Biol. Plants* **2023**, *29*, 1563–1578. [CrossRef] [PubMed]
101. Hasanuzzaman, M.; Nahar, K.; Fujita, M. Role of Tocopherol (Vitamin E) in Plants: Abiotic Stress Tolerance and Beyond. In *Emerging Technologies and Management of Crop Stress Tolerance*; Elsevier: Amsterdam, The Netherlands, 2014; pp. 267–289. [CrossRef]
102. Szarka, A.; Tomasskovics, B.; Bánhegyi, G. The Ascorbate–Glutathione–α-Tocopherol Triad in Abiotic Stress Response. *Int. J. Mol. Sci.* **2012**, *13*, 4458–4483. [CrossRef] [PubMed]
103. Surówka, E.; Potocka, I.; Dziurka, M.; Wróbel-Marek, J.; Kurczyńska, E.; Żur, I.; Maksymowicz, A.; Gajewska, E.; Miszalski, Z. Tocopherols Mutual Balance Is a Key Player for Maintaining *Arabidopsis thaliana* Growth under Salt Stress. *Plant Physiol. Biochem.* **2020**, *156*, 369–383. [CrossRef]
104. Orabi, S.A.; Abdelhamid, M.T. Protective Role of α-Tocopherol on Two *Vicia faba* Cultivars against Seawater-Induced Lipid Peroxidation by Enhancing Capacity of Anti-oxidative System. *J. Saudi Soc. Agric. Sci.* **2016**, *15*, 145–154. [CrossRef]
105. Naqve, M.; Wang, X.; Shahbaz, M.; Mahmood, A.; Bibi, S.; Fiaz, S. Alpha Tocopherol-Induced Modulations in the Morphophysiological Attributes of Okra under Saline Conditions. *Front. Plant Sci.* **2021**, *12*, 800251. [CrossRef]
106. Kobayashi, N.; DellaPenna, D. Tocopherol Metabolism, Oxidation and Recycling under High Light Stress in *Arabidopsis*. *Plant J.* **2008**, *55*, 607–618. [CrossRef]
107. Ali, Q.; Javed, M.T.; Haider, M.Z.; Habib, N.; Rizwan, M.; Perveen, R.; Ali, S.; Alyemeni, M.N.; Al-Huqail, A.A.; Ahmad, P. α-Tocopherol Foliar Spray and Translocation Mediates Growth, Photosynthetic Pigments, Nutrient Uptake, and Oxidative Defense in Maize (*Zea mays* L.) under Drought Stress. *Agronomy* **2020**, *10*, 1235. [CrossRef]
108. Diepenbrock, C.H.; Kandianis, C.B.; Lipka, A.E.; Vaillancourt, B.; Wallace, J.G.; Cepela, J.; Mesberg, A.; Bradbury, P.J.; Ilut, D.C.; Hamilton, J.; et al. Novel Loci Underlie Natural Variation in Vitamin E Levels in Maize Grain. *Plant Cell* **2017**, *29*, 2374–2392. [CrossRef]
109. Hodges, A.K.; Polutchko, S.K.; Adams, W.W. Zeaxanthin and Other Carotenoids: Roles in Abiotic Stress Defense with Implications for Biotic Defense. *Plants* **2024**, *14*, 2703. [CrossRef]
110. Swapnil, P.; Meena, M.; Singh, S.K.; Dhuldhaj, U.P.; Marwal, A. Vital Roles of Carotenoids in Plants and Humans to Deteriorate Stress with Its Structure, Biosynthesis, Metabolic Engineering and Functional Aspects. *Curr. Plant Biol.* **2021**, *26*, 100203. [CrossRef]
111. Stra, A.; Almarwaey, L.O.; Alagoz, Y.; Moreno, J.C.; Al-Babili, S. Carotenoid Metabolism: New Insights and Synthetic Approaches. *Front. Plant Sci.* **2023**, *13*, 1072061. [CrossRef] [PubMed]
112. Zulfiqar, S.; Sharif, S.; Saeed, M.; Tahir, A. Role of Carotenoids in Photosynthesis. In *Carotenoids: Structure and Function in the Human Body*; Zia-Ul-Haq, M., Dewanjee, S., Riaz, M., Eds.; Springer: Cham, Switzerland, 2021; pp. 71–88. [CrossRef]

https://doi.org/10.3390/antiox15030337

13. Uarrota, V.G.; Stefen, D.L.V.; Leolato, L.S.; Gindri, D.M.; Nerling, D. Revisiting Carotenoids and Their Role in Plant Stress Responses: From Biosynthesis to Plant Signaling Mechanisms during Stress. In *Antioxidants and Antioxidant Enzymes in Higher Plants*; Gupta, D., Palma, J., Corpas, F., Eds.; Springer: Cham, Switzerland, 2018; pp. 185–205. [CrossRef]
14. Schaub, P.; Rodriguez-Franco, M.; Cazzonelli, C.I.; Álvarez, D.; Wüst, F.; Welsch, R. Establishment of an *Arabidopsis* Callus System to Study the Interrelations of Biosynthesis, Degradation and Accumulation of Carotenoids. *PLoS ONE* **2018**, *13*, e0192158. [CrossRef] [PubMed]
15. Park, S.; Kim, S.H.; Park, S.; Lee, H.; Lee, J.S.; Park, W.S.; Ahn, M.; Kim, Y.; Jeong, J.C.; Lee, H.; et al. Enhanced Accumulation of Carotenoids in Sweetpotato Plants Overexpressing *IbOr-Ins* Gene in Purple-Fleshed Sweetpotato Cultivar. *Plant Physiol. Biochem.* **2015**, *86*, 82–90. [CrossRef]
16. Moreno, J.C.; Al-Babili, S. Are Carotenoids the True Colors of Crop Improvement? *New Phytol.* **2023**, *237*, 1946–1950. [CrossRef] [PubMed]
17. Zhu, C.; Naqvi, S.; Breitenbach, J.; Sandmann, G.; Christou, P.; Capell, T. Combinatorial Genetic Transformation Generates a Library of Metabolic Phenotypes for the Carotenoid Pathway in Maize. *Proc. Natl. Acad. Sci. USA* **2008**, *105*, 18232–18237. [CrossRef]
18. Li, F.; Vallabhaneni, R.; Yu, J.; Rocheford, T.; Wurtzel, E.T. The Maize Phytoene Synthase Gene Family: Overlapping Roles for Carotenogenesis in Endosperm, Photomorphogenesis, and Thermal Stress Tolerance. *Plant Physiol.* **2008**, *147*, 1334–1346. [CrossRef]
19. Shumbe, L.; Bott, R.; Havaux, M. Dihydroactinidiolide, a High Light-Induced β-Carotene Derivative That Can Regulate Gene Expression and Photoacclimation in *Arabidopsis*. *Mol. Plant* **2014**, *7*, 1248–1251. [CrossRef]
20. Liu, L.; Shao, Z.; Zhang, M.; Wang, Q. Regulation of Carotenoid Metabolism in Tomato. *Mol. Plant* **2015**, *8*, 28–39. [CrossRef]
21. Kumar, K.; Debnath, P.; Singh, S.; Kumar, N. An Overview of Plant Phenolics and Their Involvement in Abiotic Stress Tolerance. *Stresses* **2023**, *3*, 570–585. [CrossRef]
22. Panahirad, S.; Morshedloo, M.R.; Ali, S.; Hano, C.; Kulak, M. Secondary Metabolites and Their Potential Roles in Plant Tolerance against Abiotic and Biotic Stress. *Plant Stress* **2023**, *10*, 100292. [CrossRef]
23. Patil, J.R.; Mhatre, K.J.; Yadav, K.; Yadav, L.S.; Srivastava, S.; Nikalje, G.C. Flavonoids in Plant–Environment Interactions and Stress Responses. *Discov. Plants* **2024**, *1*, 68. [CrossRef]
24. Shri, R.; Sood, P.; Kaur, G. Exploring Different Environmental Impacts: How Abiotic Factors Shape Flavonoid Synthesis in Plants. *Biocatal. Agric. Biotechnol.* **2025**, *68*, 103726. [CrossRef]
25. Sytar, O.; Hajihashemi, S. Specific Secondary Metabolites of Medicinal Plants and Their Role in Stress Adaptation. In *Plant Stress Biology*; Wiley-VCH: Weinheim, Germany, 2023; pp. 425–479. [CrossRef]
26. Tattini, M.; Galardi, C.; Pinelli, P.; Massai, R.; Remorini, D.; Agati, G. Differential Accumulation of Flavonoids and Hydroxycinnamates in Leaves of *Ligustrum vulgare* under Excess Light and Drought Stress. *New Phytol.* **2004**, *163*, 547–561. [CrossRef]
27. Li, B.; Fan, R.; Sun, G.; Sun, T.; Fan, Y.; Bai, S.; Guo, S.; Huang, S.; Liu, J.; Zhang, H.; et al. Flavonoids Improve Drought Tolerance of Maize Seedlings by Regulating the Homeostasis of Reactive Oxygen Species. *Plant Soil* **2021**, *461*, 389–405. [CrossRef]
28. Foyer, C.H.; Noctor, G. Redox Homeostasis and Antioxidant Signaling: A Metabolic Interface between Stress Perception and Physiological Responses. *Plant Cell* **2005**, *17*, 1866–1875. [CrossRef]
29. Asada, K. Production and Scavenging of Reactive Oxygen Species in Chloroplasts and Their Functions. *Plant Physiol.* **2006**, *141*, 391–396. [CrossRef]
30. Casazza, A.P.; Tarantino, D.; Soave, C. Preparation and Functional Characterization of Thylakoids from *Arabidopsis thaliana*. *Photosynth. Res.* **2001**, *68*, 175–180. [CrossRef]
31. Kukri, A.; Rossi, M.; Freschi, L.; Czékus, Z.; Poór, P. Phytohormonal Regulation of α-Tocopherol in Plants. *Plant Stress* **2025**, *18*, 101032. [CrossRef]
32. Kocsy, G.; Kobrehel, K.; Szalai, G.; Duviau, M.; Buzás, Z.; Galiba, G. Abiotic Stress-Induced Changes in Glutathione and Thioredoxin h Levels in Maize. *Environ. Exp. Bot.* **2004**, *52*, 101–112. [CrossRef]
133. Ugalde, J.M. The Echo from Outside: Ascorbate Peroxidase 1 Modulates Cytosolic Effector-Triggered Reactive Oxygen Species. *Plant Physiol.* **2023**, *192*, 23. [CrossRef] [PubMed]
134. Castro, B.; Citterico, M.; Kimura, S.; Stevens, D.M.; Wrzaczek, M.; Coaker, G. Stress-Induced ROS Compartmentalization, Perception, and Signaling. *Nat. Plants* **2021**, *7*, 403–412. [CrossRef] [PubMed]
135. Takahama, U. Oxidation of Vacuolar and Apoplastic Phenolic Substrates by Peroxidase: Physiological Significance of the Oxidation Reactions. *Phytochem. Rev.* **2004**, *3*, 207–219. [CrossRef]
136. Postiglione, A.E.; Muday, G.K. Abscisic Acid Increases Hydrogen Peroxide in Mitochondria to Facilitate Stomatal Closure. *Plant Physiol.* **2022**, *192*, 469. [CrossRef]
137. Zandi, P.; Schnug, E. Reactive Oxygen Species, Antioxidant Responses and Implications from a Microbial Modulation Perspective. *Biology* **2022**, *11*, 155. [CrossRef]

https://doi.org/10.3390/antiox15030337

138. Noctor, G.; Foyer, C.H. Intracellular Redox Compartmentation and ROS-Related Communication in Regulation and Signaling. *Plant Physiol.* **2016**, *171*, 1581–1592. [CrossRef]
139. Liu, J.X.; Feng, K.; Duan, A.Q.; Li, H.; Yang, Q.Q.; Xu, Z.S.; Xiong, A.S. Isolation, Purification and Characterization of an Ascorbate Peroxidase from Celery and Overexpression of the *AgAPX1* Gene Enhanced Ascorbate Content and Drought Tolerance in *Arabidopsis*. *BMC Plant Biol.* **2019**, *19*, 488. [CrossRef]
140. Rezayian, M.; Ebrahimzadeh, H.; Niknam, V. Nitric Oxide Stimulates Antioxidant System and Osmotic Adjustment in Soybean under Drought Stress. *J. Soil Sci. Plant Nutr.* **2020**, *20*, 1122–1132. [CrossRef]
141. Hasanuzzaman, M.; Nahar, K.; Anee, T.; Khan, M.; Fujita, M. Silicon-Mediated Regulation of Antioxidant Defense and Glyoxalase Systems Confers Drought Stress Tolerance in *Brassica napus* L. *S. Afr. J. Bot.* **2018**, *115*, 50–57. [CrossRef]
142. Bhuiyan, T.F.; Ahamed, K.U.; Nahar, K.; Al Mahmud, J.; Bhuyan, M.B.; Anee, T.I.; Fujita, M.; Hasanuzzaman, M. Mitigation of PEG-Induced Drought Stress in Rapeseed (*Brassica rapa* L.) by Exogenous Application of Osmolytes. *Biocatal. Agric. Biotechnol.* **2019**, *20*, 101197. [CrossRef]
143. Akbarifard, Z.; Sharifi, M.; Samari, E. Antioxidant and metabolic responses of Linum album cell culture to polyethylene glycol-induced drought stress. *BMC Plant Biol.* **2026**. [CrossRef]
144. Bogoutdinova, L.R.; Lazareva, E.M.; Chaban, I.A.; Kononenko, N.V.; Dilovarova, T.; Khaliluev, M.R.; Kurenina, L.V.; Gulevich, A.A.; Smirnova, E.A.; Baranova, E.N. Salt Stress-Induced Structural Changes Are Mitigated in Transgenic Tomato Plants Over-Expressing Superoxide Dismutase. *Biology* **2020**, *9*, 297. [CrossRef]
145. Kim, Y.H.; Kim, H.S.; Park, S.C.; Ji, C.Y.; Yang, J.W.; Lee, H.U.; Kwak, S.S. Overexpression of *swpa4* Peroxidase Enhances Tolerance to Hydrogen Peroxide and High Salinity-Mediated Oxidative Stress in Transgenic Sweetpotato. *Plant Biotechnol. Rep.* **2020**, *14*, 301–307. [CrossRef]
146. Riyazuddin, R.; Bela, K.; Horváth, E.; Rigó, G.; Gallé, Á.; Szabados, L.; Fehér, A.; Csiszár, J. Overexpression of the *Arabidopsis* Glutathione Peroxidase-Like 5 Gene (*AtGPXL5*) Resulted in Altered Plant Development and Redox Status. *Environ. Exp. Bot.* **2019**, *167*, 103849. [CrossRef]
147. Saxena, S.C.; Salvi, P.; Kamble, N.U.; Joshi, P.K.; Majee, M.; Arora, S. Ectopic Overexpression of Cytosolic Ascorbate Peroxidase Gene (*Apx1*) Improves Salinity Stress Tolerance in *Brassica juncea* by Strengthening Antioxidative Defense Mechanism. *Acta Physiol. Plant.* **2020**, *42*, 45. [CrossRef]
148. Qi, Q.; Yanyan, D.; Yuanlin, L.; Kunzhi, L.; Huini, X.; Xudong, S. Overexpression of *SlMDHAR* in Transgenic Tobacco Increased Salt Stress Tolerance Involving S-Nitrosylation Regulation. *Plant Sci.* **2020**, *299*, 110609. [CrossRef] [PubMed]
149. Li, J.; Li, H.; Yang, N.; Jiang, S.; Ma, C.; Li, H. Overexpression of a Monodehydroascorbate Reductase Gene from Sugar Beet M14 Increased Salt Stress Tolerance. *Sugar Tech.* **2021**, *23*, 45–56. [CrossRef]
150. Guo, X.; He, Y.; Zhang, Y.; Wang, Y.; Huang, S.; Liu, Y.; Li, W. Kiwifruit (*Actinidia chinensis* 'Hongyang') Cytosolic Ascorbate Peroxidases (*AcAPX1* and *AcAPX2*) Enhance Salinity Tolerance in *Arabidopsis thaliana*. *J. Integr. Agric.* **2022**, *21*, 1058–1070. [CrossRef]
151. Alzahrani, S.M.; Alaraidh, I.A.; Migdadi, H.; Alghamdi, S.; Khan, M.A.; Ahmad, P. Physiological, Biochemical, and Antioxidant Properties of Two Genotypes of *Vicia faba* Grown under Salinity Stress. *Pak. J. Bot.* **2019**, *51*, 786–798. [CrossRef]
152. Zeeshan, M.; Lu, M.; Sehar, S.; Holford, P.; Wu, F. Comparison of Biochemical, Anatomical, Morphological, and Physiological Responses to Salinity Stress in Wheat and Barley Genotypes Differing in Salinity Tolerance. *Agronomy* **2019**, *10*, 127. [CrossRef]
153. Singh, M.; Singh, V.P.; Prasad, S.M. Nitrogen Alleviates Salinity Toxicity in *Solanum lycopersicum* Seedlings by Regulating ROS Homeostasis. *Plant Physiol. Biochem.* **2019**, *141*, 466–476. [CrossRef]
154. Taïbi, K.; Taïbi, F.; Ait Abderrahim, L.; Ennajah, A.; Belkhodja, M.; Mulet, J.M. Effect of Salt Stress on Growth, Chlorophyll Content, Lipid Peroxidation and Antioxidant Defence Systems in *Phaseolus vulgaris* L. *S. Afr. J. Bot.* **2016**, *105*, 306–312. [CrossRef]
155. Chiang, C.M.; Chien, H.L.; Chen, L.F.O.; Hsiung, T.C.; Chiang, M.C.; Chen, S.P.; Lin, K.H. Overexpression of the Genes Coding Ascorbate Peroxidase from *Brassica campestris* Enhances Heat Tolerance in Transgenic *Arabidopsis thaliana*. *Biol. Plant.* **2015**, *59*, 305–315. [CrossRef]
156. Nahar, K.; Hasanuzzaman, M.; Fujita, M. Physiological Roles of Glutathione in Conferring Abiotic Stress Tolerance to Plants. In *Abiotic Stress Response in Plants*; Gill, S.S., Tuteja, N., Eds.; Wiley-VCH: Weinheim, Germany, 2016; pp. 151–179.
157. Jahan, M.S.; Wang, Y.; Shu, S.; Zhong, M.; Chen, Z.; Wu, J.; Guo, S. Exogenous Salicylic Acid Increases the Heat Tolerance in Tomato (*Solanum lycopersicum* L.) by Enhancing Photosynthesis Efficiency and Improving Antioxidant Defense System through Scavenging of Reactive Oxygen Species. *Sci. Hortic.* **2019**, *247*, 421–429. [CrossRef]
158. Mei, W.; Chen, W.; Wang, Y.; Liu, Z.; Dong, Y.; Zhang, G.; Xiao, Y. Exogenous Kinetin Modulates ROS Homeostasis to Affect Heat Tolerance in Rice Seedlings. *Int. J. Mol. Sci.* **2023**, *24*, 6252. [CrossRef]
159. Che, Y.; Zhang, N.; Zhu, X.; Li, S.; Wang, S.; Si, H. Enhanced Tolerance of the Transgenic Potato Plants Overexpressing Cu/Zn Superoxide Dismutase to Low Temperature. *Sci. Hortic.* **2020**, *261*, 108949. [CrossRef]
160. He, Z.; Zhao, T.; Yin, Z.; Liu, J.; Cheng, Y.; Xu, J. The Phytochrome-Interacting Transcription Factor CsPIF8 Contributes to Cold Tolerance in Citrus by Regulating Superoxide Dismutase Expression. *Plant Sci.* **2020**, *298*, 110584. [CrossRef]

https://doi.org/10.3390/antiox15030337

161. Lin, K.H.; Sei, S.C.; Su, Y.H.; Chiang, C.M. Overexpression of the *Arabidopsis* and Winter Squash Superoxide Dismutase Genes Enhances Chilling Tolerance via ABA-Sensitive Transcriptional Regulation in Transgenic *Arabidopsis*. *Plant Signal. Behav.* **2019**, *14*, 1685728. [CrossRef] [PubMed]
162. Geng, J.; Wei, T.; Wang, Y.; Huang, X.; Liu, J. Overexpression of *PtrbHLH*, a Basic Helix-Loop-Helix Transcription Factor from *Poncirus trifoliata*, Confers Enhanced Cold Tolerance in Pummelo (*Citrus grandis*) by Modulation of H_2O_2 Level via Regulating a *CAT* Gene. *Tree Physiol.* **2019**, *39*, 2045–2054. [CrossRef] [PubMed]
163. Matsumura, T.; Tabayashi, N.; Kamagata, Y.; Souma, C.; Saruyama, H. Wheat Catalase Expressed in Transgenic Rice Can Improve Tolerance against Low Temperature Stress. *Physiol. Plant.* **2002**, *116*, 317–327. [CrossRef]
164. Guan, Z.; Chai, T.; Zhang, Y.; Xu, J.; Wei, W. Enhancement of Cd Tolerance in Transgenic Tobacco Plants Overexpressing a Cd-Induced Catalase cDNA. *Chemosphere* **2009**, *76*, 623–630. [CrossRef]
165. Yin, L.; Tanaka, K.; Wang, S.; Zhang, M.; Deng, X.; Zhang, S. High Level of Reduced Glutathione Contributes to Detoxification of Lipid Peroxide-Derived Reactive Carbonyl Species in Transgenic *Arabidopsis* Overexpressing Glutathione Reductase under Aluminum Stress. *Physiol. Plant.* **2017**, *161*, 211–223. [CrossRef]
166. Bharwana, S.A.; Ali, S.; Farooq, M.A.; Iqbal, N.; Abbas, F.; Ahmad, M.S.A. Alleviation of Lead Toxicity by Silicon Is Related to Elevated Photosynthesis, Antioxidant Enzymes Suppressed Lead Uptake and Oxidative Stress in Cotton. *J. Bioremed. Biodegrad.* **2013**, *4*, 187. [CrossRef]
167. Manne, H.; Kumari, N.; Yashveer, S.; Nain, S.; Duhan, J.; Avtar, R.; Jattan, M.; Rani, B.; Alsahli, A.A.; Ali, S. Mitigation of Lead Toxicity in *Brassica juncea* L. by Sulphur Application–Via Various Biochemical and Transcriptomic Strategies. *J. King Saud Univ. Sci.* **2024**, *36*, 103175. [CrossRef]
168. Nain, S.; Kumari, N.; Manne, H.K.; Avtar, R.; Yashveer, S.; Malik, K.; Tokas, J.; Rati, S.; Al-Anazi, K.M.; Ahmad, P. Impact of Calcium on Ascorbate–Glutathione Pool and Gene Expression under Cadmium Stress in Indian Mustard (*Brassica juncea* L.). *S. Afr. J. Bot.* **2025**, *180*, 12–20. [CrossRef]
169. Hasanuzzaman, M.; Nahar, K.; Rahman, A.; Al Mahmud, J.; Alharby, H.F.; Fujita, M. Exogenous glutathione attenuates lead-induced oxidative stress in wheat by improving antioxidant defense and physiological mechanisms. *J. Plant Interact.* **2018**, *13*, 203–212. [CrossRef]
170. Akram, T.; Ahmad, S.; Ahmad, A.; Sehrish, A.K.; Riaz, M.; Qadir, R.; Ali, S. Involvement against Drought Stress or Water Deficiency in Plants. In *Role of Antioxidants in Abiotic Stress Management*; Elsevier: Amsterdam, The Netherlands, 2025; pp. 47–67. [CrossRef]
171. Mutava, R.N.; Prince, S.J.K.; Syed, N.H.; Song, L.; Valliyodan, B.; Chen, W.; Nguyen, H.T. Understanding Abiotic Stress Tolerance Mechanisms in Soybean: A Comparative Evaluation of Soybean Response to Drought and Flooding Stress. *Plant Physiol. Biochem.* **2015**, *86*, 109–120. [CrossRef] [PubMed]
172. Anjum, S.A.; Ashraf, U.; Tanveer, M.; Khan, I.; Hussain, S.; Shahzad, B.; Zohaib, A.; Abbas, F.; Saleem, M.F.; Ali, I.; et al. Drought Induced Changes in Growth, Osmolyte Accumulation and Antioxidant Metabolism of Three Maize Hybrids. *Front. Plant Sci.* **2017**, *8*, 69. [CrossRef] [PubMed]
173. Basit, F.; Alyafei, M.; Hayat, F.; Al-Zayadneh, W.; El-Keblawy, A.; Sulieman, S.; Sheteiwy, M.S. Deciphering the Role of Glycine Betaine in Enhancing Plant Performance and Defense Mechanisms against Environmental Stresses. *Front. Plant Sci.* **2025**, *16*, 1582332. [CrossRef]
174. Wang, X.; Zhang, W.; Wang, M. The LRR-RLK Protein HSL3 Regulates Stomatal Closure and the Drought Stress Response by Modulating Hydrogen Peroxide Homeostasis. *Front. Plant Sci.* **2020**, *11*, 548034. [CrossRef]
175. Kang, S.W.; Rhee, S.G.; Chang, S.; Jeong, W.; Choi, M.H. 2-Cys Peroxiredoxin Function in Intracellular Signal Transduction: Therapeutic Implications. *Trends Mol. Med.* **2005**, *11*, 571–578. [CrossRef]
176. Zimmer, D.; Swart, C.; Graf, A.; Arrivault, S.; Tillich, M.; Proost, S.; Nikoloski, Z.; Stitt, M.; Bock, R.; Mühlhaus, T.; et al. Topology of the Redox Network during Induction of Photosynthesis as Revealed by Time-Resolved Proteomics in Tobacco. *Sci. Adv.* **2021**, *7*, eabi8307. [CrossRef]
177. Zhao, S.; Riaz, M. Contribution of Antioxidants (Enzymatic and Non-Enzymatic) in Salinity Stress Tolerance in Plants. In *Role of Antioxidants in Abiotic Stress Management*; Elsevier: Amsterdam, The Netherlands, 2025; pp. 79–97. [CrossRef]
178. Hassan, A.; Fasiha Amjad, S.; Hamzah Saleem, M.; Yasmin, H.; Imran, M.; Riaz, M.; Ali, Q.; Ahmad Joyia, F.; Ahmed, S.; Ali, S.; et al. Foliar Application of Ascorbic Acid Enhances Salinity Stress Tolerance in Barley (*Hordeum vulgare* L.) through Modulation of Morpho-Physio-Biochemical Attributes, Ion Uptake, Osmo-Protectants and Stress Response Gene Expression. *Saudi J. Biol. Sci.* **2021**, *28*, 4276–4290. [CrossRef]
179. Mishra, M.; Afzal, S.; Yadav, R.; Singh, N.K.; Zarbakhsh, S. Salinity Stress Amelioration through Selenium and Zinc Oxide Nanoparticles in Rice. *Sci. Rep.* **2025**, *15*, 12106. [CrossRef]
180. Bartoli, C.G.; Buet, A.; Gergoff Grozeff, G.; Galatro, A.; Simontacchi, M. Ascorbate–Glutathione Cycle and Abiotic Stress Tolerance in Plants. In *Ascorbic Acid in Plant Growth, Development and Stress Tolerance*; Springer: Cham, Switzerland, 2017; pp. 177–197. [CrossRef]

https://doi.org/10.3390/antiox15030337

181. Sun, T.; Rao, S.; Zhou, X.; Li, L. Plant carotenoids: Recent advances and future perspectives. *Mol. Hortic.* **2022**, *2*, 3. [CrossRef] [PubMed]
182. Gul, Z.; Tang, Z.; Arif, M.; Ye, Z. An Insight into Abiotic Stress and Influx Tolerance Mechanisms in Plants to Cope in Saline Environments. *Biology* **2022**, *11*, 597. [CrossRef] [PubMed]
183. Pirzada, A.M.; Anwar, T.; Qureshi, W.A.; Qureshi, H.; Siddiqi, E.H.; Zaman, W.; Soufan, W. Salinity Stress Mitigation in Wheat through Synergistic Application of Ascorbic Acid, Nanoparticles and *Salvadora oleoides* Extract. *Sci. Rep.* **2024**, *14*, 76194. [CrossRef] [PubMed]
184. Ahmad, S.; Sehrish, A.K.; Ahmad, A.; Akram, T.; Guo, H. Addressing Heat Stress in Plants. In *Role of Antioxidants in Abiotic Stress Management*; Elsevier: Amsterdam, The Netherlands, 2025; pp. 133–150. [CrossRef]
185. Meena, N.L.; Kumari, A.; Maheshwari, C.; Bhardwaj, R.; Dhaka, A.S.; Hasan, M. Antioxidant Defense Mechanism and High Temperature Stress Tolerance in Plants. In *Molecular Dynamics of Plant Stress and Its Management*; Springer: Singapore, 2024; pp. 193–210.
186. Sato, H.; Mizoi, J.; Shinozaki, K.; Yamaguchi-Shinozaki, K. Complex Plant Responses to Drought and Heat Stress under Climate Change. *Plant J.* **2024**, *117*, 1873–1892. [CrossRef]
187. Mahajan, S.; Thakur, P.; Das, S.; Sharma, R.P.; Manuja, S.; Jha, P.K.; Saini, A.; Sahoo, C.; Fayezizadeh, M.R. Impression of Contemporary Heat Stress Complexities in Agricultural Crops: A Review. *Plant Growth Regul.* **2025**, *105*, 1805–1823. [CrossRef]
188. Arshad, M.S.; Farooq, M.; Asch, F.; Krishna, J.S.; Prasad, P.V.V.; Siddique, K.H.M. Thermal Stress Impacts Reproductive Development and Grain Yield in Rice. *Plant Physiol. Biochem.* **2017**, *115*, 57–72. [CrossRef]
189. Eastin, J.D.; Castleberry, R.M.; Gerik, T.J.; Hultquist, J.H.; Mahalakshmi, V.; Ogunlela, V.B.; Rice, J.R. Physiological Aspects of High Temperature and Water Stress. In *Crop Reactions to Water and Temperature Stresses in Humid, Temperate Climates*; CRC Press Boca Raton, FL, USA, 2019; pp. 91–112.
190. Raja, V.; Qadir, S.U.; Alyemeni, M.N.; Ahmad, P. Impact of Drought and Heat Stress Individually and in Combination on Physio-Biochemical Parameters, Antioxidant Responses, and Gene Expression in *Solanum lycopersicum*. *3 Biotech.* **2020**, *10*, 208 [CrossRef]
191. Broin, M.; Cuiné, S.; Eymery, F.; Rey, P. The Plastidic 2-Cysteine Peroxiredoxin Is a Target for a Thioredoxin Involved in the Protection of the Photosynthetic Apparatus against Oxidative Damage. *Plant Cell* **2002**, *14*, 1417–1432. [CrossRef]
192. Maryam, A.; Khan, R.I.; Abbas, M.; Hussain, K.; Muhammad, S.; Sabir, M.A.; Ahmed, T.; Khalid, M.F. Beyond the Membrane: The Pivotal Role of Lipids in Plant Abiotic Stress Adaptation. *Plant Growth Regul.* **2025**, *105*, 1869–1887. [CrossRef]
193. Ghosh, U.K.; Hossain, M.S.; Islam, M.N.; Khan, M.A.R. Role of Tocopherol in Conferring Abiotic Stress Tolerance in Plants. In *Antioxidant Defense in Plants*; Aftab, T., Hakeem, K.R., Eds.; Springer: Singapore, 2022; pp. 193–210. [CrossRef]
194. Driedonks, N.; Xu, J.; Peters, J.L.; Park, S.; Rieu, I. Multi-Level Interactions Between Heat Shock Factors, Heat Shock Proteins, and the Redox System Regulate Acclimation to Heat. *Front. Plant Sci.* **2015**, *6*, 999. [CrossRef]
195. Mondal, S.; Karmakar, S.; Panda, D.; Pramanik, K.; Bose, B.; Singhal, R.K. Crucial Plant Processes under Heat Stress and Tolerance through Heat Shock Proteins. *Plant Stress* **2023**, *10*, 100227. [CrossRef]
196. Zhai, J.; Teng, Z.; Zheng, C.; Lv, J.; Song, T.; Zhang, J.; Ye, N. Abscisic Acid Mediates the Trade-Off between Growth and Abiotic Stress Defense. *Crop Environ.* **2025**, *5*, 100110. [CrossRef]
197. Molavian, H.; Madani Tonekaboni, A.; Kohandel, M.; Sivaloganathan, S. The Synergetic Coupling among the Cellular Antioxidants Glutathione Peroxidase/Peroxiredoxin and Other Antioxidants and Its Effect on the Concentration of H_2O_2. *Sci. Rep.* **2015**, *5*, 13620. [CrossRef]
198. Chai, J.; Wen, Z.; Chen, L.; Pu, Q.; Luo, T.; Wu, X.; Ma, Z.; Luo, Z.; Luo, J.; Wang, J. Multi-Omics Analysis of Chronic Heat Stress-Induced Biological Effects, Liver Injury, and Heat Tolerance Mechanisms via Oxidative and Anti-Inflammatory Pathways in Early-Pregnancy Sows. *Antioxidants* **2025**, *14*, 623. [CrossRef]
199. Djanaguiraman, M.; Narayanan, S.; Erdayani, E.; Prasad, P.V.V. Effects of High Temperature Stress during Anthesis and Grain Filling Periods on Photosynthesis, Lipids and Grain Yield in Wheat. *BMC Plant Biol.* **2020**, *20*, 24. [CrossRef]
200. Nizamani, M.M.; Khaskheli, M.A.; Tarafder, E.; Das, D.; Awais, M.; Leghari, U.A.; Jatoi, G.H.; Laghari, A.H.; Khaskheli, R.A. Efficacy in Cold Stress Tolerance in Plants. In *Role of Antioxidants in Abiotic Stress Management*; Elsevier: Amsterdam, The Netherlands, 2025; pp. 99–131. [CrossRef]
201. Adhikari, L.; Baral, R.; Paudel, D.; Min, D.; Makaju, S.O.; Poudel, H.P.; Acharya, J.P.; Missaoui, A.M. Cold Stress in Plants: Strategies to Improve Cold Tolerance in Forage Species. *Plant Stress* **2022**, *4*, 100081. [CrossRef]
202. Faivre, L.; Kinscher, N.; Kuhlmann, A.B.; Xu, X.; Kaufmann, K.; Schubert, D. Cold Stress Induces Rapid Gene-Specific Changes in the Levels of H3K4me3 and H3K27me3 in *Arabidopsis thaliana*. *Front. Plant Sci.* **2024**, *15*, 1390144. [CrossRef]
203. Peng, Y.; Lin, W.; Cai, W.; Arora, R. Overexpression of a *Panax ginseng* Tonoplast Aquaporin Alters Salt Tolerance, Drought Tolerance and Cold Acclimation Ability in Transgenic *Arabidopsis* Plants. *Planta* **2007**, *226*, 729–740. [CrossRef]
204. Rahman, A.; Yadav, N.S.; Byeon, B.; Ilnytskyy, Y.; Kovalchuk, I. Genomic and Epigenomic Changes in the Progeny of Cold-Stressed *Arabidopsis thaliana* Plants. *Int. J. Mol. Sci.* **2024**, *25*, 2795. [CrossRef]

https://doi.org/10.3390/antiox15030337

205. Zulfiqar, U.; Haider, F.U.; Ahmad, M.; Hussain, S.; Maqsood, M.F.; Ishfaq, M.; Shahzad, B.; Waqas, M.M.; Ali, B.; Tayyab, M.N.; et al. Chromium Toxicity, Speciation, and Remediation Strategies in Soil–Plant Interface: A Critical Review. *Front. Plant Sci.* **2023**, *13*, 1081624. [CrossRef] [PubMed]

206. Gusain, S.; Joshi, S.; Joshi, R. Sensing, Signalling, and Regulatory Mechanism of Cold-Stress Tolerance in Plants. *Plant Physiol. Biochem.* **2023**, *197*, 107646. [CrossRef]

207. Zhang, S.; Wang, N.; Gao, Z.; Gao, J.; Wang, X.; Xie, H.; Wang, C.; Zhang, S. Reductive Stress: The Key Pathway in Metabolic Disorders Induced by Overnutrition. *J. Adv. Res.* **2025**, *77*, 569–584. [CrossRef]

208. Altaf, M.A. Protective Roles against Heavy Metals or Metalloid Stress in Plants. In *Role of Antioxidants in Abiotic Stress Management*; Elsevier: Amsterdam, The Netherlands, 2025; pp. 35–45. [CrossRef]

209. Yuan, J.J.; Zhao, Y.N.; Yu, S.H.; Sun, Y.; Li, G.X.; Yan, J.Y.; Xu, J.M.; Ding, W.N.; Benhamed, M.; Qiu, R.L.; et al. The *Arabidopsis* Receptor-Like Kinase WAKL4 Limits Cadmium Uptake via Phosphorylation and Degradation of NRAMP1 Transporter. *Nat. Commun.* **2024**, *15*, 8734. [CrossRef]

210. Hu, W.; Yang, Y.; Ali, S.; Zhao, W.; Wang, S.; Zhou, Z. Optimizing Cotton Growth and Phytoremediation under Cadmium Stress through the Integration of Nanomaterials and Plant Growth-Promoting Bacteria. *J. Hazard. Mater.* **2025**, *495*, 139090. [CrossRef] [PubMed]

211. Khan, R.; Sarwar, M.J.; Shabaan, M.; Asghar, H.N.; Zulfiqar, U.; Iftikhar, I.; Aijaz, N.; Haider, F.U.; Chaudhary, T.; Soufran, W. Exploring the Synergistic Effects of Indole Acetic Acid (IAA) and Compost in the Phytostabilization of Nickel (Ni) in Cauliflower Rhizosphere. *BMC Plant Biol.* **2024**, *24*, 275. [CrossRef] [PubMed]

212. Ashraf, H.; Ghouri, F.; Zhong, M.; Cheema, S.A.; Haider, F.U.; Sun, L.; Ali, S.; Alshehri, M.A.; Fu, X.; Shahid, M.Q. *Oryza glumaepatula* and Calcium Oxide Nanoparticles Enhanced Cr Stress Tolerance by Maintaining Antioxidant Defense, Chlorophyll and Gene Expression Levels in Rice. *J. Environ. Manag.* **2024**, *368*, 122239. [CrossRef] [PubMed]

213. Ashraf, H.; Ghouri, F.; Ali, S.; Bukhari, S.A.H.; Haider, F.U.; Zhong, M.; Xia, W.; Fu, X.; Shahid, M.Q. The Protective Roles of *Oryza glumaepatula* and Phytohormone in Enhancing Rice Tolerance to Cadmium Stress by Regulating Gene Expression, Morphological, Physiological, and Antioxidant Defense System. *Environ. Pollut.* **2025**, *364*, 125311. [CrossRef]

214. Adrees, M.; Ali, S.; Rizwan, M.; Zia-ur-Rehman, M.; Ibrahim, M.; Abbas, F.; Farid, M.; Qayyum, M.F.; Irshad, M.K. Mechanisms of Silicon-Mediated Alleviation of Heavy Metal Toxicity in Plants: A Review. *Ecotoxicol. Environ. Saf.* **2015**, *119*, 186–197. [CrossRef]

215. Haider, F.U.; Wang, X.; Farooq, M.; Hussain, S.; Cheema, S.A.; Ain, N.U.; Virk, A.L.; Ejaz, M.; Janyshova, U.; Liqun, C. Biochar Application for the Remediation of Trace Metals in Contaminated Soils: Implications for Stress Tolerance and Crop Production. *Ecotoxicol. Environ. Saf.* **2022**, *230*, 113165. [CrossRef]

216. Jozefczak, M.; Remans, T.; Vangronsveld, J.; Cuypers, A. Glutathione Is a Key Player in Metal-Induced Oxidative Stress Defenses. *Int. J. Mol. Sci.* **2012**, *13*, 3145–3175. [CrossRef]

217. Krówczyńska, N.; Pietrowska-Borek, M. From Signal Perception to Adaptive Responses: A Comprehensive Review of Plant Mechanisms under Cadmium, Lead, and Aluminum Stress. *Environ. Exp. Bot.* **2025**, *238*, 106240. [CrossRef]

218. Llerena, J.P.P.; Coasaca, R.L.; Rodriguez, H.O.L.; Llerena, S.Á.P.; Valencia, Y.D.; Mazzafera, P. Metallothionein Production Is a Common Tolerance Mechanism in Four Species Growing in Polluted Cu Mining Areas in Peru. *Ecotoxicol. Environ. Saf.* **2021**, *212*, 112009. [CrossRef]

219. Wang, Q.; Zhou, Q.; Huang, L.; Xu, S.; Fu, Y.; Hou, D.; Feng, Y.; Yang, X. Cadmium Phytoextraction through *Brassica juncea* L. under Different Consortia of Plant Growth-Promoting Bacteria from Different Ecological Niches. *Ecotoxicol. Environ. Saf.* **2022**, *237*, 113541. [CrossRef]

220. Mustafa, A.; Zulfiqar, U.; Mumtaz, M.Z.; Radziemska, M.; Haider, F.U.; Holatko, J.; Hammershmiedt, T.; Naveed, M.; Ali, H.; Kintl, A.; et al. Nickel (Ni) Phytotoxicity and Detoxification Mechanisms: A Review. *Chemosphere* **2023**, *328*, 138574. [CrossRef] [PubMed]

221. Delaix, C.L.; Tomiozzo, A.; Weber, G.; Lima-Melo, Y.; Vargas, A.N.D.; Margis-Pinheiro, M.; Trenz, T.S. Interplay among Hormones, Antioxidants, and Redox Signaling in Abiotic Stress Responses. *Environ. Exp. Bot.* **2024**, *229*, 106081. [CrossRef]

222. Haider, F.U.; Liqun, C.; Coulter, J.A.; Cheema, S.A.; Wu, J.; Zhang, R.; Wenjun, M.; Farooq, M. Cadmium Toxicity in Plants: Impacts and Remediation Strategies. *J. Hazard. Mater.* **2021**, *211*, 111887. [CrossRef]

223. Hoque, M.N.; Tahjib-Ul-Arif, M.; Hannan, A.; Sultana, N.; Akhter, S.; Hasanuzzaman, M.; Akter, F.; Hossain, M.S.; Sayed, M.A.; Hasan, M.T.; et al. Melatonin Modulates Plant Tolerance to Heavy Metal Stress: Morphological Responses to Molecular Mechanisms. *Int. J. Mol. Sci.* **2021**, *22*, 11445. [CrossRef] [PubMed]

224. Shang, Q.; Wei, Z.; Fang, L.; Huang, D.; Pan, X. Integrated Physiological, Biochemical, Transcriptomic, and Metabolomic Analysis Reveals the Mechanism of Silicon in Alleviating Cadmium Stress in *Juglans sigillata*. *Ind. Crops Prod.* **2025**, *231*, 121212. [CrossRef]

225. Fedoreyeva, L.I. ROS as Signaling Molecules to Initiate the Process of Plant Acclimatization to Abiotic Stress. *Int. J. Mol. Sci.* **2024**, *25*, 11820. [CrossRef] [PubMed]

226. Krishnamurthy, H.K.; Rajavelu, I.; Pereira, M.; Jayaraman, V.; Krishna, K.; Wang, T.; Bei, K.; Rajasekaran, J.J. Inside the genome: Understanding genetic influences on oxidative stress. *Front. Genet.* **2024**, *15*, 1397352. [CrossRef] [PubMed]

https://doi.org/10.3390/antiox15030337

227. Wang, S.; Yao, Y.; Wang, J.; Ruan, B.; Yu, Y. Advancing Stress-Resilient Rice: Mechanisms, Genes, and Breeding Strategie *Agriculture* **2024**, *15*, 721. [CrossRef]
228. Zhang, H.; Zhang, J.; Wei, P.; Zhang, B.; Gou, F.; Feng, Z.; Mao, Y.; Yang, L.; Zhang, H.; Xu, N.; et al. The CRISPR/Cas9 system produces specific and homozygous targeted gene editing in rice in one generation. *Plant Biotechnol. J.* **2014**, *12*, 797–807. [CrossRef
229. Liao, S.; Qin, X.; Luo, L.; Han, Y.; Wang, X.; Usman, B.; Nawaz, G.; Zhao, N.; Liu, Y.; Li, R. CRISPR/Cas9-induced mutagenesi of *semirolled leaf1,2* confers curled leaf phenotype and drought tolerance by influencing protein expression patterns and RO scavenging in rice (*Oryza sativa* L.). *Agronomy* **2019**, *9*, 728. [CrossRef]
230. Lu, G.; Wang, C.; Wang, G.; Mao, G.; Habben, J.E.; Chen, G.; Liu, M.; Shi, Y.; Wang, W.; Wang, X.; et al. Knockouts of drought sensitive genes improve rice grain yield under both drought and well-watered field conditions. *Adv. Crop Sci. Technol.* **2020**, 444.
231. Ogata, T.; Ishizaki, T.; Fujita, M.; Fujita, Y. CRISPR/Cas9-targeted mutagenesis of *OsERA1* confers enhanced responses to abscisi acid and drought stress and increased primary root growth under nonstressed conditions in rice. *PLoS ONE* **2020**, *15*, e024337 [CrossRef]
232. Usman, B.; Nawaz, G.; Zhao, N.; Liao, S.; Liu, Y.; Li, R. Precise editing of the *OsPYL9* gene by RNA-guided Cas9 nuclease confers enhanced drought tolerance and grain yield in rice (*Oryza sativa* L.) by regulating circadian rhythm and abiotic stress responsive proteins. *Int. J. Mol. Sci.* **2019**, *21*, 7854. [CrossRef] [PubMed]
233. Shim, J.S.; Oh, N.; Chung, P.J.; Kim, Y.S.; Choi, Y.D.; Kim, J. Overexpression of *OsNAC14* improves drought tolerance in rice. *Front Plant Sci.* **2018**, *9*, 310. [CrossRef] [PubMed]
234. Qin, Q.; Wang, Y.; Huang, L.; Du, F.; Zhao, X.; Li, Z.; Wang, W.; Fu, B. A U-box E3 ubiquitin ligase *OsPUB67* is positively involved in drought tolerance in rice. *Plant Mol. Biol.* **2020**, *102*, 89–107. [CrossRef]
235. Zhang, H.; Gou, F.; Zhang, J.; Liu, W.; Li, Q.; Mao, Y.; Botella, J.R.; Zhu, K. TALEN-mediated targeted mutagenesis produces a large variety of heritable mutations in rice. *Plant Biotechnol. J.* **2016**, *14*, 186–194. [CrossRef]
236. Cantos, C.; Francisco, P.; Trijatmiko, K.R.; Slamet-Loedin, I.; Chadha-Mohanty, P.K. Identification of "safe harbor" loci in *indica* rice genome by harnessing the property of zinc-finger nucleases to induce DNA damage and repair. *Front. Plant Sci.* **2014**, *5*, 302 [CrossRef]
237. Kim, D.; Alptekin, B.; Budak, H. CRISPR/Cas9 genome editing in wheat. *Funct. Integr. Genomics* **2018**, *18*, 31–41. [CrossRef]
238. Shi, J.; Gao, H.; Wang, H.; Lafitte, H.R.; Archibald, R.L.; Yang, M.; Hakimi, S.M.; Mo, H.; Habben, J.E. *ARGOS8* variants generated by CRISPR/Cas9 improve maize grain yield under field drought stress conditions. *Plant Biotechnol. J.* **2017**, *15*, 207–216. [CrossRef]
239. Chen, S.; Zhang, N.; Zhou, G.; Hussain, S.; Ahmed, S.; Tian, H.; Wang, S. Knockout of the entire family of *AITR* genes in *Arabidopsis* leads to enhanced drought and salinity tolerance without fitness costs. *BMC Plant Biol.* **2021**, *21*, 137. [CrossRef]
240. Alam, M.S.; Kong, J.; Tao, R.; Ahmed, T.; Alamin, M.; Alotaibi, S.S.; Abdelsalam, N.R.; Xu, J. CRISPR/Cas9 mediated knockout of the *OsbHLH024* transcription factor improves salt stress resistance in rice (*Oryza sativa* L.). *Plants* **2021**, *11*, 1184. [CrossRef] [PubMed]
241. Zhang, A.; Liu, Y.; Wang, F.; Li, T.; Chen, Z.; Kong, D.; Bi, J.; Zhang, F.; Luo, X.; Wang, J.; et al. Enhanced rice salinity tolerance via CRISPR/Cas9-targeted mutagenesis of the *OsRR22* gene. *Mol. Breed.* **2019**, *39*, 47. [CrossRef] [PubMed]
242. Wang, W.; Lin, T.; Kieber, J.; Tsai, Y. Response regulators 9 and 10 negatively regulate salinity tolerance in rice. *Plant Cell Physiol* **2019**, *60*, 2549–2563. [CrossRef] [PubMed]
243. Bo, W.; Zhaohui, Z.; Huanhuan, Z.; Xia, W.; Binglin, L.; Lijia, Y.; Xiangyan, H.; Deshui, Y.; Xuelian, Z.; Chunguo, W.; et al. Targeted mutagenesis of NAC transcription factor gene, *OsNAC041*, leading to salt sensitivity in rice. *Rice Sci.* **2019**, *26*, 98–108. [CrossRef]
244. Lan, T.; Zheng, Y.; Su, Z.; Yu, S.; Song, H.; Zheng, X.; Lin, G.; Wu, W. *OsSPL10*, a SBP-box gene, plays a dual role in salt tolerance and trichome formation in rice (*Oryza sativa* L.). *G3 Genes Genomes Genet.* **2019**, *9*, 4107–4114. [CrossRef]
245. Roy, M.; Wu, R. Overexpression of *S-adenosylmethionine decarboxylase* gene in rice increases polyamine level and enhances sodium chloride-stress tolerance. *Plant Sci.* **2002**, *163*, 987–992. [CrossRef]
246. Vlčko, T.; Ohnoutková, L. Allelic variants of CRISPR/Cas9-induced mutation in an inositol trisphosphate 5/6 kinase gene manifest different phenotypes in barley. *Plants* **2020**, *9*, 195. [CrossRef]
247. Schilling, R.K.; Marschner, P.; Shavrukov, Y.; Berger, B.; Tester, M.; Roy, S.J.; Plett, D.C. Expression of the *Arabidopsis* vacuolar H^+-pyrophosphatase gene (*AVP1*) improves the shoot biomass of transgenic barley and increases grain yield in a saline field *Plant Biotechnol. J.* **2014**, *12*, 378–386. [CrossRef]
248. Wang, T.; Xun, H.; Wang, W.; Ding, X.; Tian, H.; Hussain, S.; Dong, Q.; Li, Y.; Cheng, Y.; Wang, C.; et al. Mutation of *GmAITR* genes by CRISPR/Cas9 genome editing results in enhanced salinity stress tolerance in soybean. *Front. Plant Sci.* **2021**, *12*, 779598. [CrossRef]
249. Zeng, Y.; Wen, J.; Zhao, W.; Wang, Q.; Huang, W. Rational improvement of rice yield and cold tolerance by editing the three genes *OsPIN5b*, *GS3*, and *OsMYB30* with the CRISPR–Cas9 system. *Front. Plant Sci.* **2020**, *10*, 478091. [CrossRef]
250. Shen, C.; Que, Z.; Xia, Y.; Tang, N.; Li, D.; He, R.; Cao, M. Knock out of the annexin gene *OsAnn3* via CRISPR/Cas9-mediated genome editing decreased cold tolerance in rice. *J. Plant Biol.* **2017**, *60*, 539–547. [CrossRef]

https://doi.org/10.3390/antiox15030337

251. Nawaz, G.; Han, Y.; Usman, B.; Liu, F.; Qin, B.; Li, R. Knockout of *OsPRP1*, a gene encoding proline-rich protein, confers enhanced cold sensitivity in rice (*Oryza sativa* L.) at the seedling stage. *3 Biotech* **2019**, *9*, 254. [CrossRef] [PubMed]
252. Chen, H.; Su, C.; Lin, C.; Huang, G.; Lin, Y. Expression of sweet potato cysteine protease SPCP2 altered developmental characteristics and stress responses in transgenic *Arabidopsis* plants. *J. Plant Physiol.* **2010**, *167*, 838–847. [CrossRef] [PubMed]
253. Klap, C.; Yeshayahou, E.; Bolger, A.M.; Arazi, T.; Gupta, S.K.; Shabtai, S.; Usadel, B.; Salts, Y.; Barg, R. Tomato facultative parthenocarpy results from *SlAGAMOUS-LIKE 6* loss of function. *Plant Biotechnol. J.* **2017**, *15*, 634–647. [CrossRef] [PubMed]
254. Katiyar-Agarwal, S.; Agarwal, M.; Grover, A. Heat-tolerant basmati rice engineered by over-expression of *hsp101*. *Plant Mol. Biol.* **2003**, *51*, 677–686. [CrossRef]
255. El-Esawi, M.A.; Alayafi, A.A. Overexpression of rice *Rab7* gene improves drought and heat tolerance and increases grain yield in rice (*Oryza sativa* L.). *Genes* **2018**, *10*, 56. [CrossRef]
256. Scafaro, A.P.; Atwell, B.J.; Muylaert, S.; Reusel, B.V.; Ruiz, G.A.; Rie, J.V.; Gallé, A. A thermotolerant variant of Rubisco activase from a wild relative improves growth and seed yield in rice under heat stress. *Front. Plant Sci.* **2018**, *9*, 355020. [CrossRef]
257. Baeg, G.J.; Kim, S.H.; Choi, D.M.; Tripathi, S.; Han, Y.J.; Kim, J.I. CRISPR/Cas9-mediated mutation of 5-oxoprolinase gene confers resistance to sulfonamide compounds in *Arabidopsis*. *Plant Biotechnol. Rep.* **2021**, *15*, 753–764. [CrossRef]
258. Cannea, F.B.; Padiglia, A. Antioxidant Defense Systems in Plants: Mechanisms, Regulation, and Biotechnological Strategies for Enhanced Oxidative Stress Tolerance. *Life* **2025**, *15*, 1293. [CrossRef]
259. Tavleeva, M.M.; Belykh, E.S.; Rybak, A.V.; Rasova, E.E.; Chernykh, A.A.; Ismailov, Z.B.; Velegzhaninov, I.O. Effects of Antioxidant Gene Overexpression on Stress Resistance and Malignization In Vitro and In Vivo: A Review. *Antioxidants* **2022**, *11*, 2316. [CrossRef]
260. Inal, B.; Çatkın, S.; Altıntaş, S. Oxidative stress suppression in plant system and elevation of antioxidant gene expression under diverse stress situations. In *Role of Antioxidants in Mitigating Plant Stress*; Elsevier: Amsterdam, The Netherlands, 2025; pp. 269–308. [CrossRef]
261. Oliveira Júnior, J.L.; Neto, V.G.; Santos, I.D.; Soares, P.H.; Rezende, N.T.; Dos Santos, A.N.; De Almeida, C.P.; Venâncio, T.; Ferreira, A.G.; Ribeiro, P.R.; et al. Enhanced abiotic stress tolerance in *Arabidopsis* overexpressing *Ricinus communis* FeSOD8 (RcFeSOD8). *Plant Physiol. Biochem.* **2025**, *229*, 110370. [CrossRef]
262. Xie, R.; Gao, J.; Yang, Z.; Wang, Y.; Tong, L.; Ke, Y.; Li, C.; Zheng, C.; Li, W. Improving high light tolerance of tobacco plants: Adequate magnesium supply enhances photosynthetic performance. *Agronomy* **2024**, *14*, 1396. [CrossRef]
263. Bai, Z.; Yang, X.; Zi, N.; Ren, W.; Yin, J.; Yuan, T.; Wang, M.; Yuan, F.; Liu, Y. Drought stress memory enhances the tolerance of alfalfa (*Medicago sativa* L.) in response to a subsequent drought: A physiological and omics perspective. *Environ. Exp. Bot.* **2025**, *230*, 106088. [CrossRef]
264. Zafar, S.; Afzal, H.; Ijaz, A.; Mahmood, A.; Ayub, A.; Nayab, A.; Hussain, S.; Ul-Hussan, M.; Sabir, M.A.; Zulfiqar, U.; et al. Cotton and drought stress: An updated overview for improving stress tolerance. *S. Afr. J. Bot.* **2023**, *161*, 258–268. [CrossRef]
265. Ransy, C.; Vaz, C.; Lombès, A.; Bouillaud, F. Use of H_2O_2 to cause oxidative stress, the catalase issue. *Int. J. Mol. Sci.* **2020**, *21*, 9149. [CrossRef] [PubMed]
266. Dong, J.; Li, X.; Ma, Y.; Yang, J.; Chen, J.; Yang, W.; Zhou, L.; Wang, J.; Yang, T.; Zhang, S.; et al. Overexpression of *OsGF14C* enhances salinity tolerance but reduces blast resistance in rice. *Front. Plant Sci.* **2023**, *14*, 1098855. [CrossRef] [PubMed]
267. Aleem, M.; Aleem, S.; Sharif, I.; Aleem, M.; Shahzad, R.; Khan, M.I.; Batool, A.; Sarwar, G.; Farooq, J.; Iqbal, A.; et al. Whole-genome identification of APX and CAT gene families in cultivated and wild soybeans and their regulatory function in plant development and stress response. *Antioxidants* **2022**, *11*, 1626. [CrossRef]
268. Kamran, M.; Burdiak, P.; Karpiński, S. Crosstalk between abiotic and biotic stress responses and the role of chloroplast retrograde signaling in the cross-tolerance phenomena in plants. *Cells* **2025**, *14*, 176. [CrossRef]
269. Li, H.; Luo, W.; Ji, R.; Xu, Y.; Xu, G.; Qiu, S.; Tang, H. A comparative proteomic study of cold responses in potato leaves. *Heliyon* **2021**, *7*, e06002. [CrossRef]
270. Asati, R.; Tripathi, M.K.; Tiwari, S.; Yadav, R.K.; Tripathi, N. Molecular breeding and drought tolerance in chickpea. *Life* **2022**, *12*, 1846. [CrossRef]
271. Chibane, N.; Caicedo, M.; Martinez, S.; Marcet, P.; Revilla, P.; Ordás, B. Relationship between delayed leaf senescence (stay-green) and agronomic and physiological characters in maize (*Zea mays* L.). *Agronomy* **2021**, *11*, 276. [CrossRef]
272. Hasan, N.; Choudhary, S.; Naaz, N.; Sharma, N.; Laskar, R.A. Recent advancements in molecular marker-assisted selection and applications in plant breeding programmes. *J. Genet. Eng. Biotechnol.* **2021**, *19*, 128. [CrossRef]
273. Saini, D.K.; Bardhan, K.; Somayanda, I.; Bahuguna, R.N.; Jagadish, S.K. Translational research progress and challenges for developing drought-resilient rice. *Plant Stress* **2025**, *15*, 100751. [CrossRef]
274. Xu, Y.; Zhang, B.; Ma, N.; Liu, X.; Qin, M.; Zhang, Y.; Wang, K.; Guo, N.; Zuo, K.; Liu, X.; et al. Quantitative trait locus mapping and identification of candidate genes controlling flowering time in *Brassica napus* L. *Front. Plant Sci.* **2021**, *11*, 626205. [CrossRef] [PubMed]

https://doi.org/10.3390/antiox15030337

275. Tiwari, K.; Tiwari, S.; Kumar, N.; Sinha, S.; Krishnamurthy, S.L.; Singh, R.; Kalia, S.; Singh, N.K.; Rai, V. QTLs and genes for salt stress tolerance: A journey from seed to seed continued. *Plants* **2024**, *13*, 1099. [CrossRef] [PubMed]
276. Zhang, Y.; Wang, R.; Wang, H.; Wang, H. *LuCSD3* enhances salt stress tolerance in flax: Genome-wide profiling and functional validation of the SOD gene family. *Front. Plant Sci.* **2025**, *16*, 1609085. [CrossRef] [PubMed]
277. Lu, Y.; Zhou, R.; Zeng, M.; Xu, K.; Teng, L.; Zhou, M.; Fan, X.; Cao, F. An anthocyanin acyltransferase *TaMAT1a-2B* mediates salt tolerance by regulating anthocyanin acylation in wheat. *Cell Rep.* **2025**, *44*, 116205. [CrossRef]
278. Nouraei, S.; Mia, M.S.; Liu, H.; Turner, N.C.; Khan, J.M.; Yan, G. Proteomic analysis of near-isogenic lines reveals key biomarkers on wheat chromosome 4B conferring drought tolerance. *Plant Genome* **2024**, *17*, e20343. [CrossRef]
279. Saleh, H.; Badr, A.; Zayed, E.M.; Soliman, E.R. Genetic diversity and drought stress tolerance of a global collection of pearl millet at germination and early seedling growth stages. *Biochem. Genet.* **2024**. [CrossRef]
280. Raza, A.; Mubarik, M.S.; Sharif, R.; Habib, M.; Jabeen, W.; Zhang, C.; Chen, H.; Chen, Z.H.; Siddique, K.H.M.; Zhuang, W.; et al. Developing drought-smart, ready-to-grow future crops. *Plant Genome* **2023**, *16*, e20279. [CrossRef]
281. Lee, K.P.; Kim, C. Photosynthetic ROS and retrograde signaling pathways. *New Phytol.* **2024**, *244*, 1183–1198. [CrossRef]
282. Singh, V.K.; Singh, B.D.; Kumar, A.; Maurya, S.; Krishnan, S.G.; Vinod, K.K.; Singh, M.P.; Ellur, R.K.; Bhowmick, P.K.; Singh, A.K. Marker-assisted introgression of *Saltol* QTL enhances seedling stage salt tolerance in the rice variety "Pusa Basmati 1". *Int. J. Genom.* **2018**, *2018*, 8319879. [CrossRef]
283. Alafari, H.A.; Freeg, H.; Abdelrahman, M.; Attia, K.A.; Jalal, A.S.; El-Banna, A.; Fiaz, S. Integrated analysis of yield response and early stage biochemical, molecular, and gene expression profiles of pre-breeding rice lines under water deficit stress. *Sci. Rep.* **2024**, *14*, 17855. [CrossRef]
284. Singh, G.; Raigar, O.P.; Kaur, S.; Bishnoi, R.; Mondal, K.; Abreha, K.B.; Nayak, A.K.; Athar, T.; Sharma, V.; Aziz, D.; et al. Advances in genomics-assisted breeding strategies for enhancing nutrient uptake and use efficiency in cereals: A pathway toward sustainable agriculture. *Plant Stress* **2025**, *18*, 101002. [CrossRef]
285. Verma, S.; Gupta, A.R.S.S.H.; Yalla, S.; Shreya Patel, P.J.; Sharma, R.; Donga, A.A.V. Integrating marker-assisted (MAS) and genomic selection (GS) for plant functional trait improvement. In *Plant Functional Traits for Improving Productivity*; Kumar, N. Singh, H., Eds.; Springer: Singapore, 2024. [CrossRef]
286. Gulcin, İ. Antioxidants: A comprehensive review. *Arch. Toxicol.* **2025**, *99*, 1893. [CrossRef] [PubMed]
287. Yu, T.; Ma, X.; Zhang, J.; Cao, S.; Li, W.; Yang, G.; He, C. Progress in transcriptomics and metabolomics in plant responses to abiotic stresses. *Curr. Issues Mol. Biol.* **2025**, *47*, 421. [CrossRef] [PubMed]
288. Shahzaib, M.; Aprile, D.; Squillaro, T.; Alessio, N.; Peluso, G.; Di Bernardo, G.; Galderisi, U. The interactome era: Integrating RNA-seq, proteomics, and network biology to decode cellular senescence. *Ageing Res. Rev.* **2025**, *113*, 102916. [CrossRef] [PubMed]
289. Li, X.; Li, Y.; Sun, Y.; Li, S.; Cai, Q.; Li, S.; Sun, M.; Yu, T.; Meng, X.; Zhang, J. Characterization of the soybean *GPAT* gene family identifies *GmGPAT1* as a key protein in salt stress tolerance. *Plants* **2024**, *14*, 2862. [CrossRef]
290. Guo, X.; Liang, J.; Lin, R.; Zhang, L.; Zhang, Z.; Wu, J.; Wang, X. Single-cell transcriptome reveals differentiation between adaxial and abaxial mesophyll cells in *Brassica rapa*. *Plant Biotechnol. J.* **2022**, *20*, 2233. [CrossRef]
291. Burian, M.; Podgórska, A.; Szal, B. Respiratory Burst Oxidase Homolog D as a Modulating Component of Oxidative Response under Ammonium Toxicity. *Antioxidants* **2022**, *11*, 703. [CrossRef]
292. Chen, X.; Pan, Z.; Li, N.; Chen, J.; Du, N.; Xu, Y.; Wang, Y. From invasion to remediation: *Phytolacca americana* outcompetes congeners through superior cadmium hyperaccumulation and stress resilience in contaminated soils. *Ecotoxicol. Environ. Saf.* **2025**, *304*, 119135. [CrossRef]
293. Jiang, M.; Zhang, J. Water stress-induced abscisic acid accumulation triggers the increased generation of reactive oxygen species and up-regulates the activities of antioxidant enzymes in maize leaves. *J. Exp. Bot.* **2002**, *53*, 2401–2410. [CrossRef]
294. Yoshida, K.; Hisabori, T. Current insights into the redox regulation network in plant chloroplasts. *Plant Cell Physiol.* **2023**, *64*, 704. [CrossRef]
295. Michelet, L.; Zaffagnini, M.; Morisse, S.; Sparla, F.; Pérez-Pérez, M.E.; Francia, F.; Danon, A.; Marchand, C.H.; Fermani, S.; Trost, P.; et al. Redox regulation of the Calvin–Benson cycle: Something old, something new. *Front. Plant Sci.* **2013**, *4*, 470. [CrossRef] [PubMed]
296. Fan, L.; Chen, H.; Chen, L.; Guo, H. Integrative multi-omics approaches for identifying and characterizing biological elements in crop traits: Current progress and future prospects. *Int. J. Mol. Sci.* **2025**, *26*, 1466. [CrossRef] [PubMed]
297. Chachar, Z.; Fan, L.; Chachar, S.; Ahmed, N.; Narejo, M.U.N.; Ahmed, N.; Lai, R.; Qi, Y. Genetic and genomic pathways to improved wheat (*Triticum aestivum* L.) yields: A review. *Agronomy* **2024**, *14*, 1201. [CrossRef]
298. Sahoo, A.; Yadav, G.; Mehta, T.; Meena, M.; Swapnil, P. Omics-driven insights into plant growth-promoting microorganisms for sustainable agriculture. *Discov. Sustain.* **2025**, *6*, 659. [CrossRef]
299. Hasanuzzaman, M.; Fotopoulos, V.; Nahar, K.; Fujita, M. Oxidative stress and antioxidant defense in plants exposed to ultraviolet radiation. In *Environmental Stress Physiology of Plants*; Wiley: Hoboken, NJ, USA, 2017; pp. 371–420. [CrossRef]

00. Thingujam, D.; Gouli, S.; Cooray, S.P.; Chandran, K.B.; Givens, S.B.; Gandhimeyyan, R.V.; Tan, Z.; Wang, Y.; Patam, K.; Greer, S.A.; et al. Climate-Resilient Crops: Integrating AI, Multi-Omics, and Advanced Phenotyping to Address Global Agricultural and Societal Challenges. *Plants* **2025**, *14*, 2699. [CrossRef]
01. Padilla, Y.G.; Álvarez, S.; Calatayud, Á.; Hernández, J.A.; Penella, C.; Tasa, M. Advancements in Water-Saving Strategies and Crop Adaptation to Drought: A Comprehensive Review. *Physiol. Plant.* **2025**, *177*, e70332. [CrossRef]
02. Maury, G.L.; Rodríguez, D.M.; Hendrix, S.; Escalona Arranz, J.C.; Boix, Y.F.; Pacheco, A.O.; Díaz, J.G.; Morris-Quevedo, H.J.; Dubois, A.F.; Aleman, E.I.; et al. Antioxidants in Plants: A Valorization Potential Emphasizing the Need for the Conservation of Plant Biodiversity in Cuba. *Antioxidants* **2020**, *9*, 1048. [CrossRef]
03. Birben, E.; Sahiner, U.M.; Sackesen, C.; Erzurum, S.; Kalayci, O. Oxidative Stress and Antioxidant Defense. *World Allergy Organ. J.* **2012**, *5*, 9. [CrossRef]
04. Meulmeester, F.L.; Luo, J.; Martens, L.G.; Mills, K.; Noordam, R. Antioxidant Supplementation in Oxidative Stress-Related Diseases: What Have We Learned from Studies on Alpha-Tocopherol? *Antioxidants* **2022**, *11*, 2322. [CrossRef]
05. Mollo, L.; Petrini, A.; Norici, A.; Ferrante, A.; Cocetta, G. Enhanced Growth and Photosynthetic Efficiency in Wild Rocket (*Diplotaxis tenuifolia* L.) Following Multi-Species Microalgal Biostimulant Application. *Plant Sci.* **2025**, *359*, 112643. [CrossRef]
06. Yin, Z.P.; Shang, Z.W.; Wei, C.; Ren, J.; Song, X.S. Foliar Sprays of Photosynthetic Bacteria Improve the Growth and Anti-Oxidative Capability on Chinese Dwarf Cherry Seedlings. *J. Plant Nutr.* **2012**, *35*, 840–853. [CrossRef]
07. Zhang, C.; Ge, Y. Response of Glutathione and Glutathione S-Transferase in Rice Seedlings Exposed to Cadmium Stress. *Rice Sci.* **2008**, *15*, 73–76. [CrossRef]
08. Choudhary, A.; Kumar, A.; Kaur, N. ROS and Oxidative Burst: Roots in Plant Development. *Plant Divers.* **2019**, *42*, 33. [CrossRef] [PubMed]
09. Johnson, R.; Puthur, J.T. Seed Priming as a Cost-Effective Technique for Developing Plants with Cross Tolerance to Salinity Stress. *Plant Physiol. Biochem.* **2021**, *162*, 247–257. [CrossRef] [PubMed]
10. Saha, D.; Choyal, P.; Mishra, U.N.; Dey, P.; Bose, B.; Gupta, N.K.; Mehta, B.K.; Kumar, P.; Pandey, S.; Chauhan, J.; et al. Drought Stress Responses and Inducing Tolerance by Seed Priming Approach in Plants. *Plant Stress* **2022**, *4*, 100066. [CrossRef]
11. Rhaman, M.S. Seed Priming Before the Sprout: Revisiting an Established Technique for Stress-Resilient Germination. *Seeds* **2025**, *4*, 29. [CrossRef]
12. Hasanović, M.; Karalija, E. Seed Priming Beyond Stress Adaptation: Broadening the Agronomic Horizon. *Agronomy* **2025**, *15*, 1829. [CrossRef]
13. Alam, M.U.; Fujita, M.; Nahar, K.; Rahman, A.; Anee, T.I.; Masud, A.A.C.; Amin, A.R.; Hasanuzzaman, M. Seed Priming Upregulates Antioxidant Defense and Glyoxalase Systems to Confer Simulated Drought Tolerance in Wheat Seedlings. *Plant Stress* **2022**, *6*, 100120. [CrossRef]
14. Edo, G.I.; Mafe, A.N.; Ali, A.B.; Akpoghelie, P.O.; Yousif, E.; Isoje, E.F.; Igbuku, U.A.; Zainulabdeen, K.; Owheruo, J.O.; Essaghah, A.E.A.; et al. Eco-Friendly Nanoparticle Phytosynthesis via Plant Extracts: Mechanistic Insights, Recent Advances, and Multifaceted Uses. *Nano TransMed* **2025**, *4*, 100080. [CrossRef]
15. Haider, F.U.; Zulfiqar, U.; Ul Ain, N.; Hussain, S.; Maqsood, M.F.; Ejaz, M.; Yong, J.W.H.; Li, Y. Harnessing Plant Extracts for Eco-Friendly Synthesis of Iron Nanoparticles (Fe-NPs): Characterization and Their Potential Applications for Ameliorating Environmental Pollutants. *Ecotoxicol. Environ. Saf.* **2024**, *281*, 116620. [CrossRef]
16. Putri, G.E.; Labanni, A.; Arief, S.; Dafriani, P.; Darma, I.Y.; Handayani, S.; Jaffar, N.; Mahmud, M.; Ritonga, A.H. Green Synthesis of Cerium Oxide Nanoparticles Using *Citrus nobilis* Lour. Peel Extract and Evaluation of Their Potential as Antibacterial and Antioxidant Agents. *Case Stud. Chem. Environ. Eng.* **2025**, *11*, 101062. [CrossRef]
17. Soleymani, S.; Piri, S.; Aazami, M.A.; Salehi, B. Cerium Oxide Nanoparticles Alleviate Drought Stress in Apple Seedlings by Regulating Ion Homeostasis, Antioxidant Defense, Gene Expression, and Phytohormone Balance. *Sci. Rep.* **2025**, *15*, 11805. [CrossRef] [PubMed]
18. Pietrzak, M.; Skiba, E.; Wolf, W.M. Root-Applied Cerium Oxide Nanoparticles and Their Specific Effects on Plants: A Review. *Int. J. Mol. Sci.* **2023**, *25*, 4018. [CrossRef] [PubMed]
19. Siddique, A.B.; Shabala, S.; Li, C.; Chen, Z.; Varshney, R.K.; Zhao, C.; Zhou, M. Reducing Heat Stress Damage in Cereal Crops Through Agronomic Management and Breeding Strategies. *Plant Stress* **2025**, *16*, 100888. [CrossRef]
20. Mahra, S.; Tripathi, S.; Tiwari, K.; Sharma, S.; Mathew, S.; Kumar, V.; Sharma, S. Harnessing Nanotechnology for Sustainable Agriculture: From Seed Priming to Encapsulation. *Plant Nano Biol.* **2025**, *11*, 100124. [CrossRef]
21. Salam, A.; Afridi, M.S.; Javed, M.A.; Saleem, A.; Hafeez, A.; Khan, A.R.; Zeeshan, M.; Ali, B.; Azhar, W.; Sumaira; et al. Nano-Priming Against Abiotic Stress: A Way Forward Towards Sustainable Agriculture. *Sustainability* **2021**, *14*, 14880. [CrossRef]
22. Bindyalaxmi, K.; Siril, S.; Aswath, K.S.; Binu, N.K.; Shiran, K. From Seed to Seedling: Impact of ZnO Nano-Priming to Enhance Seed Vigour and Stress Tolerance. In *Advances Seed Quality Evaluation Improvement*; Roy, B., Shukla, G., Dunna, V., Sharma, P., Shukla, P.S., Eds.; Springer: Singapore, 2025. [CrossRef]

https://doi.org/10.3390/antiox15030337

323. Ali, A.; Zafar, S.; Mehmood, K.; Hasnain, Z.; Nawaz, M.; Gilani, S.M.J.J.; Siddiquai, M.H.; Shahbaz, K. Silicic acid seed pre-treatment modulates growth and antioxidant responses in maize under drought stress. *Sci. Nat.* **2025**, *112*, 70. [CrossRef]
324. Chen, Q.; Song, Y.; An, Y.; Lu, Y.; Zhong, G. Soil Microorganisms: Their Role in Enhancing Crop Nutrition and Health. *Diversity* **2024**, *16*, 734. [CrossRef]
325. Saeed, Q.; Xiukang, W.; Haider, F.U.; Kučerik, J.; Mumtaz, M.Z.; Holatko, J.; Naseem, M.; Kintl, A.; Ejaz, M.; Naveed, M.; et al. Rhizosphere Bacteria in Plant Growth Promotion, Biocontrol, and Bioremediation of Contaminated Sites: A Comprehensive Review of Effects and Mechanisms. *Int. J. Mol. Sci.* **2021**, *22*, 10529. [CrossRef]
326. Zulfiqar, U.; Haider, F.U.; Maqsood, M.F.; Shabaan, M.; Ahmad, M.; Kaleem, M.; Ishfaq, M.; Aslam, Z.; Shahzad, B. Recent Advances in Microbial-Assisted Remediation of Cadmium-Contaminated Soil. *Plants* **2023**, *12*, 3147. [CrossRef]
327. El-Saadony, M.T.; Saad, A.M.; Mohammed, D.M.; Fahmy, M.A.; Elesawi, I.E.; Ahmed, A.E.; Algopishi, U.B.; Elrys, A.S.; Desoky, E.-S.M.; Mosa, W.F.A.; et al. Drought-Tolerant Plant Growth-Promoting Rhizobacteria Alleviate Drought Stress and Enhance Soil Health for Sustainable Agriculture: A Comprehensive Review. *Plant Stress* **2024**, *14*, 100632. [CrossRef]
328. Chieb, M.; Gachomo, E.W. The Role of Plant Growth Promoting Rhizobacteria in Plant Drought Stress Responses. *BMC Plant Biol.* **2023**, *23*, 407. [CrossRef] [PubMed]
329. Chattaraj, S.; Samantaray, A.; Ganguly, A.; Thatoi, H. Employing Plant Growth-Promoting Rhizobacteria for Abiotic Stress Mitigation in Plants: With a Focus on Drought Stress. *Discov. Appl. Sci.* **2025**, *7*, 68. [CrossRef]
330. Hasanuzzaman, M.; Raihan, M.R.; Nowroz, F.; Fujita, M. Insight into the Mechanism of Salt-Induced Oxidative Stress Tolerance in Soybean by the Application of *Bacillus subtilis*: Coordinated Actions of Osmoregulation, Ion Homeostasis, Antioxidant Defense, and Methylglyoxal Detoxification. *Antioxidants* **2022**, *11*, 1856. [CrossRef]
331. Valente, I.D.; Wancura, J.H.; Zabot, G.L.; Mazutti, M.A. Endophytic and Rhizospheric Microorganisms: An Alternative for Sustainable, Organic, and Regenerative Bioinput Formulations for Modern Agriculture. *Microorganisms* **2025**, *13*, 813. [CrossRef]
332. García-Villaraco, A.; Boukerma, L.; Lucas, J.A.; Gutierrez-Mañero, F.J.; Ramos-Solano, B. Tomato Bio-Protection Induced by *Pseudomonas fluorescens* N21.4 Involves ROS Scavenging Enzymes and PRs, Without Compromising Plant Growth. *Plants* **2021**, *10*, 331. [CrossRef]
333. Kour, D.; Negi, R.; Khan, S.S.; Kumar, S.; Kaur, S.; Kaur, T.; Sharma, B.; Dasila, H.; Kour, H.; Ramniwas, S.; et al. Microbes-Mediated Induced Systemic Response in Plants: A Review. *Plant Stress* **2024**, *11*, 100334. [CrossRef]
334. Postel, S.; Kemmerling, B. Plant Systems for Recognition of Pathogen-Associated Molecular Patterns. *Semin. Cell Dev. Biol.* **2009**, *20*, 1025–1031. [CrossRef]
335. Zhang, S.; Li, C.; Si, J.; Han, Z.; Chen, D. Action Mechanisms of Effectors in Plant–Pathogen Interaction. *Int. J. Mol. Sci.* **2021**, *23*, 6758. [CrossRef]
336. Nautiyal, C.S.; Srivastava, S.; Chauhan, P.S.; Seem, K.; Mishra, A.; Sopory, S.K. Plant Growth-Promoting Bacteria *Bacillus amyloliquefaciens* NBRISN13 Modulates Gene Expression Profile of Leaf and Rhizosphere Community in Rice During Salt Stress. *Plant Physiol. Biochem.* **2013**, *66*, 1–9. [CrossRef]
337. Han, L.; Zhang, M.; Du, L.; Zhang, L.; Li, B. Effects of *Bacillus amyloliquefaciens* QST713 on Photosynthesis and Antioxidant Characteristics of Alfalfa (*Medicago sativa* L.) Under Drought Stress. *Agronomy* **2022**, *12*, 2177. [CrossRef]
338. Wu, S.; Shan, L.; He, P. Microbial Signature-Triggered Plant Defense Responses and Early Signaling Mechanisms. *Plant Sci.* **2014**, *118*. [CrossRef]
339. Li, Q.; Lin, F.; Su, Z. Endophytic fungi—Big player in plant-microbe symbiosis. *Curr. Plant Biol.* **2025**, *42*, 100481. [CrossRef]
340. Mehmood, S.; Ahmed, W.; Ikram, M.; Imtiaz, M.; Mahmood, S.; Tu, S.; Chen, D. Chitosan Modified Biochar Increases Soybean (*Glycine max* L.) Resistance to Salt-Stress by Augmenting Root Morphology, Antioxidant Defense Mechanisms and the Expression of Stress-Responsive Genes. *Plants* **2020**, *9*, 1173. [CrossRef] [PubMed]
341. Kamarudin, Z.S.; Yusop, M.R.; Ismail, M.R.; Harun, A.R. Growth Performance and Antioxidant Enzyme Activities of Advanced Mutant Rice Genotypes under Drought Stress Condition. *Agronomy* **2018**, *8*, 279. [CrossRef]
342. Tripathi, S.; Dabral, S.; Kundu, S.; Saini, D.K.; Jamal, H.; Meena, R.K.; Somayanda, I.; Varma, A.; Bahuguna, R.N.; Jagadish, S.K. Harnessing the Plant-Associated Microbiome: A Sustainable Solution for Enhancing Crop Resilience to Abiotic Stresses and Problematic Soils. *Plant Stress* **2025**, *18*, 101033. [CrossRef]
343. Khaskheli, M.A.; Nizamani, M.M.; Tarafder, E.; Das, D.; Jatoi, G.H.; Leghari, U.A.; Laghari, A.H.; Khaskheli, R.A.; Awais, M.; Wang, Y. Potential of Soil Associated Plant Growth-Promoting Microbes in Improving the Abiotic-Stress Resilience of Agricultural Crops. In *Role of Antioxidants in Abiotic Stress Management*; Elsevier: Amsterdam, The Netherlands, 2024; pp. 333–362. [CrossRef]
344. Naik, B.; Kumar, V.; Rizwanuddin, S.; Mishra, S.; Kumar, V.; Saris, P.E.J.; Khanduri, N.; Kumar, A.; Pandey, P.; Gupta, A.K.; et al. Biofortification as a Solution for Addressing Nutrient Deficiencies and Malnutrition. *Heliyon* **2024**, *10*, e30595. [CrossRef]
345. Zulfiqar, U.; Khokhar, A.; Maqsood, M.F.; Shahbaz, M.; Naz, N.; Sara, M.; Maqsood, S.; Sahar, S.; Hussain, S.; Ahmad, M. Genetic Biofortification: Advancing Crop Nutrition to Tackle Hidden Hunger. *Funct. Integr. Genom.* **2024**, *24*, 34. [CrossRef]

https://doi.org/10.3390/antiox15030337

46. Amaya, I.; Osorio, S.; Martinez-Ferri, E.; Lima-Silva, V.; Doblas, V.G.; Fernández-Muñoz, R.; Fernie, A.R.; Botella, M.A.; Valpuesta, V. Increased Antioxidant Capacity in Tomato by Ectopic Expression of the Strawberry D-Galacturonate Reductase Gene. *Biotechnol. J.* **2015**, *10*, 490–500. [CrossRef]
47. Broad, R.C.; Bonneau, J.P.; Beasley, J.T.; Roden, S.; Sadowski, P.; Jewell, N.; Brien, C.; Berger, B.; Tako, E.; Glahn, R.P.; et al. Effect of Rice GDP-L-Galactose Phosphorylase Constitutive Overexpression on Ascorbate Concentration, Stress Tolerance, and Iron Bioavailability in Rice. *Front. Plant Sci.* **2020**, *11*, 595439. [CrossRef]

https://doi.org/10.3390/antiox15030337

antioxidants

MDPI

Review

Geochemical Speciation, Uptake, and Transportation Mechanisms of Arsenic, Cadmium, and Lead in Soil–Rice Systems: Additional Aspects and Challenges

Chaw Su Lwin †, Ha-il Jung *,†, Myung-Sook Kim, Eun-Jin Lee and Tae-Gu Lee

Soil and Water Environment Division, National Institute of Agricultural Sciences, Rural Development Administration, Wanju 55365, Republic of Korea; chlwin1990@korea.kr (C.S.L.); msk74@korea.kr (M.-S.K.); eunjin0219@korea.kr (E.-J.L.); leetg7942@korea.kr (T.-G.L.)

* Correspondence: hj255@korea.kr

† These authors contributed equally to this work.

Abstract: Potentially toxic elements (PTE), such as cadmium (Cd), lead (Pb), and arsenic (As), threaten rice (*Oryza sativa* L.) crop productivity and pose significant risks to human health when they are present in soil. This review summarizes the current understanding of soil and rice contamination with As, Cd, and Pb to provide an in-depth understanding of the dynamics of these contaminants and the mechanisms regulating their flow from soil to plants. It focuses on the following aspects: (1) these metals' geochemical distribution and speciation in soil–rice systems; (2) factors influencing the transformation, bioavailability, and uptake of these metals in paddy soils; (3) metal uptake, transport, translocation, and accumulation mechanisms in rice grains; and (4) the roles of transporters involved in metal uptake, transport, and accumulation in rice plants. Moreover, this review contributes to a clearer understanding of the environmental risks associated with these toxic metals in soil–rice ecosystems. Furthermore, it highlights the challenges in simultaneously managing the risks of As, Cd, and Pb contamination in rice. The study findings may help inspire innovative methods, biotechnological applications, and sustainable management strategies to mitigate the accumulation of As, Cd, and Pb in rice grains while effectively addressing multi-metal contamination in paddy soils.

Keywords: soil–rice systems; uptake; transporters; As; Cd; Pb

1. Introduction

The contamination of agricultural soils with potentially toxic elements (PTE) is a serious global concern [1–4]. PTE primarily enter agricultural soil ecosystems via atmospheric deposition from mining and industrial activities; the application of synthetic fertilizers (e.g., phosphate fertilizers), farmyard manure, and agrochemicals (e.g., pesticides); the introduction of sewage sludge and wastewater; and irrigation using contaminated groundwater [2,5,6]. Among PTE, arsenic (As), cadmium (Cd), and lead (Pb) are of particular concern because they frequently occur in the environment, which increases the likelihood for human exposure [7]. In fact, the Agency for Harmful Substances and Disease Registry and the U.S. Environmental Protection Agency list these three toxic elements among the top 10 PTE [8,9]. Their toxicity levels are dose-dependent, and chronic exposure to these PTE through food consumption has been linked to various health risks, including cancer [10,11]. They are non-biodegradable, persistent, and devoid of biological functions once they enter the environment.

Moreover, these PTE significantly degrade soil health and fertility by disrupting soil ecosystem functions, negatively affecting natural soil microbial communities, and altering soil chemical and physical properties [12,13]. Upon entering the edible parts of crops, these PTE interfere with various physiological and biochemical processes, including nutrient uptake, membrane stability, photosynthesis (through decreased chlorophyll production), ion homeostasis, and lipid peroxidation, which generate reactive oxygen species, ultimately leading to decreased plant biomass and grain yield [14–16]. Furthermore, edible crops with elevated concentrations of PTE pose significant risks to consumers and are unsuitable for consumption because of compromised food quality and safety.

Therefore, the mechanism behind the transfer of PTE from the soil to edible plant parts is a critical research area because it is directly related to food safety and security, with potential implications for human health. For instance, rice, which contains both dietary calories and toxic metals [17], is the primary staple food for approximately half of the global human population [18]. Rice-based agroecosystems play a critical role in global food security. The uptake and accumulation of toxic elements in rice grains is directly related to food safety and public health. Flooded paddy soils, in particular, create unique chemical conditions that can increase the mobility and bioavailability of certain PTE, especially As. According to the World Health Organization (WHO), Food and Agricultural Organization (FAO), and European Commission legislation, the maximum allowable limits for total As, Cd, and Pb are 300 μg kg^{-1}, 200–400 μg kg^{-1}, and 100–200 μg kg^{-1}, respectively [19–21]. Therefore, understanding the behavior of PTE in rice-growing environments and their uptake pathways and accumulation in rice is essential for developing effective strategies to ensure the safety and sustainability of rice production.

This review focuses on three PTE—As, Cd, and Pb—owing to their high potential transfer from soil to the food chain. These toxic elements exhibit distinct biogeochemical behaviors, complicating the development of remediation strategies that can simultaneously decrease their toxicity and accumulation in agricultural products. Previous studies have introduced a simple and cost-effective method (chemical immobilization technology with soil amendments) for reducing metal bioavailability [4,22,23], with detailed explanations of its mechanisms for the remediation of contaminated arable soils. However, a single technique is often insufficient for simultaneously and effectively decreasing the bioavailability and uptake of As, Cd, and Pb in plants. Therefore, understanding the mechanisms of their uptake and translocation mechanisms in plants is essential for developing effective strategies to decrease their risks in soil–rice systems.

This review summarizes and discusses the geochemical distribution and speciation of As, Cd, and Pb in the soil, their uptake and transportation mechanisms from the soil to plant roots, and their subsequent movement into aerial plant parts. Furthermore, this review provides an in-depth understanding of the dynamics of As, Cd, and Pb in the soil; their controlling factors; their interactions with other metals, cations, and anions; and the mechanisms regulating their flow from soil to plants.

2. Literature Search Strategy

This review is based on a comprehensive search of the published literature using several electronic databases, including PubMed, Scopus, Web of Science, and Google Scholar. The search was conducted using a combination of specific keywords and corresponding terms found in titles and/or abstracts: "arsenic", "cadmium", "lead", "rice", "paddy soil", "metal uptake", "bioavailability", "speciation", and "metal transporters".

Articles were included if they focused on rice systems and examined the behavior, transformation, bioavailability, or uptake of these toxic elements (As, Cd, and Pb) in paddy soils and rice plants. The initial screening process involved removing duplicate records, followed by a review of titles, abstracts, and full texts to exclude studies that were not relevant to the objectives of this review. No formal inclusion or exclusion criteria or systematic data synthesis methods were applied. Instead, the goal was to provide a broad overview and a critical discussion of key findings and emerging trends in the literature.

3. Geochemical Distribution and Speciation of As, Cd, and Pb in Paddy Soils

3.1. *As*

Arsenic (As) has an atomic weight of 74.9 and a specific gravity of 5.73. This semi-metallic element exists in the +5, +3, 0, and −3 oxidation states, depending on the prevailing physiochemical conditions of the environment. In soils, As can occur in the form of inorganic species, such as arsenate oxyanions [As(V)] and arsenite [As(III)], as well as methylated organic species, such as monomethylarsonic acid (MMA), dimethylarsinic acid (DMA), and trimethylarsinic acid (TMA) [24,25]. In contaminated soils, As exists primarily in two inorganic forms: As(V) and As(III). In addition, trace amounts of methylated organic As species, such as MMA and DMA, are also present in soils, often resulting from extensive pesticide, herbicide, and defoliant applications, as well as the microbial methylation of As [26,27]. Under aerobic conditions, As(V) is the predominant As species [28]. It can undergo hydrolysis and bind strongly to iron (oxy)hydroxides through inner-sphere bidentate and monodentate surface complexes [26]. However, under anaerobic conditions in flooded paddy soils, As(V) is reduced to As(III) through microbial activity. Moreover, As(III) is ten times more toxic and soluble than As(V) [29,30]. High concentrations of As(III) are dissolved in the soil solution under anaerobic conditions, leading to higher As concentrations in rice grains [29,31], as rice is typically cultivated under anaerobic conditions [32]. It was reported that As(V) accounts for approximately 73–96% of the total As content of aerobic soils [33], whereas As(III) accounts for approximately 87–94% of As in flooded soils. Although As(V) is less toxic than As(III), it is thermodynamically more stable under normal conditions and remains a major contaminant, owing to its toxicity and carcinogenicity to humans. The anthropogenic sources that contribute to elevated As concentrations in the soil include atmospheric deposition from metal mining; smelting; fossil fuel combustion; industrial dust and waste accumulation; the irrigation of wastewater and As-contaminated water originating from geogenic sources; and phosphate fertilizer, manure, pesticide, and herbicide application [34–37].

3.2. *Cd*

Cadmium (Cd) is a transition metal that has an atomic number of 48, an atomic weight of 112.4 g mol^{-1}, and a density of 8.65 g cm^{-3}. It predominantly exists as divalent Cd^{2+} ions and is mobilized under oxic and acidic conditions. Cd is more mobile and soluble than other toxic elements in the soil, making it readily available for plant uptake [38]. The behavior of Cd^{2+} in soil largely depends on the soil pH. At pH levels below 6.5, Cd^{2+} tends to form complexes with inorganic ligands (e.g., Cl^-, SO_4^{2-} etc.), while at pH levels above 6.5, it binds to iron oxides [39]. Additionally, a significant portion of Cd^{2+} binds to humic acids. In some soil types, $CdCO_3$ is the dominant species; meanwhile, small quantities of Cd^{2+} sulfide species may also occur in other soil types. The primary sources of Cd^{2+} input into agricultural soils include the application of phosphate fertilizers, pesticides, wastewater irrigation, and biosolids, as well as the atmospheric deposition of dust and fumes from industrial activities, incineration, and mining [40–44].

3.3. Pb

Lead (Pb) has an atomic number of 82, an atomic weight of 207.19 g mol^{-1}, and a specific gravity of 11.34 g cm^{-3}. It primarily exists in the +2 oxidation state rather than in the +4 oxidation state. Pb^{2+} ranks second after As in terms of occurrence, human exposure potential, and toxicity. Inorganic Pb^{2+} is less mobile and soluble than As and Cd; therefore, rice absorbs minute concentration of Pb^{2+} [45,46]. Most of the absorbed Pb^{2+} ions are deposited in rice roots rather than being translocated to the aerial parts of rice. Pb^{2+} strongly binds to humic matter in organic soils and to iron oxides in mineral-rich soils. In flooded paddy soils, Pb^{2+} can bind to sulfur to form PbS precipitates; however, under oxidizing conditions, PbS releases Pb^{2+} ions. Pb^{2+} is released into agricultural soil ecosystems through various anthropogenic activities, including the use of leaded gasoline, fertilizers, and pesticides; the disposal of municipal solid waste; and the deposition of industrial dust and fumes [47].

4. Factors Influencing As, Cd, and Pb Distribution in Paddy Soils and Their Translocation in Rice

Not all PTE in the soil are available for uptake by rice; their availability depends on the type, concentration, and chemical forms of metals. Additionally, As, Cd, and Pb exhibit dynamic behaviors in paddy soils. Their bioavailability and uptake mechanisms are influenced by various factors, including soil physicochemical properties (e.g., soil pH and redox potential); plant root characteristics; As, Cd, and Pb forms and concentrations in the soil solution; environmental conditions; plant genotype; growth stages; soil amendment applications; and soil microbial dynamics [48–51]. Such a wide range of factors can significantly influence the bioavailability and uptake of toxic ions by rice, thereby affecting their accumulation in edible plant parts [52].

4.1. Soil pH

Soil pH significantly influences the concentrations of available As, Cd, and Pb in the soil, as well as their translocation to rice roots. As soil pH increases, the adsorption of cationic metals (Cd and Pb) tends to increase due to hydrolysis and the formation of MOH^+ (M = Cd^{2+} or Pb^{2+}), which, in turn, decreases their solubility and bioavailability. At high pH levels, several factors contribute to the decrease in bioavailable Cd and Pb in paddy soils: (1) an increase in negative charges on the soil surface, which enhances the adsorption of cationic metals (Cd and Pb); (2) the production of $M(OH)^+$ precipitates due to hydrolysis; and (3) the formation of stable phases that bind to carbonate-bound, phosphate-bound, and Fe–Mn oxide-bound fractions [53]. It was also reported that increasing soil pH from 5.0 to 6.5 decreased the extractable phase of Cd by approximately 2.6 to 3 times [54]. In contrast, soil acidification increases the solubility of Cd, transforming it from a stable to a more bioavailable form [55]. Furthermore, Pb translocates to shoots when Pb ions cannot precipitate and are retained in plant cell walls due to low pH [56]. Conversely, oxyanionic As has the opposite solubility pattern. For example, the desorption of As increases with increasing soil pH, mainly because of proton consumption, which increases the number of anions on the exchange sites of the soil surface, along with both As(III) and As(V). It was reported that high soil pH (pH 8.5) increases negative surface charges [57], such as hydroxyl ions, thereby facilitating the desorption of As from Fe oxides. This, in turn, increases the mobilization of As at the root surface, leading to greater As accumulation in rice. Consequently, this condition enhances the desorption and mobilization of As at the root surface, ultimately increasing As accumulation in polished rice [58]. Therefore, the contrasting behaviors of oxyanionic As and cationic metals (Cd and Pb) in response to soil pH in paddy soils pose a significant challenge for simultaneously controlling the accumulation of these toxic elements in plants and rice.

4.2. Oxidation–Reduction Reactions (Redox Potential)

Redox potential is also a key driving factor that significantly and directly affects the solubility of As, Cd, and Pb in flooded paddy soils. Flooding and draining cycles during rice growing seasons cause substantial fluctuations in the redox potential, pH, and solubility of these toxic elements. The mobility and toxicity of As are highly dependent on the soil redox environment. The solubility of As increases dramatically as the redox potential decreases. For example, reduced redox potential facilitates the transformation of As(V) into As(III), which is more soluble and bioavailable than As(V) [59], because As(III) has a lower affinity for the soil solid phase. Conversely, As(V) readily adsorbs onto mineral components such as iron (hydr)oxides [60]. During drainage, As mobility decreases as As(V) becomes the predominant species [31].

By contrast, Cd and Pb are redox-insensitive toxic elements that do not change their oxidation states with respect to their solubility. However, the reductive dissolution of Fe and Mn hydr(oxide)s under decreasing redox potential in flooded soils can significantly increase the release and mobility of Cd and Pb through desorption reactions. Additionally, sulfide-rich conditions in reduced paddy soils promote the precipitation of Cd and Pb with sulfide minerals. For instance, the solubility of Cd decreases with lower redox potential due to the formation of CdS [61]. During drainage, increasing redox potential causes the dissolution of CdS [62] and the formation of water-soluble Cd sulfate ($CdSO_4$), which, in turn, increases the solubility of Cd^{2+} in the soil solution, making it more available for rice roots to adsorb [63].

Sulfate reduction may also decrease the solubility of As, particularly in paddy soils that significantly produce Fe^{2+} under reducing conditions, which is likely due to the co-precipitation or sorption of As(III) by the newly formed FeS minerals [64]. Flooding conditions can also alter soil pH. Typically, soil pH in flooded paddy soils stabilizes within a neutral range. In flooded acidic soils, pH increases to a neutral level, owing to proton consumption during various reduction reactions. By contrast, alkaline soils tend to exhibit decreased soil pH due to CO_2 accumulation during flooding [65]. The solubility of Cd and Pb decreases markedly as the redox potential decreases under flooded conditions because an increase in soil pH enhances the sorption of Cd^{2+} and Pb^{2+} ions.

Therefore, the solubility of As increases dramatically as the redox potential decreases, primarily because of the reduction of As(V) to As(III) and the attainment of neutral pH levels. When paddy soil is drained, the opposite processes occur rapidly, leading to increased Cd^{2+} and Pb^{2+} solubility and decreased As solubility [66–68].

4.3. Fe Plaque Formation and Radial O_2 Loss

As rice is a semi-aquatic plant (moist or wet conditions), its roots have extensive longitudinal aerenchyma (a spongy tissue that forms spaces/air channels in plants) [69], which facilitates the transport of O_2 from the shoots to the root tips, allowing it to thrive and extend its roots into flooded soils. Oxygen diffusion through the aerenchyma into the surrounding rhizosphere is referred to as radial O_2 loss (ROL). ROL oxygenates the rhizosphere and promotes the oxidation of ferrous Fe (Fe^{2+}) to ferric Fe (Fe^{3+}). Consequently, yellow Fe (hydr) oxide precipitates, known as iron plaques, form on the root surface. These plaques primarily comprise amorphous or crystalline Fe (hydr)oxides [70]. As strong sorbents, Fe plaques hinder the uptake of As, Cd, and Pb by rice [70–72]. Several studies have reported that Fe plaques decrease metal concentrations in rice seedlings, with a negative correlation observed between As levels in rice grains and amorphous Fe oxide-bound As concentrations in the soil [73–75]. The role of Fe plaques in influencing As, Cd, and Pb concentrations varies depending on several factors, including the extent of plaque formation on the root surface, which differs among rice genotypes. In addition, Fe plaque formation is influenced by waterlogging conditions, O_2 availability in the soil, the growth stages of rice plants, and the background concentration of Fe in paddy soils [76,77].

4.4. Rice Root Activity

As the first organ exposed to As, Cd, and Pb in the soil, rice roots serve as the first barrier against their toxicity. The root hairs and epidermal cells in the mature zone of the root tip absorb toxic metal ions from the soil. Root activity significantly influences the concentration and speciation of metal ions in the rhizosphere, thereby affecting the ability of roots to take up metal ions. To maintain the charge balance resulting from the absorption of nutrient ions, roots compensate by excreting H^+ or OH^- ions [78]. For instance, metal ions are exchanged with H^+ and OH^- ions that are produced during plant respiration and are adsorbed onto the surface of root epidermal cells [79]. Additionally, plant roots produce certain low-molecular-weight compounds and organic (e.g., dissolved organic C) or inorganic ligands (e.g., NO_3^- and Cl^-) that form soluble metal–ligand complexes [80]. Consequently, these complexes enhance the mobility of metal ions and facilitate their entry into the root epidermis as chelates. Other factors, such as root surface area, mycorrhizal relationships in the rhizosphere, transpiration pull, and the extent of the root system, can influence the solubility of toxic metal ions in the soil. Moreover, rice cultivars may affect the solubility of metals in the rhizosphere and their accumulation in grains. Different rice cultivars exhibit varying capacities for As, Cd, and Pb uptake, transport, and accumulation in the reproductive tissues [81].

4.5. Interaction with Cations (Ca^{2+}, Mg^{2+}, K^+, Na^+, and Mn^{2+}) and Anions (NO_3^-, SO_4^{2-}, and PO_4^{2-})

The transport pathways of Cd and Pb chemically resemble those of other divalent cations. Consequently, the presence of competing cations, such as Ca^{2+}, Mg^{2+}, or K^+, in the soil medium can alter the amount of Cd^{2+} or Pb^{2+} concentrations taken up by rice roots because they share the same transporter proteins [82]. Na^+ has been reported to enhance Cd uptake by plants [83,84]. Conversely, other metallic elements, such as Ca^{2+}, Mg^{2+}, and K^+, inhibit Cd^{2+} and Pb^{2+} uptake by blocking their transport into rice roots [85]. Other divalent metal ions, such as Mn^{2+}, Fe^{2+}, Zn^{2+}, and Si, also enter rice root cells through ion channels and carrier proteins and share the same pathways as Cd and Pb. These cations can inhibit Cd and Pb uptake through competitive interactions with these channels [41,86–90]. For instance, the excessive release of Mn owing to the dissolution of Mn oxides inhibits Cd absorption by rice roots. Mn oxides, which have strong oxidation and adsorption capacities, can influence the availability of Cd by interfering with its adsorption onto other metal oxides [41,91].

By contrast, anionic As primarily interacts with elemental S in paddy soils, especially under reducing conditions in which reduced S species bind to arsenite [As(III)]. In flooded soils, sulfate is readily reduced to dissolved sulfide (S^{2-}) by sulfate-reducing bacteria. It was also reported that reduced S^{2-} immobilizes As through the formation of amorphous As–sulfide precipitates, such as AsS and As_2S_3 [92]. For instance, the high-sulfate treatment of rice plants has been reported to decrease the translocation of As from the roots to the shoots. Additionally, in the presence of Fe, S^{2-} can bind with Fe(II) to form FeS minerals, such as Fe_3S_4 and FeS_2, thereby enhancing the adsorption of As(III) and reducing its mobilization [93]. In paddy soils, N predominantly exists in the form of ammonium (NH_4^+), with lower concentrations of nitrate (NO_3^-). The release of O_2 from rice roots into the rhizosphere can promote the microbial nitrification of NH_4^+ near the root surface [94]. NO_3^- concentrations can enhance the oxidation of As(III) and decrease its mobilization in paddy soils by increasing the abundance and activity of NO_3^--dependent oxidizers [80,95]. The microbial oxidation of Fe(II) can occur simultaneously, forming insoluble Fe(III) precipitates that immobilize As(III) through co-precipitation and adsorption [96]. Furthermore, denitrification (NO_3^- reduction) and Fe(II) oxidation can occur concurrently, facilitated by specific microbes that enhance As(III) adsorption, thereby decreasing As (III) availability under flooded conditions [97,98].

5. As, Cd, and Pb Uptake and Transport Mechanisms in Rice

Understanding the uptake and transport mechanisms of As, Cd, and Pb is essential for producing tolerant rice varieties with low metal accumulation through breeding or genetic engineering. Various transporter proteins are critical in facilitating the transport of free As, Cd, and Pb ions from soil solutions into rice. Transporters and intracellular high-affinity binding sites regulate metal uptake across cell membranes. Several protein families, including heavy metal P-type ATPases (HMA), natural resistance-associated macrophage proteins (NRAMP), cation diffusion facilitators (CDF), Zn–Fe permease proteins (ZIP), and ATP-binding cassettes (ABC), have been implicated in metal ion transport in plants [99]. The transport of As, Cd, and Pb involves five major processes: (i) uptake by the roots; (ii) translocation to the xylem; (iii) transport to the shoot via the xylem; (iv) redistribution through intervascular transfer at the node; and (v) transport to rice grains from leaf blades via the phloem. Additionally, rice plants can absorb metal ions through their aerial parts, such as leaves. These aforementioned processes determine the accumulation of As, Cd, and Pb in rice grains [100].

Rice roots act as the primary entry point for metal ion uptake and serve as the first barrier to the absorption of toxic elements [101,102]. Metal ions adsorbed onto the surface of the root epidermis enter root cells via rhizosphere interactions, root membrane transporters, and ion channels. Once inside the root, metal ions move from the epidermis to the cortex, stele, and xylem for upward transport to the shoots. This movement occurs via two primary pathways: the apoplastic pathway (via extracellular spaces, including the cell wall, interstitial spaces, and conduit cavities) and the symplastic pathway (cell-to-cell movement mediated by polarized influx and efflux transporters) [103,104]. In the apoplastic pathway, Casparian strips act as barriers in the endodermis, limiting metal ion movement to the xylem. Meanwhile, the symplastic pathway is slower and involves intracellular flow through the plasmodesmata. Metal ions cross the endodermis, enter the stele, move into the parenchyma cells through the pericycle sheath, and finally enter the xylem [105]. During this process, a fraction of metal ions is sequestered into root cell vacuoles via specific transporter proteins before entering the xylem. Once in the xylem, metal ions are transported to aboveground plant parts, including leaves, stems, and nodes, driven by transpiration pull and root pressure. The efficiency of metal ion translocation depends on the sequestration capacity of the vacuole and efficiency of xylem loading. High levels of metal ion accumulation in rice grains are primarily associated with phloem transport, as the phloem redistributes metal ions within the aboveground parts of the plant [106,107]. Thus, metal ions absorbed by the roots are first translocated to the aboveground parts through the xylem and are subsequently transported to rice grains via the phloem [108,109]. The journey of As, Cd, and Pb from soil to rice grain is presented in Figure 1.

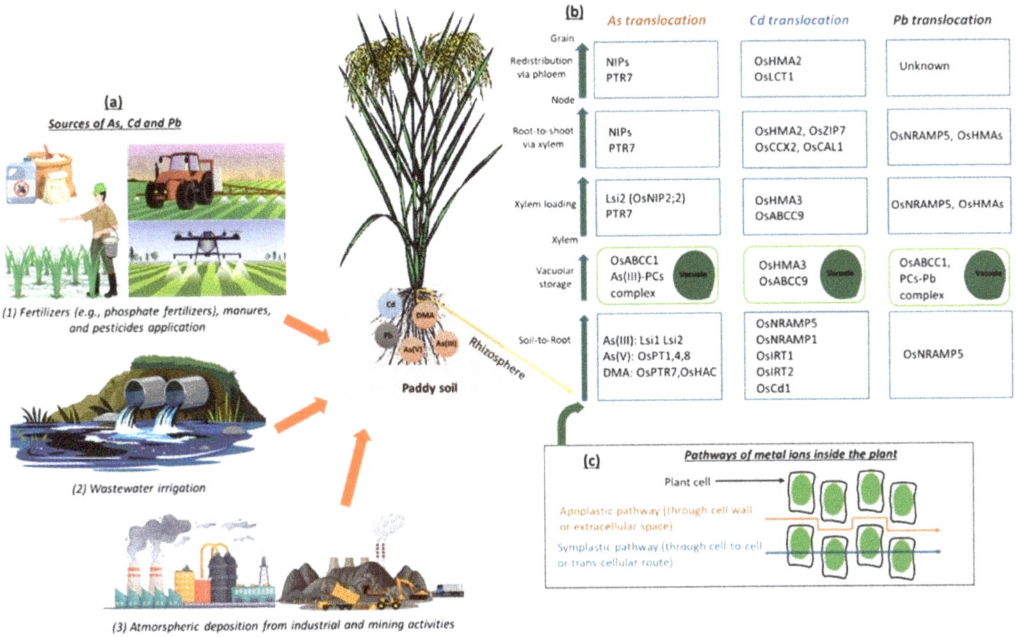

Figure 1. Summary of the distribution and direction of As, Cd, and Pb transportation in soil–rice systems. (**a**) Possible anthropogenic sources of As, Cd, and Pb in paddy soil. (**b**) Summary of the contribution of transporter genes in As, Cd, and Pb translocation from the roots to the aboveground parts of rice and, finally, accumulation in the grain. (**c**) Pathways of As, Cd, and Pb transportation in rice plant first occurs through root absorption via apoplastic and symplastic routes. Once metal ions enter the root cells, they are sequestered in root vacuoles or transferred to the xylem and then transported upward via xylem loading, transferred from the root to the shoot via the xylem, and redistributed to the grain via the phloem after reaching the node.

5.1. As Transport in Soil–Rice Systems

Under flooded conditions, As(III) is the most dominant As species in the soil, and rice roots rapidly take up As(III), primarily through the influx transporter OsNIP2;1 [50,101,110]. OsNIP2;1, also known as Si influx transporter (Lsi1), primarily functions as a Si transporter and serves as the major pathway for As(III) entry into rice roots. Lsi1 is strongly expressed in rice roots, and mutations in this transporter significantly decrease As(III) influx into rice roots [110]. Lsi1 is localized in the plasma membrane on the distal side of both endodermal and exodermal cells, where Casparian strips occur. Although other NIP family members, such as OsNIP1;1, OsNIP2;2, and OsNIP3;1, are expressed at very low levels, they are unlikely to play a significant role in As(III) influx [50].

After entering the root cells, a portion of As(III) is detoxified via sequestering into vacuoles following complexation with thiol compounds, such as phytochelatins (PCs). This process occurs predominantly in the pericycle and endodermal cells of rice roots [71]. The transport of As(III)–PC complexes into vacuoles for further detoxification is catalyzed by *Oryza sativa* ABC transporter (OsABCC1) [111], which are crucial in limiting As translocation by sequestering As(III)–PC complexes into vacuoles, wherein As is stored and distributed in plants. As OsABCC1 is expressed in various plant parts, including the roots, leaves, and nodes, removing it will significantly increase As accumulation in rice grains [112].

While Lsi1 is responsible for As(III) uptake into rice roots, the Si efflux transporter (Lsi2) mediates As(III) translocation to the xylem [110,113]. Lsi2 is localized on the plasma membrane of exodermal and endodermal cells in the roots; however, unlike Lsi1, it is situated on the proximal side [114]. The pathway of As from the soil medium to the stele involves uptake through the aquaporin channel Lsi1 and efflux toward the stele mediated by Lsi2 [113,114]. Lsi2 mutation significantly decreases As accumulation in rice shoots by 66–75% [110]. Similarly, knocking out Lsi1 decreases As concentration in the straw, although it does not significantly affect the grains. By contrast, knocking out Lsi2 leads to a significant decrease in As accumulation in both the straw and grains compared to wild-type rice. These findings suggest that Lsi2 is more critical than Lsi1 in regulating As transport and accumulation in rice shoots and grains. However, Lsi2 disruption also impairs Si uptake, which negatively affects rice growth and can decrease grain yield by up to 60% [113].

When paddy soils are flooded, certain concentrations of As(V) are present due to the oxidation of As(III) coupled with the denitrification and release of oxygen by plant roots [99]. In this case, the As(V) species is taken up by rice roots through phosphate transporters, such as OsPT1, OsPT4, and OsTP8. Once inside the root cells, As(V) is rapidly reduced to As(III) via the activity of transporters such as OsHAC1;1, OsHAC1;2, and OsHAC4 [110]. The resulting As(III) is transported to the xylem sap via Lsi2. Consequently, As(III) is the primary form of As in plant cells [115,116]. Notably, the overexpression of OsHAC1;1, OsHAC1;2, and OsHAC4 significantly increases the efflux of As(III) from the root cells, thereby reducing As accumulation in rice.

Aside from inorganic As species, several methylated As species, including methylated As species such as MMA and DMA, are present in the soil. While rice plants lack the As methylation ability, methylated As species in rice are derived from the soil by microbial activity after soil flooding and additions of organic matter [117]. Although rice roots can absorb methylated As species mediated by certain transporters like Lsi1 [118], their uptake by rice plants is slower than that of As(III). However, organic As species are much easier and 10 times more efficiently translocated to the shoots and reproductive parts than inorganic As species via the xylem and phloem [119,120]. Additionally, DMA does not form complexes with PCs or become sequestered in vacuoles. Consequently, DMA remains highly mobile within plants [121]. The putative peptide transporter (OsPTR7) in rice has been identified as being involved in DMA accumulation in rice grains [122].

5.2. *Cd Transport in Soil–Rice Systems*

Cd is highly mobile in the soil solution, and the root uptake of Cd begins when Cd ions reach the root surface [123]. Rice roots primarily take up Cd ions in the forms of Cd^{2+} and $CdCl^{-}$. The uptake of Cd^{2+} is mainly facilitated by the Mn transporter *Oryza sativa* NRAMP5 (OsNRAMP5) [79]. The NRAMP family, known as the transporters of divalent transition metals, is present across all living organisms [124]. OsNRAMP5 is predominantly expressed in rice roots and is polarly localized to the plasma membranes on the distal sides of both exodermal and endodermal cells [79]. The expression of OsNRAMP5 promotes Cd^{2+} uptake, leading to increased Cd accumulation in rice grains [125]. Hence, knocking out OsNRAMP5 significantly decreases Cd^{2+} uptake by rice roots and decreases Cd accumulation in shoots and grains [126]. Other transporters, such as OsNRAMP1, *Oryza sativa* iron-regulated transporters (OsIRT1 and OsIRT2), and OsCd1 (a member of a major facilitator family), are also involved in Cd uptake; however, their contributions are minor compared to those of OsNRAMP5 [127,128]. Studies have shown that OsIRT1 and OsIRT2 influence the root absorption of Cd [129], whereas OsCd1 also contributes to the uptake of Cd by rice roots [130].

After being transported into rice roots, a fraction of the Cd is sequestered into root vacuoles via the *Oryza sativa* heavy metal ATPase family protein (OsHMA3) [131]. Variations in the coding sequence of OsHMA3 across rice cultivars affect Cd sequestration and transportation, resulting in differences in Cd accumulation among cultivars [132]. OsHMA3, which is predominantly expressed in root cells, sequesters Cd in vacuoles and regulates its loading into the xylem [131]. Additionally, the ABC family member OsABCC9 contributes to Cd sequestration in root vacuoles [43,133]. As OsABCC9 is localized in the tonoplast of parenchyma cells, knocking it out results in increased Cd sensitivity in rice plants and elevated Cd levels in the xylem sap and grains [130,134].

Cd translocation from roots to shoots is facilitated by transporters such as OsHMA2 and OsZIP7. OsHMA2, which is localized in the plasma membrane of the root pericycle and parenchyma cells of vascular bundles in the nodes, is essential for delivering Cd from roots to developing tissues [135–137]. Hence, knocking out these genes decreases the inward flow of Cd [138]. Additionally, OsHMA2 is localized in the phloem of the vascular bundle and mediates Cd transport through the phloem into growing structures. Knocking out OsHMA2 significantly decreases Cd translocation to rice grains; moreover, it also disrupts Zn transport and affects plant growth [139]. Similarly, OsZIP7 is involved in xylem loading and transporting Cd upward to the phloem [137].

Node-expressed transporters such as OsCCX2 also mediate Cd translocation. OsCCX2, a plasma membrane-localized Ca^{2+} transporter, regulates Cd transport between roots and stems. Thus, knocking out this gene decreases Cd concentrations in rice grains [140]. Additionally, OsCAL1, a defensin-like protein expressed mainly in the root exodermis and xylem parenchyma cells, chelates Cd for secretion into the extracellular spaces, thereby reducing cytosolic Cd levels and facilitating root-to-shoot transport via xylem vessels [141]. Cd transport into rice grains occurs primarily via phloem transport, which is regulated by low-affinity cation transporters (OsHMA2 and OsLCT1). OsHMA2 is responsible for reloading Cd from parenchymal cells into the phloem of diffuse vascular bundles [125]. OsLCT1, a phloem plasma membrane protein localized in the uppermost node of the vascular bundle, is involved in Cd translocation through the phloem [142,143]. OsLCT1 expression increases significantly during grain maturity, which is a critical period for Cd accumulation in rice grains [142].

5.3. *Pb Transport in Soil–Rice Systems*

Rice roots serve as the initial point for facilitating the entry of Pb into the plant. Adsorbed Pb passively penetrates the roots and follows the water transport channels [51]. Once Pb is absorbed and passively accumulated in the roots, it utilizes the apoplastic pathway and is subsequently translocated through the vascular flow system. However, the specific transport mechanisms, including the roles of the transporter proteins responsible for stimulating Pb uptake in rice, remain unclear. Some studies have pointed out that plants can take up Pb via Ca^{2+}-permeable channels [144]. Additionally, similar to divalent Cd^{2+}, Pb^{2+} uptake is linked to the NRAMPs family of transporters in rice [90]. OsNRAMP5 has been identified as a major transporter for Pb^{2+} uptake, similar to its role in Mn^{2+} and Cd^{2+} uptake. Short-term hydroponic studies have shown that knocking out OsNRAMP5 Pb accumulation in rice grains by 50% significantly decreases its accumulation in roots and shoots. Moreover, studies have confirmed that the expression of transporter genes, including OsNRAMP5 and OsHMA6, is induced by Pb exposure in rice [145].

HMA facilitates metal detoxification. For example, OsHMA9 is predominantly expressed in vascular tissues (phloem and xylem) and contributes to Pb uptake and transport within and from these tissues [90]. Additionally, OsHMA2, OsHMA3, and OsHMA4 are implicated in Pb loading. These genes exhibit significantly higher expression in rice roots than in shoots and leaves, which likely explains increased Pb accumulation in the roots and decreased translocation to the shoots [146]. Furthermore, the ABCC1 transporter gene is expressed in rice, particularly in the roots of Pb-stressed plants, where it facilitates PC–Pb complex formation and vacuolar sequestration [146].

As Pb^{2+} is a divalent metal cation, it may also enter plant cells through channels or transporters associated with other divalent transition metals such as Cd, Co, Fe, Ni, and Zn. However, further studies are required to elucidate the precise mechanisms underlying Pb uptake in rice. Pb^{2+} is less mobile in the soil–rice system than Cd and As. Most Pb is retained in the root tissues, with only a small portion translocated to the aerial parts through transporters [147]. Several factors limit Pb transport from the roots to the aerial plant parts, including immobilization by negatively charged pectins in the cell wall [48,148,149], the precipitation of insoluble Pb salts in intercellular spaces [150], accumulation in the plasma membrane [151], and sequestration in the vacuoles of rhizodermal and cortical cells.

6. Key Challenges and Future Directions

Several studies have been conducted on the uptake, toxicity, and oxidative stress in rice exposed to high As, Cd, and Pb concentrations. Additionally, other researchers have attempted to develop various mitigation strategies, such as incorporating soil amendments to decrease metal bioavailability and limit plant uptake by modifying the soil environment, releasing essential plant nutrients, and developing rice varieties that can tolerate or avoid the accumulation of HM ions by enhancing or modifying transporter proteins. Numerous studies have demonstrated that agricultural soils are exposed to multiple PTE. Thus, the toxicological consequences of exposure to a mixture of metals differ significantly from those of exposure to individual metals. Therefore, the process and need for specific adaptations that simultaneously minimize metal uptake and translocation to the edible parts of rice are complicated. Information and research on developing new rice varieties that can tolerate multi-metal contamination is still lacking, which is one of the future challenges in mitigation strategies and research.

Metal(loid)s behave differently in the soil, affecting their availability, uptake, and transport to plants. In plants, the absorption of metal ions is facilitated by their respective transporters, and once inside the cells, different organelles utilize distinct transport proteins according to their metal needs and biological functions. Due to the chemical similarities between PTE such as As, Cd, and Pb and essential minerals, several transporters that manage nutrient uptake may also inadvertently allow for heavy metal uptake. Thus, creating or modifying transporter proteins that selectively exclude As, Cd, and Pb without interfering with the uptake of essential minerals is difficult.

Additionally, the bioavailability of As (metalloid), Cd, and Pb (metals) in flooded paddy soils varies widely because of factors such as pH, redox potential, organic matter content, and microbial interactions. A rice variety that is tolerant to metals in one type of soil may not perform well under different soil conditions. Therefore, specific soil and water management practices must be selected under metal-enriched conditions.

Effectively addressing multi-metal contamination in soils and preventing harmful translocation in plants may require a combination of genetic engineering, microbial support, and soil amendment strategies. Balancing HM tolerance with crop productivity and nutritional quality is a long-term goal, particularly under diverse environmental conditions. Developing new rice varieties that can tolerate multi-metal contamination presents significant challenges owing to the complexity of metal uptake and translocation pathways, environmental variability, and the need to maintain crop yield and quality. One major hurdle is designing transporter proteins that can exclude toxic metals, such As, Cd, and Pb, while allowing essential nutrients to enter the plant. Additionally, achieving tissue-specific exclusion, in which metals are sequestered in roots or shoots rather than in grains, requires advanced gene-editing techniques and detailed knowledge of regulatory networks. Multi-metal tolerance also demands balance; tolerance to one metal may increase sensitivity to others or decrease nutrient uptake, making it difficult to engineer effective broad-spectrum resistance. Furthermore, rigorous field trials across diverse environmental conditions are needed to validate laboratory results, whereas statutory approval and public acceptance of genetically modified or gene-edited rice can delay development. These complexities underscore the need for continued research, collaboration, and innovation in breeding and biotechnological approaches to develop safe, high-yield, and resilient rice varieties for growth in contaminated soils.

7. Conclusions

The contamination of paddy soils with As, Cd, and Pb, as well as the uptake and translocation of these PTE to rice, have raised significant environmental and food safety concerns. This review highlights the complex interactions among soil properties, metal speciation, and uptake and translocation mechanisms that govern the mobility and accumulation of As, Cd, and Pb in rice. While considerable progress has been made in understanding the uptake, translocation, and detoxification of these potentially toxic elements (PTE), managing multi-metal contamination remains a significant challenge. Future strategies should integrate soil and water management, the genetic improvement of rice varieties with selective metal exclusion or sequestration traits, and effective regulations to decrease the sources of these toxic elements to the rice ecosystems. Advancements in biotechnology and breeding approaches can offer promising tools for developing resilient and high-yielding rice varieties suited to contaminated environments, but sustained interdisciplinary efforts will be critical to ensure food safety and security under growing soil environmental pressure.

Author Contributions: Conceptualization: C.S.L. and H.-i.J.; methodology: C.S.L. and H.-i.J.; formal analysis: C.S.L., H.-i.J., M.-S.K., E.-J.L. and T.-G.L.; visualization: C.S.L.; writing—original draft: C.S.L.; writing—review and editing: H.-i.J. All authors have read and agreed to the published version of the manuscript.

Funding: This study was supported by the "Cooperative Research Program for Agriculture Science and Technology Development (Project No. PJ017441)", RDA, South Korea.

Conflicts of Interest: The authors declare that the research was conducted in the absence of any commercial or financial relationships that could be construed as a potential conflict of interest.

References

1. Pourret, O.; Hursthouse, A. It's Time to Replace the Term "Heavy Metals" with "Potentially Toxic Elements" When Reporting Environmental Research. *Int. J. Environ. Res. Public Health* **2019**, *16*, 4446. [CrossRef] [PubMed]
2. Wuana, R.A.; Okieimen, F.E. Heavy metals in contaminated soils: A review of sources chemistry risks and best available strategies for remediation. *ISRN Ecol.* **2011**, *2011*, 402647. [CrossRef]
3. Hussain, B.; Ashraf, M.N.; Rahman, S.U.; Abbas, A.; Li, J.M.; Farooq, M. Cadmium stress in paddy fields: Effects of soil conditions and remediation strategies. *Sci. Total Environ.* **2021**, *754*, 142188. [CrossRef] [PubMed]
4. Lwin, C.S.; Kim, Y.N.; Lee, M.; Jun, H.I.; Kim, K.R. In situ immobilization of potentially toxic elements in arable soil by adding soil amendments and the best ways to maximize their use efficiency. *J. Soil. Sci. Plant Nutr.* **2024**, *24*, 115–134. [CrossRef]
5. Arao, T.; Ishikawa, S.; Murakami, M.; Abe, K.; Maejima, Y.; Makino, T. Heavy metal contamination of agricultural soil and countermeasures in Japan. *Paddy Water Environ.* **2010**, *8*, 247–257. [CrossRef]
6. Hasnine, M.T.; Huda, M.E.; Khatun, R.; Saadat, A.H.M.; Ahasan, M.; Akter, S.; Uddin, M.F.; Monika, A.N.; Rahman, M.A.; Ohiduzzaman, M. Heavy metal contamination in agricultural soil at DEPZA, Bangladesh. *Environ. Ecol. Res.* **2017**, *5*, 510–516. [CrossRef]

7. Wang, H.; Chen, P.; Kopittke, P.M.; Zhao, F.J. Cadmium contamination in agricultural soils of China and the impact on food safet *Environ. Pollut.* **2019**, *249*, 1038–1048. [CrossRef]
8. Gall, J.E.; Boyd, R.S.; Rajakaruna, N. Transfer of heavy metals through terrestrial food webs: A review. *Environ. Monit. Asses* **2015**, *187*, 201. [CrossRef]
9. Mansoor, S.; Kour, N.; Manhas, S.; Zahid, S.; Wani, O.A.; Sharma, V.; Wijaya, L.; Alyemeni, M.N.; Alsahli, A.A.; El-Serehy, H.A. Biochar as a tool for effective management of drought and heavy metal toxicity. *Chemosphere* **2020**, *271*, 129458. [CrossRef]
10. Hunter, P. A toxic brew we cannot live without. Micronutrients give insights into the interplay between geochemistry an evolutionary biology. *EMBO Rep.* **2008**, *9*, 15–18. [CrossRef] [PubMed] [PubMed Central]
11. Koyama, H.; Kamogashira, T.; Yamasoba, T. Heavy Metal Exposure: Molecular Pathways, Clinical Implications, and Protectiv Strategies. *Antioxidants* **2024**, *13*, 76. [CrossRef] [PubMed]
12. Wang, Y.; Shi, I.; Wang, H.; Lin, Q.; Chen, X.; Chen, Y. The influence of soil heavy metals pollution on soil microbial biomas enzyme activity, and community composition near a copper smelter. *Ecotoxicol. Environ. Saf.* **2007**, *67*, 75–81. [CrossRef] [PubMed]
13. Bakshi, S.; Banik, C.; He, Z. The impact of heavy metal contamination on soil health. In *Managing Soil Health for Sustainabl Agriculture*; Burleigh Dodds Science Publishing Ltd.: Cambridge, UK, 2018; Volume 2, pp. 63–95. [CrossRef]
14. Aslam, M.; Aslam, A.; Sheraz, M.; Ali, B.; Ulhassan, Z.; Najeeb, U.; Zhou, Z.; Gill, R.A. Lead toxicity in cereals: Mechanisti insight into toxicity, mode of action, and management. *Front. Plant Sci.* **2021**, *11*, 587785. [CrossRef] [PubMed]
15. Alsherif, E.A.; Al-Shaikh, T.M.; Almaghrabi, O.; AbdElgawad, H. High Redox Status as the Basis for Heavy Metal Tolerance o *Sesuvium portulacastrum* L. Inhabiting Contaminated Soil in Jeddah, Saudi Arabia. *Antioxidants* **2022**, *11*, 19. [CrossRef]
16. Bano, K.; Kumar, B.; Alyemeni, M.N.; Ahmad, P. Exogenously-Sourced Salicylic Acid Imparts Resilience towards Arsenic Stres by Modulating Photosynthesis, Antioxidant Potential and Arsenic Sequestration in Brassica napus Plants. *Antioxidants* **2022**, *11*, 2010. [CrossRef]
17. Sohn, E. Contamination: The toxic side of rice. *Nature* **2014**, *514*, S62–S63. [CrossRef]
18. Song, Y.; Wang, Y.; Mao, W.; Sui, H.; Yong, L.; Yang, D.; Jiang, D.; Zhang, L.; Gong, Y. Dietary cadmium exposure assessmen among the Chinese population. *PLoS ONE* **2017**, *12*, e0177978. [CrossRef]
19. Sombatjinda, P.; Sunphorka, S. Hongsibsong, Heavy metals contamination in paddy fields and rice crops in the vicinity of a industrial estate. *Environ. Monit. Assess.* **2021**, *193*, 1–11.
20. FAO/WHO Codex Alimentarius Commission. Joint FAO/WHO Food Standards Programme Codex Committee on Contaminant in Foods. Fifth Session. In Proceedings of the Working Document for Information and Use in Discussions Related to Contaminant and Toxins in the GSCTFF (Prepared by Japan and the Netherlands) CF/5 INF/1, The Hague, The Netherlands, 21–25 March 2011; 89p.
21. FAO/WHO Codex Alimentarius Commission. Joint FAO/WHO Food Standards Programme CODEX Committee on Contaminants in Foods. Eighth Session. In Proceedings of the Editorial Amendments to the General Standard for Contaminants and Toxins in Food and Feed. CX/CF 14/8/12, The Hague, The Netherlands, 31 March–4 April 2014.
22. Lwin, C.S.; Seo, B.H.; Kim, H.U.; Owens, G.; Kim, K.R. Application of soil amendments to contaminated soils for heavy meta immobilization and improved soil quality—A critical review. *Soil. Sci. Plant Nutr.* **2018**, *64*, 156–167. [CrossRef]
23. Khalid, S.; Shahid, M.; Niazi, N.K.; Murtaza, B.; Bibi, I.; Dumat, C. A comparison of technologies for remediation of heavy meta contaminated soils. *J. Geochem. Explor.* **2017**, *182*, 247–268. [CrossRef]
24. Vithanage, M.; Herath, I.; Almaroai, Y.A.; Rajapaksha, A.U.; Huang, L.; Sung, J.K. Effects of carbon nanotube and biochar on bioavailability of Pb, Cu and Sb in multi-metal contaminated soil. *Environ. Geochem. Health* **2017**, *39*, 1409–1420. [CrossRef] [PubMed]
25. Bali, A.S.; Sidhu, G.P.S. Arsenic acquisition, toxicity and tolerance in plants-From physiology to remediation: A review. *Chemosphere* **2021**, *283*, 131050. [CrossRef]
26. Meharg, A.A.; Williams, P.N.; Adomako, E.; Lawgali, Y.Y.; Deacon, D.; Villada, A.; Cambell, R.C.J.; Sun, G.; Zhu, Y.G.; Feldmann, J.; et al. Geographical variation in total and inorganic As content of polished (white) rice. *Environ. Sci. Technol.* **2009**, *43*, 1612–1617. [CrossRef] [PubMed]
27. Chen, J.; Rosen, B.P. The arsenic methylation cycle: How microbial communities adapted methylarsenicals for use as weapons in the continuing war for dominance. *Front. Environ. Sci.* **2020**, *8*, 43. [CrossRef]
28. Takahashi, Y.; Minamikawa, R.; Hattori, K.H.; Kurishima, K.; Kihou, N.; Yuita, K. Arsenic behavior in paddy fields during the cycle of flooded and non-flooded periods. *Environ. Sci. Technol.* **2004**, *38*, 1038–1044. [CrossRef] [PubMed]
29. Xu, X.Y.; McGrath, S.P.; Meharg, A.A.; Zhao, F.J. Growing rice aerobically markedly decreases arsenic accumulation. *Environ. Sci. Technol.* **2008**, *42*, 5574–5579. [CrossRef]
30. Amin, M.N.; Kaneco, S.; Kitagawa, T.; Begum, A.; Katsumata, H.; Suzuki, T.; Ohta, K. Removal of arsenic in aqueous solutions by adsorption onto waste rice husk. *Ind. Eng. Chem. Res.* **2006**, *45*, 8105–8110. [CrossRef]
31. Arao, T.; Kawasaki, A.; Baba, K.; Mori, S.; Matsumoto, S. Effects of water management on cadmium and arsenic accumulation and dimethylarsinic acid concentrations in Japanese rice. *Environ. Sci. Technol.* **2009**, *43*, 9361–9367. [CrossRef]

Williams, P.N.; Villada, A.; Deacon, C.; Raab, A.; Figuerola, J.; Green, A.J.; Feldmann, J.; Meharg, A.A. Greatly enhanced As shoot assimilation in rice leads to elevated grain levels compared to wheat and barley. *Environ. Sci. Technol.* **2007**, *41*, 6854–6859. [CrossRef]

3. Das, S.; Chou, L.M.; Jean, J.S.; Liu, C.C.; Yang, H.J. Water management impact on arsenic behaviour and rhizosphere bacterial communities and activities in a rice agro-ecosystem. *Sci. Total Environ.* **2016**, *542*, 442–452. [CrossRef]

Basu, B.; Kundu, M.; Hedyatullah, M.; Kundu, C.K.; Bandyopadhyay, P.; Bhattacharya, K.; Sarkar, S. Mitigation of arsenic in rice through deficit irrigation in field and use of filtered water in kitchen. *Int. J. Environ. Sci. Technol.* **2015**, *12*, 2065–2070. [CrossRef]

5. Bhardwaj, S.; Soni, R.; Gupta, S.K.; Shukla, D.P. Mercury, arsenic, lead and cadmium in waters of the Singrauli coal mining and power plants industrial zone, Central East India. *Environ. Monit. Assess.* **2020**, *192*, 251. [CrossRef] [PubMed]

Hossain, M.F. Arsenic contamination in Bangladesh—An overview. *Agric. Ecosyst. Environ.* **2006**, *113*, 1–16. [CrossRef]

7. Brammer, H.; Ravenscroft, P. Arsenic in groundwater: A threat to sustainable agriculture in South and South-east Asia. *Environ. Int.* **2009**, *35*, 647–654. [CrossRef]

8. Zou, M.; Zhou, S.; Zhou, Y.; Jia, Z.; Guo, T.; Wang, J. Cadmium pollution of soil-rice ecosystems in rice cultivation dominated regions in China: A review. *Environ. Pollut.* **2021**, *280*, 116965. [CrossRef]

9. Wang, Z.; Zeng, X.; Yu, X.; Zhang, H.; Li, Z.; Jin, D. Adsorption behaviors of Cd_{2+} on Fe_2O_3/MnO_2 and the effects of coexisting ions under alkaline conditions. *Chin. J. Geochem.* **2010**, *29*, 197–203. [CrossRef]

0. Yuan, Z.; Luo, T.; Liu, X.; Hua, H.; Zhuang, Y.; Zhang, X.; Ren, J. Tracing anthropogenic cadmium emissions: From sources to pollution. *Sci. Total Environ.* **2019**, *676*, 87–96. [CrossRef] [PubMed]

Kubier, A.; Wilkin, R.T.; Pichler, T. Cadmium in soils and groundwater: A review. *Appl. Geochem.* **2019**, *108*, 104388. [CrossRef] [PubMed] [PubMed Central]

2. Wang, M.; Wang, L.; Zhao, S.; Li, S.; Lei, X.; Qin, L.; Sun, X.; Chen, S. Manganese facilitates cadmium stabilization through physicochemical dynamics and amino acid accumulation in rice rhizosphere under flood-associated low pe+pH. *J. Hazard. Mater.* **2021**, *416*, 126079. [CrossRef]

3. Yang, H.; Xiong, Z.T.; Xu, Z.R.; Liu, R.X. Interactive effects of lanthanum and calcium on cadmium accumulation in wheat with special reference to TaNramp5 expression regulated by calmodulin. *J. Agric. Food Chem.* **2021**, *69*, 6870–6878. [CrossRef]

4. Liu, H.B.; Qu, M.K.; Chen, J.; Guang, X.; Zhang, J.L.; Liu, M.S.; Kang, J.F.; Zhao, Y.; Huang, B. Heavy metal accumulation in the surrounding areas affected by mining in China: Spatial distribution patterns, risk assessment, and influencing factors. *Sci. Total Environ.* **2022**, *825*, 154004. [CrossRef] [PubMed]

5. Penn, C.J.; Camberato, J.J. A critical review on soil chemical processes that control how soil pH affects phosphorus availability to plants. *Agriculture* **2019**, *26*, 120. [CrossRef]

6. Palansooriya, K.N.; Shaheen, S.M.; Chen, S.S.; Tsang, D.C.W.; Hashimoto, Y.; Hou, D.; Bolan, N.S.; Rinklebe, J.; Ok, Y.S. Soil amendments for immobilization of potentially toxic elements in contaminated soils: A critical review. *Environ. Int.* **2020**, *134*, 105046. [CrossRef]

7. Obeng-Gyasi, E. Sources of lead exposure in various countries. *Rev. Environ. Health* **2019**, *34*, 25–34. [CrossRef] [PubMed]

8. Arias, J.A.; Peralta-Videa, J.R.; Ellzey, J.T.; Ren, M.H.; Viveros, M.N.; Gardea-Torresdey, J.L. Effects of Glomus deserticola inoculation on prosopis: Enhancing chromium and lead uptake and translocation as confirmed by X-ray mapping, ICP-OES and TEM techniques. *Environ. Exp. Bot.* **2010**, *68*, 139–148. [CrossRef]

9. Ashraf, U.; Kanu, A.S.; Mo, Z.; Hussain, S.; Anjum, S.A.; Khan, I.; Abbas, R.N.; Tang, X. Lead toxicity in rice: Effects, mechanisms, and mitigation strategies–A mini review. *Environ. Sci. Pollut. Control Ser.* **2015**, *22*, 18318–18332. [CrossRef]

0. Zhao, F.J.; Wang, P. Arsenic and cadmium accumulation in rice and mitigation strategies. *Plant Soil.* **2020**, *446*, 1–21. [CrossRef]

1. Ashraf, U.; Mahmood, M.H.R.; Hussain, S.; Abbas, F.; Anjum, S.A.; Tang, X. Lead (Pb) distribution and accumulation in different plant parts and its associations with grain Pb contents in fragrant rice. *Chemosphere* **2020**, *248*, 126003. [CrossRef]

2. Vega, F.; Andrade, M.; Covelo, E. Influence of soil properties on the sorption and retention of cadmium, copper and lead, separately and together, by 20 soil horizons: Comparison of linear regression and tree regression analyses. *J. Hazard. Mater.* **2010**, *174*, 522–533. [CrossRef]

3. Mao, C.; Song, Y.; Chen, L.; Ji, J.; Li, J.; Yuan, X.; Yang, Z.; Ayoko, G.A.; Frost, R.L.; Theiss, F. Human health risks of heavy metals in paddy rice based on transfer characteristics of heavy metals from soil to rice. *CATENA* **2019**, *175*, 339–348. [CrossRef]

4. Chen, H.; Zhang, W.; Yang, X.; Wang, P.; McGrath, S.P.; Zhao, F.J. Effective methods to reduce cadmium accumulation in rice grain. *Chemosphere* **2018**, *207*, 699–707. [CrossRef] [PubMed]

5. Li, W.; Xu, B.; Song, Q.; Liu, X.; Xu, J.; Brookes, P.C. The identification of 'hotspots' of heavy metal pollution in soil-rice systems at a regional scale in eastern China. *Sci. Total Environ.* **2014**, *472*, 407–420. [CrossRef]

6. Singh, S.; Parihar, P.; Singh, R.; Singh, V.P.; Prasad, S.M. Heavy metal tolerance in plants: Role of transcriptomics, proteomics, metabolomics, and ionomics. *Front. Plant Sci.* **2016**, *6*, 1143. [CrossRef] [PubMed]

7. Mitra, A.; Chatterjee, S.; Moogouei, R.; Gupta, D.K. Arsenic accumulation in rice and probable mitigation approaches: A review. *Agronomy* **2017**, *7*, 67. [CrossRef]

58. Yamaguchi, N.; Nakamura, T.; Dong, D.; Takahashi, Y.; Amachi, S.; Makino, T. Arsenic release from flooded paddy soils influenced by speciation, Eh, pH, and iron dissolution. *Chemosphere* **2011**, *83*, 925–932. [CrossRef]
59. Li, R.Y.; Stroud, J.L.; Ma, J.F.; McGrath, S.P.; Zhao, F.J. Mitigation of arsenic accumulation in rice with water management an silicon fertilization. *Environ. Sci. Technol.* **2009**, *43*, 3778–3783. [CrossRef]
60. Singh, A.P.; Dixit, G.; Mishra, S.; Dwivedi, S.; Tiwari, M.; Mallick, S.; Pandey, V.; Trivedi, P.K.; Chakrabarty, D.; Tripathi, R.I Salicylic acid modulates arsenic toxicity by reducing its root to shoot translocation in rice (*Oryza sativa* L.). *Front. Plant Sci.* **201** *6*, 340. [CrossRef]
61. Furuya, M.; Hashimoto, Y.; Yamaguchi, N. Time-course changes in speciation and solubility of cadmium in reduced and oxidize paddy soils. *Soil. Sci. Soc. Am. J.* **2016**, *80*, 870. [CrossRef]
62. El-Naggar, A.; Shaheen, S.M.; Ok, Y.S.; Rinklebe, J. Biochar affects the dissolved and colloidal concentrations of Cd, Cu, Ni, an Zn and their phytoavailability and potential mobility in a mining soil under dynamic redox-conditions. *Sci. Total Environ.* **201** *624*, 1059–1071. [CrossRef]
63. Han, H.; Wang, Q.; He, L.; Sheng, X. Increased biomass and reduced rapeseed Cd accumulation of oilseed rape in the presence c Cd-immobilizing and polyamine-producing bacteria. *J. Hazard. Mater.* **2018**, *353*, 280–289. [CrossRef]
64. Xu, X.; Wang, P.; Zhang, J.; Chen, C.; Wang, Z.M.; Kopittke, P.; Kretzschmar, R.; Zhao, F.J. Microbial sulfate reduction decrease arsenic mobilization in flooded paddy soils with high potential for microbial Fe reduction. *Environ. Pollut.* **2019**, *251*, 952–96 [CrossRef]
65. Kirk, G.J.D. *The Biogeochemistry of Submerged Soils*; John Wiley and Sons: Chichester, UK, 2004.
66. de Livera, J.; McLaughlin, M.J.; Hettiarachchi, G.M.; Kirby, J.K.; Beak, D.G. Cadmium solubility in paddy soils: Effects of so oxidation, metal sulfides and competitive ions. *Sci. Total Environ.* **2011**, *409*, 1489–1497. [CrossRef] [PubMed]
67. Khaokaew, S.; Chaney, R.L.; Landrot, G.; Ginder-Vogel, M.; Sparks, D.L. Speciation and release kinetics of cadmium in an alkalin paddy soil under various flooding periods and draining conditions. *Environ. Sci. Technol.* **2011**, *45*, 4249–4255. [CrossRef]
68. Fulda, B.; Voegelin, A.; Kretzschmar, R. Redox-controlled changes in cadmium solubility and solid-phase speciation in a padd soil as affected by reducible sulfate and copper. *Environ. Sci. Technol.* **2013**, *47*, 12775–12783. [CrossRef] [PubMed]
69. Steffens, B.; Geske, T.; Sauter, M. Aerenchyma formation in the rice stem and its promotion by H_2O_2. *New Phytol.* **2011**, *19* 369–378. [CrossRef]
70. Liu, W.J.; Zhu, Y.G.; Hu, Y.; Williams, P.N.; Gault, A.G.; Meharg, A.A.; Charnock, J.M.; Smith, F.A. Arsenic sequestration i iron plaque, its accumulation and speciation in mature rice plants (*Oryza sativa* L.). *Environ. Sci. Technol.* **2006**, *40*, 5730–573 [CrossRef] [PubMed]
71. Moore, K.L.; Schroder, M.; Wu, Z.; Martin, B.G.H.; Hawes, C.R.; McGrath, S.P.; Hawkesford, M.J.; Ma, J.F.; Zhao, F.J.; Groveno C.R.M. High-resolution secondary ion mass spectrometry reveals the contrasting subcellular distribution of arsenic and silicon i rice roots. *Plant Physiol.* **2011**, *156*, 913–924. [CrossRef]
72. Liu, J.; Leng, X.; Wang, M.; Zhu, Z.; Dai, Q. Iron plaque formation on roots of different rice cultivars and the relation with lea uptake. *Ecotoxicol. Environ. Saf.* **2011**, *74*, 1304–1309. [CrossRef]
73. Liu, W.J.; Zhu, Y.G.; Smith, F.A.; Smith, S.E. Do iron plaque and genotypes affect arsenate uptake and translocation by ric seedlings (*Oryza sativa* L.) grown in solution culture? *J. Exp. Bot.* **2004**, *55*, 1707–1713. [CrossRef]
74. Liu, W.J.; Zhu, Y.G.; Smith, F.A. Effects of iron and manganese plaques on arsenic uptake by rice seedlings (*Oryza sativa* L.) grow in solution culture supplied with arsenate and arsenite. *Plant Soil.* **2005**, *277*, 127–138. [CrossRef]
75. Yu, H.Y.; Li, F.B.; Liu, C.S.; Huang, W.; Liu, T.-X.; Yu, W.-M. Iron redox cycling coupled to transformation and immobilizatio of heavy metals: Implications for paddy rice safety in the red soil of South China. In *Advances in Agronomy*; Sparks, D.L., Ed Elsevier: Amsterdam, The Netherlands, 2016; Volume 137, pp. 279–317.
76. Colmer, T.D.; Cox, M.C.H.; Voesenek, L.A.C.J. Root aeration in rice (*Oryza sativa*), evaluation of oxygen, carbon dioxide, an ethylene as possible regulators of root acclimatization. *New Phytol.* **2006**, *170*, 767–778. [CrossRef] [PubMed]
77. Li, R.; Zhou, Z.; Zhang, Y.; Xie, X.; Li, Y.; Shen, X. Uptake and accumulation characteristics of arsenic and iron plaque in rice a different growth stages. *Commun. Soil. Sci. Plant Anal.* **2015**, *46*, 2509–2522. [CrossRef]
78. Hinsinger, P.; Plassard, C.; Tang, C.; Jaillard, B. Origins of root-mediated pH changes in the rhizosphere and their responses t environmental constraints: A review. *Plant Soil.* **2003**, *248*, 43–59. [CrossRef]
79. Sasaki, A.; Yamaji, N.; Yokosho, K.; Ma, J.F. Nramp5 is a major transporter responsible for manganese and cadmium uptake i rice. *Plant Cell* **2012**, *24*, 2155–2167. [CrossRef]
80. Yuan, Z.; Lin, J.; Pan, Y.; Hu, Y.; Zhang, J.; Shakouri, M.; Chen, N.; Chernikov, R.; Wiens, E.; Jia, Y. Effects of nitrate concentration on As(III) immobilization via new ferric arsenite hydroxynitrate precipitates. *Geoderma* **2023**, *432*, 116423. [CrossRef]
81. He, Y.B.; Huang, D.Y.; Zhu, Q.H.; Wang, S.; Liu, S.L.; He, H.B.; Zhu, H.H.; Xu, C. A three-season field study on the in sit remediation of Cd-contaminated paddy soil using lime, two industrial by-products, and a low-Cd-accumulation rice cultiva *Ecotoxicol. Environ. Saf.* **2017**, *136*, 135–141.

82. Zhang, Y.C.; Fan, J.J.; Fu, M.L.; Ok, Y.S.; Hou, Y.; Cai, C. Adsorption antagonism and synergy of arsenate [As(V)] and cadmium(II) onto Fe-modified rice straw biochars. *Environ. Geochem. Health* **2019**, *41*, 1755–1766. [CrossRef]
83. Manousaki, E.; Kadukova, J.; Papadantonakis, N.; Kalogerakis, N. Phytoextraction and phytoexcretion of Cd by the leaves of Tamarix smyrnensis growing on contaminated non-saline and saline soils. *Environ. Res.* **2008**, *106*, 326–332. [CrossRef]
84. Chai, M.W.; Shi, F.C.; Li, R.L.; Liu, F.C.; Qiu, G.Y.; Liu, L.M. Effect of NaCl on growth and Cd accumulation of halophyte *Spartina alterniflora* under $CdCl_2$ stress. *S. Afr. J. Bot.* **2013**, *85*, 63–69. [CrossRef]
85. Kim, Y.Y.; Yang, Y.Y.; Lee, Y. Pb and Cd uptake in rice roots. *Physiol. Plant* **2002**, *116*, 368–372. [CrossRef]
86. Takahashi, R.; Ishimaru, Y.; Nakanishi, H.; Nishizawa, N.K. Role of the iron transporter osnramp1 in cadmium uptake and accumulation in rice. *Plant Signal. Behav.* **2011**, *6*, 1813–1816. [CrossRef]
87. Sasaki, A.; Yamaji, N.; Xia, J.X.; Ma, J.F. OsYSL6 is involved in the detoxification of excess manganese in rice. *Plant Physiol.* **2011**, *157*, 1832–1840. [CrossRef]
88. Li, Z.; Yang, W.; Zou, J.; Zhou, H.; Zeng, Q.; Liao, B. Effects of calcium magnesium phosphate fertilizer on Cd bioavailability in soil and Cd contents in rice. *Acta Sci. Circumst.* **2017**, *37*, 2322–2330. [CrossRef]
89. Shao, J.F.; Che, J.; Yamaji, N.; Shen, R.F.; Ma, J.F. Silicon reduces cadmium accumulation by suppressing expression of transporter genes involved in cadmium uptake and translocation in rice. *J. Exp. Bot.* **2017**, *68*, 5641–5651. [CrossRef]
90. Chang, J.D.; Gao, W.; Wang, P.; Zhao, F. J OsNRAMP5 is a major transporter for lead uptake in rice. *Environ. Sci. Technol.* **2022**, *56*, 17481–17490. [CrossRef] [PubMed]
91. Wang, J.; Wang, P.-M.; Gu, Y.; Kopittke, P.M.; Zhao, F.J.; Wang, P. Iron–manganese oxy(hydr)oxides, rather than oxidation of sulfides, determine mobilization of Cd during soil drainage in paddy soil systems. *Environ. Sci. Technol.* **2019**, *53*, 2500–2508. [CrossRef] [PubMed]
92. Burton, E.D.; Johnston, S.G.; Planer-Friedrich, B. Coupling of arsenic mobility to sulfur transformations during microbial sulfate reduction in the presence and absence of humic acid. *Chem. Geol.* **2013**, *343*, 12–24. [CrossRef]
93. Wang, H.Y.; Göttlicher, J.; Byrne, J.M.; Guo, H.M.; Benning, L.G.; Norra, S. Vertical redox zones of Fe–S–As coupled mineralogy in the sediments of Hetao Basin–constraints for groundwater As contamination. *J. Hazard. Mater.* **2020**, *408*, 124924. [CrossRef]
94. Srivastava, S.; Pathare, V.S.; Sounderajan, S.; Suprasanna, P. Nitrogen supply influences arsenic accumulation and stress responses of rice (*Oryza sativa* L.) seedlings. *J. Hazard. Mater.* **2019**, *367*, 599–606. [CrossRef]
95. Zhang, J.; Zhao, S.; Xu, W.; Huang, K.; Tang, Z.; Zhao, F.-J. Nitrate stimulates anaerobic microbial arsenite oxidation in paddy soils. *Environ. Sci. Technol.* **2017**, *51*, 4377–4386. [CrossRef]
96. Hohmann, C.; Winkler, E.; Morin, G.; Kappler, A. Anaerobic Fe(II)-oxidizing bacteria show as resistance and immobilize as during Fe(III) mineral precipitation. *Environ. Sci. Technol.* **2010**, *44*, 94–101. [CrossRef] [PubMed]
97. Smith, R.L.; Kent, D.B.; Repert, D.A.; Böhlke, J.K. Anoxic nitrate reduction coupled with iron oxidation and attenuation of dissolved arsenic and phosphate in a sand and gravel aquifer. *Geochim. Cosmochim. Acta* **2017**, *196*, 102–120. [CrossRef]
98. Kumarathilaka, P.; Seneweera, S.; Meharg, A.; Bundschuh, J. Arsenic speciation dynamics in paddy rice soil-water environment: Sources, physicochemical, and biological factors-a review. *Water Res.* **2018**, *140*, 403–414. [CrossRef]
99. Broadley, M.R.; White, P.J.; Hammond, J.P.; Zelko, I.; Lux, A. Zinc in plants. *New Phytol.* **2007**, *173*, 677–702. [CrossRef] [PubMed]
100. Uraguchi, S.; Fujiwara, T. Rice breaks ground for cadmium-free cereals. *Curr. Opin. Plant Biol.* **2013**, *16*, 328–334. [CrossRef]
101. Zhang, X.; Yang, M.; Yang, H.; Pian, R.; Wang, J.; Wu, A.-M. The Uptake, Transfer, and Detoxification of Cadmium in Plants and Its Exogenous Effects. *Cells* **2024**, *13*, 907. [CrossRef] [PubMed]
102. Zhang, J.L.; Zhu, Y.C.; Yu, L.J.; Yang, M.; Zou, X.; Yin, C.X.; Lin, Y.J. Research advances in cadmium uptake, transport and resistance in rice (*Oryza sativa* L.). *Cells* **2022**, *11*, 569. [CrossRef]
103. White, P. Ion uptake mechanisms of individual cells and roots: Short-distance transport. In *Mineral Nutrition of Higher Plants*, 3rd ed.; Marschner, P., Ed.; Elsevier Ltd.: Amsterdam, The Netherlands, 2012; pp. 41–43.
104. Barberon, M.; Geldner, N. Radial transport of nutrients: The plant root as a polarized epithelium. *Plant Physiol.* **2014**, *166*, 528–537. [CrossRef]
105. Cai, H.; Huang, S.; Che, J.; Yamaji, N.; Ma, J.F. The tonoplast-localized transporter oshma3 plays an important role in maintaining Zn homeostasis in rice. *J. Exp. Bot.* **2019**, *70*, 2717–2725. [CrossRef]
106. Uraguchi, S.; Mori, S.; Kuramata, M.; Kawasaki, A.; Arao, T.; Ishikawa, S. Root-to-shoot Cd translocation via the xylem is the major process determining shoot and grain cadmium accumulation in rice. *J. Exp. Bot.* **2009**, *60*, 2677–2688. [CrossRef]
107. Kutrowska, A.; Szelag, M. Low-molecular weight organic acids and peptides involved in the long-distance transport of trace metals. *Acta Physiol. Plant* **2014**, *36*, 1957–1968. [CrossRef]
108. Tao, J.; Lu, L. Advances in genes-encoding transporters for cadmium uptake, translocation, and accumulation in plants. *Toxics* **2022**, *10*, 411. [CrossRef]
109. Awasthi, S.; Chauhan, R.; Srivastava, S.; Tripathi, R.D. The journey of arsenic from soil to grain in rice. *Front. Plant Sci.* **2017**, *8*, 1007. [CrossRef] [PubMed]

110. Ma, J.F.; Yamaji, N.; Mitani, N.; Xu, X.Y.; Su, Y.H.; McGrath, S.P.; Zhao, F.J. Transporters of arsenite in rice and their role in arsen accumulation in rice grain. *Proc. Nati. Acad. Sci. USA* **2008**, *105*, 9931–9935. [CrossRef]
111. Song, W.Y.; Yamaki, T.; Yamaji, N.; Ko, D.; Jung, K.H.; Fujii-Kashino, M.; An, G.; Martinoia, E.; Lee, Y.S.; Ma, J.F. A rice AB transporter, OsABCC1, reduces arsenic accumulation in the grain. *Proc. Natl. Acad. Sci. USA* **2014**, *111*, 15699–15704. [CrossRe [PubMed]
112. Hayashi, S.; Kuramata, M.; Abe, T.; Takagi, H.; Ozawa, K.; Ishikawa, S. Phytochelatin synthase OsPCS1 plays a crucial role i reducing arsenic levels in rice grains. *Plant J.* **2017**, *91*, 840–848. [CrossRef]
113. Ma, J.F.; Yamaji, N.; Mitani, N.; Tamai, K.; Konishi, S.; Fujiwara, T.; Katsuhara, M.; Yano, M. An efflux transporter of silicon in ric *Nature* **2007**, *448*, 209–212. [CrossRef]
114. Ma, J.F.; Tamai, K.; Yamaji, N.; Mitani, N.; Konishi, S.; Katsuhara, M.; Ishiguro, I.; Murata, Y.; Yano, M. A silicon transporter in ric *Nature* **2006**, *440*, 688–691. [CrossRef]
115. Shi, S.; Wang, T.; Chen, Z.; Tang, Z.; Wu, Z.; Salt, D.E.; Chao, D.Y.; Zhao, F.J. OsHAC1;1 and OsHAC1;2 function as arsenat reductases and regulate arsenic accumulation. *Plant Physiol.* **2016**, *172*, 1708–1719. [CrossRef]
116. Xu, J.; Shi, S.; Wang, L.; Tang, Z.; Lv, T.; Zhu, X.; Ding, X.; Wang, Y.; Zhao, F.J.; Wu, Z. OsHAC4 is critical for arsenate toleranc and regulates arsenic accumulation in rice. *New Phytol.* **2017**, *215*, 1090–1101. [CrossRef]
117. Zhao, F.J.; Zhu, Y.G.; Meharg, A.A. Methylated arsenic species in rice: Geographical variation, origin, and uptake mechanism *Environ. Sci. Technol.* **2013**, *47*, 3957–3966. [CrossRef] [PubMed]
118. Li, R.Y.; Ago, Y.; Liu, W.J.; Mitani, N.; Feldmann, J.; McGrath, S.P.; Ma, J.F.; Zhao, F.-J. The rice aquaporin Lsi1 mediates uptake o methylated arsenic species. *Plant Physiol.* **2009**, *150*, 2071–2080. [CrossRef]
119. Carey, A.M.; Scheckel, K.G.; Lombi, E.; Newville, M.; Choi, Y.; Norton, G.J.; Charnock, J.M.; Feldmann, J.; Price, A.H.; Mehar A.A. Grain unloading of arsenic species in rice. *Plant Physiol.* **2010**, *152*, 309–319. [CrossRef] [PubMed]
120. Lomax, C.; Liu, W.J.; Wu, L.; Xue, K.; Xiong, J.; Zhou, J.; McGrath, S.P.; Meharg, A.A.; Miller, A.J.; Zhao, F.J. Methylated arseni species in plants originate from soil microorganisms. *New Phytol.* **2012**, *193*, 665–672. [CrossRef] [PubMed]
121. Tang, Z.; Kang, Y.; Wang, P.; Zhao, F.J. Phytotoxicity and detoxification mechanism differ among inorganic and methylated arseni species in *Arabidopsis thaliana*. *Plant Soil.* **2016**, *401*, 243–257. [CrossRef]
122. Tang, Z.; Chen, Y.; Chen, F.; Ji, Y.; Zhao, F.-J. OsPTR7 (OsNPF8. 1), a putative peptide transporter in rice, is involved i dimethylarsenate accumulation in rice grain. *Plant Cell Physiol.* **2017**, *58*, 904–913. [CrossRef]
123. Huang, L.; Li, W.C.; Tam, N.F.Y.; Ye, Z. Effects of root morphology and anatomy on cadmium uptake and translocation in ric (*Oryza sativa* L.). *J. Environ. Sci.* **2019**, *75*, 296–306. [CrossRef]
124. Bozzi, A.T.; Gaudet, R. Molecular mechanism of Nramp-family transition metal transport. *J. Mol. Biol.* **2021**, *433*, 16699 [CrossRef]
125. Ma, J.F.; Shen, R.F.; Shao, J.F. Transport of cadmium from soil to grain in cereal crops: A review. *Pedosphere* **2021**, *31*, 3–1 [CrossRef]
126. Yang, M.; Zhang, Y.; Zhang, L.; Hu, J.; Zhang, X.; Lu, K.; Dong, H.; Wang, D.; Zhao, F.J.; Huang, C.F.; et al. OsNRAMP5 contribute to manganese translocation and distribution in rice shoots. *J. Exp. Bot.* **2014**, *65*, 4849–4861. [CrossRef]
127. Lee, S.; An, G. Over-expression of OsIRT1 leads to increased iron and zinc accumulations in rice. *Plant Cell Environ.* **2009**, *3* 408–416. [CrossRef] [PubMed]
128. Jiang, M.; Jiang, J.; Li, S.; Li, M.; Tan, Y.; Song, S.; Shu, Q.; Huang, J. Glutamate alleviates cadmium toxicity in rice via suppressin cadmium uptake and translocation. *J. Hazard. Mater.* **2020**, *384*, 121319. [CrossRef]
129. Nakanishi, H.; Ogawa, I.; Ishimaru, Y.; Mori, S.; Nishizawa, N. K Iron deficiency enhances cadmium uptake and translocatio mediated by the Fe^{2+} transporters OsIRT1 and OsIRT2 in rice. *Soil. Sci. Plant Nutr.* **2006**, *52*, 464–469. [CrossRef]
130. Moons, A. Ospdr9, which encodes a PDR-type ABC transporter, is induced by heavy metals, hypoxic stress and redox perturba tions in rice roots. *FEBS Lett.* **2003**, *553*, 370–376. [CrossRef]
131. Miyadate, H.; Adachi, S.; Hiraizumi, A.; Tezuka, K.; Nakazawa, N.; Kawamoto, T.; Katou, K.; Kodama, I.; Sakurai, K.; Takahash H.; et al. OsHMA3, a P1B-type of ATPase affects root-to-shoot cadmium translocation in rice by mediating efflux into vacuole *New Phytol.* **2011**, *189*, 190–199. [CrossRef] [PubMed]
132. Sui, F.Q.; Zhao, D.K.; Zhu, H.T.; Gong, Y.; Tang, Z.; Huang, X.Y.; Zhang, G.; Zhao, F.J. Map-based colonising of a new tota loss-of-function allele of OsHMA3 causes high cadmium accumulation in rice grain. *J. Exp. Bot.* **2019**, *70*, 2857–2871. [CrossRef
133. Yang, G.Z.; Fu, S.; Huang, J.J.; Li, L.Y.; Long, Y.; Wei, Q.X.; Wang, Z.G.; Chen, Z.W.; Xia, J. The tonoplast-localized transporte OsABCC9 is involved in cadmium tolerance and accumulation in rice. *Plant Sci.* **2021**, *307*, 110894–110902. [CrossRef]
134. Oda, K.; Otani, M.; Uraguchi, S.; Akihiro, T.; Fujiwara, T. Rice ABCG43 is Cd inducible and confers Cd tolerance on yeast. *Biosc Biotechnol. Biochem.* **2011**, *75*, 1211–1213. [CrossRef]
135. Takahashi, R.; Ishimaru, Y.; Shimo, H.; Ogo, Y.; Senoura, T.; Nishizawa, N.K.; Nakanishi, H. The OsHMA2 transporter is involvec in root-to-shoot translocation of Zn and Cd in rice. *Plant Cell Environ.* **2012**, *35*, 1948–1957. [CrossRef]

6. Li, Z.; Liang, Y.; Hu, H.; Shaheen, S.M.; Zhong, H.; Tack, F.M.G.; Wu, M.J.; Li, Y.F.; Gao, Y.; Rinklebe, J.; et al. Speciation, transportation, and pathways of cadmium in soil-rice systems: A review on the environmental implications and remediation approaches for food safety. *Environ. Int.* **2021**, *156*, 106749. [CrossRef]
7. Yamaji, N.; Xia, J.; Mitani-Ueno, N.; Yokosho, K.; Feng Ma, J. Preferential delivery of zinc to developing tissues in rice is mediated by P-type heavy metal ATPase OsHMA2. *Plant Physiol.* **2013**, *162*, 927–939. [CrossRef]
8. Tan, L.; Zhu, Y.; Fan, T.; Peng, C.; Wang, J.; Sun, L.; Chen, C. OsZIP7 functions in xylem loading in roots and inter-vascular transfer in nodes to deliver Zn/Cd to grain in rice. *Biochem. Biophys. Res. Commun.* **2019**, *512*, 112–118. [CrossRef] [PubMed]
39. Yamaji, N.; Takemoto, Y.; Miyaji, T.; Mitani-Ueno, N.; Yoshida, K.T.; Ma, J.F. Reducing phosphorus accumulation in rice grains with an impaired transporter in the node. *Nature* **2017**, *541*, 92–95. [CrossRef]
40. Hao, X.; Zeng, M.; Wang, J.; Zeng, Z.; Dai, J.; Xie, Z.; Yang, Y.; Tian, L.; Chen, L.; Li, D. A node-expressed transporter OsCCX2 is involved in grain cadmium accumulation of rice Front. *Plant Sci.* **2018**, *9*, 476. [CrossRef]
41. Luo, J.S.; Huang, J.; Zeng, D.L.; Peng, J.S.; Zhang, G.B.; Ma, H.L.; Guan, Y.; Yi, H.Y.; Fu, Y.L.; Han, B.; et al. A defensin-like protein drives cadmium efflux and allocation in rice. *Nat. Commun.* **2018**, *9*, 645. [CrossRef] [PubMed]
42. Uraguchi, S.; Kamiya, T.; Sakamoto, T.; Kasai, K.; Sato, Y.; Nagamura, Y.; Yoshida, A.; Kyozuka, J.; Ishikawa, S.; Fujiwara, T. Low-affinity cation transporter (OsLCT1) regulates cadmium transport into rice grains. *Proc. Natl. Acad. Sci. USA* **2011**, *108*, 20959–20964. [CrossRef] [PubMed]
43. Uraguchi, S.; Kamiya, T.; Clemens, S.; Fujiwara, T. Characterization of OsLCT1, a cadmium transporter from indica rice (*Oryza sativa*). *Physiol. Plant* **2014**, *151*, 339–347. [CrossRef]
44. Moon, J.Y.; Belloeil, C.; Ianna, M.L.; Shin, R. Arabidopsis cngc family members contribute to heavy metal ion uptake in plants. *Int. J. Mol. Sci.* **2019**, *20*, 413. [CrossRef] [PubMed]
45. Ghouri, F.; Shahid, M.J.; Zhong, M.; Zia, M.A.; Alomrani, S.O.; Liu, J.; Sun, L.; Ali, S.; Liu, X.; Shahid, M.Q. Alleviated lead toxicity in rice plant by co-augmented action of genome doubling and TiO_2 nanoparticles on gene expression, cytological and physiological changes. *Sci. Total Environ.* **2024**, *911*, 168709. [CrossRef]
46. Salavati, J.; Fallah, H.; Niknejad, Y.; Barari Tari, D. Methyl jasmonate ameliorates lead toxicity in *Oryza sativa* by modulating chlorophyll metabolism, antioxidative capacity and metal translocation. *Physiol. Mol. Biol. Plants* **2021**, *27*, 1089–1104. [CrossRef]
47. Kumar, B.; Smita, K.; Flores, L.C. Plant mediated detoxification of mercury and lead. *Arab. J. Chem.* **2017**, *10* (Suppl. 2), S2335–S2342. [CrossRef]
48. Islam, E.; Yang, X.; Li, T.; Liu, D.; Jin, X.; Meng, F. Effect of Pb toxicity on root morphology, physiology and ultrastructure in the two ecotypes of *Elsholtzia argyi*. *J. Hazard. Mater.* **2007**, *147*, 806–816. [CrossRef] [PubMed]
49. Kopittke, P.M.; Asher, C.J.; Kopittke, R.A.; Menzies, N.W. Toxic effects of Pb2+ on growth of cowpea (*Vigna unguiculata*). *Environ. Pollut.* **2007**, *150*, 280–287. [CrossRef] [PubMed]
50. Meyers, D.E.R.; Auchterlonie, G.J.; Webb, R.I.; Wood, B. Uptake and localisation of lead in the root system of *Brassica juncea*. *Environ. Pollut.* **2008**, *153*, 323–332. [CrossRef]
51. Jiang, W.; Liu, D. Pb-induced cellular defense system in the root meristematic cells of *Allium sativum* L. *BMC Plant Biol.* **2010**, *10*, 40. [CrossRef]

MDPI AG
Grosspeteranlage 5
4052 Basel
Switzerland
Tel.: +41 61 683 77 34

Antioxidants Editorial Office
E-mail: antioxidants@mdpi.com
www.mdpi.com/journal/antioxidants

www.ingramcontent.com/pod-product-compliance
Lightning Source LLC
Chambersburg PA
CBHW040804250926
51850CB00007BA/1074
9783725878192